Please remember that this is a library book,
and that it belongs only temporarily to each
person who uses it. Be considerate. Do
not write in this, or any, library book.

ENVIRONMENTAL SCIENCE

sixth edition

ENVIRONMENTAL SCIENCE

A Study of Interrelationships

ELDON D. ENGER

Delta College

BRADLEY F. SMITH

Western Washington University

WCB
McGraw-Hill

Boston, Massachusetts   Burr Ridge, Illinois   Dubuque, Iowa
Madison, Wisconsin   New York, New York   San Francisco, California   St. Louis, Missouri

WCB/McGraw-Hill

*A Division of The **McGraw·Hill** Companies*

ENVIRONMENTAL SCIENCE: A STUDY OF INTERRELATIONSHIPS

1 2 3 4 5 7 8 9 0 QPD/QPD 9 0 9 8 7

ISBN 0-697-28656-8 (paper)

Publisher: *Kevin T. Kane*
Sponsoring editor: *Margaret J. Kemp*
Developmental editor: *Thomas C. Lyon*
Project manager: *Sue Dillon*
Production supervisor: *Laura Fuller*
Designer: *Kaye Farmer*
Photo research coordinator: *John Leland*
Art editor: *Brenda Ernzen*
Compositor: *Graphic World Inc.*
Typeface: *10/12 Goudy*
Printer: *Quebecor Printing Group/Dubuque*

Front cover: © Fred Felleman/Tony Stone Images
Back cover: © Alan Levenson/Tony Stone Images
The credits section for this book begins on page 447 and is considered
an extension of the copyright page.

Library of Congress Cataloging-in-Publication Data

Enger, Eldon D.
    Environmental science: a study of interrelationships / Eldon D.
Enger, Bradley F. Smith. — 6th ed.
        p.   cm.
    Includes bibliographical references and index.
    ISBN 0-697-28656-8 (pbk.).
    1. Environmental sciences.   I. Smith, Bradley Fraser.   II. Title.
GE105.E54   1997
363.7—dc21   97-13135
    CIP

http://www.mhhe.com

DEDICATION

To my parents, Iver and Helen Enger, who instilled in their
children a respect for people and nature.

For my parents, Brent and Catherine Smith, who taught me
that perseverance was the longest part of the journey,
and the pride of loved ones was the reward.

B R I E F

EXPANDED

contents

chapter 17
Air Pollution 348

chapter 18
Solid Waste Management and Disposal 371

chapter 19
Regulating Hazardous Materials 386

chapter 20
Environmental Policy and Decision Making 404

BOXED

*New boxes

List of Boxed Readings

PREFACE

Environmental science is an interdisciplinary field. Because environmental disharmonies occur as a result of the interaction between humans and the natural world, we must include both when seeking solutions to environmental problems. It is important to have a historical perspective, appreciate economic and political realities, recognize the role of different social experiences and ethical backgrounds, and integrate these with the science that describes the natural world and how we affect it. *Environmental Science: The Study of Interrelationships* incorporates all of these sources of information when discussing any environmental issue. Furthermore, the authors have endeavored to present a balanced view of issues, diligently avoiding personal biases and fashionable philosophies.

Environmental Science: The Study of Interrelationships is intended as a text for a one-semester, introductory course for students with a wide variety of career goals. They will find it interesting and informative. The central theme is interrelatedness. No text of this nature can cover all issues in depth. What we have done is to identify major issues and give appropriate examples that illustrate the complex interactions that are characteristic of all environmental problems. There are many facts present in charts, graphs, and figures that help to illustrate the scope of environmental issues. However, this is not the core of the text, since the facts will change. The core is the central theme: interrelatedness.

Organization and Content

This book is divided into five parts and twenty chapters. It is organized to provide an even, logical flow of concepts and to provide clear illustrations of the major environmental issues of today.

Part 1 establishes the theme of the book in chapter 1 by looking at the kinds of environmental issues typical of different regions of North America. In each region, the specific issues selected involve scientific, social, political, and economic components typical of environmental problems. Chapter 2 focuses on the philosophical base needed to examine environmental issues by discussing various ethical and moral stands that shape how people approach environmental issues.

Part 2 provides an understanding of the ecological principles that are basic to organism interactions and the flow of matter and energy in ecosystems. The nature of food chains and how they affect the flow of matter and energy are discussed. Other topics included are: the efficiency of energy flow through ecosystems, the intricacies of organism to organisms interaction, and the creative role of natural selection in shaping ecological relationships. Principles of population structure and organization are also developed in this section, with particular attention to the implications of these principles to growth and impact of human populations.

Part 3 focuses on energy. A major emphasis is on the historically important, nonrenewable fossil fuels that have stimulated economic success of the developed economies of the world. Renewable sources of energy are discussed, but with the recognition that they currently are a small part of the world energy picture. Weapons production and nuclear power plants use the enormous amounts of energy that can be released from the nucleus of the atom. Both of these uses have caused fear among the public related to the dangers of radiation and the adequacy of waste disposal. These issues are discussed in this section.

Part 4 emphasizes the impact of human activity on natural ecosystems. As human populations grow, and technology changes, the magnitude of human actions becomes more apparent. The natural ecosystems on land and water are modified to meet human needs. The heavy use of pesticides in agriculture is discussed in this section.

Part 5 deals with the major types of pollution. Pollution affects the health and welfare of humans and other organisms. Air pollution, solid waste, and hazardous and toxic substances are discussed in this section. The cost of pollution cannot always be measured in financial terms, but may be reflected in the mental and physical health of the populace. Government often uses risk assessment and analysis to help set policy regarding pollution and other environmental issues. Separate chapters deal with the ways in which government sets policy and evaluates risk.

New to This Edition

1. The text has been edited throughout and rewritten where needed to include the most recent data and ways of thinking about environmental issues.
2. Many new illustrations were developed and many others were modified to improve their ability to convey information.
3. Chapter 10 has been retitled "Nuclear Energy: Benefits and Risks." Although it still includes discussion of nuclear power plants, it was significantly rewritten to address the problems associated with the cleanup of sites contaminated by activities related to the building of nuclear weapons in the United States and Russia.
4. Chapter 19 has been retitled "Regulating Hazardous Materials," and now includes discussions of hazardous materials and hazardous wastes. New sections clarify the

difference between hazardous materials, toxic materials, and hazardous wastes. Additional sections address the process of establishing regulations, pollution prevention, and waste minimization.
5. Many new topics or boxed readings have been introduced into chapters that did not require major changes in chapter organization:
   - Corporate attention to environmental issues and international trade in endangered species in chapter 2,
   - A discussion of lighting efficiency in chapter 3,
   - A discussion of the Department of Interior's experiment on restoring the Colorado River by causing an artificial flood in chapter 4,
   - The changing nature of the concept of climax communities in chapter 5,
   - An exploration of the relationship between hunger, food production, and environmental degradation and a discussion of the relationship between population and poverty in chapter 7,
   - The politics and economics of the Artic National Wildlife Refuge in chapter 9,
   - Over-exploitation of ocean fisheries and possible recovery of rhinoceros population in chapter 11,
   - China's improving standard of living and its relationship to increased demands on the soil and water resources of the country in chapter 14,
   - A comparison of the ways that water is used and abused in industrialized and developing countries, and some improved water use examples from Mexico in chapter 15,
   - Expanded coverage of risk assessment and an analysis of economic decision making and the biophysical world in chapter 16,
   - The changes in thinking about the likelihood of global climate change and probable consequences in chapter 17,
   - Expanded coverage of the economics of recycling in chapter 18,
   - An entire new section on environmental policy and regulation in chapter 20, along with expanded coverage of international environmental concerns.

Special Features and Learning Aids

1. Each of the five parts of the text begins with an **introduction** that places the upcoming chapters in context for the reader by recalling previously discussed material and by describing the organization of the chapters to come. A world map with political boundaries, land occupied by each country and continent, and populations of each country and continent can be found immediately following this preface. We believe that this will help the reader to more fully understand and appreciate global environmental issues.

2. Each chapter begins with a set of learning **objectives,** an **outline,** and a **conceptual diagram**—all of which give the student a broad overview of the interrelated forces that are involved in the material to be discussed. The student is encouraged to refer to these resources while reading and reviewing the chapter.

3. Chapters conclude with an **Issues and Analysis** case study, **summary,** a list of **key terms,** and **review questions.** The case studies have been specifically selected to allow the reader to apply the chapter concepts to actual situations. Review questions are related to the chapter objectives, and thus serve to reinforce understanding of basic concepts and principles.

4. To dramatize and clarify text material, each chapter includes a number of **tables, charts, graphs, maps, drawings,** or **photographs.** Each illustration has been carefully chosen to provide a pictorial image or an organized format for showing detailed information, which helps the reader comprehend the chapter material.

5. Each chapter also includes one or more **boxed items.** These provide an in-depth consideration of a specific situation that is relevant to the content, an alternative viewpoint, or a wider world view of the issues discussed in the chapter.

6. The text concludes with a section on **environmental careers,** a **list of environmental, work, study, and travel opportunities,** a **list of active environmental organizations,** a **list of U.S. governmental agencies,** a **metric conversion chart,** a thorough **glossary,** and an **index.**

About Internet URLs in this Text

The authors and publisher have prepared this text with the best of their abilities; however, no guarantees can be made, implicit or inferred, with regard to the information contained within. The authors and publisher therefore disclaim, without limitation, any implied guarantees of appropriateness or usefulness for a specific purpose with respect to listings in the book, or the techniques depicted therein. No responsibility shall be placed upon the authors or publisher for loss or damages of any kind, including consequential or incidental damages, special applications or other usage not limited to the foregoing in conjunction with, or resulting from, the supply, application, or use of this text.

Photographs and illustrations obtained on-line bear captions identifying their on-line sources. Text and images accessed through the Internet or through other on-line sources may be copyrighted and/or subject to rights owned by third parties; therefore, permission issues should be explored before an attempt is made to reproduce the material.

Words or terms believed to carry a trademark, service mark, or other proprietary rights have been printed with an initial capital, however, not all personal-computer words or terms in which ownership rights may exist have been indicated as such. The validity or legal status of any right or ownership claimed in a word or term is not meant to be challenged by including, excluding, or defining that word or term.

Useful Ancillaries

1. An **Instructor's Manual** accompanies the text. It includes chapter outlines, objectives, and key terms; a range of test and discussion questions; suggestions for demonstrations; and suggestions for audiovisual materials and other teaching aids. The Instructor's Manual also provides **additional case studies** for instructors who wish to use additional concrete examples of how the concepts in the chapter can be applied to the real world. It is available on disk only, in either Mac or IBM format.

2. A set of **one hundred transparencies** is also available to users of the text. The transparencies duplicate text figures that clarify essential ecological, political, economic, social, and historical concepts.

3. A completely revised **Field and Laboratory Activities** manual is available that features thirty-seven exercises on environmental topics. The manual can be adapted to a variety of situations, including rural and urban, northern and southern, desert and forestland, coastal and inland. A **Laboratory Resource Guide** can also be ordered that specifies procedures, objectives, and equipment for each exercise.

4. **Computerized Testing Software** rounds out the supplementary materials. Available for either Windows® or MacIntosh®, this software allows for easy test generation using the questions found in the printed testbank.

Related Titles of Interest Available from Dushkin/McGraw Hill

Annual Editions: Environment 97/98 (0-697-37266-9) © 1997. Editor: John L. Allen, University of Connecticut, Storrs— 34 articles that address the current state of Earth and the changes it faces; world population and hunger; present and future energy needs and problems; endangered species; natural resources; and pollution.

Taking Sides: Clashing Views on Controversial Environmental Issues, 7th edition (0-697-37536-6) © 1997. Editor: Theodore D. Goldfarb, State University of New York, Stony Brook—18 issues debating general philosophical and political issues; the environment and technology; disposing of wastes; and the environment and the future.

Sources: Notable Selections in Environmental Studies, 1st edition (0-697-32894-5) © 1997. Editor: Theodore D. Goldfarb, State University of New York, Stony Brook—18 chapters that provide an overview of environmental issues and focus specifically on energy, environmental degradation, population issues and the environment, human health and the environment, and environment and society.

The Dushkin Student Atlas of Environmental Issues, 1st edition (0-697-36520-4) © 1997. Editor: John Allen, University of Connecticut—covers recent agricultural, industrial, and demographic changes in every world region. 48 full-color maps illustrate global patterns in the physical and human environments; and presents human impact on the air, fresh water and the oceans, the biosphere, and land.

Acknowledgements

We would like to thank our many colleagues who have reviewed all or part of *Environmental Science: A Study of Interrelationships* over the years, along with those who responded to our market research requests. Their valuable input has contributed significantly to the quality of this text.

John W. Adams, University of Texas at San Antonio
Terry C. Allison, University of Texas-Pan American
Mark Aronson, Scott Community College
Elizabeth Ayre, Whatcom Community College
Loretta M. Bates, Concordia College at Moorhead
Loren Baumbach, Minneapolis Community College
Glenn Bellah, Bethany College
Jeffrey H. Black, Oklahoma Baptist University
Larry G. Blackburn, Davidson County Community College
Marion E. Bontrager, Hesston College
Richard D. Bowden, Allegheny College
Kathleen Brown, Georgia Military College
Scott Byington, Nash Community College
Dominic Calvetti, National College
Jay P. Clymer III, Marywood College
Richard D. Coleman, Volunteer State Community College
James Connors, University of Idaho
Theodore J. Crovello, California State University, Los Angeles
Sam Crowley, National College
George A. Damoff, East Texas Baptist University
Darlene Coleman Deecher, Mohawk Valley Community College
Sara B. Dodd, Davidson County Community College
David Dubose, Shasta College
Mark Finley, Heartland Community College
Bob Galbraith, Crafton Hills College
Lowell L. Getz, University of Illinois

William H. Gilbert, Simpson College
John H. Green, Nicholls State University
John Lee Griffis, Southwest State University
Charles Hart, University of Wisconsin Center-Manitowoc County
Roald Hazelhoff, Birmingham Southern College
Harry L. Holloway, Jr., University of North Dakota
Jerry F. Howell, Jr., Morehead State University
John A. Jones, Miami-Dade Community College
Wayne E. Kiefer, Central Michigan University
Ronald R. Keiper, Penn State University-Mont Alto
Allan H. Lee, Seneca College (New York, Ontario)
R.G. Litchford, University of Tennessee at Chattanooga
Jane Maler, Austin Community College
David McCalley, University of Northern Iowa
Bernard McGonigle, Community College of Philadelphia
Eric J. Meyer, Iowa Western Community College
Robert Miller, Moorpark College
Mercedes C. Mondecar, Morehouse College
Muthena Naseri, Moorpark College
Lisa H. Newton, Fairfield University
Victor J. Newton, Fairfield University
G.L. Peterson, Central Community College
Christine Richard, Southwestern Michigan College
Ingrid Ritchie, Indiana University-Indianapolis
Jarl Roine, Northern Michigan University
Anna E. Ross, Christian Brothers College
Lynette Rushton, Centralia College
Charles Scarborough, Michigan State University
Mark Secord, Bee County College
Mel Seifert, Sheldon Jackson College
Robert M. Shealy, University of South Carolina-Spartanburg
Philip C. Shelton, University of Virginia-Clinch Valley College
Doris M. Shoemaker, Dalton College
Linda Silva, Heald College
Richard Slavich, Butte College
Barbara W. Smigel, Community College of S. Nevada
Wayne L. Smith, University of Tampa
Susan P. Speece, Anderson University
David A. Stewart, Ferris State University
Deborah S. Temperley, Northwood Institute
Theresa Hoffman-Till, Northern Virginia Community College
John B. Topp, Gaston College
Jack A. Turner, University of South Carolina, Spartanburg
Deborah F. Verfaillie, Grossmont College
Linda L. Wallace, University of Oklahoma
Phillip L. Watson, Ferris State University
Clint Westervelt, Chapman College
James M. Willard, Cleveland State University
Ray E. Williams, Rio Hondo College
Jim Yoder, Hesston College

Eldon D. Enger

Eldon D. Enger is a professor of biology at Delta College, a community college near Saginaw, Michigan. He received his B.A. and M.S. degrees from the University of Michigan. Professor Enger has over 30 years of teaching experience, during which he has taught biology, zoology, environmental science, and several other courses. He has been very active in curriculum and course development. Recent activities include the development of a learning community course in stream ecology, which involves students in two weekend activities including canoeing and camping, and a plant identification course that incorporates weekend field activities with backpacking and camping. In addition, he was involved in the development of an environmental regulations course and an environmental technician curriculum.

Professor Enger is an advocate for variety in teaching methodology. He feels that if students are provided with varied experiences, they are more likely to learn. In addition to the standard textbook assignments, lectures, and laboratory activities, his classes are likely to include writing assignments, student presentation of lecture material, debates by students on controversial issues, field experiences, individual student projects, and discussions of local examples and relevant current events. Textbooks are very valuable for presenting content, especially if they contain accurate, informative drawings and visual examples. Lectures are best used to help students see themes and make connections, and laboratory activities provide important hands-on activities.

Professor Enger has been a Fulbright Exchange Teacher to Australia and Scotland, received the Bergstein Award for Teaching Excellence and the Scholarly Achievement Award from Delta College, and participated as a volunteer in an Earthwatch Research Program in Costa Rica, studying the behavior of a bird known as the long-tailed manakin. He has also visited New Zealand, New Guinea, Fiji, Puerto Rico, Mexico, Canada, many areas in Europe, and much of the United States. During these travels he has spent considerable time visiting coral reefs, ocean coasts, mangrove swamps, alpine tundra, prairies, tropical rainforests, cloud forests, deserts, temperate rainforests, coniferous forests, deciduous forests, and many other special ecosystems. This extensive experience provides the background to look at environmental issues from a broad perspective.

Professor Enger is married, has two college-aged sons, and enjoys a variety of outdoor pursuits such as cross-country skiing, hiking, hunting, fishing, camping and gardening. Other interests include reading a wide variety of periodicals, playing soccer, bee-keeping, singing in a church choir, and preserving garden produce.

Bradley F. Smith

Bradley F. Smith is the Dean of Huxley College of Environmental Studies at Western Washington University in Bellingham, Washington. Prior to assuming the position as Dean in 1994, he served as the first Director of the Office of Environmental Education for the U.S. Environmental Protection Agency in Washington, D.C. from 1991 to 1994. Dean Smith also served as the Acting President of the National Environmental Education and Training Foundation in Washington, D.C. and as a Special Assistant to the EPA Administrator.

Before moving to Washington, D.C., Dean Smith was a professor of political science and environmental studies for fifteen years, and the executive director of an environmental education center and nature refuge for five years.

Dean Smith has considerable international experience. He was a Fulbright Exchange Teacher to England and worked as a research associate for Environment Canada in New Brunswick, Canada. He is a frequent speaker on environmental issues worldwide and serves on the International Scholars Program for the U.S. Information Agency. He also served as a U.S. representative on the Tri-Lateral Commission on environmental education with Canada and Mexico. In 1995, he was awarded a NATO Fellowship to study the environmental problems associated with the closure of former Soviet military bases in Eastern Europe. Dean Smith is an Adjunct Professor at Far Eastern State University in Vladivostok, Russia and is a member of the Russian Academy of Transport.

Nationally, Dean Smith serves as a member/advisor for many environmental organizations' boards of directors, advisory councils, and executive committees, including President Clinton's Council for Sustainable Development (Education Task Force).

Dean Smith holds B.A. and M.A. degrees in Political Science and Public Administration and a Ph.D. from the School of Natural Resources and Environment at the University of Michigan.

Dean Smith lives with his wife Daria, daughter Morgan, son Ian, and English Setter Skye, along Puget Sound south of Bellingham. He is an avid outdoor enthusiast.

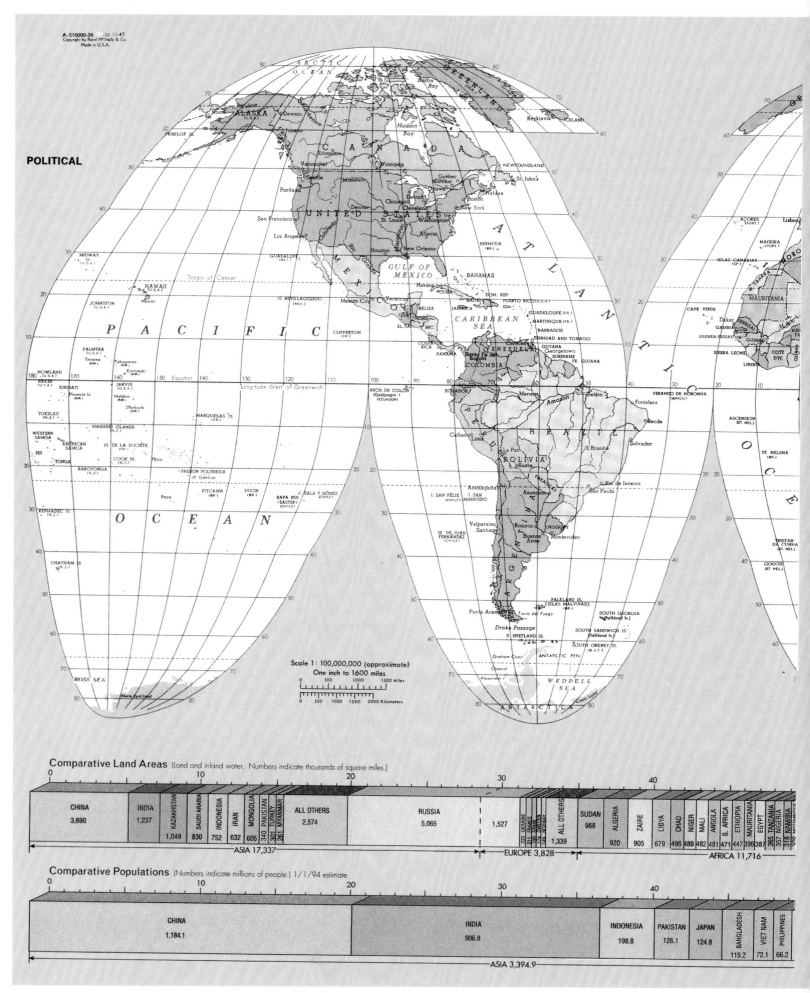

POLITICAL

Scale 1 : 100,000,000 (approximate)
One inch to 1600 miles

Comparative Land Areas (land and inland water. Numbers indicate thousands of square miles.)

| CHINA 3,690 | INDIA 1,237 | KAZAKHSTAN 1,049 | SAUDI ARABIA 830 | INDONESIA 752 | IRAN 632 | MONGOLIA 605 | PAKISTAN 340 | TURKEY 301 | MYANMAR 261 | ALL OTHERS 2,574 | RUSSIA 5,065 | 1,527 | UKRAINE 233 | FRANCE 211 | SPAIN 195 | SWEDEN 174 | NORWAY 149 | ALL OTHERS 1,339 | SUDAN 968 | ALGERIA 920 | ZAIRE 905 | LIBYA 679 | CHAD 496 | NIGER 489 | MALI 482 | ANGOLA 481 | S. AFRICA 471 | ETHIOPIA 447 | MAURITANIA 396 | EGYPT 387 | TANZANIA 365 | NIGERIA 357 | NAMIBIA 318 | MOZAMBIQUE 309 |

ASIA 17,337 — EUROPE 3,828 — AFRICA 11,716

Comparative Populations (Numbers indicate millions of people.) 1/1/94 estimate

| CHINA 1,184.1 | INDIA 906.8 | INDONESIA 198.8 | PAKISTAN 126.1 | JAPAN 124.8 | BANGLADESH 115.2 | VIET NAM 72.1 | PHILIPPINES 66.2 |

ASIA 3,394.9

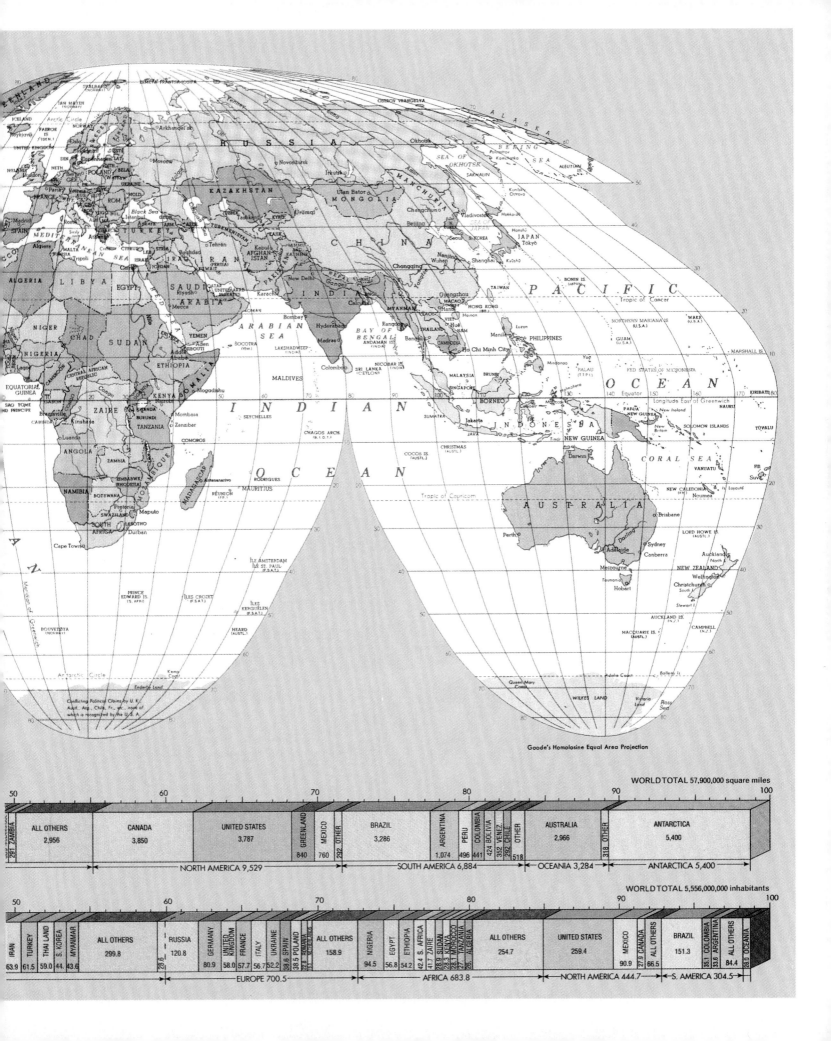

Goode's Homolosine Equal Area Projection

WORLD TOTAL 57,900,000 square miles

50		60						70			80									90		100

| ZAMBIA 291 | ALL OTHERS 2,956 | CANADA 3,850 | UNITED STATES 3,787 | GREENLAND 840 | MEXICO 760 | OTHER 292 | BRAZIL 3,286 | ARGENTINA 1,074 | PERU 496 | COLOMBIA 441 | BOLIVIA 424 | VENEZ. 352 | CHILE 292 | OTHER 518 | AUSTRALIA 2,966 | OTHER 318 | ANTARCTICA 5,400 |

NORTH AMERICA 9,529 —— SOUTH AMERICA 6,884 —— OCEANIA 3,284 —— ANTARCTICA 5,400

WORLD TOTAL 5,556,000,000 inhabitants

50		60							70			80									90		100

| IRAN 63.9 | TURKEY 61.5 | THAILAND 59.0 | S. KOREA 44. | MYANMAR 43.6 | ALL OTHERS 299.8 | RUSSIA 120.8 | GERMANY 80.9 | UNITED KINGDOM 58.0 | FRANCE 57.7 | ITALY 56.7 | UKRAINE 52.2 | SPAIN 38.8 | POLAND 38.5 | ROMANIA 22.8 | NETHERLANDS 15.3 | ALL OTHERS 158.9 | NIGERIA 94.5 | EGYPT 56.8 | ETHIOPIA 54.2 | S. AFRICA 42.9 | ZAIRE 41.7 | SUDAN 28.9 | KENYA 28.8 | MOROCCO 28.6 | TANZANIA 28.4 | ALGERIA 26. | ALL OTHERS 254.7 | UNITED STATES 259.4 | MEXICO 90.9 | CANADA 27.9 | ALL OTHERS 66.5 | BRAZIL 151.3 | COLOMBIA 35.1 | ARGENTINA 33.6 | ALL OTHERS 84.4 | OCEANIA 28.0 |

EUROPE 700.5 —— AFRICA 683.8 —— NORTH AMERICA 444.7 —— S. AMERICA 304.5

PART
one

Environmental science is an interdisciplinary study that describes problems caused by human use of the natural world. It also seeks remedies for these problems. To learn about this complex field of study, it helps to understand three things: First, it is important to understand the natural processes (both physical and biological) that operate in the world. Second, it is important to appreciate the role that technology plays in our society and its capacity to alter natural processes as well as solve problems caused by human impact. Third, it helps to understand the complex social processes that characterize human populations. When we integrate that understanding with a knowledge of technology and natural processes, we can fully appreciate our role in the natural world.

Chapter 1 introduces the central theme of interrelatedness by analyzing some environmental issues in North America region by region. Chapter 2 discusses the differences that can exist among individuals in a society and the different behaviors people exhibit, depending on whether they are acting as individuals, as part of a corporation, or as part of government.

chapter 1

Environmental Interrelationships

Objectives

After reading this chapter, you should be able to:

- Understand why environmental problems are complex and interrelated.
- Realize that environmental problems involve social, political, and economic issues, not just scientific issues.
- Understand that acceptable solutions to environmental problems are not often easy to achieve.
- Understand that all organisms have an impact on their surroundings.
- Understand what is meant by an ecosystem approach to environmental problem solving.
- Recognize that different geographic regions have somewhat different environmental problems, but the process for resolving them is the same and involves compromise.

Chapter Outline

The Field of Environmental Science

The Interrelated Nature of Environmental Problems

Environmental Close-Up: *Science Versus Policy*

Global Perspective: *Fish, Seals, and Jobs*

An Ecosystem Approach

Regional Environmental Concerns

Environmental Close-Up: *The Greater Yellowstone Ecosystem*

Environmental Close-Up: *Ecosystem Size*

   The Wilderness North

   The Agricultural Middle

   The Dry West

   The Forested West

   The Great Lakes and Industrial Northeast

   The Diverse South

The Field of Environmental Science

Environmental science is an interdisciplinary area of study that includes both applied and theoretical aspects of human impact on the world. Since humans are generally organized into groups, environmental science must deal with politics, social organization, economics, ethics, and philosophy. Thus, environmental science is a mixture of traditional science, societal values, and political awareness. (See figure 1.1.)

Environmental science as a field of study is evolving, but it is rooted in the early history of civilization. Many ancient cultures expressed a reverence for the plants, animals, and geographic features that provided them with food, water, and transportation. These features are still appreciated by modern people. The following quote from Henry David Thoreau (1817–1862) is over a century old but is consistent with current environmental philosophy:

> I wish to speak a word for Nature,
> for absolute freedom and wildness,
> as contrasted with a freedom and
> culture merely civil . . . to regard
> man as an inhabitant, or a part and
> parcel of Nature, rather than a
> member of society.

The current interest in the state of the environment began with philosophers like Thoreau and received an additional push by the organization of the first Earth Day on April 22, 1970. The second Earth Day on April 22, 1990 reaffirmed this commitment, as have similar Earth Days since then. As a result of this continuing interest in the state of the world and how people affect it, environmental science is now a standard course on many college campuses and is also a part of most high school course offerings. Most of the concepts covered by environmental science courses had previously been taught in ecology, conservation, or geography courses. Environmental science incorporates the scientific aspects of these courses with input from the social

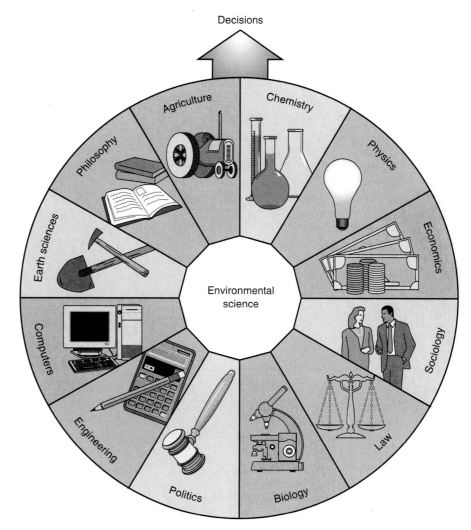

Figure 1.1

Environmental Science   The field of environmental science involves an understanding of scientific principles, economic influences, and political action. Environmental decisions often involve compromise. A decision that may be supportable from a scientific or economic point of view may not be supportable from a political point of view without modification. Often political decisions relating to the environment may not be supported by economic analysis.

sciences, such as economics, sociology, and political science, creating a new interdisciplinary field.

The Interrelated Nature of Environmental Problems

Environmental science is by nature an interdisciplinary field. The word *environment* is usually understood to mean the surrounding conditions that affect people and other organisms. In a broader definition, **environment** is everything that affects an organism during its lifetime. From a human perspective, environmental issues involve concerns about science, nature, health, employment, profits, politics, ethics, and economics.

Most social and political decisions are made with respect to political jurisdictions, but environmental problems do not necessarily coincide with these artificial, political boundaries. For example, air pollution may involve several local units of government, several states or provinces, and even different nations. Air

environmental
close-up

Science Versus Policy

Scientific knowledge and government policy do not always agree. The scientific community can advise governments but cannot insist that certain policies be adopted. Governments may halt some scientific research because they control funding sources, or they may introduce regulations that make continuing the research difficult. For example, much federal money was spent on alternative energy research during the Carter presidency, but many of these projects were not in favor during the Reagan and Bush presidencies. Funding was reduced and much of the research into alternative fuels stopped. Conversely, the passage of the 1990 Clean Air Act during the Bush presidency mandated that alternative-fuel

automobiles be used in some cities with severe air-pollution problems.

Government policy may be contrary to prevailing scientific opinion for economic or political reasons. For many years during the Reagan administration, most scientists in the United States and Canada agreed that the burning of high-sulfur coal and other acid-producing fuels was responsible for acid rain, which was leading to the deaths of lakes in parts of Canada and the northeastern United States. The administration continued to insist that the information was not conclusive and that the problem should be studied in greater detail.

GLOBAL PERSPECTIVE

Fish, Seals, and Jobs

In 1995, the Canadian government announced a moratorium on cod fishing along the east coast of Canada. The cod industry contributes $700 million a year and 31,000 jobs to the Canadian economy, primarily in Newfoundland. At the same time, the government announced that it would begin a program to encourage the harvesting of harp seals by helping develop markets for seal products. Environmental groups oppose the harvesting of the seals.

How do all these pieces fit together? It is thought that the low numbers of cod in the North Atlantic are partially the result of an increasing population of harp seals. While overfishing, larger nets, and other factors have also contributed to the decline of the cod, it is true that harp seals feed on fish that could have been harvested. The current harp seal population of the Atlantic Coast is estimated at 3.5 million, which is an increase of 75 percent from the 1985 estimate of 2 million animals. The increase in harp seals is at least partly the result of actions during the 1970s by environmental groups that sought to stop the killing of seals because they considered the harvesting method inhumane. The traditional method involves clubbing the young seals to death. In 1996, the Canadian government increased the seal hunt quota to 287,000 from 186,000.

It appears that what might be seen initially as isolated, unrelated factors are really interrelated in a way that affects the economy of an entire province.

pollution problems in Juarez, Mexico, are also problems in El Paso, Texas. But the issue is more than air quality and human health. Lower wage rates and less strict environmental laws have influenced some industries to move to Mexico for economic reasons. Mexico and many other developing nations are struggling to improve their environmental image and need the money generated by foreign invest-

ment to improve the conditions and the environment in which their people live.

Air pollutants produced in the major industrial regions of the United States drift across the border into Canada, where acid rain damages lakes and forests. A long-standing dispute exists between the United States and Canada over this issue. Canada claims that the United States should be doing more to reduce emissions

that cause acid rain, and the United States claims it is doing as much as it can. In another example, farmers who use water from the Colorado River for irrigation reduce the quality and quantity of water entering Mexico. This causes political friction between Mexico and the United States.

Because of all these political, economic, ethical, and scientific links, solving

Figure 1.2

The Earth Summit   The United Nations Conference on Environment and Development held in Rio de Janeiro in 1992 was the first worldwide conference at which heads of state were asked to commit their countries to specific environmental policies.

environmental problems is complex. The problems seldom have simple solutions. However, international organizations such as the International Joint Commission have had major bearing on the quality of the environment over broad regions of the world.

The International Joint Commission was established in 1909 when the Boundary Waters Treaty was signed between the United States and Canada. The treaty was established in part to provide that the "boundary waters and waters flowing across the boundary shall not be polluted on either side to the injury of health or property of the other." The Commission has been instrumental in identifying areas of concern and encouraging the cleanup of polluted sites that affect the quality of the Great Lakes and other boundary waters. In general, the two governments have listened to the Commission's advice and have responded by initiating cleanup activities.

The first worldwide meeting of heads of state directed to concern for the environment took place at the Earth Summit, formally known as the United Nations Conference on Environment and Development (UNCED) in Rio de Janeiro in 1992. (See figure 1.2.) Most countries signed agreements on sustainable development and biodiversity. Many nations had signed previous agreements on global warming and depletion of the ozone layer. It may be years before we will know if all countries that signed these agreements will meet their commitments to environmental improvement, but they have at least stated their intention to do so.

The United Nations, through the United Nations Educational, Scientific, and Cultural Organization (UNESCO) and the United Nations Environment Programme (UNEP), has supported many environmental programs. A recent undertaking is the International Environmental Education Programme (IEEP). This program recognizes the need for both formal environmental education in schools and the informal education that occurs through the media and groups of interested citizens. Conferences on environmental education were first held during the 1970s and continue to the present.

An Ecosystem Approach

The natural world is organized into interrelated units called ecosystems. An **ecosystem** is a region in which the organisms and the physical environment form an interacting unit. Weather affects plants, plants use minerals in the soil, and affect animals, animals spread plant seeds, plants secure the soil, and plants evaporate water, which affects weather.

Ecosystems sometimes have fairly discrete boundaries, as is the case with a lake or island. Sometimes the boundaries are indistinct, as in the transition from grassland to desert. Grassland gradually becomes desert depending on the historical pattern of rainfall in an area.

An ecosystem approach requires a look at the way the natural world is organized. Where do the rivers flow? What are the prevailing wind patterns? What are the typical plants and animals in the area? How does human activity affect nature? The task of an environmental scientist is to recognize and understand the natural interactions that take place and to integrate these with the uses humans must make of the natural world.

To illustrate the interrelated nature of environmental issues, we will look at several regions of North America and highlight some of the key features and issues of each.

Regional Environmental Concerns

No region is free of environmental concerns. Most regions tend to focus on specific, local environmental issues that apply directly to them. For example, protecting endangered species is a concern in many parts of the world. It has become a matter of particular concern in the Pacific Northwest where an endangered species known as the northern spotted owl appears to require undisturbed, mature forests for its survival. In most metropolitan areas the problem of endangered species is purely intellectual, since the construction of cities has essentially destroyed the previously existing

The Greater Yellowstone Ecosystem

In 1872, the U.S. government established Yellowstone National Park as the world's first national park. It was a huge area that protected unique natural features such as geysers, hot springs, rivers, lakes, and mountains. It was also a preserve for many kinds of wildlife such as grizzly bears, elk, moose, and bison. At the time it was established, the park was thought to be of adequate size to protect the scenic resources and the wildlife. Since that time, the lands surrounding the park have been converted to a variety of uses, including cattle grazing, timber production, hunting, and mining.

Fortunately, most of the lands surrounding Yellowstone National Park and the adjacent Grand Teton National Park are still under government control as national forests, national wildlife refuges, and other state, local, or federal bodies. Some of the park wildlife, particularly the grizzly bear and bison, often wander across the park boundaries. The grizzly in particular needs large regions of wilderness to survive as a species.

Many people assert that it is essential that these lands be integrated into a Greater Yellowstone Ecosystem management plan encompassing about 7.3 million hectares. The plan is based on more natural boundaries than the original boundaries established in 1872. This would require changes in the way much of the land surrounding Yellowstone is currently being used. The trade-offs are significant. Logging, mining, hunting, and grazing would be stopped or significantly reduced. This would result in a loss of jobs in those industries. Proponents argue that additional jobs would be created in the tourist and related service industries. The advantages, they argue, would be equal to or greater than the economic losses caused by stopping current uses.

Ecosystem Size

Many people interested in songbirds have documented a significant decrease in the numbers of certain species of songbirds. Species particularly affected are those that migrate from North America to South America and require relatively large areas of undisturbed forest. Many of these birds are being negatively affected by human activities that fragment large patches of forest into many smaller patches. This creates more edges between forested and nonforested habitats. Bird and animal species that use edge habitats, which have a mixture of closed forest and open areas, supplant the songbirds, which require large patches of undisturbed forest. Species are affected most severely when both their northern and southern habitats are disturbed.

A survey of migrating sharp-shinned hawks indicates greatly reduced numbers in recent years. It is thought that since sharp-shinned hawks use small songbirds as their primary source of food, the reduction in their numbers is directly related to the reduction in migratory songbirds.

Indigo bunting

Sharp-shinned hawk

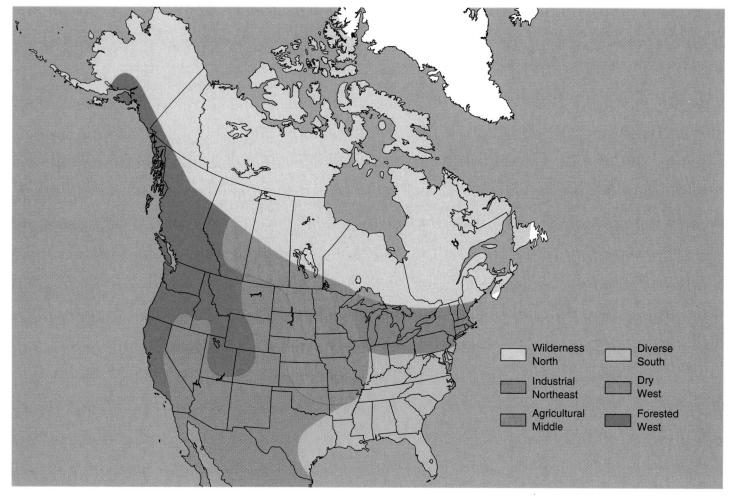

Figure 1.3

Regions of North America Because of natural features of the land and the uses people make of the land, different regions of North America face different kinds of environmental issues. Certainly within each region people face a large number of specific issues, but certain kinds of issues are more important in some regions than in others.

ecosystem. Here we present a number of regional vignettes to illustrate the complexity and interrelatedness of environmental issues. (See figure 1.3.)

The Wilderness North

Much of Alaska and Northern Canada can be characterized as **wilderness**—areas with minimal human influence. Much of this land is owned by governments, not by individuals, so government policies have a large effect on what happens in these regions. These areas have important economic values in their trees, animals, scenery, and other natural resources. Exploitation of the region's natural resources involves significant

trade-offs. Usually, a portion of the natural world is altered permanently, but the area altered is so small that many people consider it insignificant. Because of the severe climate, northern wilderness areas tend to be very sensitive to insults and take a long time to repair damage done by unwise exploitation. Mining, oil exploration, development of hydroelectric projects, and harvesting of timber all require roads and other human artifacts, involve the insertion of new technologies into native culture, and generate economic benefits.

In the past, many short-term political and economic decisions failed to look at long-term environmental implications. Today, however, people are concerned

about these remaining wilderness areas. Politicians are more willing to look at the scientific and recreational values of wilderness as well as the economic value of exploitation.

Native people, who consider much of this region to be their land, have become increasingly sophisticated in negotiating with state, provincial, and federal governments to protect rights they feel they were granted in treaties. They are sensitive to changes in land use or government policy that would force changes in their traditional way of life.

Native people, other concerned citizens, business interests, and environmental activists have become increasingly sophisticated in influencing decisions made by government. The process of compromise is often difficult and does not always assure wise decisions, but most governments now realize they must listen to the concerns of their citizens and balance economic benefits with social and cultural benefits. (See figure 1.4.)

The Agricultural Middle

The middle of the North American continent is dominated by intensive agriculture. This means that the original, natural ecosystems have been replaced by managed, agricultural enterprise. It is important to understand that this area was at one time wilderness. Today, you would need to search very hard to find regions of true wilderness in Iowa, Missouri, or southern Manitoba. Some special areas have been set aside to preserve fragments of the original natural plant and animal associations, but most of the land has been converted to agriculture wherever practical.

The economic value generated by this use of a rich soil resource is tremendous, and consequently most of the land is privately owned. Governments cannot easily control what happens on these privately held lands. But governments indirectly encourage certain activities through departments of agriculture that encourage agricultural research, grant special subsidies to farmers in the form of guaranteed prices for their products and other special payments, and develop markets for products. Yet because the economic risks involved in farming are great, the number of farmers constantly declines. There are a number of reasons for farm failures, including drought, disease, and lack of markets.

One of the major, nonpoint pollution sources (pollution that does not have an easily identified point of origin) is agriculture. Air pollution in the form of dust is an inevitable result of tilling the land. Soil erosion occurs when soil is exposed to wind and moving water and leads to siltation of rivers, impoundments, and lakes. Fertilizers and other agricultural chemicals blow or are washed from the areas where they are applied. Nutrients washed from the land enter rivers and lakes where they encourage the growth of algae, lowering water quality. The use of pesticides causes concern about human exposure,

Walrus harvesting

Grizzly bear fishing for salmon

A clear-cut forest

Discussions with government officials

Figure 1.4

The Wilderness North Protection of wilderness is a major issue of this region. The major points of conflict involve the government role in managing these lands, the protection of the rights and beliefs of native people, and the desire of many to exploit the mineral and other resources of the region.

effects on wild animals that are accidentally exposed, and residues in farm products.

Since many communities in this region rely on groundwater for drinking water, the use of fertilizers and pesticides, and their potential for entering the groundwater as a result of unwise or irresponsible use, is a concern. In addition, many farmers use groundwater for irrigation, which lowers the water table and leaves less groundwater for other purposes.

In an effort to stay in business and preserve their way of life, farmers must use modern technology. Careful use of these tools can reduce their impact; irresponsible use causes increased erosion, water pollution, and risk to humans. (See figure 1.5.)

The Dry West

Where rainfall is inadequate to support agriculture, ranching and raising livestock are possible. This is true in much of the drier portions of western North America. Because much of the land is of low economic value, most is still the property of government, which encourages its use by providing water for livestock and irrigation at minimal cost, offering low rates for grazing rights, and encouraging mining and other development.

Many people believe that government agencies have seriously mismanaged these lands. They assert that the agencies are controlled by special interest groups and powerful politicians sensitive to the needs of ranchers, that they subsidize ranchers by charging too little for grazing rights, and that they allow destructive overgrazing because of the economic needs of ranchers. Ranchers argue that they need access to government-owned land, cannot afford significantly increased grazing fees, and that changing government policies would destroy a way of life that is important to the regional economy.

A well-kept farm

Farm auction

Agricultural chemicals

Ship loaded with grain

Figure 1.5

The Agricultural Middle   The rich soil resource of this region has been converted to managed agricultural activity. The use of pesticides and fertilizer and exposure of the land to erosion cause concern about pollution of surface and groundwater. Farmers maintain that these practices are essential in modern agriculture and that they can be used safely and with minimal pollution.

Water is an extremely valuable resource in this region. It is needed for municipal use and for agriculture. Many areas, particularly the river valleys, have fertile soils that can be used for intensive agriculture. Cash crops such as cotton, fruits, and vegetables can be grown if water is available for irrigation. Because water tends to evaporate from the soil rapidly, long-term use of irrigated lands often results in the buildup of salts in the soil, thus reducing fertility. Irrigation water flowing from fields is polluted by agricultural chemicals that make it unsuitable for other uses such as drinking. As cities in the region grow, an increasing conflict arises between urban dwellers who need water for drinking and other purposes, and ranchers and farmers who need the water for livestock and agriculture. Increased demand for water will result in shortages, and decisions will have to be made about who will ultimately get the water. If the urban areas get the water they want, some farmers and ranchers will go out of business. If the agricultural interests get the water, urban growth and development will have to be limited and expensive changes will have to be made to conserve domestic water use.

Because population density is low in most of this region, much of the land has a wilderness character. Increasingly, a conflict has developed between the economic management of the land for livestock production and the desire on the part of many to preserve the "wilderness." Designating an area as wilderness means that certain uses are no longer permitted. This offends individuals and groups who have traditionally used the area for grazing, hunting, and other pursuits. A long history of use and abuse of this land by overgrazing, modification to encourage plants valuable for livestock, and the introduction of grasses for livestock has significantly altered the region so that it cannot truly be called wilderness. The low population density does, however, provide a remoteness and wild character that many seek to preserve. (See figure 1.6.)

The Forested West

The coastal areas and mountain ranges of the western United States and Canada receive sufficient rainfall that coniferous forests are the dominant vegetation. Since most of these areas are not suitable for farmland, they have been maintained as forests with some grazing activity in the more open forests. Governments and large, commercial timber companies own large sections of these lands. Government forest managers (U.S. Forest Service, Bureau of Land Management, Environment Canada, and various state and provincial departments) historically have sold timber-cutting rights at a loss and are thought by many to be too interested in the production of forest products at the expense of other, less tangible values. In 1993, the U.S. Forest Service was directed to stop below-cost timber sales.

This policy change has become a major issue in the old-growth forests of the Pacific Northwest where timber interests maintain that they must have access to government-owned forests in order to remain in business. Many of these areas have significant wilderness, scenic, and recreational value. Environmental interests point out that it makes no sense to complain about the destruction of tropical rainforests in South America while North America makes plans to cut large areas of previously uncut, temperate rainforest. Are the intangible values of preserving an ancient forest ecosystem as important as the economic values provided by timber and jobs?

Environmental interests are concerned about the consequences logging would have on organisms that require mature, old-growth forests for their survival. Grizzly bear habitat in Alaska and British Columbia could be altered significantly by logging; the northern spotted owl has become a symbol of the conflict between logging and preservation in Oregon and Washington; and

Overgrazed land

Wilderness area

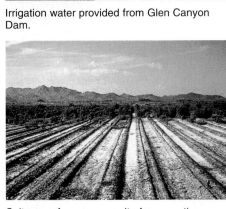
Irrigation water provided from Glen Canyon Dam.

Bryce Canyon

Salt on surface as a result of evaporation on irrigated land

Figure 1.6

The Dry West Water is a key issue in this region. Both city dwellers and rural ranchers and farmers need water, and conflict results when there is not enough water to satisfy the desires of all. In addition, much of the land in this region is owned by government. This raises concerns about how government manages the land and how government policy affects the people of the region.

preservation of coastal redwood forests has become an issue in northern California. (See figure 1.7.)

The Great Lakes and Industrial Northeast

While much of the West and Central region of North America is characterized by low population densities and small towns, major portions of the Great Lakes and Northeast are dominated by large metropolitan complexes that generate social and resource needs that are difficult to satisfy. Many of these older cities formed around industrial centers that have declined, leaving behind poverty, environmental problems in abandoned industrial sites, and difficulties with solid waste disposal, air quality, and land-use priorities. Interspersed among the major metropolitan areas are small towns, farmland, and forests.

One of the major resources of the region is its access to water transport. The Great Lakes and eastern seacoast are extremely important to commerce, since ships can travel throughout the area by way of the St. Lawrence Seaway and the Great Lakes through a series of locks and canals that bypass natural barriers. Because of the importance of shipping in this region, harbors have been constructed and waterways have been deepened by dredging. The waterways are maintained at considerable government expense.

One of the greatest problems associated with the industrial uses of the Great Lakes and East Coast is contamination of the water with toxic materials.

Old-growth forest

U.S. Forest Service ownership

Logging

Wildlife

Figure 1.7

The Forested West   The cutting of forested areas for timber production destroys the previous ecosystem. Some see the trees as a valuable resource that provides jobs and building materials. Others see the forest ecosystem as a natural resource that should be preserved. In addition, government ownership of much of this land has generated considerable political debate about what the appropriate use of the land should be.

In some cases, unthinking or unethical individuals have dumped toxins directly into the water. In other cases, small, accidental spills or leaks over long periods of time have contaminated the sediments in harbors and bays.

A major concern about these pollutants is that they bioaccumulate (see chapter 14) in the food chain. The concentrations of some chemicals in the fat tissue of top predators, such as lake trout and fish-eating birds, can be a million times higher than the concentration in the water. Because of this, government agencies have issued consumption advisories for some fish and shellfish in contaminated areas. Since many kinds of fish can swim great distances, advisories for the Great Lakes warn against eating certain fish taken anywhere within the lakes, not just from the site of contamination. Similarly, Chesapeake Bay has been subjected to years of thoughtless pollution, resulting in reduced fish and shellfish populations and advisories against consuming some organisms taken from the bay.

Water always generates considerable recreational value. Consequently, conflicts arise between those who want to use the water for industrial and shipping purposes and those who wish to use it for recreation. Since so much of the North American population is concentrated in this region, the economic value of recreational use is extremely high. Pressure is great to clean up contaminated sites and prevent the pollution of new ones. Contaminated areas do not enhance tourism.

Most of these older, large cities had no plan to shape their growth. As a result, open space for people is limited and urban dwellers have few opportunities

Inner-city decay in Chicago

Central Park in New York City

Harbor in Superior, Wisconsin

Figure 1.8

The Great Lakes and the Industrial Northeast Industry, waterways, and population centers are the defining elements of this region. The historically extensive use of the Great Lakes and coastal areas of the Northeast for industry, because of the ease of providing water transportation, has resulted in many older cities with poor land-use practices. Rebuilding cities, providing recreational opportunities for urban dwellers, and repairing previous environmental damage are important issues. The water resources of the region provide transportation, recreation, and industrial opportunities.

Recreational fishing along Lake Erie

to interact with the natural world. Children who grow up in these cities often do not know that milk comes from a cow—they have never seen, smelled, or touched a cow. Consequently, urban people have difficulty understanding the feeling rural people have for the land. These urban dwellers may never have an opportunity to experience wilderness. Their major environmental priorities are cleaning up contaminated sites, providing more parks and recreation facilities, reducing air and water pollution, and improving transportation. (See figure 1.8.).

The Diverse South

In many ways, the South is a microcosm of all the regions previously discussed. The petrochemical industry dominates the economies of Texas and Louisiana, and forestry and agriculture are significant elements of the economy in other parts of the region. Major metropolitan areas thrive, and much of the area is linked to the coast either directly or by the Mississippi River and its tributaries. The environmental issues faced in the South are as diverse as those in all the other regions.

Some areas of the South (particularly Florida) have had extremely rapid population growth, which has led to groundwater problems, transportation problems, and concerns about regulating the rate of growth. Growth means money to developers and investors, but it requires municipal services, which are the responsibility of local governments. Too many people and too much development also threaten remaining natural ecosystems.

Poverty has been a problem in many areas of the South. This creates a climate that encourages state and local governments to accept industrial development at the expense of other values. Often, jobs are more important than the environmental consequences of the jobs; low-paying jobs are better than no jobs.

The use of the coastline is of major concern in many parts of the South. The coast is a desirable place to live, which encourages unwise development on barrier islands and in areas that are subject to flooding during severe weather. In addition, industrial activity along the coast has resulted in the loss of wetlands. (See figure 1.9.)

Barge on the Mississippi River

Everglades

Miami metropolitan area

Coal miners exiting mine

Chemical plant on lower Mississippi

Figure 1.9

The Diverse South Poverty has been a historically important problem in the region. Often the creation of jobs was considered more important than the environmental consequences of those jobs. The use of coastal areas for industry has resulted in pollution of coastal waters. The heavy use of the Mississippi River for transportation and industry has caused pollution problems. In addition, the desirable climate in the South has resulted in intense pressure to develop new housing for those who want to move to the region. Unwise development of housing on fragile coastal sites has resulted in damage to buildings by storms and the actions of the oceans. This causes intense debate on land use.

SUMMARY

Artificial political boundaries create difficulties in managing environmental problems because most environmental units, ecosystems, do not coincide with political boundaries. Therefore, a regional approach to solving environmental problems, one that incorporates natural geographic units, is ideal. Each region of the world has certain environmental issues that are of primary concern because of the mix of population, resource use patterns, and culture.

Environmental problems become issues when someone finds a situation offensive. This inevitably leads to a confrontation between groups that have different views on what constitutes an environmental problem. Many social, economic, ethical, and scientific inputs shape a person's opinions. The process of environmental decision-making must take all of these inputs into account and arrive at an acceptable compromise.

Environmental problems are people problems. They occur because the uses of natural resources, which some people feel are justified, result in a diminished environment for others in the region. Environmental problems are defined by the person who perceives the problem. When perceptions differ, conflict occurs. Environmental decisions inevitably involve economic considerations because someone is receiving value from the resources being used or someone perceives an economic loss because a use has been withdrawn.

- Some argue that economic considerations should not be important when making environmental decisions; others argue that economic considerations can resolve all environmental issues.
- Some argue that regulation is necessary to protect resources; others argue that regulation hinders valuable use of resources.
- Some consider nonhuman organisms as important as humans; others feel that humans have a special place in nature.
- Some are against change; others recognize that change must occur.

With all these differing opinions, compromise is the only way to resolve the conflicts. The social institution of government must play a role. Economic evaluation is important. Recogni- tion of the validity of opposing points of view is essential. The field of environmental science seeks to find that middle ground.

REVIEW QUESTIONS

1. Describe why finding solutions to environmental problems is so difficult. Do you think it has always been as complicated?
2. Describe what is meant by an ecosystem approach to environmental problem-solving. Is this the right approach?
3. List two key environmental issues for each of the following regions: the wilderness North, the agricultural middle, the forested West, the dry West, the Great Lakes and industrial Northeast, and the South. How are the issues changing?
4. Define *environment* and *ecosystem* and provide examples of these terms from your region.
5. Describe how environmental conflicts are resolved.
6. Select a local environmental issue and write a short essay presenting all sides of the question. Is there a solution to this problem?

KEY TERMS

ecosystem   7

environment   5

environmental science   5

wilderness   9

RELATED INTERNET SITES

Web directory search engine for environmental organizations. Categories range from *auto* to *'zines*.
    http://www.webdirectory.com
CEPA and the globalization of environmental protection.
    http://www.doe.ca/cepa/ip19/e19_03.html
Coastal ecosystems programs in the United States.
    http://www.fws.gov/~cep/cepcode.html
U.S. Fish and Wildlife Service detailing the seven environmental regions of the United States. Lots of great links.
    http://www.fws.gov/
Fact sheet for the Northwestern region of the United States.
    http://www.fws.gov/~r5ws/facts.html

Award-winning site with great up-to-date links on environment and technology.
    http://www.gnet.org/
Searchable index of a great variety of environmental issues. Great starting point.
    http://www.nerdworld.com/nw1185.html
Cears information on climate. Another good starting point to connect to a wide variety of sites.
    http://www2.waikato.ac.nz/cears/index.html
Save Harp Seals. Great photos and contacts for saving the baby harp seals.
    http://www.bhs.mq.edu.au/~laurie/seal.html

chapter 2

Environmental Ethics

Objectives

After reading this chapter, you should be able to:

- Differentiate between ethics and morals.
- Define personal ethics.
- Explain the connection between material wealth and resource exploitation.
- Describe how industry must exploit resources and consume energy to produce goods.
- Explain how corporate behavior is determined.
- Describe the influential power that corporations wield because of their size.
- Explain why governmental action was necessary to force all companies to meet environmental standards.
- Describe the factors associated with environmental justice.
- Describe what has been the general attitude of society toward the environment.
- Explain the relationship between economic growth and environmental degradation.
- List three conflicting attitudes toward nature.

Chapter Outline

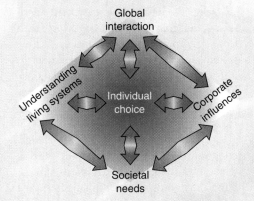

Views of Nature

The most beautiful object I have ever seen in a photograph in all my life, is the planet Earth seen from the distance of the moon, hanging in space, obviously alive. Although it seems at first glance to be made up of innumerable separate species of living things, on closer examination every one of its things, working parts, including us, is interdependently connected to all the other working parts. It is, to put it one way, the only truly closed ecosystem any of us know about.

Lewis Thomas

One of the marvels of recent technology is that we can see the Earth from the perspective of space, a blue sphere unique among all the planets in our solar system. (See figure 2.1.) Looking at ourselves from space, it becomes obvious, says ecologist William Clark of Harvard University, that only as a global species, "pooling our knowledge, coordinating our actions, and sharing what the planet has to offer—do we have any prospect for managing the planet's transformation along pathways of sustainable development."

Many people see little value in an undeveloped river and feel it is unreasonable to leave it flowing in a natural state. Many rivers throughout the world have been "controlled" to provide power, irrigation, and navigation at the expense of the natural world. Many people think that to not use these resources would be wasteful.

In the U.S. Pacific Northwest, there is a conflict over the value of old-growth forests. Economic interests want to use the forests for timber production and feel that to not do so would cause economic hardship. They argue that the trees are going to die anyway and they might as well be used for the betterment of the community. Others feel that all the living things that make up the forest have a value we do not yet appreciate. Removing the trees would destroy something that took hundreds of years to develop and may never be replaced.

Interactions between people and their environment are as old as human

Figure 2.1

The Earth as Seen from Space   Political, geographical, and nationalistic differences among humans do not seem so important from this perspective. In reality, we all share the same "home."

civilization. The problem of managing those interactions, however, has been transformed today by unprecedented increases in the rate, scale, and complexity of the interactions. At one time, pollution was a local, temporary event. Today, pollution may involve several countries—as with the concern over acid deposition in Europe and in North America—and affect multiple generations. The debates over chemical and radioactive waste disposal are examples of the increasingly international nature of pollution. For example, many European countries are concerned about the transportation of radioactive and toxic wastes across their borders. What were once straightforward confrontations between ecological preservation and economic growth now involve multiple linkages that blur the distinction between right and wrong. For example, the greenhouse

effect is thought to result from energy consumption, agricultural practices, and climatic change.

Many people believe that we have entered an era characterized by global change that stems from the interdependence between human development and the environment. They argue that self-conscious, intelligent management of the earth is one of the greatest challenges facing humanity as we approach the twenty-first century. To meet this challenge, they believe, a new environmental ethic must evolve.

Ethical issues dealing with the environment are different from other kinds of ethical problems. Depending on your perspective, an environmental ethic could encompass differing principles and beliefs. Perhaps ecologist and writer Aldo Leopold summed up his understanding of an environmental ethic best in his essay "The

environmental
close-up

What is Ethical?

Ethics is one branch of philosophy. Ethics seeks to define fundamentally what is right and what is wrong, regardless of cultural differences. For example, most cultures have a reverence for life and hold that all individuals have a right to live. It is considered unethical to deprive an individual of life.

Morals differ somewhat from ethics because morals reflect the predominant feelings of a culture about ethical issues. For example, in almost all cultures, it is certainly unethical to kill someone; however, when a country declares war, most of its people accept the necessity of killing the enemy. Therefore, it is a moral thing to do even though ethics says that killing is wrong. No nation has ever declared an immoral war.

Environmental issues require a consideration of ethics and morals. For example, because there is currently enough food in the world to feed everyone adequately, it is unethical to allow some people to starve while others have more than enough. However, the predominant mood of those in the developed world is one of indifference. They don't feel morally bound to share what they have with others. In reality, this indifference says that it is permissible to allow people to starve. This moral stand is not consistent with a purely ethical one.

As we can see, ethics and morals are not always the same; thus, it is often difficult to clearly define what is right and what is wrong. Some individuals view the world's energy situation as serious and have reduced their consumption. Others do not believe there is a problem and so have not modified their energy use. Still others do not care what the situation is. They will use energy as long as it is available. Other issues are population and pollution. Is it ethical to have more than two children when the world faces overpopulation? Should an industry persuade the public to vote "no" on a particular legislative bill because it might reduce profits, even though its passage would improve the environment? The stand we take on such issues often depends on our position. An industrial leader, for example, would probably not look upon pollution as negatively as someone who is active outdoors. In fact, many business leaders view the behavior of active preservationists as immoral because it restricts growth and, in some cases, causes unemployment.

Most ethical questions are very complex. Ethical issues dealing with the environment are no different. It is important to explore environmental issues from several points of view before taking a stand. One point to consider is the difference between the short-term and long-term effects of a course of action.

When we take an ethical stand, we become open to attack from those who disagree with our stand. Often, individuals are portrayed as villains for pursuing a course of action they consider righteous.

Land Ethic" from *A Sand County Almanac and Sketches Here and There* (1949):

> All ethics so far evolved rest upon a single premise: that the individual is a member of a community of interdependent parts. The land ethic simply enlarges the boundaries of the community to include soils, waters, plants, and animals, or collectively, the land...a land ethic changes the role of *Homo sapiens* from conqueror of the land-community to plain member and citizen of it...It implies respect for his fellow-members, and also respect for the community as such.

Leopold also wrote that "a thing is right when it tends to preserve the integrity, stability, and beauty of the biotic community. It is wrong when it tends otherwise.... We abuse land because we regard it as a commodity belonging to us." When we see land as a community to which we belong, we may begin to use it with love and respect."

Some environmental ethics are founded on an awareness that humanity is part of nature and that nature's many parts are interdependent. In any natural community, the well-being of the individual and of each species is tied to the well-being of the whole. In a world increasingly without environmental borders, nations, like individuals, should have a fundamental ethical responsibility to respect nature and to care for the Earth, protecting its life-support systems, biodiversity, and beauty and caring for the needs of other countries and future generations.

Environmental ethicists argue that to consider environmental protection as a "right" of the planet is a natural extension of the concept of human rights. Many also argue that an environmental ethic considers one's actions toward the environment as a matter of right and wrong, rather than one of self-interest.

Environmental Attitudes

There are many different attitudes about the environment, most of which fall under one of three headings: (*a*) the development ethic, (*b*) the preservation ethic, and (*c*) the conservation ethic. Each of these ethical positions has its own code of conduct against which ecological mortality may be measured. (See figure 2.2.)

environmental
close-up

Naturalist Philosophers

The philosophy behind the environmental movement had its roots in the last century. Among many notable conservationist philosophers, several stand out: Ralph Waldo Emerson, Henry David Thoreau, John Muir, Aldo Leopold, and Rachel Carson.

In Emerson's first essay, *Nature*, published in 1836, he claimed that "behind nature, throughout nature, spirit is present." Emerson was an early critic of rampant economic development, and he sought to correct what he considered to be the social and spiritual errors of his time. In his *Journals*, published in 1840, Emerson stated that "a question which well deserves examination now is the Dangers of Commerce. This invasion of Nature by Trade with its Money, its Credit, its Steam, its Railroads, threatens to upset the balance of Man and Nature."

A naturalist who held beliefs similar to Emerson's was Henry David Thoreau. Thoreau's bias fell on the side of "truth in nature and wilderness over the deceits of urban civilization." The countryside around Concord, Massachusetts, fascinated and exhilarated him as much as the commercialism of the city depressed him. It was near Concord that Thoreau wrote his classic, *Walden*, which describes a year in

which he lived in the country to have direct contact with nature's "essential facts of Life." In his later writings and journals, Thoreau summarized his feelings toward nature with prophetic vision:

> But most men, it seems to me, do not care for Nature and would sell their share in all her beauty, as long as they may live, for a stated sum—many for a glass of rum. Thank God, man cannot as yet fly, and lay waste the sky as well as the earth! We are safe on that side for the present. It is for the very reason that some do not care for these things that we need to continue to protect all from the vandalism of a few. (1861).

John Muir combined the intellectual ponderings of a philosopher with the hard-core, pragmatic characteristics of a leader. Muir believed that "wilderness mirrors divinity, nourishes humanity, and vivifies the spirit." Muir tried to convince people to leave the cities for a while to enjoy the wilderness. However, he felt that the wilderness was threatened. In the 1876 article entitled, "God's First Temples: How Shall We Preserve Our Forests," published in the Sacramento *Record Union*, Muir argued that only government control

Ralph Waldo Emerson

Henry David Thoreau

John Muir

The **development ethic** is based on individualism or egocentrism. It assumes that the human race is and should be the master of nature and that the earth and its resources exist for our benefit and pleasure. This view is reinforced by the work ethic, which dictates that humans should

be busy creating continual change and that things that are bigger, better, and faster represent "progress," which itself is good. This philosophy is strengthened by the idea that, "if it can be done, it should be done," or that our actions and energies are best harnessed in creative work.

Examples of the development ethic abound. The notion that bigger is better is certainly not new to us, nor is the belief that if something can be done or built it should be. The dream of upward mobility is embodied in this ethic. In some circles, questioning growth is considered almost could save California's finest sequoia groves from the "ravages of fools." In the early 1890s, Muir organized the Sierra Club to "explore, enjoy, and render accessible the mountain regions of the Pacific Coast" and to enlist the support of the government in preserving these areas. His actions in the West convinced the federal government to restrict development in the Yosemite Valley, which preserved its beauty for generations to come.

Another thinker as well as a doer in the early conservation field was Aldo Leopold. As a philosopher, Leopold summed up his feelings in *A Sand County Almanac*:

> Wilderness is the raw material out of which man has hammered the artifact called civilization. No living man will see again the long grass prairie, where a sea of prairie flowers lapped at the stirrups of the pioneer. No living man will see again the virgin pineries of the Lake States, or the flatwoods of the coastal plain, or the giant hardwoods.

Leopold founded the field of game management. In the 1920s, while serving in the Forest Service, he worked for the development of a wilderness policy and pioneered his concepts of game management. He wrote extensively in the *Bulletin* of the American Game Association and stated that the amount of space and the type of forage of a wildlife habitat determine the number of animals that can be supported in an area. Furthermore, he said that regulated hunting can maintain a proper balance of wildlife.

While most people talk about what's wrong with the way things are, few actually go ahead and change it. Rachel Carson ranks among those few. A distinguished naturalist and best-selling nature writer, Rachel Carson published a series of articles in the *New Yorker* in 1960, which generated widespread discussion about pesticides. In 1962, she published *Silent Spring*, which dramatized the potential dangers of pesticides to food, wildlife, and humans and eventually led to changes in pesticide use in the United States.

Although some technical details of her book have been shown to be in error by later research, her basic thesis that pesticides can contaminate and cause widespread damage to the ecosystem has been established. Unfortunately, Carson's early death from cancer came before her book was recognized as one of the most important events in the history of environmental awareness and action in this century.

Aldo Leopold

Rachel Carson

unpatriotic. In the development ethic, nature has only instrumental value; that is, the environment has value only insofar as human beings economically utilize it. Only in the past fifty to one hundred years have the by-products and waste associated with development been considered.

The **preservation ethic** considers nature special in itself. Nature, it is argued, has intrinsic value or inherent worth apart from human appropriation. Preservationists have different reasons for wanting to preserve nature. Some have a religious belief regarding nature. They

hold a reverence for life and respect the right of all creatures to live, no matter what the social and economic costs. Some preservationists' interest in nature is primarily aesthetic or recreational. They believe that nature is beautiful and refreshing and should be available for

Preservation

Development

Conservation

Recreation

Figure 2.2

The Views of Nature   Individuals envision the same resources used differently.

picnics, hiking, camping, fishing, or just peace and quiet.

In addition to the religious and recreational preservationists, there are also preservationists whose reasons are essentially scientific. They argue that the human species depends on and has much to learn from nature. Rare and endangered species and ecosystems, as well as the more common ones, must be preserved because of their known or assumed long-range, practical utility. In this view, natural diversity, variety, complexity, and wilderness is thought to be superior to humanized uniformity, simplicity, and domesticity. Scientific preservationists want to lock up not all the land but only what they consider important to future generations.

The third environmental ethic is referred to as the **conservation ethic.** It is related to the scientific preservationist view but extends the rational consideration to the entire earth and for all time. It recognizes the desirability of decent living standards, but it works toward a balance of resource use and resource availability.

The conservation ethic stresses a balance between total development and absolute preservation. It stresses that rapid and uncontrolled growth in population and economics is self-defeating in the long run. The goal of the conservation ethic is one people living together in one world, indefinitely.

Societal Environmental Ethics

Society is composed of a great variety of people with differing viewpoints. This variety can be distilled into a set of ideas that reflect the prevailing attitudes of society. The collective attitudes can be analyzed from an ethical point of view. Western, developed societies have long acted as if the earth has unlimited reserves of natural resources, an unlimited ability to assimilate wastes, and a limitless ability to accommodate unchecked growth.

The economic direction and rationale of developed nations have been that of continual growth. Unfortunately, this growth has not always been carefully planned or even desired. This "growth-mania" has resulted in the use of our nonrenewable resources for comfortable homes, well-equipped hospitals, convenient transportation, fast-food outlets, VCRs, home computers, and battery-operated toys, among many other things. In economic statistics, such "growth" measures out as "productivity," and life looks rosy for all. But the question arises, "What is enough?" Many poor societies have too little, but rich societies never say, "Halt! We have enough." As the Indian philosopher and statesman Mahatma Gandhi said, "The earth provides enough to satisfy every person's need, but not every person's greed."

Growth, expansion, and domination remain the central sociocultural objectives of most advanced societies. **Economic growth** and **resource exploitation** are attitudes shared by developing societies. We continue to consume natural resources as if the supplies were never ending. All of this is reflected in our increasingly unstable relationship with the environment, which grows out of our tendency to take from the "common good" without regard for the future.

This attitude is deeply embedded in the fabric of our society. Since the first settlers arrived in North America, nature has been considered an enemy. Frequently, the colonists expressed their relation to the wilderness in military terms. They viewed nature as an enemy to be "conquered," "subdued," or "vanquished" by a pioneer "army." Any qualms the pioneers may have felt about invading and exploiting the wilderness were justified by religious beliefs. This attitude toward nature is still popular today. Many view wilderness solely as underdeveloped land and see value in land only if it is farmed, built upon, or in some way developed. The notion that land and wilderness should be preserved is incomprehensible to some. The thought of purposely opting to not develop a resource is considered almost a sin by many.

environmental *close-up*

Pigeon River

In 1968, oil was discovered in a remote area of northern Michigan known as the Pigeon River Forest. The Pigeon River Forest is unique in many ways. Its 37,600 hectares is regarded as the wildest country in the state's Lower Peninsula. Because of its wilderness qualities, grouse, black bear, bobcats, deer, beaver, and many other animals inhabit the region. Its streams provide excellent conditions for healthy populations of native brook trout. Hunting, fishing, cross-country skiing, camping, and hiking are major forms of recreation in the forest. It is also home to the largest North American elk herd east of the Mississippi.

In the early 1970s, five successful oil wells were drilled in the area. Then, a series of lawsuits filed by environmental groups halted further exploration until 1981. Oil companies, environmental groups, and the state eventually reached a compromise to restrict and closely monitor future oil drilling. The agreement limited drilling to the southern 11,700 hectares of the forest, protecting what then was the elk herd's primary range. The argument in 1980 was that the oil from the Pigeon River area was needed to combat the energy crisis and that the state of Michigan would get several hundred million dollars in royalties from it, which could then be used to buy other recreational lands. By the early 1980s, it was clear that most of the glowing promises on which the compromise was based had not been kept. Much less oil had been found beneath the forest than was originally predicted, so royalties were a fraction of what had been estimated. More important was that the Pigeon River Forest had changed.

Even wtih all the safeguards in the compromise, pollution problems did arise. Groundwater became contaminated with chloride from the drilling when pits meant to store brine began to leak. Part of the forest's most prized commodity—solitude—had been lost, despite the best efforts of state officials, citizen watchdogs, and the oil companies to minimize the disruption.

Now even an untrained observer notices a winding two-track road change into a straight, flat, wide gravel-bed highway when it reaches the southernmost facility where gas and oil from wells in the forest are pumped and stored. At the forest's northern pumping site, a heavy odor of natural gas fumes is evident.

The Pigeon River was not true wilderness, but it was the closest thing to it, the largest single roadless area in the Lower Peninsula of Michigan, and the legacy of a half-century of careful management aimed at restoring what the early loggers had ruined. Trade-offs are a way of life in the world today. The case of the Pigeon River Forest was also a trade-off. A barrel of oil has a straightforward economic value. What value do we place on wilderness? More importantly, what value do you place on wilderness?

a.

b.

(*a*) In economic terms, it is easy to set a value on a barrel of oil. (*b*) There is no good way to place a value on some things, such as scenery or solitude. We often recognize their value only after they are lost.

A Corporate Perspective

The polarization that has separated the environmental and economic sectors is being replaced with increasing cooperation. As a society, we are beginning to understand that it is not a question of the economy *or* the environment but a question of *both*. We cannot have a healthy economy without a healthy environment or a healthy environment without a healthy economy. Many of the leading multinational corporations are adopting this new attitude as a philosophy of business. One example is General Motors Corporation.

General Motors leadership recognizes that as companies pursue their strategies worldwide, they must accept social and environmental responsibility for their actions. These responsibilities include promoting a sustainable global economy and recognizing their accountability to the economies, environments, and communities where they do business. GMC recognizes the importance of balancing environmental protection with economic objectives and the need to establish rational policies to foster development that can be sustained over the long term. A key General Motors philosophy is that the company can achieve both economic growth and environmental protection through cooperative efforts.

"We've found some new ways to help improve the environment. Their names are Elena, Andrew, Miesha, and Brian." General Motors is a major contributor to the Global Rivers Environmental Education Network (GREEN). GREEN works on watershed education programs in over 100 nations.

Corporate Environmental Ethics

Many tasks of industry, such as procuring raw materials, manufacturing and marketing, and disposing of wastes, are in large part responsible for pollution. This is not because any industry or company has adopted pollution as a corporate policy.

Industry is naturally dirty because it consumes energy and resources. When raw materials are processed, some waste (useless material) is inevitable. It is usually not possible to completely control the dispersal of all by-products of a manufacturing process. Also, some of the waste material may simply be useless.

For example, the food-service industry uses energy to prepare meals. Much of this energy is lost as waste heat. Smoke and odors are released into the atmosphere, and bone, fat, and discolored food items must be discarded.

The cost of controlling waste can be very important in determining a company's profit margin. **Corporations** are legal entities designed to operate at a profit, which is not in itself harmful. The corporation has no ethics, but the people who make up the corporation are faced with ethical decisions. Ethics are involved when a corporation cuts corners in production quality or waste disposal to maxi-

While General Motors continues to address the challenges of past environmental practices, it is working to enhance environmental management systems and continually improve its environmental performance. General Motors' environmental strategy consists of three objectives:

- Implement environmental policy and strategy worldwide;
- Integrate environmental issues into all business decisions;
- Continue to develop global environmental management systems.

GMC is not alone in the corporate world in its growing attention to the environment. Most major corporations now publish an annual report on their environmental progress. Examples include DOW Chemical, which publishes its annual *Continuing the Responsible Care Journey: Steps Toward Sustainability*; Weyerhaeuser, which publishes its *Annual Environmental Performance Report*; and Boeing, which publishes its annual *Progress Report on Environmental Affairs*. Perhaps you could obtain similar reports from local companies and investigate whether they are living up to what they profess.

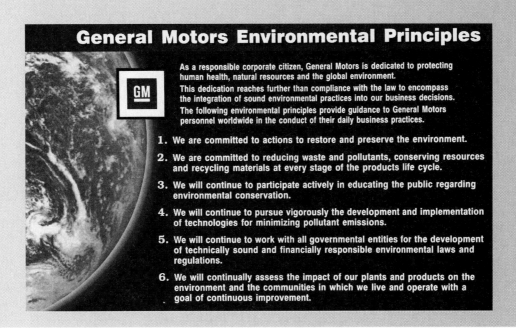

General Motors Environmental Principles

As a responsible corporate citizen, General Motors is dedicated to protecting human health, natural resources and the global environment.

This dedication reaches further than compliance with the law to encompass the integration of sound environmental practices into our business decisions. The following environmental principles provide guidance to General Motors personnel worldwide in the conduct of their daily business practices.

1. We are committed to actions to restore and preserve the environment.

2. We are committed to reducing waste and pollutants, conserving resources and recycling materials at every stage of the products life cycle.

3. We will continue to participate actively in educating the public regarding environmental conservation.

4. We will continue to pursue vigorously the development and implementation of technologies for minimizing pollutant emissions.

5. We will continue to work with all governmental entities for the development of technically sound and financially responsible environmental laws and regulations.

6. We will continually assess the impact of our plants and products on the environment and the communities in which we live and operate with a goal of continuous improvement.

mize profit. The cheaper it is to produce an item, the greater the possible profit. It is cheaper to dump wastes into a river than to install a wastewater treatment facility, and it is cheaper to release wastes into the air than it is to trap them in filters. Many people consider such pollution unethical and immoral, but corporations think of it as just one of the factors that determines **profitability.** (See figure 2.3.) Because stockholders expect an immediate return on their investment, corporations often make decisions based on short-term profitability rather than long-term benefit to society.

The amount of profit a corporation realizes determines how much it can expand. To expand continually, a corporation increases the demand for its products through advertising. The more it expands, the more power it attains. The more power it has, the greater its influence over decision makers who can create conditions favorable to its expansion plans. The process becomes a seemingly never-ending spiral.

Nations of the world must confront the problem of corporate irresponsibility toward the environment. In business, incorporation allows for the organization and concentration of wealth and power far surpassing that of individuals or partnerships. Some of the most important decisions affecting our environment are made not by governments or the public but by executives who wield massive

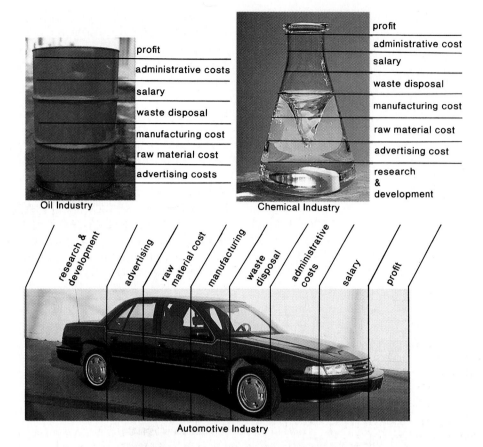

Figure 2.3

Corporate Decision Making Corporations must make a profit. When they look at pollution control, they view its cost like any other cost: any reductions in cost increase profits.

a.

b.

Figure 2.4

CERES Principles The 1989 oil spill in Alaska led to the development of the CERES Principles. (*a*) These waterfowl were victims of the spill. (*b*) The *Exxon Valdez*.

corporate power. Often, these executives make only minimal concessions to the public interest, while they make every effort to maximize profits.

Business decisions and technological developments have increased the exploitation of natural resources. In addition, many political and legal institutions have generally supported the development of private enterprise. They have also defended and promoted private property rights rather than social and environmental concerns. Businesses and individuals typically use loopholes, political pressure, and the time-consuming nature of legal action to circumvent or delay compliance with social or environmental regulations.

Is industry becoming more environmentally concerned? Corporations have certainly made more frequent references to worldwide environmental issues over the past several years. Is such concern only rhetoric and social marketing, or is it the beginning of a new corporate ethic? The "Valdez Principles," a tool that industry and the public can use to evaluate corporate environmental responsibility, were developed as a result of the 1989 oil spill in Alaska. (See figure 2.4). In 1992, the Valdez Principles were proposed and adopted by the CERES Organization (Coalition for Environmentally Responsible Economics). This organization is an independent group of environmentalists and social investors. The Valdez Principles have been renamed the CERES Principles.

The CERES Principles are a set of codes that businesses may adopt voluntarily. The codes provide environmental standards against which all companies can be assessed and compared. The ten principles encompass a wide range of goals that include minimizing pollutants, making sustainable use of renewable re-

sources, reducing health and safety risks for employees and communities, and representing environmental interests on corporate boards. The CERES Principles are looked upon as a guide for corporate environmentalism. The goal, some argue, should be to make compliance with the CERES Principles a prerequisite for doing business.

Practicing an environmental ethic should not interfere with corporate and other social responsibilities or obligations, though this is not always the case. It must be integrated into overall systems of belief and coordinated with economic

26

Chico Mendes and Extractive Reserves

Chico Mendes, whose real name was Francisco Alves Mendes Filho, was a Brazilian rubber tapper active in the rubber tappers' union. He and many other peasants made a living by extracting latex from rubber trees and selling it. Rubber tappers also collect and sell other natural products of the forest, such as Brazil nuts, fruits, and native medicines. Mendes was interested in preserving the portion of the Brazilian rainforest that provided their livelihood, and he supported the concept of "extractive reserves."

Extractive reserves involves setting aside land for rubber tappers, who would continue their traditional lifestyles and use the rainforest in its natural state for generations to come. This idea put Mendes in conflict with powerful people interested in clearing the rainforest to raise cattle. Most cattle-ranching operations show short-term economic gains but ultimately become uneconomical.

On 22 December 1988, as he walked from his home, Chico Mendes was shot by members of a vigilante group who supported local ranchers. In 1990, Darli Alves da Silva, and his son were convicted of the murder.

Before his death, Mendes had said, "I want to live to defend the Amazon." His life and death appear to have made a difference in the way the Brazilian rainforest is being used. Due to the circumstances that surrounded his death and his role as a leader in the rubber tapers' union, his murder received international notice and caused many people to ask if the natural rainforest perhaps has as much to offer as ranches do.

During 1990, the Brazilian government established four refuges. The first, named the Chico Mendes Reserve in the Juruá River valley, covers about 6 percent of the state of Acre in northwest Brazil. Unfortunately, by 1993, economic realities were once again surfacing in the area. While confrontations between the rubber tappers and ranchers have decreased, there have been new problems relating to the mismanagement of funds in the Chico Mendes Foundation and mismanagement of the rubber tappers' cooperative that runs the Brazil-nut factory. By 1994 it was still costing $1.50 per pound to process the nuts, which sell for $1.00 per pound. The future of the cooperative and the foundation are still in question.

systems. Environmental advocates, in turn, need to consider others' objectives just as they demand that others consider the environmental consequences in decision making. It makes little sense to preserve the environment if that objective produces national economic collapse. Nor does it make sense to maintain stable industrial productivity at the cost of breathable air, drinkable water, wildlife species, parks, and wilderness. But to maintain profitability, influence, and freedom, businesses must be sensitive to their impact on current and future citizens, not just in terms of the price and quality of the goods they produce but also in terms of public approval of their social and political influence. A 1995 Harris Poll, for example, found that 70 percent of Americans wanted increased government spending to control acid rain and toxic waste dumping, even if they had to pay higher taxes. In another poll, eight of ten Americans said they would be willing to pay extra for a product packaged with recyclable materials.

In the middle 1990s a concept emerged called industrial ecology that reflects the link between the economy and the environment. This concept argues that good ecology is also good economics and that alternatives exist for corporations to provide goods and services in ways that do not destroy the environment.

One of the most important elements of **industrial ecology** is that, as in biological systems, it accounts for waste. Dictionaries define waste as useless or worthless material. In nature, however, nothing is eternally discarded; in various ways, all materials are reused. In our industrial world, discarding materials taken from the Earth at great cost is also generally unwise. Perhaps materials and products that are obsolete should be termed *residues* rather than *wastes;* wastes are merely residues that our economy has not yet learned to use efficiently. A simpler way of saying this is to view a pollutant as a resource out of place. Such a statement forces us to view pollution and waste in a new way.

Environmental Justice

At its core, **environmental justice** means fairness. It speaks to the impartiality that should guide the application of laws designed to protect the health of human beings and the productivity of ecological systems on which all human activity, economic activity included, depends. It is emerging as an issue because studies show that certain groups of North Americans and citizens of other nations may suffer disproportionately from the effects of pollution.

Governments have established numerous laws, mandates, and directives to eliminate discrimination in housing, education, and employment, but few attempts have been made to address discriminatory environmental practices. In the United States, people of color have borne a disproportionate burden in the siting of municipal landfills, incinerators, and hazardous-waste treatment, storage, and disposal facilities.

Hazardous waste sites and incinerators are not randomly located. While waste generation is correlated directly with per capita income, few toxic waste sites are located in affluent suburbs. Waste facilities are often located in communities that have high percentages of poor, elderly, young, and minority residents. Often such facilities are deliberately sited in these communities because they are seen as providing the path of least resistance.

Questions of environmental justice extend beyond the siting of toxic waste sites. Exposure to harmful pesticides and other toxic agricultural substances is a major health issue among hired farm workers, the majority of whom are people of color. There is also concern that because some Native American communities consume much greater amounts of fish from certain areas such as the Great Lakes than does the general population, they are at greater risk for dietary exposure to toxic chemicals.

Historically, the environmental movement has been a concern of middle-class whites, but there is a growing level of activism by people of color. Minority participation has broadened the debate to include many issues that were being ignored. It has also forced a dialogue about race, class, discrimination, and equity. Minorities have pushed the plight of their communities to the forefront. They have also brought a new perspective to the environmental movement and will be a part of any future environmental agenda.

Individual Environmental Ethics

The environmental movement has effectively influenced public opinion and moved the business community toward an environmental ethic. The result of this changing view of business's responsibilities will complicate business decision making into the next century. More complex demands by the public and a broadening of horizons on the part of business will be the dominant theme of corporate life during the next decade. As human populations and economic activity continue to grow, we are facing a number of environmental problems that threaten not only human health and the productivity of ecosystems, but in some cases the very habitability of the globe.

If we are to respond to those problems successfully, our environmental ethic must express itself in broader and more fundamental ways. We have to recognize that each of us is individually responsible for the quality of the environment we live in and that our personal actions affect environmental quality, for better or worse. The recognition of individual responsibility must then lead to changes in individual behavior. In other words, our environmental ethic must begin to express itself not only in national laws, but also in subtle but profound changes in the ways we all live our daily lives.

A Roper Poll in 1994 indicated that Americans think environmental problems can often be given a quick technological fix. Says the Roper organization, "They believe that cars, not drivers, pollute, so business should invent pollution-free autos. Coal utilities, not electricity consumers, pollute, so less environmentally dangerous generation methods should be found." It appears that many individuals want the environment cleaned up, but they do not want to make major lifestyle changes to make that happen.

Decisions and actions by individuals faced with ethical choices collectively determine the hopes and quality of life for everyone. As ecological knowledge and awareness begin to catch up with good intentions, people in all walks of life will need to live by an environmental ethic.

Global Environmental Ethics

In 1990, Noel Brown, the Director of the United Nations Environmental Program, stated:

> Suddenly and rather uniquely the world appears to be saying the same thing. We are approaching what I have termed a consensual moment in history, where suddenly from most quarters we get a sense that the world community is now agreeing that the environment has to become a matter of global priority and action.

International Trade In Endangered Species

At a recent meeting of the United Nations Convention on International Trade in Endangered Species (CITES), 120 nations raised the issue of trade sanctions against China and Taiwan if they do not halt their trade in rhinoceros and tiger parts sold for medicinal purposes. The animals are killed to provide material for a growing international market. Rhino horns and tiger parts are still used frequently in traditional Chinese medicine to treat fevers, impotence, and joint and bone maladies such as rheumatism and arthritis. Recent investigations have shown that rhino horn is available for sale in pharmacies in Taiwan and through government-run companies in China and is a lucrative export commodity.

It is estimated that fewer than 7,000 tigers and 10,000 rhinos remain in the world. The fact that CITES has proposed trade sanctions is a positive step toward stemming the trade. Overcoming traditional customs and beliefs, however, may not be so simply accomplished.

This new sense of urgency and common cause about the environment is leading to unprecedented cooperation in some areas. Despite their political differences, Arab, Israeli, Russian, and American environmental professionals have been working together for several years. Ecological degradation in any nation almost inevitably impinges on the quality of life in others. For years, acid rain has been a major irritant in relations between the United States and Canada. Drought in Africa and deforestation in Haiti have resulted in waves of refugees. From the Nile to the Rio Grande, conflicts flare over water rights. The growing megacities of the Third World are time bombs of civil unrest.

Much of the current environmental crisis is rooted in and exacerbated by the widening gap between rich and poor nations. Industrialized countries contain only 20 percent of the world's population, yet they control 80 percent of the world's goods and create most of its pollution. The developing countries are hardest hit by overpopulation, malnutrition, and disease. As these nations struggle to catch up with the developed world and improve the quality of life for their people, a vicious circle begins: Their efforts at rapid industrialization poison their cities, while their attempts to boost agricultural production often result in the destruction of their forests and the depletion of their soils, which lead to greater poverty. (See figure 2.5).

Perhaps one of the most important questions for the future is, "Will the nations of the world be able to set aside their political differences to work toward a global environmental course of action?" The United Nations Conference on Human Environment held in Stockholm, Sweden, in 1972 was a step in the right direction. Out of that international conference was born the U.N. Environment Programme, a separate department of the United Nations that deals with environmental issues. A second world environmental conference was held in 1992 in Brazil. It followed up the Stockholm conference with many new international initiatives. (See Global Perspective: Earth Summit.) Through organizations and conferences such as these, nations can work together to solve common environmental problems.

At the individual level, people have begun to respond to increased awareness of global environmental change by altering their values, beliefs, and actions. Changes in individual behavior are necessary but are not enough. As a global species, we are changing the planet. By pooling our knowledge, coordinating our actions, and sharing what the planet has to offer, we may achieve a global environmental ethic.

Figure 2.5

Lifestyle and Environmental Impact   Significant differences in lifestyles and their environmental impact exist between the rich and poor nations of the world. What would be the environmental impact on the Earth if the citizens of China and India and other less developed countries enjoyed the standard of living of North Americans? Can we deny them that opportunity?

Earth Summit

In June 1992, representatives from 178 countries, including 115 heads of state, met in Rio de Janeiro, Brazil, at the Earth Summit. Officially, the meeting was titled the United Nations Conference on Environment and Development (UNCED), and it was the largest gathering of world leaders ever held. The first Earth Summit had been held 20 years earlier in Stockholm, Sweden. At that time, the planet was divided into rival East and West blocs and was preoccupied with the perils of the nuclear arms race. With the collapse of the East bloc and the thawing of the cold war, a fundamental shift in the global base of power had occurred.

Today, the more important diversion, especially on environmental issues, is not between East and West but between "North" (Europe, North America, and Japan) and "South" (most of Asia, Africa, and Latin America). And, though the immediate threat of nuclear destruction has lifted, the planet is still at risk.

The idea behind the Earth Summit was that the relaxation of cold war tensions, combined with the growing awareness of ecological crises, offered a rare opportunity to persuade countries to look beyond their national interests and agree to some basic changes in the way they treat the environment. The major issues are clear: The developed countries of the North have grown accustomed to lifestyles that are consuming a disproportionate share of natural resources and generating the bulk of global pollution. Many of the developing countries of the South are consuming irreplaceable global resources to provide for their growing populations.

In June 1992 representatives from 178 countries, including 115 heads of state, met in Rio de Janeiro, Brazil, at the Earth Summit. In addition to the official governmental representatives, people representing nongovernmental organizations gathered to express their opinions, encourage cooperation, and join in ecumenical religious services. (This photograph was taken at an ecumenical service associated with UNCED.)

The Earth Summit was intended to promote better integration of nations' environmental goals with their economic aspirations. Although the hopes of some developing nations for large commitments of new foreign assistance did not fully materialize, much was accomplished during the Summer.

- The *Rio Declaration on Environment and Development* sets out 27 principles to guide the behavior of nations toward more environmentally sustainable patterns of development. The declaration, a compromise between developing and industrialized countries that was crafted at preparatory meetings, was adopted in Rio without negotiation due to fears that further debate would jeopardize any agreement.

- States at UNCED also adopted a voluntary action plan called *Agenda 21*, named because it is intended to provide an agenda for local, national, regional, and global action into the 21st century. UNCED Secretary General Maurice Strong called Agenda 21 "the most comprehensive, the most far-reaching and, if implemented, the most effective program of international action ever sanctioned by the international community." Agenda 21 includes hundreds of pages of recommended actions to address environmental problems and promote sustainable development. It also represents a process of building consensus on a "global work-plan" for the economic, social, and environmental tasks of the United Nations as they evolve over time.

- The third official product of UNCED was a *"non-legally binding authoritative statement of principles for a global consensus on the management, conservation, and sustainable development of all types of forests."* Negotiations on the forest statement, begun as negotiations for a legally binding convention on forests, were among the most difficult of the UNCED process. Many states and experts, dissatisfied with the end result, came away from UNCED seeking further negotiations toward agreement on a framework convention on forests.

Two international conventions were presented and opened for signature at UNCED, each of which attracted signatures of representatives of more than 150 countries.

- A *Framework Convention on Climate Change* requires signatories to take steps to reduce their emissions of gases believed to contribute to global warming, although no mandatory targets and timetables for such actions were met, largely at the insistence of U.S. negotiators. In Rio, then president George Bush signed the climate change convention. President Clinton later pledged that the United States would reduce its emissions of greenhouse gases to their 1990 levels by the year 2000 and would take the lead in addressing global warming.

- A *Framework Convention on Biological Diversity* prescribes steps for the protection and sustainable use of the world's diverse plant and animal species. Then president Bush refused to sign the biodiversity convention, citing concerns about protection of intellectual property rights and the treaty's financing arrangements. President Clinton later reported that his administration had worked out an interpretive statement addressing some business and environmental groups' concerns, and he announced that the United States would sign the biodiversity convention.

Antarctica—Resource or Refuge?

Few places on earth have not been exploited by humans. One such place is Antarctica. It is as close to an unpolluted environment as there is on earth, but it is not without its problems.

Seals and whales were the earliest exploited resource in Antarctica. There was money to be made, and this "opportunity" resulted in the near extinction of the southern fur seal, the elephant seal, and the blue whale.

By the 1950s, aboveground nuclear testing had spread radioactive particles around the planet, including Antarctica. Pesticides like DDT were turning up in the tissues and blood of certain Antarctic bird and marine mammal species. A growing hole in the ozone layer above the Antarctic continent is caused by the use of chlorofluorocarbons throughout the world. Fossil-fuel combustion contributes to the greenhouse effect, which in turn threatens to melt the ice in Antarctica's Western Peninsula.

During the past several decades, Antarctica has been the site of extensive scientific exploration. Much of this exploration has been economically motivated. For example, government scientists, with the aid of satellites, are advising oil and mineral prospectors. Much of the so-called scientific research is conducted with geopolitical or military objectives in mind.

Antarctica is also being proposed as a tourist attraction. Australia has suggested building a hotel, while Argentina is considering chartering a vessel to transport six hundred tourists from South America seven times a year. Several sites are also being viewed as potential ski resorts.

From an ecological perspective, Antarctica is fragile. The thin layer on the surface of the ocean, nourished by the sun, supports the tiny shrimp-like krill, which sustain fish, whales, seals, and penguins. These short, simple food chains are extremely sensitive to environmental insults.

In the mid-1970s, New Zealand proposed designating the continent an Antarctic World Park. This would turn Antarctica into an international wilderness area, a region on earth where we recognize that humanity does not belong.

In 1991, 24 countries signed an agreement to ban mineral and oil exploration in Antarctica for 50 years. The agreement, which was hailed as historic by governments and environmental groups, includes new regulations for wildlife protection, waste disposal, marine pollution, and continued monitoring of the Antarctic, which covers nearly one-tenth of the world's land surface. The signing of the agreement in Madrid, Spain, was the result of two years of negotiations. The protocol protects Antarctica's delicate flora and fauna and sets procedures to assess environmental effects of all human activities on the continent.

- Should we turn a continent into a world park?
- Would humanity be better served by developing the natural resources of Antarctica, such as oil and minerals?
- Should the natural beauty of Antarctica be opened up to tourism so it can be enjoyed by many?
- Is it possible to strike a balance between preservation and development in a fragile ecosystem? Can you give examples?

Should Antarctica be preserved or utilized for human use?

SUMMARY

People of different cultures view their place in the world from different perspectives. Among the things that shape their views are religious understandings, economic pressures, and fundamental knowledge of nature. Because of this diversity of backgrounds, different cultures put different values on the natural world and the individual organisms that compose it.

Three prevailing attitudes toward nature are the development ethic, which assumes that nature is for people to use for their own purposes; the preservation ethic, which assumes that nature has value in itself and should not be disturbed; and the conservation ethic, which recognizes that we will use nature but that it should be used in a sustainable manner.

Ethical issues can be examined at several levels. Growth and exploitation have been the prevailing priorities of our society for generations. This does not mean that everyone in society has the same opinions, but the general attitude has been one of the development rather than preservation. Most environmental decisions have really been economic decisions, and the rationale has been: If a resource is available for use, it should be used.

Corporate ethics are even more strongly influenced by economics. Corporations exist to make a profit. Any way that they can reduce costs makes them more profitable. Unfortunately, pollution and exploitation of rare resources may be costly to individuals or society while being profitable to corporations. In addition, corporations wield tremendous economic power and can sway public opinion and political will. Many corporations have begun to openly acknowledge their responsibilities to carefully examine their impact on the natural world.

Society and corporations are composed of individuals. An increasing sensitivity of individual citizens to environmental concerns can change the political and economic climate for society and corporations. However, people often do not have a clear idea of what should be done and often do not act in a way that supports their stated beliefs.

Global environmental concerns have become more important. The world is getting "smaller" and more interrelated. As more people are added to the world's population each year, there is increasing competition for the resources needed to live a decent life. An environmental disaster is no longer a local problem but affects us globally. The increasing economic difference between rich and poor nations affects the global environment, since the poor aspire to have what the rich take for granted. All peoples and nations need to work together to solve environmental problems.

REVIEW QUESTIONS

1. How does personal wealth relate to ethics? Can you provide personal examples?
2. Why do industries pollute?
3. Why would normal economic forces work against pollution control? Do you feel that this is changing?
4. Is it reasonable to expect a totally unpolluted environment? Why or why not?
5. What has been the dominant societal attitude toward resource use?
6. Describe the differences between development, preservation, and conservation ethics. Must there always be conflict among these ethics?
7. What is a major motivating force of corporate management?
8. Why do decision makers view the actions of corporations differently from the way they view the actions of individuals?

KEY TERMS

conservation ethic 22
corporation 24
development ethic 20
economic growth 22
environmental justice 27
ethics 19
industrial ecology 27
morals 19
preservation ethic 21
profitability 25
resource exploitation 22

RELATED INTERNET SITES

Department of Philosophy in Sweden discusses philosophical questions related to environmental issues and ethics.
http://www.phil.gu.se/Environment.html

Indigenous treaties versus mining rights.
http://www.menominee.com/nomining/home.htm

PART

two

Y ou can better understand many of the environmental problems that have surfaced over the past 20 years if you clearly understand how organisms interact with their physical environment and with each other. Chapter 3 deals with energy principles and the structure of matter. Chapters 4, 5, and 6 present principles of ecology and population biology that can be applied to specific problems we face as a result of a growing world population putting demands on shrinking resources. Chapter 7 focuses on human population issues. It specifically relates population growth to social and economic problems.

Ecological Principles and Their Application

chapter 3

Interrelated Scientific Principles: Matter, Energy, and Environment

Objectives

After reading this chapter, you should be able to:

- Understand that science is exact because information is gathered in a manner that requires evaluation and revision.
- Understand that environmental science is a new discipline that includes both applied and theoretical aspects of traditional science, and that social, economic, and political aspects are also involved.
- Understand that matter is made up of atoms that have a specific subatomic structure of protons, neutrons, and electrons.
- Recognize that different elements have different atomic structures and that different atoms of the same element may differ in the number of neutrons present.
- Recognize that atoms may be combined and held together by chemical bonds to produce molecules.
- Understand that rearranging chemical bonds results in chemical reactions and that these reactions are associated with energy changes.
- Recognize that matter may be solid, liquid, or gas, depending on the amount of kinetic energy contained in the molecules.
- Realize that energy can be neither created nor destroyed, but when energy is converted from one form to another, some energy is converted into a useless form.
- Understand that energy can be of different qualities.

Chapter Outline

Scientific Thinking

Limitations of Science

The Structure of Matter

    Atomic Structure

Environmental Close-Up: *The Periodic Table of the Elements*

Environmental Close-Up: *Typical Household Chemicals*

    Molecules and Mixtures

    Acids, Bases, and pH

    Inorganic and Organic Matter

    Chemical Reactions

    Chemical Reactions in Living Things

Energy Principles

    States of Matter

    Kinds of Energy

    First and Second Laws of Thermodynamics

    Environmental Implications of Energy Flow

Issues and Analysis: *Improvements in Lighting Efficiency*

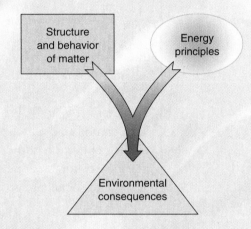

Scientific Thinking

Since environmental science involves the analysis of data, it is useful to understand how scientists gather and evaluate information. It is also important to understand some chemical and physical principles as a background for evaluating environmental issues. An understanding of these scientific principles will also help you appreciate the ecological concepts in the chapters that follow.

The word *science* creates a variety of images in the mind. Some people feel that it is a powerful word and are threatened by it. Others are baffled by scientific topics and have developed an unrealistic belief that scientists are brilliant individuals who can solve any problem. Neither of these images accurately portrays what science is really like. **Science** is a body of knowledge characterized by the requirement that information be gathered and evaluated by impartial testing of hypotheses and that it be shared so that it can be evaluated by others. The **scientific method** of gathering information generally involves observation, hypothesis formation, hypothesis testing, critical evaluation of results, and the publishing of findings. (See figure 3.1.) Underlying all of these activities is constant attention to accuracy and freedom from bias.

Observation simply means the ability to notice something. Sometimes, the observation is made with the unaided senses—we see, feel, or smell something. Often, machines such as microscopes, chemical analyzers, or radiation detectors may be used to extend our senses. Because these machines are complicated, we might get the feeling that science is incredibly complex, when in reality the questions being asked are relatively easy to understand.

A microscope has several knobs to turn and a specially designed light source. It requires considerable skill to use properly, but it is essentially a fancy magnifying glass that allows small objects to be seen more clearly. The microscope has enabled scientists to answer some relatively simple questions such as, Are there living things in pond water? and Are living things made up of smaller subunits? Similarly, a meter stick allows us to measure distance or a pH

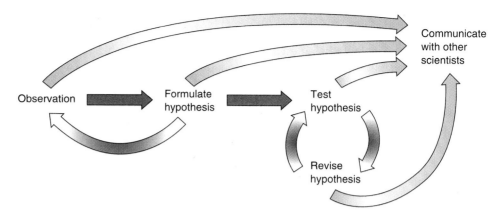

Figure 3.1

Elements of the Scientific Method The scientific method consists of several kinds of activities. Observation of a natural phenomenon is usually the first step. This is followed by the construction of a hypothesis that explains why the phenomenon occurred. The hypothesis is then tested to see if it is supported. Often this involves experimentation. If the hypothesis is not substantiated, it is modified and tested in its new form. It is important at all times that others in the scientific community be informed by publishing observations of unusual events, their probable cause, and the results of experiments that test hypotheses.

meter to measure a chemical property of a solution. Both are simple activities, but if we are not familiar with the units of measure, we might consider the processes hard to understand.

When you have formulated a question that needs scientific investigation, your first step is to form a hypothesis. A **hypothesis** is a logical statement that explains an event or answers a question. A good hypothesis should be as simple as possible, while taking all of the known facts into account. Furthermore, a hypothesis must be testable. In other words, you must be able to support it or prove it incorrect. The construction and testing of hypotheses is one of the most difficult (creative) aspects of the scientific method. Often, artificial situations must be constructed to test hypotheses. These are called **experiments.** A standard kind of experiment is one called a **controlled experiment.** Two groups are created that are identical in all respects except one. The control group has nothing out of the ordinary done to it. The experimental group has one thing different. If the experimental group gives different results from the control group, those results must be caused by the single difference (variable) between the two groups.

The results of a well-designed experiment should be able to support or disprove a hypothesis. However, this does not always occur. Sometimes the results of an experiment are inconclusive. This means that a new experiment has to be conducted or that more information has to be collected. Often, it is necessary to have large amounts of information before a decision can be made about the validity of a hypothesis. The public often finds it difficult to understand why it is necessary to perform experiments on so many subjects, or why it is necessary to repeat experiments again and again.

The concept of **repeatability** is important to the scientific method. Because it is often not easy for scientists to eliminate unconscious bias, independent investigators must repeat the experiment to see if they get the same results. To do this, they must have a complete and accurate written document to work from. That means the scientists must publish the results of their experiment. This process of publishing results for others to examine and criticize is one of the most important steps in the process of scientific discovery. If a hypothesis is supported by many experiments and by different investigators, it is considered reliable.

A hypothesis that has survived repeated examination by many investigators over a long time and that has central importance to an area of science may become known as a **law.** For example, the **biogenetic law** states that all living things are produced by previously living organisms. This has been tested repeatedly over hundreds of years and there have been no exceptions. Broadly written statements that cover large bodies of scientific knowledge are often called **theories.** For example, the **theory of evolution** holds that the characteristics of plants and animals and their kinds change over time. A theory is generally accepted by scientists to be true, but it cannot be proved true in *every* case because it is impossible to test *every* case. It is important to recognize that the word *theory* is often used in a much less restrictive sense. Often it is used to describe a vague idea or a hunch. This is not a theory in the scientific sense. So when you see or hear the word *theory* you must look at the context to see if the speaker or writer is referring to a theory in the scientific sense.

Now that we have some idea of how the scientific method works, let's look at an example. In many rivers in industrial parts of the world, it is possible to notice tumors of the skin and liver in the fish that live in the rivers (*observation*). This raises the question of what causes the tumors. Many people feel that the tumors are caused by the toxic chemicals that have been released into the rivers by industrial plants (*hypothesis*). Now, how could an experiment be conducted to test the hypothesis? If an industrial plant is suspected of releasing toxic chemicals that cause tumors, resident species of fish that do not migrate can be collected upstream and downstream from the plant's wastewater discharge pipes (*outfall*). Fish collected above the outfall constitute the control group, and those collected below the outfall constitute the experimental group. Large numbers of fish would have to be collected and examined. If the fish below the outfall have significantly more tumors than those above the outfall, it is because of where they live in the river and so the toxic chemicals from the industrial plants are probably the cause of the tumors. This is particularly true if the chemicals are already known to cause

tumors. After the data were evaluated, the results of the experiment would be published. Certainly, the owners of the industrial plants would want to look at the data and might want to repeat the experiment to see if they get the same results.

Limitations of Science

Science is a powerful tool for developing an understanding of the natural world, but it cannot analyze international politics, decide if family-planning programs should be instituted, or evaluate the significance of a beautiful landscape. These tasks are beyond the scope of scientific investigation. This does not mean that scientists cannot comment on such issues. They often do. But they should not be regarded as more knowledgeable on these issues just because they are scientists. Scientists may know more about the scientific aspects of these issues, but they struggle with the same moral and ethical questions that face all people, and their judgments on these matters can be just as faulty as anyone else's.

It is important to differentiate between the scientific data collected and the opinions scientists have about what the data mean. Scientists form and state opinions that may not always be supported by fact, just as other people do. Equally reputable scientists commonly state opinions that are in direct contradiction. This is especially true in environmental science, where predictions about the future must be based on inadequate or fragmentary data.

It is important to recognize that some knowledge can be used in both a scientific and a nonscientific manner. For example, it is a fact that many of the kinds of chemicals used in modern agriculture are toxic to humans and other animals. This does not mean that foods grown with the use of these chemicals are less nutritious or that "organically grown" foods are necessarily more nutritious because they have been grown without agricultural chemicals. The idea that something that is artificial is neces-

sarily bad and something natural is necessarily good is false. After all, tobacco, poison ivy, and the prickly cactus are natural, while chemical fertilizers account for a large proportion of the food grown in the world. It is often very easy to jump to conclusions or confuse fact with hypothesis. This is particularly true when we generalize.

The Structure of Matter

Now that we have an appreciation for the methods of science, it is time to explore some basic information and theories about the structure and function of various kinds of matter. **Matter** is anything that takes up space and has weight. Air, water, trees, cement, and gold are all examples of matter. A central theory that describes the structure and activity of matter is the **kinetic molecular theory.** This theory states that all matter is made up of tiny particles that are in constant motion. Although different kinds of matter have different properties, they are similar in one fundamental way. They are all made up of one or more kinds of smaller subunits called atoms.

Atomic Structure

Atoms are the fundamental subunits of matter. They are made up of protons, neutrons, and electrons. There are 92 kinds of atoms found in nature. (See the Periodic Table of Elements on next page.) Each kind forms a specific type of matter known as an **element.** Gold (Au), oxygen (O), and mercury (Hg) are examples of elements. All atoms are composed of a central region known as a **nucleus,** which is composed of two kinds of relatively heavy particles: positively charged particles called **protons** and uncharged particles called **neutrons.** Surrounding the nucleus are clouds of relatively lightweight, fast-moving, negatively charged particles called **electrons.** As mentioned earlier, each kind of element is composed of a different kind of atom. The elements differ from one another in the number of protons, neutrons, and electrons

environmental *close-up*

The Periodic Table of the Elements

Traditionally, elements are represented in a shorthand form by letters. For example, the formula for water, H_2O, shows that a molecule of water consists of two atoms of hydrogen and one atom of oxygen. These chemical symbols for each of the atoms can be found on any periodic table of the elements. Using the periodic table, we can determine the number and position of the various parts of atoms. Notice that atoms number 3, 11, 19, and so on are in column one. The atoms in this column act in a similar way since they all have one electron in their outermost layer. In the next column, Be,

Mg, Ca, and so on act alike because these metals all have two electrons in their outermost electron layer. Similarly, atoms number 9, 17, 35, and so on all have seven electrons in their outer layer. Knowing how fluorine, chlorine, and bromine act, you can probably predict how iodine will act under similar conditions. At the far right in the last column, argon, neon, and so on all act alike. They all have eight electrons in their outer electron layer. Atoms with eight electrons in their outer electron layer seldom form bonds with other atoms.

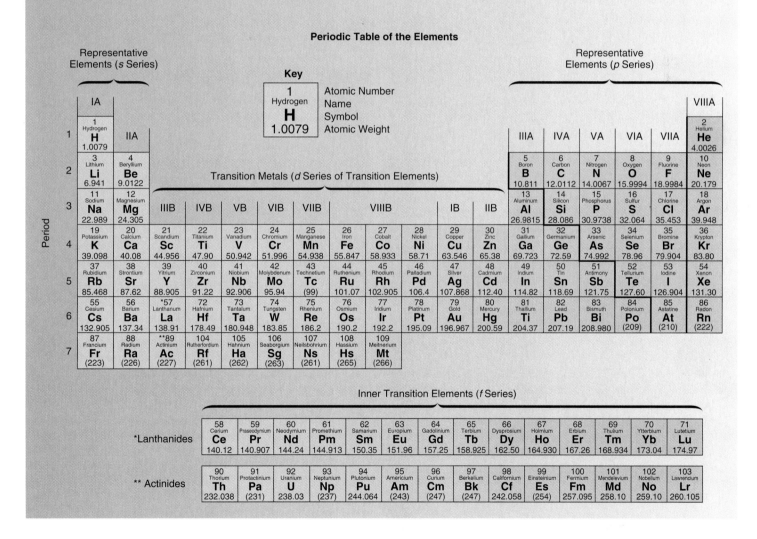

Periodic Table of the Elements

environmental
close-up

Typical Household Chemicals

Modern society uses many different kinds of chemicals. A survey of a typical household would probably yield the following inorganic chemicals:

Common name	Chemical name	Use
Table salt	Sodium chloride, $NaCl$	Flavor
Saltpeter	Potassium nitrate, KNO_3	Preservative
Baking soda	Sodium bicarbonate, $NaHCO_3$	Leavening agent
Ammonia	Ammonia, NH_3	Disinfectant
Bleach	Sodium hypochlorite, $NaHClO$	Bleaching
Caustic soda	Sodium hydroxide, $NaOH$	Drain cleaner

Other products we use contain mixtures of inorganic chemicals. Fertilizers are good examples. They usually contain a nitrate such as ammonium nitrate (NH_4NO_3), a phosphate compound (PO_4^{3+}), and potash, which is potassium oxide (K_2O).

In addition, we use a vast array of organic chemicals: ethyl alcohol in alcoholic beverages, acetic acid in vinegar, methyl alcohol for fuel, and cream of tartar (tartaric acid) for flavoring. We also use many complex mixtures of organic molecules in flavorings, pesticides, cleaners, and other applications.

Most of us know very little about the activities of the molecules we use. Many of them can be dangerous if used improperly. Fertilizer is poisonous, caustic soda can cause severe burns, and bleach or ammonia in high enough concentrations can damage skin or other tissues. The disposal of unused or unwanted household chemicals is a problem. Many of them should not just be dumped down the sink but should be disposed of in such a way that the material is converted to a harmless product or stored in a secure place. Unfortunately, most people do not know how to dispose of unwanted chemicals. For this reason, many manufacturers of household chemicals that

CARPET BEETLES — Thoroughly apply as a spot treatment. Spray along baseboards and edges of carpeting, under carpeting, rugs and furniture, in closets and on shelving, or wherever these insects are seen or suspected. **FLEAS, BROWN DOG TICKS** — Remove soiled bed bedding and clean thoroughly or destroy. Spray sleeping quarters of pets, along baseboards, windows, door frames, cracks and crevices, carpets, rugs, floors where these pests may be found. Put fresh bedding in pet quarters after spray has dried. **DO NOT SPRAY ANIMALS.** Pets should be treated with FLEA-B-GON® Flea Killer (aerosol) or ORTHO Pet Flea & Tick Spray Formula II (pump spray).
STORAGE: To store, rotate nozzle to closed position. Keep pesticide in original container. Do not put concentrate or dilute into food or drink containers. Avoid contamination of feed and foodstuffs. Store in a cool, dry place, preferably in a locked storage area.
DISPOSAL: PRODUCT — Partially filled bottle may be disposed of by securely wrapping original container in several layers of newspaper and discard in trash. **CONTAINER** — Do not reuse empty bottle. Rinse thoroughly before discarding in trash.

NOTICE: Buyer assumes all responsibility for safety and use not in accordance with directions.

Chevron Chemical Company © 1984
Ortho Consumer Products Division
P.O. Box 5047 San Ramon CA 94583-0947
Form 10152-N Product 5466 Made in U.S.A.
EPA Reg. No. 239-2490-AA
EPA Est. 239-IA-3

0   8
7 1549 01980

have a potential to cause harm print statements on the containers explaining how to properly dispose of the unused product and the container. In addition, many communities have regular clean-up efforts for household hazardous waste, in which volunteers who know the contents of such products help determine how to dispose of them properly.

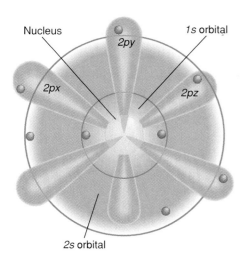

Figure 3.2

Diagrammatic Oxygen Atom Most oxygen atoms are composed of a nucleus containing eight positively charged protons and eight neutrons without charges. Eight negatively charged electrons spin around the nucleus.

in their atoms. For example, a typical atom of mercury contains 80 protons and 80 electrons; gold has 79, and oxygen only eight. (See figure 3.2.) An atom always has the same number of protons and electrons, but the number of neutrons may vary from one atom to the next. Atoms of the same element that differ from one another in the number of neutrons they contain are called **isotopes.**

Molecules and Mixtures

Atoms can be bonded to one another into stable units called **molecules.** When two or more different kinds of atoms are bonded to one another, the kind of matter formed is called a **compound.** While only 92 kinds of atoms are commonly found, there are millions of ways atoms can be combined to form compounds. Water (H_2O), sugar ($C_6H_{12}O_6$), salt (NaCl), and methane gas (CH_4) are examples of compounds.

Many other kinds of matter are **mixtures,** variable combinations of atoms or molecules. Honey is a mixture of several sugars and water; concrete is a mixture of cement, sand, gravel, and reinforcing rod; and air is a mixture of several gases of which the most common are nitrogen and oxygen.

Acids, Bases, and pH

Acids and bases are two classes of compounds that are of special interest. Their characteristics are determined by the nature of their chemical bonds. When acids are dissolved in water, hydrogen ions (H^+) are set free. An **ion** is an atom or molecule that has gained or lost one or more electrons and, therefore, has either a positive charge or a negative charge. A *hydrogen ion* is positive because it has lost its electron and now has only the positive charge of the proton. An **acid** is any ionic compound that releases hydrogen ions in a solution. One other way of thinking of an acid is that it is a substance able to donate a proton to a solution. This is only part of the definition of an acid. We also think of acids as compounds that act like the hydrogen ion: they attract negatively charged particles. An example of a common acid is the sulfuric acid (H_2SO_4) in our automobile batteries.

A **base** is the opposite of an acid in that it is an ionic compound that releases a group known as a **hydroxyl ion,** or OH^- group. This group is composed of an oxygen atom and a hydrogen atom bonded together but with an additional electron. The hydroxyl ion is negatively charged. It is a base because it is able to donate electrons to the solution. A base can also be thought of as any substance able to attract positively charged particles. A very strong base used in oven cleaners is NaOH, or sodium hydroxide.

The strength of an acid or base is represented by a number called its **pH** number. The pH scale is a measure of hydrogen ion concentration. A pH of seven indicates that the solution is neutral and has an equal number of H^+ ions and OH^- ions to balance each other. As the number of hydrogen ions in the solution increases, the pH number gets smaller. The lower the number, the stronger the acid. A number higher than seven indicates that the solution has more OH^- than H^+. As the number of hydroxyl ions increases, the pH number gets larger. The higher the number, the stronger the base.

Inorganic and Organic Matter

Inorganic and organic matter are usually distinguished from one another by one fact: organic matter consists of molecules that contain carbon atoms bonded to form chains or rings. Consequently, organic molecules are usually large. Many different kinds of organic molecules are possible. Inorganic molecules are generally small and are of relatively few kinds. All living things contain organic molecules. They must either be able to manufacture organic molecules from inorganic molecules or be able to modify organic molecules they obtain from eating organic material. Typically, organic molecules contain a large amount of chemical energy that can be released when they are broken down to inorganic molecules. Salt, water, metals, sand, and oxygen are examples of inorganic matter. Sugars, proteins, fats, natural gas, gasoline, and coal are all examples of organic matter.

Chemical Reactions

The atoms within a molecule are held together by chemical bonds. (See figure 3.3.) **Chemical bonds** are physical attractions between atoms resulting from the interaction of their electrons. When chemical bonds are broken or formed, a chemical reaction occurs. When it does, the amount of energy within the chemical bonds changes, and some of the energy may be released as heat and light. A common example is the burning of natural gas. The primary ingredient in

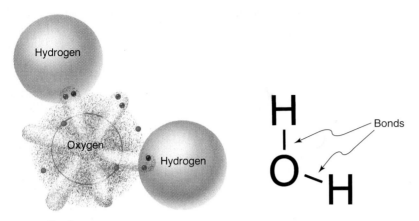

Figure 3.3

Water Molecule Water molecules are an atom of oxygen bonded to two atoms of hydrogen, with the hydrogen atoms located on one side of the oxygen.

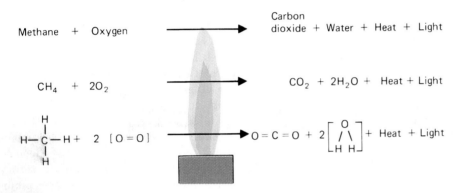

Figure 3.4

A Chemical Reaction When methane is burned, chemical bonds are changed, and the excess chemical bond energy is released as light and heat. The same atoms are present, but they are bonded in different ways, resulting in different molecules.

natural gas is the compound methane. When methane and oxygen are mixed together and a small amount of energy is used to start the reaction, the chemical bonds in the methane and oxygen (reactants) are rearranged to form two new compounds, carbon dioxide and water (products). In this kind of reaction, some chemical-bond energy is left over; it is released as light and heat. (See figure 3.4.) In every reaction, the amount of energy in the reactants and in the products can be compared and the differences accounted for by energy loss or gain. Even energy-yielding reactions usually need an input of energy to get the reaction started. This initial input of

energy is called **activation energy.** In certain cases, the amount of activation energy required to start the reaction can be reduced by the use of a catalyst. A **catalyst** is a substance that alters the rate of a reaction, but the catalyst itself is not altered in the process. Catalysts are used in catalytic converters, which are attached to automobile exhaust systems. The purpose of the catalytic converter is to bring about more complete burning of the fuel, thus resulting in less air pollution. Most of the materials that are not completely burned by the engine require high temperatures to react further; with the presence of catalysts, these reactions can occur at lower temperatures.

Chemical Reactions in Living Things

Living things contain catalysts called **enzymes,** which are composed of proteins. Most chemical reactions that occur in organisms are assisted by enzymes that reduce the activation energy needed to start the reactions. This is important since the high temperatures required to start these reactions without enzymes would destroy living organisms. Many of these enzymes are arranged in such a way that they cooperate in controlling a chain of reactions, as in photosynthesis and respiration.

Photosynthesis is the process plants use to convert inorganic material into organic matter, using the assistance of light energy. Light energy enables the smaller inorganic molecules (water and carbon dioxide) to be converted into organic sugar molecules. In the process, oxygen is released. Photosynthesis takes place in the green portions of the plant, usually the leaves. (See figure 3.5.) The organic molecules produced as a result of photosynthesis can be used as a source of energy by the plants and by organisms that eat the plants.

Respiration involves the use of oxygen to break down large, organic molecules (sugars, fats, and proteins) into smaller, inorganic molecules (carbon dioxide and water). This process releases energy the organisms can use. (See figure 3.6.) All organisms must carry on some form of respiration, since all need a source of energy to maintain life.

Energy Principles

The previous section started out with a description of matter, yet it used the concepts of energy to describe chemical bonds, chemical reactions, and the movement of atoms and molecules. That is because energy and matter are inseparable. It is difficult to describe one without the other. **Energy** is the ability to do work, which typically results in matter being moved over a distance. This occurs even at the molecular level.

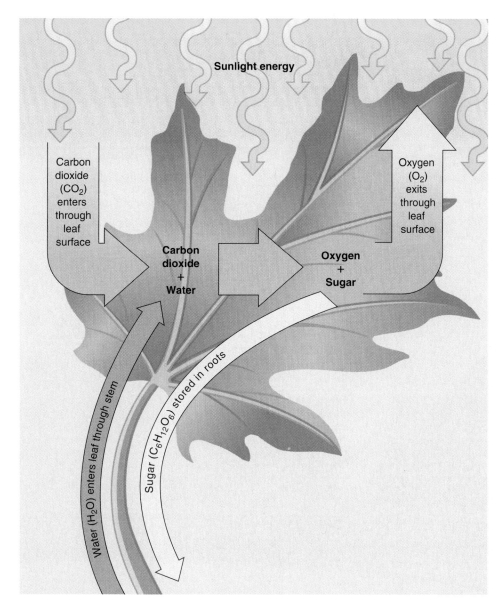

Sunlight energy

Carbon dioxide (CO_2) enters through leaf surface

Carbon dioxide + Water

Oxygen (O_2) exits through leaf surface

Oxygen + Sugar

Water (H_2O) enters leaf through stem

Sugar ($C_6H_{12}O_6$) stored in roots

Figure 3.5

Photosynthesis This reaction is an example of one that requires an input of energy (sunlight) to combine low-energy molecules (CO_2 and H_2O) to form sugar with a greater amount of chemical bond energy. Oxygen is also produced.

States of Matter

Depending on the amount of energy present, matter can occur in three states: solid, liquid, or gas. The physical nature of matter changes when the amount of kinetic energy its molecules contain changes, but the chemical nature of matter remains the same. For example, water vapor, liquid water, and ice all have the same chemical composition but differ in the arrangement and activity of their molecules. The amount of kinetic energy molecules have determines how rapidly they move. (See figure 3.7.) In solids, the molecules have low amounts of energy, and they vibrate in place very close to one another. In liquids, the higher-energy molecules are farther apart and will roll, tumble, and flow over each other. The molecules of gases move very rapidly and are very far apart. All that is necessary to change the physical nature of a type of matter is an energy change. Heat energy must be added or removed.

Kinds of Energy

There are several kinds of energy. Heat, light, electricity, and chemical energy are common forms. The energy contained by moving objects is called **kinetic energy.** The moving molecules in air have kinetic energy, as does water running downhill or a dog chasing a ball. **Potential energy** is in a special category; it is the energy matter has because of its position. The water behind a dam has potential energy by virtue of its elevated position. (See figure 3.8.) An electron at some distance from the nucleus has potential energy due to the distance between the electron and the nucleus.

First and Second Laws of Thermodynamics

Energy can exist in several different forms, and it is possible to convert one kind of energy into another. However, the total amount of energy remains constant. The **first law of thermodynamics** states that energy can neither be created nor destroyed; it can only be changed from one form into another. From a human perspective, some forms of energy are more useful than others . We tend to make extensive use of electrical energy for a variety of purposes, but there is very little electrical energy present in nature. Therefore, we convert other forms of energy into electrical energy. When converting energy from one form to another, some of the useful energy is lost. This is the **second law of thermodynamics.** There is no loss of *total* energy, but there is a loss of *useful* energy. For example, coal can be burned in a power plant to produce electrical energy. However, large amounts of useless heat energy are also produced. Therefore, the amount of useful energy (electricity) is much less than the total amount of chemical energy present in the coal. (See figure 3.9.)

Energy is being converted from one form to another continuously within the

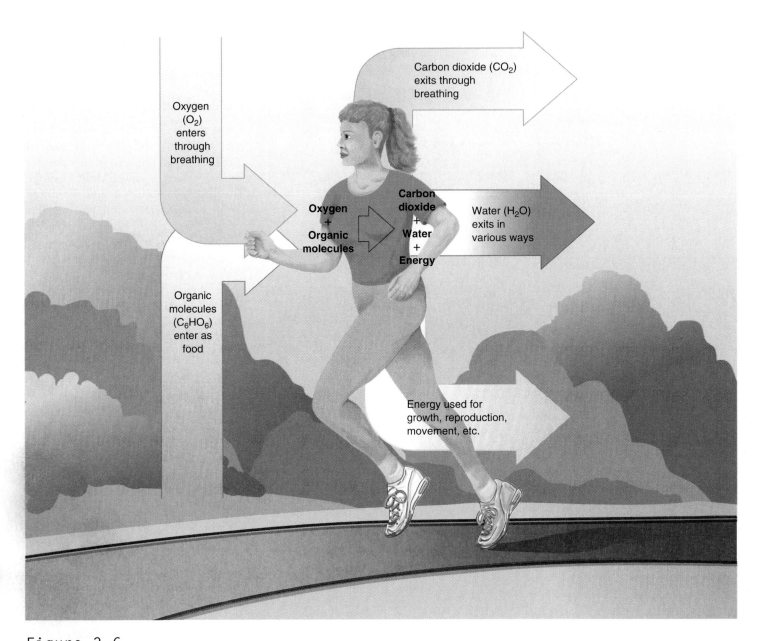

Figure 3.6

Respiration   Respiration involves the release of energy from organic molecules when they react with oxygen. In addition to providing energy in a usable form, respiration produces carbon dioxide and water.

universe. Stars are converting nuclear energy into heat and light. Animals are converting the chemical energy found in food into the kinetic energy of motion. Plants are converting sunlight energy into the chemical bond energy of sugar molecules. In each of these cases, some useless energy is produced, generally in the form of heat.

Environmental Implications of Energy Flow

The heat produced when energy conversions occur is dissipated throughout the universe. This is a common experience. Valuable things always disintegrate unless we work to maintain them. Houses fall into ruin, automobiles rust, and appliances wear out. In reality, all of these phenomena involve the loss of heat. The organisms that decompose the wood in our houses release heat. The chemical reaction that causes rust releases heat. Friction, caused by the movement of parts of a machine against each other, generates heat and causes the parts to wear.

Orderly arrangements of matter, such as clothing, automobiles, or living organisms, always tend to become disordered.

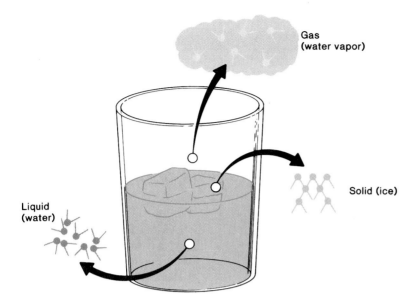

Gas
(water vapor)

Solid (ice)

Liquid
(water)

Figure 3.7

States of Matter Matter exists in one of three states, depending on the amount of kinetic energy the molecules have. The higher the amount of energy, the greater the distance between molecules and the greater their degree of freedom of movement.

Figure 3.8

Kinetic and Potential Energy Kinetic and potential energy are interconvertible. The potential energy possessed by the water behind a dam is converted to kinetic energy as the water flows to a lower level.

Eventually, nonliving objects wear out and living things die and decompose. This process of becoming more disordered coincides with the constant flow of energy toward a dilute form of heat. This dissipated, low-quality heat has little value to us.

It is important to understand that different energy forms are of different quality. Some are of high quality, such as electrical energy, which can be easily used to perform a variety of useful actions. Some are of low quality, such as the

heat in the water of the ocean. Although the total *quantity* of heat energy in the ocean is much greater than the total amount of electrical energy in the world, little can be done with the heat energy in the ocean because it is of low *quality*. Therefore, it is not as valuable as other forms of energy that can be used to do work for us.

The reason the heat of the ocean is of little value involves the small temperature difference between two sources of heat. When two objects differ in temperature, heat will flow from the warmer to the cooler object. The greater the temperature difference, the more useful the work that can be done. For example, fossil-fuel power plants burn fuel to heat water and convert it to steam. Cold cooling-water provides a steep temperature gradient, and heat energy flows from the steam to the cold water. That causes a turbine to turn, which generates electricity. Because the average temperature of the ocean is not high, and it is difficult to find another object that has a greatly lower temperature than the ocean, the huge heat content of the ocean cannot do useful work for us.

These quantitative and qualitative factors are also evident in the energy expended by a stream as the water runs downhill. The steeper the slope, the greater the amount of energy expended per kilometer of its length. If there is no point along the stream where the slope is very steep, the stream has low-quality energy, because the energy is dissipated along the entire length of the stream. To make this a high-quality (concentrated) source of energy, the water must be dammed so that it will drop a long distance at one point. This means that it will give up much of its energy over a short distance. With damming, the *quantity* of energy has not changed but the *quality* has.

Because of the second law of thermodynamics, all organisms, including humans, are in the process of converting high-quality energy into low-quality energy. Waste heat is produced when the chemical-bond energy in food is converted into the energy needed to move, grow, or respond. The process of releasing

Figure 3.9

Second Law of Thermodynamics   Whenever energy is converted from one form to another, some of the useful energy is lost, usually in the form of heat. The conversion of fuel to electricity produces heat, which is lost to the atmosphere. As the electricity moves through the wire, resistance generates some additional heat. When the electricity is converted to light in a light bulb, heat is produced as well. All of these steps produce useless heat as a result of the second law of thermodynamics.

chemical-bond energy from food by organisms is known as cellular respiration. From an energy point of view, it is similar to the process of **combustion,** which is the burning of fuel to obtain heat, light, or some other form of useful energy. The efficiency of cellular respiration is relatively high. About 40 percent of the energy contained in food is released in a useful form. The rest is dissipated as low-quality heat. Table 3.1 lists the efficiencies of many common energy conversion systems.

An unfortunate consequence of energy conversion is pollution. The heat lost from most energy conversions is a pollutant. The wear of the brakes used to stop cars results in pollution. The emissions from power plants pollute. All of these are examples of the effect of the second law of thermodynamics. If each individual on earth used less energy, there would be less waste heat and other forms of pollution that result from energy conversion. The amount of energy in the universe is limited. Only a small portion of that energy is of high quality. The use of high-quality energy decreases the amount of useful energy available, as more low-quality heat is generated. All life and all activities are subject to these important physical principles known as the first and second laws of thermodynamics.

Table 3.1

The Efficiency of Some Energy Conversion Systems

Energy conversion system	% efficiency*
Electric motor	93
Hydroelectric power plant	85
Home oil furnace	65
Fluorescent lamp	65
Steam-power plant	47
High-intensity lamp	32
Automobile engine	25
Incandescent lamp	4

*Efficiency with which the energy of the power or fuel source is converted to a useful form.

Source: Data, in part, from Robert H. Romer, Energy: An Introduction to Physics, 1976, W.H. Freeman and Company Publishers, New York, NY.

Improvements in Lighting Efficiency

In the United States, 36 percent of the energy consumed is electricity. Of that, nearly 40 percent is used for lighting. Improvements in the efficiency of lighting would significantly reduce the demand for electrical energy. All forms of lighting involve the conversion of electricity into light, but some systems are much more efficient than others. The incandescent light bulb is extremely inefficient yet is still used in many homes and commercial buildings. Compact fluorescent lights use about 25 percent of the energy of an incandescent bulb to produce the same amount of light and can be used in the standard incandescent light bulb socket. In many situations, such as commercial buildings, newer high-efficiency fluorescent lights with electronic ballasts can reduce energy consumption by about 15 percent over standard fluorescent bulbs.

In some cases, fluorescent lighting is not practical. It does not work well in the cold and in most situations cannot be used with dimmer switches. Other kinds of higher efficiency lighting are available, however. Halogen lights are incandescent lights with somewhat better efficiency than standard incandescent lights. Sodium vapor, mercury vapor, and metal halide lights are very efficient but produce a light of a different color than normal daylight. These lights also require several minutes to come up to full lighting power. They are used in places where color is not important and where they are not turned on and off repeatedly, such as exterior lighting in parking lots. The U. S. Department of Energy has helped develop a new sulfur lamp that is even more efficient than fluorescent lighting and has better color that other high-efficiency lamps.

Any improvement in the efficiency with which electricity is converted to light also reduces the amount of waste heat produced, which reduces the amount of electricity needed to cool buildings. Often the cost of replacing inefficient incandescent light bulbs is offset by subsidies from local electric utilities, since increased lighting efficiency reduces the demand for electricity and allows utilities to put off building expensive new power plants.

How many incandescent light bulbs do you have in your home? Why haven't they been replaced?

SUMMARY

Science is a method of gathering and organizing information. It involves observation, hypothesis formation, the testing of hypotheses, and publication of the results for others to evaluate. A hypothesis is a logical prediction about how things work that must account for all the known information and be testable. The process of science attempts to be careful, unbiased, and reliable in the way information is collected and evaluated. This often involves conducting experiments to test the validity of a hypothesis. If a hypothesis is continually supported by the addition of new facts, it may become known as a law. A theory is a broadly written, widely accepted generalization that ties together large bodies of information.

The fundamental unit of matter is the atom, which is made up of protons and neutrons in the nucleus and electrons circling the nucleus. The number of protons for any one type of atom is constant, but the number of neutrons in different atoms of the same type may vary. The number of electrons is equal to the number of protons. Protons have a positive charge, neutrons lack a charge, and electrons have a negative charge.

When two or more atoms combine with one another, they form stable units known as molecules. Chemical bonds are physical attractions between atoms resulting from the interaction of their electrons. When chemical bonds are broken or formed, a chemical reaction occurs, and the amount of energy within the chemical bonds is changed. Chemical reactions require activation energy to get the reaction started.

Matter that is composed of only one kind of atom is known as an element. Matter that is composed of molecules containing atoms bonded in specific ratios is known as a compound. Atoms or molecules that have gained or lost electrons so that they have an electric charge are known as ions.

Matter can occur in three states: solid, liquid, and gas. These three differ in the amount of energy the molecules contain and the distance between the molecules. Kinetic energy is the energy contained by moving objects. Potential energy is the energy an object has because of its position.

The first law of thermodynamics states that the amount of energy in the universe is constant, that energy can neither be created nor destroyed. The second law of thermodynamics states that when energy is converted from one form to another, some of the useful energy is lost. Some forms of energy are more useful than others. The quality of the energy determines how much useful work can be accomplished by expending the energy. Low-temperature heat sources are of poor quality, since they cannot be used to do useful work.

REVIEW QUESTIONS

1. How do scientific disciplines differ from nonscientific disciplines?
2. What is a hypothesis? Why is it an important part of the way scientists think?
3. Why are events that happen only once difficult to analyze from a scientific point of view?
4. What is the scientific method, and what processes does it involve?
5. How are the second law of thermodynamics and pollution related?
6. Diagram an atom of oxygen and label its parts.
7. What happens to atoms during a chemical reaction?
8. State the first and second laws of thermodynamics.
9. How do solids, liquids, and gases differ from one another at the molecular level?
10. List five kinds of energy.
11. Are all kinds of energy equal in their capacity to bring about changes? Why or why not?

KEY TERMS

acid   41
activation energy   42
atom   38
base   41
biogenetic law   38
catalyst   42
chemical bond   41
combustion   46
compound   40
controlled experiment   37
electron   38

element   38
energy   42
enzyme   42
experiment   37
first law of
    thermodynamics   43
hydroxyl ion   41
hypothesis   37
ion   41
isotope   41
kinetic energy   43

kinetic molecular theory   38
law   38
matter   38
mixture   41
molecule   41
neutron   38
nucleus   38
observation   37
pH   41
photosynthesis   42

potential energy   43
proton   38
repeatability   37
respiration   42
science   37
scientific method   37
second law of
    thermodynamics   43
theory   38
theory of evolution   38

RELATED INTERNET SITES

Department of Energy's home page. Links to a wide assortment of energy topics and issues.
   http://www.doe.gov/
Clean Energy Basics. Covers energy issues dealing with water, sun, wind, plant, and geothermal power.
   http://www.nrel.gov/ceb.html
U.S. Department of Energy site dealing with efficiency and renewable energy networks.
   http://www.eren.doe.gov/
University of Minnesota database on chemical compounds.
   http://dragon.labmed.umn.edu/~lynda/index.html

Award-winning site from the Department of Energy's Alternative Fuels Data Center.
   http://www.afdc.doe.gov
Award-winning site about Horticulture in a virtual perspective.
   ***http://www.hes.ohio.state.edu/hes.html
Chemical Science Technology Division at Los Alamos Laboratory.
   http://www.cst.lanl.gov

Interactions: Environment and Organisms

Objectives

After reading this chapter, you should be able to:

- Identify the abiotic and biotic factors in an ecosystem.
- Define *niche*.
- Describe the process of natural selection as it operates to refine the fit between organism, habitat, and niche.
- Describe predator–prey, parasite–host, competitive, mutualistic, and commensalistic relationships.
- Differentiate between a community and an ecosystem.
- List some of the components of an ecosystem.
- Define the roles of producer, herbivore, carnivore, omnivore, scavenger, parasite, and decomposer.
- Describe energy flow in an ecosystem.
- Relate the concepts of food webs and food chains to trophic levels.
- Explain the cycling of nutrients such as nitrogen, carbon, and phosphorus through an ecosystem.

Chapter Outline

Ecological Concepts
  Environment
  Limiting Factors
  Habitat and Niche
The Role of Natural Selection and Evolution
  Species Definition
  The Mechanism of Natural Selection
Kinds of Organism Interactions
  Predation
  Competition
  Symbiotic Relationships
Community and Ecosystem Interactions
  Major Roles of Organisms
  Energy Flow Through Ecosystems
Environmental Close-Up: *Name That Relationship*
  Food Chains and Food Webs
  Nutrient Cycles in Ecosystems
Environmental Close-Up: *Colorado River Restoration*
Environmental Close-Up: *Organic Contaminants in Great Lakes Fish*
Issues and Analysis: *The Reintroduction of the Moose into Michigan*

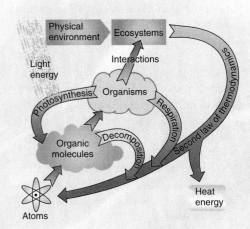

Ecological Concepts

The science of **ecology** is the study of the way organisms interact with each other and with their nonliving surroundings. It deals with the ways in which organisms are molded by their surroundings, how they make use of these surroundings, and how an area is altered by the presence and activities of organisms. These interactions involve energy and matter. Living things require a constant flow of energy and matter to assure their survival. If the flow of energy and matter ceases, the organisms die.

All organisms are dependent on other organisms in some way. One organism may eat another and use it for energy and raw materials. One organism may temporarily use another without harming it. One organism may provide a service for another, such as when animals distribute plant seeds or bacteria break down dead organic matter for reuse. The study of ecology can be divided into many specialties and be looked at from several levels of organization. (See figure 4.1.) Before we can explore the field of ecology in greater depth, we must become familiar with some of the standard vocabulary used.

Environment

Everything that affects an organism during its lifetime is collectively known as its **environment.** (See figure 4.2.) Environment is a very broad concept. For example, during its lifetime an animal such as a raccoon is likely to interact with millions of other organisms (bacteria, food organisms, parasites, mates, predators), drink copious amounts of water, breathe huge quantities of air, and respond to daily changes in temperature and humidity. This list only begins to describe the various components that make up the raccoon's environment. Because of this complexity, it is useful to subdivide the concept of environment into **abiotic** (nonliving) and **biotic** (living) **factors.**

Abiotic factors include the flow of energy necessary to maintain any organism, the physical factors that affect it, and the supply of molecules required for its life functions. The ultimate source of energy for almost all organisms is the sun; in the case of plants, the sun directly supplies the energy necessary for them to maintain themselves. Animals obtain their energy by eating plants or other animals that eat plants. Ultimately, the amount of living material that can exist in an area is determined by the plants and the amount of energy they can trap.

Other physical factors include such things as climate (average weather patterns over a number of years); temperature (average annual temperature as well as daily variations); precipitation, including its type (rain, snow, hail), amount, and seasonal distribution; type of soil present (sandy or clay, dry or wet, fertile or unfertile); and even the three-dimensional shape of the space the organism inhabits.

All forms of life require atoms such as carbon, nitrogen, and phosphorus, and molecules such as water, to construct and maintain themselves. Organisms constantly obtain these materials from their environment by eating food or taking them up through the process of photosynthesis. The atoms are used for a period of time as part of the organism's body structure, and eventually are returned to the environment through respiration, excretion, or death and decay.

The biotic factors influencing an organism include all forms of life with which it interacts. Plants that carry on photosynthesis; animals that eat other organisms; bacteria and fungi that cause decay; and bacteria, viruses, and other parasitic organisms that cause disease are all part of an organism's biotic environment.

Limiting Factors

Although organisms interact with their surroundings in many ways, certain factors may be recognized as key to a particular species' success. A shortage or absence of a key factor restricts the success of the species; thus the factor is known as a **limiting factor.** Limiting factors may be either abiotic or biotic and can be quite different from one species to another. Many plants are limited by scarcity of water or specific soil nutrients. Animals may be limited by climate or the availability of a specific food. For example, many snakes and lizards are limited to the warmer parts of the world, and Monarch butterflies require milkweed plants as food for their developing caterpillars.

The limiting factor for many species of fish is the amount of dissolved oxygen in the water. In a swiftly flowing, tree-lined mountain stream, the level of dissolved oxygen is high and so provides a favorable environment for trout. (See figure 4.3.) As the stream continues down the mountain, the steepness of the slope decreases, which results in fewer rapids to oxygenate the water. In addition, the canopy of trees over the stream usually is thinner, allowing more sunlight to reach the stream and warm the water. Warm water cannot hold as much dissolved oxygen as cool water. Therefore, slower-flowing, warm-water streams contain less oxygen than rapidly moving, cool streams. Fish such as black bass and walleye replace the trout, since they are able to tolerate lower oxygen concentrations and higher water temperatures. These species have a greater **range of tolerance** to oxygen concentration and water temperature. Thus, low levels of oxygen and high water temperatures are limiting factors for the distribution of trout.

Other factors such as the abundance of silt may influence the ability of water to support certain species of fish. Silt reduces visibility, making it difficult for fish to find food, and covers gravel beds fish need for spawning. Under these conditions, the bass and walleye may be replaced by such species as carp and catfish, which have an even greater ability to withstand high temperatures and low oxygen concentrations and are better able to survive in water with a high amount of silt.

The amount of grass available for food is a limiting factor for grazing animals; the saltiness (salinity) of water is limiting for many kinds of ocean animals and prevents their migration into freshwater; and the amount of available sunlight is a limiting factor for many kinds of plants.

Figure 4.1

Ecology The study of ecology is pursued at various levels, ranging from atoms and molecules to individual organisms to groups of interacting organisms to ecosystems.

Figure 4.2

Environment Everything that affects an organism during its lifetime is collectively called its environment. Physical factors, such as weather, soil type, altitude, living space, and the amount of sunlight, have a significant effect on the kinds of organisms that can live in an area. Likewise, how the organism interacts with other organisms, such as the types of plants used for food and shelter, parasites, and predators, is part of the environment.

Habitat and Niche

As we have just seen, it is impossible to understand an organism apart from its environment. The environment influences the organism, and organisms affect the environment. To focus attention on specific elements of this interaction, ecologists have developed two concepts that need careful explanation: habitat and niche.

The **habitat** of an organism is the space that the organism inhabits, the place where it lives. We tend to identify an organism's habitat with particular physical environmental characteristics like soil type, availability of water, or climatic con-

ditions, or we identify it with the predominant plant species that exist in the area. For example, mosses are small plants that dry out and die if they are exposed to sunlight, wind, and drought. Therefore, the habitat of moss is likely to be cool, moist, and shady. (See figure 4.4.) Likewise, a rapidly flowing, cool, well-oxygenated stream with many bottom-dwelling insects is good trout habitat, while open prairie with lots of grass is preferred by bison, prairie dogs, and many kinds of hawks and falcons. Elm bark beetles will reside only in areas where elm trees are found. The particular biological requirements of an organism determine the kind of habitat in which it is likely to be found.

The **niche** of an organism is the functional role it has in its surroundings. An organism's niche includes everything that affects the organism and everything affected by the organism during its lifetime. For example, beavers frequently flood areas by building dams of mud and sticks across steams. (See figure 4.5.) The flooding has several effects. It provides beavers with a larger area of deep water, which they need for protection; it provides a pond habitat for many other species of animals like ducks and fish; and it kills trees that cannot live in saturated soil. The animals attracted to the pond and the beavers often fall to predators. After the beavers have eaten all the suitable food, such as aspen, they will abandon the pond and migrate to other areas along the stream and begin the whole process over again.

In this recitation of beaver characteristics, we have listed several effects that the animal has. It changes the physical environment by flooding, it kills trees, it enhances the environment for other animals, and it is a food source for predators. This is only a superficial glimpse of the many aspects of the beaver's interaction with its environment. A complete catalog of all aspects of its niche would make up a separate book.

Another familiar organism is the dandelion. (See figure 4.6.) It is an opportunistic plant that rapidly becomes established in sunny, disturbed sites. In a few days it can produce thousands of parachutelike seeds that are easily carried by the wind over long distances. (You have probably helped this process by blowing on the fluffy white head of a mature dandelion flower.) Furthermore, it often produces several sets of flowers per year. Since there are so many seeds and they are so easily distributed, the plant can easily establish itself in any sunny, disturbed site, including lawns. Since it is a plant, one major aspect of its niche is the ability to carry on photosynthesis and grow. Dandelions need a large amount of sunlight to grow successfully and do not do well in shady areas. Mowing lawns helps provide just the right conditions for dandelions, since the vegetation is never allowed to get so tall that dandelions are shaded. Many kinds of animals, including humans, use the plant for food. The young leaves may be eaten in a salad, and the blossoms can be used to

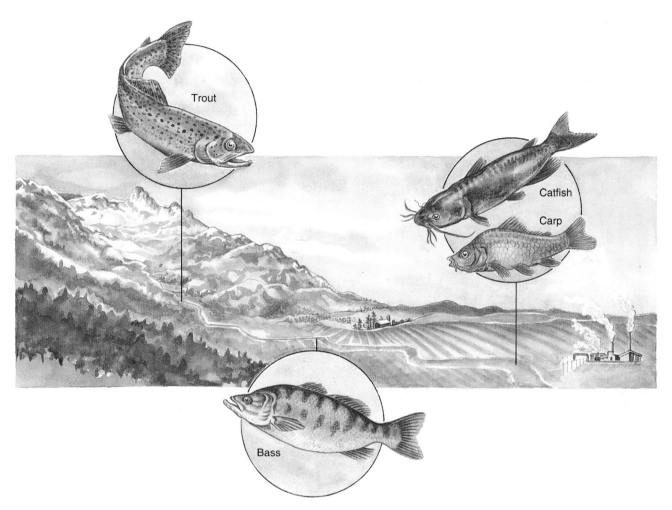

Figure 4.3

Limiting Factors   In aquatic habitats, the amount of oxygen dissolved in the water is often a limiting factor for many species of fish. Cool, highly oxygenated water, which is typical of the rapidly flowing upper sections of river systems, supports trout, but warmer, less oxygenated water is unsuited for trout. Other fish, which are more tolerant of low levels of oxygen, such as bass and carp, occupy the lower sections of the river, where the water is warmer, there is less oxygen, and the river contains much silt and other soil particles.

Figure 4.4

Moss Habitat   The habitat of mosses must be cool and shady, since mosses tend to dry out and die if they are exposed to the drying action of the sun.

make dandelion wine. Bees visit the flowers regularly to obtain nectar and pollen.

The Role of Natural Selection and Evolution

As we have seen, each organism is finely tuned to a particular habitat and has a very specific role (niche) within its habitat. But how is it that each plant, animal, fungus, or bacterium fits into its environment in such a precise way? The process that leads to this close fit between organisms and their environment is known as natural selection. **Natural selection** is the process of more successful individuals surviving and reproducing larger numbers of offspring than those that are less successful. Those that reproduce more successfully pass on to the next generation the characteristics that made them successful. Thus each kind of organism is continually refined as each generation is subjected to the same environmental conditions.

Species Definition

It is important to recognize that natural selection happens only within a **species,** which is a population of organisms in which the individuals are potentially able to interbreed and produce fertile offspring. Individual organisms are members

Figure 4.5

Ecological Niche   The ecological niche of an organism involves everything that affects it during its lifetime as well as all the effects it has. The ecological niche of beaver includes the need for streams with aspen trees nearby, the building of dams and flooding of forested areas, the killing of trees, the providing of ponds for ducks and other animals, and many other effects.

Figure 4.6

The Niche of a Dandelion   A dandelion is a common plant that does well in disturbed sites with lots of sunlight. It is able to invade these areas easily because it produces many seeds that are blown easily to new areas. Lawn mowing benefits dandelions since it prevents shading by tall grasses and other plants.

yellow flowers. Other species are not as easy to recognize. Most of us cannot tell one species of mosquito from another or identify different species of grasses. Because of this we tend to lump organisms into large categories and do not recognize the many subtle niche differences that exist among the similar-appearing species. However, different species of mosquitoes are quite distinct from one another. Only certain ones carry the human disease malaria. Other species carry dog heartworm. Each species is active during certain portions of the day or night. And each species requires specific conditions to reproduce.

The Mechanism of Natural Selection

In 1859, Charles Darwin, a British naturalist, introduced the theory of natural selection in his book *On the Origin of Species by Means of Natural Selection, or the Preservation of Favored Races in the Struggle for Life.* As a young man Darwin had the opportunity to travel around the world as the naturalist on a British survey ship. Wherever Darwin visited, he saw many different kinds of plants and animals and noted how well they were adapted to their particular situations. This caused him to wonder about the mechanism that

of a species. This definition of species contains two points that require explanation. Obviously, there are individuals in any population that never reproduce, and many pairs of individuals that will never meet one another. However, they still have the potential to interbreed and are considered members of the same species. The second point is the ability to produce fertile offspring. In some instances, two kinds of organisms may interbreed and

produce offspring, but the offspring are sterile and never reproduce. For example, horses and donkeys can breed and produce offspring called mules, but since the mules are sterile, the horse and donkey are considered separate species.

Some species are easy to recognize. We easily recognize humans as a distinct species. Most people recognize a dandelion when they see it and do not confuse it with other kinds of plants that have caused this high degree of adaptation. He concluded that several factors could interact to allow for natural selection:

1. Individuals within a species showed variation; some of the variations were very useful and others were not. For example, individual animals that are part of the same species show color variations. Some colors make the animal more conspicuous while others make it less conspicuous.

2. Organisms within a species typically produce many more offspring than can survive. This means that there is not enough suitable habitat for all of the offspring to grow to maturity. One apple tree may produce hundreds of apples with several seeds in each apple, or a pair of rabbits may have three to four litters of offspring each summer, with several young in each litter.

3. The excess number of individuals results in a struggle for survival. Individuals within the population must compete with each other for food, space, mates, or other requirements that are in limited supply. If you plant 100 bean seeds in a pot, many of them will begin to grow, but eventually some will become taller and shade the remaining plants. Great horned owls typically produce two young at a time, but if food is in short supply, the larger of the two young will get the majority of the food.

4. Because of variation among individuals, some have a greater chance of surviving and reproducing than others. The competition for resources often results in the less fit individuals dying; therefore, they do not get a chance to reproduce themselves. Even if they do not die, they may mature more slowly and not be able to reproduce as many times as the more fit members of the species.

5. As time passes and each generation is subjected to the same process, the percentage of individuals showing favorable variations will increase and those having unfavorable variations will decrease. Thus, the species will become better and better adapted to its environment.

Many people have tested the theory of natural selection, and it continues to be tested today. In a classic experiment, H.B.D. Kettlewell studied the processes that resulted in some areas of England having mostly dark-colored forms of the peppered moth, while other areas had mostly light-colored forms. He hypothesized that the color of tree trunks was a major influence. Peppered moths normally rest on tree trunks during the day. If they are noticed by birds, they are eaten. In industrialized regions of England, the trees were blackened by the presence of soot from the smokestacks of coal-burning industrial plants. Kettlewell discovered that, in these areas, the light-colored moths were more conspicuous and were more frequently located and eaten by birds. (See figure 4.7.) In areas with less air pollution the trees were light colored and were covered with light-colored lichens. Birds had greater difficulty locating the light-colored moths, and the more conspicuous dark-colored moths were rare. In these two different locations, natural selection, through the act of birds killing and eating moths, was selecting for those moths that were able to blend in with their surroundings. Each generation of moths had a greater proportion of individuals that were able to blend in than the previous generation had.

Since Darwin's time, many discoveries have helped to explain how the process of natural selection works. Today, we know that each organism has DNA that determines the characteristics it will develop. We also know that the genes contained in DNA control, in addition to structural characteristics, behavior and how organs and body parts function. We also have a better appreciation for how subtle competition can be and how the behavior of organisms contributes to their successful reproduction. We better understand the age of the Earth and appreciate how it has changed over millions of years. In fact, studies of recent fossils and other geologic features show that only thousands of years ago, huge glaciers covered much of Europe and the northern parts

Figure 4.7

Natural Selection in Peppered Moths   Air pollution in industrial regions of England resulted in blackened tree trunks. The dark form of peppered moths was less conspicuous and, therefore, they were not seen as readily by their bird predators. The light form was more easily seen and more frequently eaten. The number of dark moths increased and the number of light moths decreased, because the predators ate more of the light form of the moth, preventing them from breeding and passing on their genes for light color.

of North America, affecting the lives of humans who ate mammoths and encountered saber-toothed tigers and giant cave bears. As the climate became warmer and the glaciers receded, new pressures affected the world's organisms. Some, including the mammoths, saber-toothed tigers, and giant cave bears, did not adapt and became extinct. Others, such as humans, horses, and most kinds of plants, adapted to the new conditions and so survive to the present. Natural selection is constantly at work shaping organisms to fit a changing environment.

Scientists have continuously shown that this theory of natural selection can explain the development of most aspects of the structure, function, and behavior of organisms. It is the central idea that helps explain how species adapt to their surroundings. When we discuss environmental problems, it is helpful to understand that species change, and that as the environment is changed either naturally or by human action, some species will adapt to the new conditions while others will not.

When natural selection occurs over long periods of time (thousands to millions of years), considerable change can be seen in the nature of organisms. This change in the kinds of organisms that exist and in their characteristics is called **evolution.** During the process of evolution, which occurs as a result of natural selection, new species can be produced. The production of new species from previously existing species is known as **speciation** and is thought to occur as a result of a species dividing into two isolated subpopulations. If this separation occurs and the two subpopulations must adapt to somewhat different environments, they will probably develop different characteristics. Eventually the differences may be so great that the two subpopulations are not able to interbreed. At this point they are two different species.

Kinds of Organism Interactions

Ecology looks at organisms and how they interact with their surroundings, including other organisms. We will discuss several kinds of interactions recognized as distinct types. The concept of natural selection allows us to see how interacting organisms can result in populations that are better adapted to their environment. As you read this section, notice how different species have characteristics that fit them for their specific roles (niches).

Predation

One common kind of interaction occurs when one animal, known as a **predator,** kills and eats another, known as the **prey.** (See figure 4.8.) Some examples of predator–prey relationships are lions and zebras, birds and worms, wolves and moose, and frogs and insects. Predator–prey relationships are often thought to be one-sided. The predator seems to reap all the benefits. It has the meal and goes on living after killing and eating the prey. However, this relationship does have value for the prey as well. Prey species have a higher reproductive rate than predator species. For example, field mice may have 10 to 20 offspring per year, while

Figure 4.8

Predator/Prey Relationship   Lions are predators on zebras. The quicker lions are more likely to get food, and the slower, sickly, or weaker zebras are more likely to become prey.

hawks typically have two to three. Because of this high reproductive rate, prey species can endure a high mortality rate. Certainly, the individual organism that has been killed and eaten did not benefit, but the species does since the prey organisms that die are likely to be the old, the slow, the sick, and the less fit members of the population. The healthier, quicker, and more fit individuals are more likely to survive. When these survivors reproduce, their offspring are more likely to have characteristics that help them survive; they are better adapted to their environment. At the same time, a similar process is taking place in the predator population. Since poorly adapted individuals are less likely to capture prey, they are less likely to survive and reproduce. The predator is a participant in the natural selection process and so is the prey. This dynamic relationship between predator and prey species is a complex one that continues to intrigue ecologists.

Competition

A second type of interaction between species is **competition,** in which two organisms strive to obtain the same limited resource. In the process, both organisms are harmed to some extent. (See figure 4.9.) However, this does not mean that there is no winner. If two robins are competing for the same worm, only one gets it. Both organisms were harmed because they had to expend energy in fighting for the worm, but one got some food and was harmed less than the one that fought and got nothing. This example of competition, which is between members of the same species, is known as **intraspecific competition.** Other examples of intraspecific competition include corn plants in a field competing for water and nutrients, male elk competing with one another for the right to mate with the females, and certain species of woodpeckers competing for the holes in dead trees to use for nesting sites.

Competition among members of the same species does not have only a negative impact. As was the case with predation, competition has its good side when seen from the point of view of the species as a whole. When resources are limited, less fit organisms are more likely to die or be denied mating privileges. Consequently, the next generation of organisms will contain more of the genetic characteristics that are favorable for survival in

Figure 4.9

Competition Whenever a needed resource is in limited supply, organisms compete for it. This competition may be between members of the same species and is called intraspecific competition, or it may be between different species and is called interspecific competition. This photograph shows several vultures competing for a food source.

Flea (external parasite)

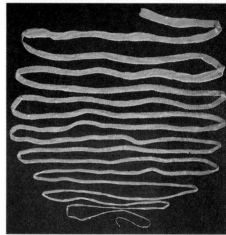

Tapeworm (internal parasite)

Figure 4.10

Parasitism Fleas are small insects that live in the feathers of birds or the fur of mammals, where they bite their hosts to obtain blood. Since they live on the outside of their hosts, they are called ectoparasites. Tapeworms live inside the intestines of their hosts, where they absorb food from their hosts' intestines. Since they live inside their hosts, they are called endoparasites.

that particular environment. Since individuals of the same species have similar needs, competition among them is usually very intense. A slight advantage on the part of one individual may mean the difference between survival and death.

Competition may also occur between organisms of different species. This is called **interspecific competition.** In a forest, very little light reaches the forest floor. The different species of plants must compete for the available light. Mosses and ferns can tolerate shade while grasses cannot; therefore, the grasses lose in this competitive interaction.

The more similar two species are, the more intense the competition between them. If one of the two competing species is better adapted to live in the area than the other, the less fit species must evolve into a slightly different niche, migrate to a different geographic area, or become extinct. As with intraspecific competition, one of the effects of interspecific competition is that the successful species emerge from the interaction better adapted to their environment.

Symbiotic Relationships

Symbiosis is a close, long-lasting, physical relationship between two different species. In other words, the two species are usually in physical contact and at least one of them derives some sort of benefit from this contact. There are three different categories of symbiotic relationships: parasitism, commensalism, and mutualism.

Parasitism

Parasitism is a relationship in which one organism, known as the **parasite,** lives in or on another organism, known as the **host,** from which it derives nourishment. Generally, the parasite is much smaller than the host. Although the host is harmed by the interaction, it is generally not killed by the parasite. Because the evolution of a parasitic way of life involves a long-standing interaction between two species—the parasite and the host—the two species generally evolve in such a way that they can accommodate one another. It is not in the parasite's best interest to kill its host. If it does, it must find another. Likewise, the host evolves defenses against the parasite, often reducing the harm done by the parasite to something the host can tolerate.

Parasites that live on the surface of their hosts are known as **ectoparasites.**

Fleas, lice, and some molds and mildews are examples of ectoparasites. (See figure 4.10.) Many other parasites, like tapeworms, malaria parasites, many kinds of bacteria, and some fungi, are called **endoparasites.** They live inside the bodies of their hosts. The tapeworm lives in the intestines of its host where it is able to resist being digested and makes use of the nutrients in the intestine. If the host has only one or two tapeworms, it can live for some time with little discomfort, supporting itself and its parasites. If the number of parasites is large, the host may die.

Even plants can be parasites. Mistletoe is a flowering plant that is parasitic on trees. It establishes itself on the surface of a tree when a bird transfers the seed to the tree. It then grows down into the tissues of the tree and uses it as a source of nutrients.

Parasitism is a very common niche. If we were to categorize all the organisms in the world, we would find many more parasitic species than nonparasitic species. Each organism, including you, has many others that use it as a host.

Commensalism

If the relationship between organisms is one in which one organism benefits while the other is not affected, it is called **commensalism.** It is possible to visualize a parasitic relationship evolving into a commensal one. Since parasites generally evolve to do as little harm to their host as possible and the host is combating the negative effects of the parasite, they might eventually evolve to the point where the host is not harmed at all. There are many examples of commensal relationships. Many orchids use trees as a surface upon which to grow. The tree is not harmed or helped, but the orchid needs a surface upon which to establish itself and also benefits by being close to the top of the tree, where it can get more sunlight and rain. Some mosses, ferns, and many vines also make use of the surfaces of trees in this way.

In the ocean, many sharks have a smaller fish known as a remora attached to them. Remoras have a sucker on the top of their heads that they can use to attach to the shark. In this way, they can hitchhike a ride as the shark swims along. When the shark feeds, the remora obtains small bits of food that the shark misses. The shark does not appear to be positively or negatively affected by the remoras. (See figure 4.11.)

Mutualism

Some symbiotic relationships are actually beneficial to both species involved. This kind of a relationship is called **mutualism.**

Figure 4.11

Commensalism Remoras hitchhike a ride on sharks and feed on the scraps of food lost by the sharks. This is a benefit to the remoras. The sharks do not appear to be affected by the presence of the remoras.

In many mutualistic relationships, the relationship is obligatory. Flowers of the yucca plant are pollinated by a specific insect known as the yucca moth. The moth, in turn, lays its eggs in the flower, where the immature moth larvae feed on the developing seeds. The yucca is dependent on the moth for pollination, and the moth is dependent on the seeds as a source of food for its larvae. Neither organism can exist without the other. The number of eggs laid by the moth is small enough that some of the yucca seeds grow to maturity to provide new yucca plants for future generations of yucca moths. Many other flowering plants are also dependent on specific insects as pollinators.

One plant nutrient that is usually in short supply in soil is nitrogen. Many kinds of plants, such as beans, clover, and alder trees, have bacteria that live in their roots in little nodules. The roots form these nodules when they are infected with certain kinds of bacteria. The bacteria do not cause disease but provide the plants with nitrogen-containing molecules that the plants can use for growth. The nitrogen-fixing bacteria benefit from the living site and nutrients that the plants provide, and the plants benefit from the nitrogen they receive. (See figure 4.12.)

Figure 4.12

Mutualism The growths on the roots of this plant contain bacteria that are beneficial to the plant because they make nitrogen available to it. The relationship is also beneficial to the bacteria because the plant makes necessary raw materials available to the bacteria. It is a mutually beneficial relationship.

Table 4.1

Roles in an Ecosystem

Category	Major role or action	Examples
Producer	*Converts simple inorganic molecules into organic molecules by the process of photosynthesis*	*Trees, flowers, grasses, ferns, mosses algae*
Consumer	*Uses organic matter as a source of food*	*Animals, fungi, bacteria*
Herbivore	Eats plants directly	Grasshopper, elk, human vegetarian
Carnivore	Kills and eats animals	Wolf, pike, dragonfly
Omnivore	Eats both plants and animals	Rats, raccoons, most humans
Scavenger	Eats meat, but often gets it from animals that died by accident or illness, or that were killed by other animals	Coyote, vulture, blowflies
Parasite	Lives in or on another living organism and gets food from it	Tapeworm, many bacteria, some insects
Decomposer	*Returns organic material to inorganic material; completes recycling of atoms*	*Fungi, bacteria, some insects and worms*

Community and Ecosystem Interactions

Thus far, we have discussed specific ways in which individual organisms interact with one another and with their physical surroundings. We have come to think of interacting groups of species as **communities** in which each kind of organism has a specific niche or role to play. Some organisms play minor roles while others play major roles, but all are part of the community. For example, the grasses of the prairie have a major role since, without them, there would be no prairie. Grasshoppers, prairie dogs, and bison are important consumers of grass. A meadowlark, though a conspicuous and colorful part of the prairie scene, has a relatively minor role and has little to do with maintaining a prairie community.

Communities consist of interacting species, but these species do not interact in a vacuum. They must interact with the physical world around them as well. The physical world has a major impact on what kinds of plants and animals can live in an area. We do not expect to see a banana tree in the arctic or a walrus in the Mississippi River. Banana trees are adapted to warm, moist, tropical areas, and walruses require cold ocean waters. Periodic fires are often important to maintain grasslands and certain kinds of forests. The kind of soil and the amount

of moisture also influence the kinds of organisms found in an area.

Organisms can also impact their physical surroundings. Trees break the force of the wind, grazing animals form paths, and earthworms create holes in the soil. This system of interacting organisms and their nonliving surroundings is frequently referred to as an **ecosystem.** While the concepts of community and ecosystem are closely related, an ecosystem is a broader concept because it involves the physical as well as the biological realm.

Every system has parts that are related to one another in specific ways. A bicycle has wheels, a frame, handlebars, brakes, pedals, and a seat. These parts must be organized in a certain way or the system known as a bicycle will not function. Similarly, an ecosystems have parts that must be organized in specific ways or the systems will not operate. In order to more fully develop the concept of ecosystem, we will look at ecosystems from three points of view: the major roles played by organisms, the way energy is utilized within ecosystems, and the way atoms are cycled from one organism to another.

Major Roles of Organisms

Several categories of organisms are found in any ecosystem. **Producers** are able to make new, complex, organic material from the atoms in their environment.

To do this, they must have a source of energy. In nearly all ecosystems, this energy is supplied by the sun, and the organisms that utilize this energy are plants that carry on photosynthesis. Since producers are the only organisms in an ecosystem that can trap energy and make new organic material from inorganic material, all other organisms rely on producers as a source of food, either directly or indirectly. These other organisms are called **consumers** because they consume organic matter to provide themselves with energy and the organic molecules necessary to build their own bodies. **Primary consumers** eat producers (plants) as a source of food. They are also known as **herbivores. Secondary consumers** or **carnivores** eat other animals. Some carnivores eat herbivores, while others may eat other carnivores. In addition, many animals, called **omnivores,** have mixed diets that include both plants and animals.

A final category of consumer is the decomposer. **Decomposers** use nonliving organic matter as a source of food. Many small animals, fungi, and bacteria fill this niche. (See table 4.1.)

Energy Flow Through Ecosystems

An ecosystem is a stable, self-regulating unit. To maintain itself, it must have a

environmental *close-up*

Name That Relationship

Sometimes it is not easy to categorize the relationships that organisms have with each other. For example, it is not always easy to say whether a relationship is a predator–prey relationship or a host–parasite relationship. How would you classify a mosquito or a tick? Both of these animals require blood meals to live and reproduce. They don't kill and eat their prey. Neither do they live in or on a host. This question points out the difficulty encountered when we try to place all organisms into a few categories.

Another relationship that doesn't fit well is the relationship that certain birds like cowbirds and cuckoos have with other birds. Cowbirds and cuckoos do not build nests but lay their eggs in the nests of other species of birds, who are left to care for a foster nestling at the expense of their own nestlings, who generally die. This could be called parasitism, but it doesn't fit the standard textbook definition.

What about grazing animals? Are they predators on the plants that they eat? And what kind of relationships do humans have with their domesticated plants and animals? Do we have a mutualistic relationship with cows and chickens? We certainly derive benefit from the milk and eggs we get, and cows and chickens benefit from the care they are given. Or are we primarily parasites on these species? Certainly, we are sometimes predators when we kill and eat them.

It would be convenient to unambiguously classify every kind of relationship into a clear category. But this is not possible. Moreover, each relationship may change in character over time.

Is this cowbird nestling being fed by a red-eyed vireo a parasite?

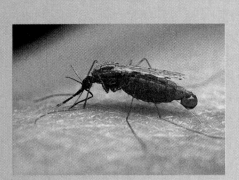

Is this mosquito a parasite?

Are bison grass predators?

---

continuous input of energy. The only significant source of energy for most ecosystems is sunlight energy. Producers are the only organisms that are capable of trapping this energy through the process of photosynthesis and making it available to the ecosystem. The energy is stored in the form of chemical bonds in large organic molecules such as carbohydrates (sugars, starches), fats, and proteins. The energy stored in the molecules of plants can be transferred to other organisms when they consume plants. Each step in the flow of energy through an ecosystem is known as a **trophic level.** Producers (plants) constitute the first trophic level, and herbivores constitute the second trophic level. Carnivores that eat herbivores are the third

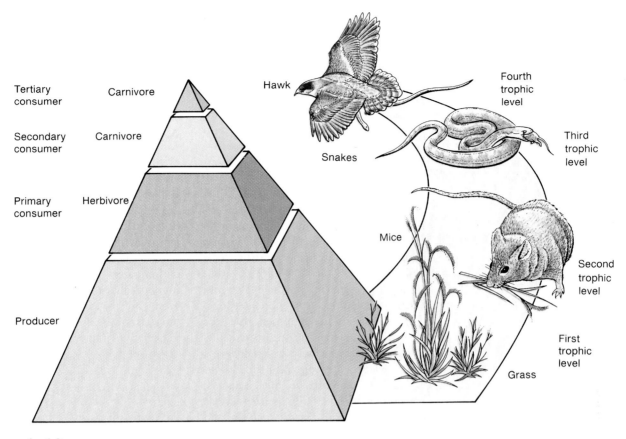

Figure 4.13

Energy Flow Through an Ecosystem As energy flows through an ecosystem, it passes through several levels known as trophic levels. Each time energy moves to a new trophic level, approximately 90 percent of the useful energy is lost. Therefore, higher trophic levels contain less energy and fewer organisms in most ecosystems.

trophic level, and carnivores that eat other carnivores are the fourth trophic level. Omnivores, parasites, and scavengers occupy different trophic levels, depending on what they happen to be eating at the time. If we eat a piece of steak, we are at the third trophic level; if we eat celery, we are at the second trophic level. (See figure 4.13.)

As energy passes from one trophic level to the next, some of the useful energy is lost due to the second law of thermodynamics. Much of this loss is in the form of low-quality heat, which is dissipated to the surroundings and warms the air, water, or soil. In addition to this loss of heat, organisms must expend energy to maintain their own life processes. It takes energy to chew food, defend nests, walk to waterholes, or raise offspring. Therefore, the amount of energy contained in higher trophic levels is con-

siderably less than that at lower levels. Approximately 90 percent of the useful energy is lost with each transfer to the next highest trophic level. So in any ecosystem, the amount of energy contained in the herbivore trophic level is only about 10 percent of the energy contained in the producer trophic level. The amount of energy at the third trophic level is approximately 1 percent of that found in the first trophic level.

Because it is difficult to actually measure the amount of energy contained in each trophic level, ecologists often use other measures to approximate the relationship between the amounts of energy at each level. One of these is the biomass. The **biomass** is the weight of living material in a trophic level. It is often possible in a simple ecosystem to collect and weigh all the producers, herbivores, and carnivores. The weights often show the

same 90 percent loss from one trophic level to the next as happens with the amount of energy.

Food Chains and Food Webs

The passage of energy from one trophic level to the next as a result of one organism consuming another is known as a **food chain.** For example, willow trees grow well in very moist soil, perhaps near a pond. The trees' leaves capture sunlight and convert carbon dioxide and water into sugars and other organic molecules. The leaves serve as a food source for insects, such as caterpillars and leaf beetles, that have chewing mouth parts and a digestive system adapted to plant food. Some of these insects are eaten by spiders, which fall from the trees into the pond below, where they are consumed by a frog. As the frog swims from one lily pad

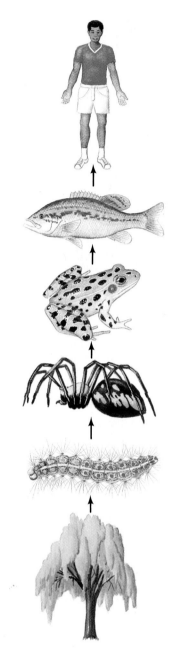

Figure 4.14

Food Chain As one organism feeds on another organism, energy flows through the series. This is called a food chain.

niche, and each organism in the food chain is involved in converting energy and matter from one form to another.

When a plant or animal dies, the chemical energy contained within its body is ultimately released as heat by organisms that decompose the body into smaller molecules, such as water and carbon dioxide. These decomposers—bacteria, fungi, and some small consumers, such as insects and worms—play a significant role in the ultimate recycling of the atoms that make up the bodies of living things. They are involved in the breakdown of small bits of organic material called **detritus.** Detritus food chains are found in a variety of situations. The bottoms of the deep lakes and oceans are too dark for photosynthesis. The animals and decomposers that live there rely on a steady rain of small bits of organic matter from the upper layers of the water where photosynthesis does take place. Similarly, in most streams, leaves and other organic debris serve as the major source of organic material and energy. A sewage treatment plant is also a detritus food chain in which particles and dissolved organic matter are constantly supplied to a series of bacteria and protozoa that use this material for food.

In another example, the soil on a forest floor receives leaves, which fuel a detritus food chain. In detritus food chains, a mixture of insects, crustaceans, worms, bacteria, and fungi cooperate in the breakdown of the large pieces of organic matter, while at the same time feeding on one another. When a leaf dies and falls to the forest floor it is colonized by bacteria and fungi, which begin the breakdown process. An earthworm will also feed on the leaf and at the same time consume the bacteria and fungi. If that earthworm is eaten by a bird, it becomes part of a larger food chain that includes material from both a detritus food chain and a photosynthesis-driven food chain. When several food chains overlap and intersect, they make up a **food web.** (See figure 4.15.)

Notice in the upper-left-hand corner of figure 4.15 that the Cooper's and sharp-shinned hawks use many different kinds of birds as a source of food. These hawks

fit into several food chains. If one source of prey is in short supply, they can switch to something else without too much trouble. These kinds of complex food webs tend to be more stable than simple food chains with few cross-links.

Nutrient Cycles in Ecosystems

Organisms are composed of molecules and atoms that are cycled from one living organism to another. Some atoms are more common in living things than others. Carbon, nitrogen, oxygen, hydrogen, and phosphorus are found in important molecules like proteins, DNA, carbohydrates, and fats, which are found in all kinds of living things. These atoms are passed from one organism to another when one organism consumes another, and they are recycled when an organism dies.

Carbon Cycle

All living things are composed of organic molecules that contain the atom carbon. The way in which carbon is converted into organic molecules that are used and released from organisms is called the **carbon cycle** (See figure 4.16.) Carbon and oxygen combine to form the molecule carbon dioxide, which is a gas present in the atmosphere in small quantities. During photosynthesis, carbon dioxide from the atmosphere is taken into plant leaves and combined with hydrogen from water molecules, which are absorbed from the soil by the roots and transported to the leaves. Complex organic molecules such as carbohydrates (sugars) are formed. Oxygen molecules are released from the leaves into the atmosphere during the process of photosynthesis when water molecules are split to provide hydrogen atoms for the organic molecules. In this process, light energy is converted to chemical-bond energy in organic molecules such as sugar. Plants use these sugars for growth and to provide energy for other processes.

Herbivores can use these complex organic molecules as food. When an herbivore eats a plant, it breaks down the complex organic molecules into simpler molecules, which can be incorporated into the chemical structure of the animal.

to another, a large bass consumes the frog. A human may use an artificial frog as a lure to entice the bass from its hiding place. A fish dinner is the final step in this chain of events that began with the leaves of a willow tree. (See figure 4.14.) This food chain has five trophic levels. Each organism occupies a specific niche and has special abilities that fit it for its

PART TWO Ecological Principles and Their Application

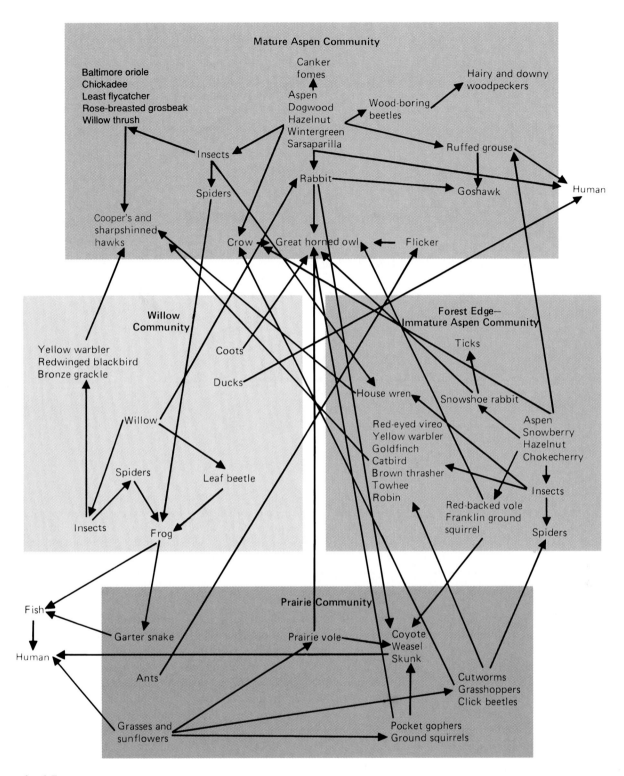

Figure 4.15

Food Web   The many kinds of interactions among organisms in an ecosystem constitute a food web. In this network of interactions, several organisms would be affected if one key organism were reduced in number. Look at the rabbit in the mature aspen community and note how many organisms use it as food.

Source: Adapted from Ralph D. Bird, "Biotic Communities of the Aspen Parkland of Central Canada" in Ecology 11:410, April 1930.

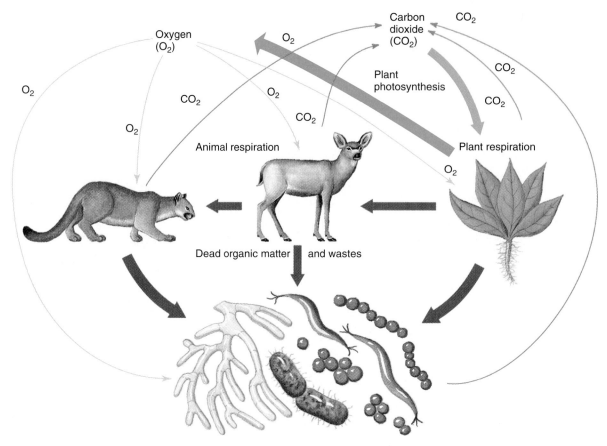

Figure 4.16

Carbon Cycle Carbon atoms are cycled through ecosystems. Plants can incorporate carbon atoms from carbon dioxide into organic molecules when they carry on photosynthesis. The carbon-containing organic molecules are passed to animals when they eat plants or other animals. Organic wastes or dead organisms are consumed by decay organisms. All organisms, plants, animals, and decomposers return carbon atoms to the atmosphere when they carry on respiration. Oxygen atoms are being cycled at the same time that carbon atoms are being cycled.

The carbon atom, which was once part of the plant, is now part of the herbivore. All organisms also carry on the process of respiration, in which oxygen from the atmosphere is used to break down large organic molecules into carbon dioxide and water. Much of the chemical-bond energy is released by respiration and is lost as heat, but the remainder is used by the herbivore for movement, growth, and other activities.

When an herbivore is eaten by a carnivore, the carbon-containing molecules of the herbivore become incorporated into the body of the carnivore. Carnivores also carry on respiration and release carbon dioxide and water in the process of obtaining energy from organic molecules.

The waste products of all kinds of organisms, and any organisms that die, are acted upon by decomposers so that even the naturally occurring organic molecules in nonliving matter are used by decomposers to provide energy. Carbon dioxide is produced in the process. The carbon atom begins as part of a carbon dioxide molecule in the atmosphere, becomes part of an organic molecule in one or more organisms, and then is released back into the atmosphere by the process of respiration.

Nitrogen Cycle

Another very important cycle, the **nitrogen cycle,** involves the flow of nitrogen atoms through organisms in an ecosystem.

Seventy-eight percent of the gas in the air we breathe is made up of molecules of nitrogen gas. However, very few organisms are able to use it in this form. Since plants are at the base of nearly all food chains, they must make new nitrogen-containing molecules, such as proteins and DNA. Plants are unable to use the nitrogen in the atmosphere and must get it in the form of nitrate (NO_3^-) or ammonia (NH_3). The amount of nitrogen available to plants controls their growth and is often a limiting factor. The primary way in which plants obtain nitrogen compounds they can use is with the help of bacteria.

Some bacteria are able to convert the nitrogen gas (N_2) in the atmosphere into forms that plants can use. These bacteria are called **nitrogen-fixing bacteria.**

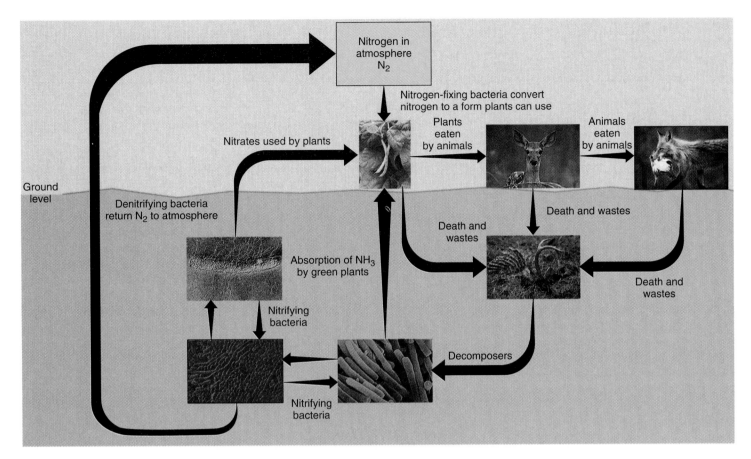

Figure 4.17

Nitrogen Cycle   Nitrogen atoms are cycled in ecosystems. Atmospheric nitrogen is converted by nitrogen-fixing bacteria to a form that plants can use to make protein and other compounds. Proteins are passed to other organisms when one organism is eaten by another. Dead organisms and waste products are acted on by decay organisms to form ammonia, which may be reused by plants or converted to other nitrogen compounds by other kinds of bacteria. Denitrifying bacteria are able to convert inorganic nitrogen compounds into atmospheric nitrogen.

Some kinds of these bacteria live in the soil and are called **free-living nitrogen-fixing bacteria.** Others live in nodules in the roots of plants known as legumes (peas, beans, and clover) and are known as **symbiotic nitrogen-fixing bacteria.** Some grasses and evergreen trees appear to have a similar relationship with certain root fungi that seem to improve the nitrogen-fixing capacity of the plant.

Other kinds of bacteria can also make nitrogen available to plants. Dead organisms and their waste products contain molecules that contain nitrogen. Many kinds of soil bacteria decompose these large nitrogen-containing organic molecules, releasing ammonia, which can be used directly by many kinds of plants or be converted to nitrate by other kinds of bacteria. Additional kinds of bacteria, known as **denitrifying bacteria,** are, under certain conditions, able to convert a molecule known as nitrite to atmospheric nitrogen gas. The nitrogen atoms in the gas can enter the cycle again with the aid of nitrogen-fixing bacteria. (See figure 4.17.)

In naturally occurring soil, nitrogen is often a limiting factor of plant growth. To increase yields, farmers often provide extra sources of nitrogen in several ways. Inorganic fertilizers are a primary method of increasing the nitrogen available. These fertilizers may contain ammonia, nitrate, or both.

Since the manufacture of nitrogen fertilizer takes a lot of energy, fertilizer is expensive. Farmers have looked for alternative methods to reduce this cost. A farmer might alternate nitrogen-demanding crops like corn with nitrogen-yielding crops like soybeans. Soybeans are legumes that have symbiotic nitrogen-fixing bacteria in their roots. If soybeans are planted one year, the excess nitrogen left in the soil can be used by the corn plants grown the next year. A slightly different technique involves growing a nitrogen-fixing crop for a short period of time and then plowing the crop under and letting the organic matter decompose. Farmers can also add nitrogen to the soil by spreading manure on the field and relying on the soil bacteria to decompose the organic matter and release the nitrogen for plant use.

Phosphorus Cycle

Phosphorus is another kind of atom common in the structure of living things. It is present in many important biological molecules such as DNA and in the membrane structure of cells. In addition, the bones and teeth of animals contain significant

Colorado River Restoration

The Colorado River originally was a warm water river that fluctuated widely in flow and delivered large amonts of sediment to the river system. The building of dams on the river changed its nature significantly. The sediments were deposited above the dams, and since water was discharged from the bottom of the dam, the water was colder. Furthermore, since the river was dammed to provide water resources for cities and agriculture, the amount of water flowing through the river system was reduced.

In order to restore some of the original character to the portion of the river that flows through the Grand Canyon, the U.S. Department of the Interior in March 1996 released large amounts of water from the Glen Canyon Dam. This action was intended to simulate the normal flooding that used to occur before the dam was built. The flooding restored beaches, marshes, and backwaters that are important for many species native to the Grand Canyon portion of the Colorado River. The humpback chub, for example, requires backwaters for spawning. Other fish species benefit from the restoration as well. The flooding increased the amount of organic material in the water. This is important as a source of food for insects that are in turn eaten by fish and other organisms.

Information gathered from this test release will be important in assessing future management of the Colorado River ecosystem and will have application to other river systems where dams alter flow.

quantities of phosphorus. The ultimate source of phosphorus atoms is rock. In nature, new phosphorus compounds are released by the erosion of rock. Plants use the dissolved phosphorus compounds to construct the molecules they need. Animals consume plants or other animals and obtain needed phosphorus. When an organism dies or excretes waste products, decomposer organisms recycle the phosphorus compounds back into the soil. (See figure 4.18.) In many soils, phosphorus is in short supply and must be provided to crop plants to get maximum yields.

Fertilizers usually contain nitrogen, phosphorus, and potassium compounds. The numbers on a fertilizer bag indicate the percentage of each in the fertilizer. For example, a 6–24–24 fertilizer has 6 percent nitrogen, 24 percent phosphorus, and 24 percent potassium compounds. In addition to carbon, nitrogen, and phosphorus, potassium and other elements are cycled within ecosystems. In an agricultural ecosystem, these elements are removed when the crop is harvested. Farmers must return these elements to the soil with fertilizer.

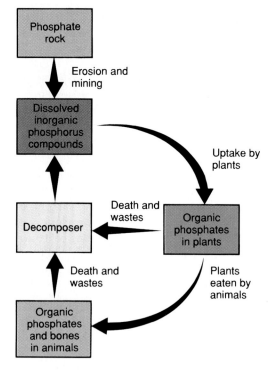

Figure 4.18

Phosphorus Cycle The source of phosphorus is rock that, when dissolved, provides a source of phosphate used by plants and animals.

environmental close-up

Organic Contaminants in Great Lakes Fish

The Great Lakes area has a concentration of industry because of the availability of water for manufacturing processes, and because water transportation is an efficient way to move raw materials and products. In the past, many of these industries released heavy metals and organic molecules into the water as an accidental by-product of the manufacturing process, or because it was a cheap way to get rid of unwanted material. Many of the organic molecules are products of modern organic synthesis and, therefore, are not something that bacteria and fungi are able to decay. Since these molecules are of recent origin, the decomposers have not had time to evolve mechanisms to break them down.

Approximately five hundred different organic compounds that scientists think are contaminants have been identified in the bodies of Great Lakes fish. Most are present in extremely small amounts and probably do not represent a serious hazard, but others are present in high enough concentrations to cause public health officials to be concerned. The contaminants are a problem because they cannot be broken down and eliminated. They persist. Fish that are carnivores tend to accumulate more of these toxic materials in their bodies as they get older. It just so happens that most of the fish that people catch and eat are carnivores. Furthermore, the older a fish is, the greater can be its concentration of contaminants.

It is not possible to check every fish caught to see if it is fit to eat. The cost of doing so would be on the order of several hundred to several thousands of dollars per individual fish, depending on which contaminants are looked for. Therefore, the states and provinces surrounding the Great Lakes have developed advisory statements to help people avoid unsafe fish.

1. Some species of fish, such as carp and catfish, feed on the bottom and tend to accumulate contaminants more quickly than others. In many areas of the Great Lakes, people are advised to not eat these fish.
2. Larger fish have generally consumed more food and have had an opportunity to accumulate more contaminants. Therefore, people are advised to eat only smaller fish.
3. Since many of the organic contaminants are fat soluble, removal of the fat or cooking in such a way that the fat is allowed to separate from the flesh is also advised in many areas.
4. Because the amount of contamination a person is exposed to is directly related to the number of fish a person eats, people are advised to limit the number of fish consumed, and women of childbearing age and young children are often advised not to eat the fish at all.

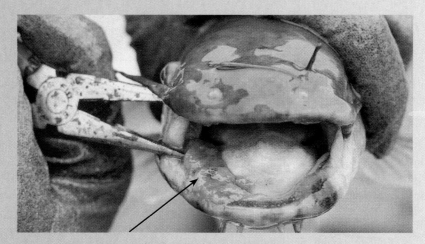

The lip cancer on this bullhead indicates carcinogens in Wisconsin's Fox River. In some tributaries of the Great Lakes, fish cancer rates may reach 84 percent.

The Reintroduction of the Moose into Michigan

At one time, much of the American continent was covered by forests. The Upper Peninsula of Michigan was the southern portion of a spruce and fir forest, which originally supported such animals as wolves, moose, the martin, and the fisher. During the 1800s, mining was the main occupation in the region, and trees were cut for lumber, to heat homes, to provide mining timbers, and to provide charcoal for smelting. This destroyed much of the wilderness character of the forest, and wolves, moose, and many other animals disappeared due to the combination of reduced habitat and hunting. Moose are browsing animals that require large numbers of small trees as a source of food and travel over large areas to get what they need.

Over many years, the mining industry diminished, people were less dependent on fuelwood for heating, and the human population declined, allowing many areas to regrow to something approximating the original forest. In 1986, the state of Michigan and the province of Ontario worked out a swap of animals. Michigan provided wild turkeys for a restocking effort in Ontario, and Ontario provided moose for restocking in the Upper Peninsula. The moose were checked by veterinarians, tagged with radio collars so they could be followed, and then released. Currently, the animals seem to be establishing a viable population in the area.

- What is the value of such a reintroduction?
- What kinds of relationships existed between moose and humans?
- What made reintroduction possible?

SUMMARY

Everything that affects an organism during its lifetime is collectively known as its environment. The environment of an organism can be divided into biotic (living) and abiotic (nonliving or physical) components.

The space an organism occupies is known as its habitat, and the role it plays in its environment is known as its niche. The niche of an organism is the result of natural selection directing the adaptation of the organism to a specific set of environmental conditions.

Organisms interact with one another in a variety of ways. Predators kill and eat prey. Organisms that have the same needs compete with one another and do mutual harm, but one is usually harmed less and survives. Symbiotic relationships are those in which organisms live in physical contact with one another. Parasites live in or on another organism and derive benefit from the relationship, harming the host in the process. Commensal organisms derive benefit from another organism but do not harm the host. Mutualistic organisms both derive benefit from their relationship.

A community is a set of interacting groups of organisms. Those organisms and their abiotic environment constitute an ecosystem. In an ecosystem, energy flows from producers through various trophic levels of consumers (herbivores, carnivores, omnivores, and decomposers). About 90 percent of the energy is lost as it passes from one trophic level to the next. This means that the amount of biomass at higher trophic levels is usually much less than that at lower trophic levels. The sequence of organisms through which energy flows is known as a food chain. Several interconnecting food chains constitute a food web.

The flow of atoms through an ecosystem involves all the organisms in the community. The carbon, nitrogen, and phosphorus cycles are examples of how these materials are cycled in ecosystems.

REVIEW QUESTIONS

1. Define *environment*.
2. Describe, in detail, the niche of a human.
3. How is natural selection related to the concept of niche?
4. List five predators and their prey organisms.
5. How is an ecosystem different from a community?
6. Humans raising cattle for food is what kind of relationship?
7. Give examples of organisms that are herbivores, carnivores, and omnivores.
8. What are some different trophic levels in an ecosystem?
9. Describe the carbon cycle, the nitrogen cycle, and the phosphorus cycle.
10. Analyze an aquarium as an ecosystem. Identify the major abiotic and biotic factors. List members of the producer, primary consumer, secondary consumer, and decomposer trophic levels.

KEY TERMS

abiotic factors 50
biomass 61
biotic factors 50
carbon cycle 62
carnivore 59
commensalism 58
community 59
competition 56
consumer 59
decomposer 59
denitrifying bacteria 65
detritus 62

ecology 50
ecosystem 59
ectoparasite 57
endoparasite 57
environment 50
evolution 56
food chain 61
food web 62
free-living nitrogen-fixing bacteria 65
habitat 52
herbivore 59

host 57
interspecific competition 57
intraspecific competition 56
limiting factor 50
mutualism 58
natural selection 53
niche 52
nitrogen cycle 64
nitrogen-fixing bacteria 64
omnivore 59
parasite 57
parasitism 57

predator 56
prey 56
primary consumer 59
producer 59
range of tolerance 50
secondary consumer 59
speciation 56
species 53
symbiosis 57
symbiotic nitrogen-fixing bacteria 65
trophic level 60

RELATED INTERNET SITES

Environmentally sensitive residential area living in harmony with nature.

    http://www.eagle-lake.com

Searchable index and global change directory covering the atmosphere, hydrosphere, and biosphere.

    http://gcmd.gsfc.nasa.gov/

International geosphere and biosphere site, focusing on the study of global environmental change.

    http://www.igbp.kva.se/

Award-winning museum tour site specializing in mollusks, marine mammals, and birds.

    http://www.rain.org

Animal Diversity Web, specializing in species accounts, zoology, and more.

    ***http://www.oit.itd.umich.edu/projects/ADW/

Award-winning site filled with Sierra Club news and information on conservation, preservation, and restoration.

    ***http://www.sierraclub.org/john_muir_exhibit/

Worldwide Fund for Nature focuses on 200 key "eco-regions" for conserving endangered species.

    http://www.livingplanet.org/

Kinds of Ecosystems and Communities

Objectives

After reading this chapter, you should be able to:

- Recognize the difference between primary and secondary succession.
- Describe the process of succession from pioneer to climax community in both terrestrial and aquatic situations.
- Associate typical plants and animals with the various terrestrial biomes.
- Recognize the physical environmental factors that determine the kind of climax community that will develop.
- Differentiate the forest biomes that develop based on temperature and rainfall.
- Describe the various kinds of aquatic ecosystems and the factors that determine their characteristics.

Chapter Outline

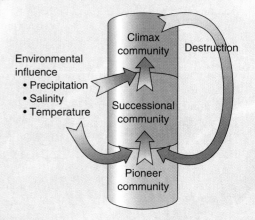

Succession

Ecosystems are dynamic, changing units. On a daily basis, plants grow and die, animals feed on plants and on one another, and decomposers recycle the chemical elements that make up the biotic portion of any ecosystem. Since all organisms are linked together in a community, any change in the community affects many organisms within it. Certain conditions within a community are keys to the kinds of organisms that are found associated with one another. Grasshoppers need grass for food, robins need trees to build nests, and herons need shallow water to find food. Each organism has specific requirements that must be met in a community, or it will not survive.

Over long periods of time, it is possible to see trends in the way the structure of a community changes. This series of regular, predictable changes in the structure of a community over time is called **succession.** Succession occurs because organisms cause changes in their surroundings that make the environment less suitable for themselves and more suitable for other kinds of organisms. One community of organisms is replaced by a slightly changed community, which itself is replaced by a subsequent collection of organisms. Ecologists recognize two different kinds of succession: **primary succession,** which begins with bare mineral surfaces or water, and **secondary succession,** which begins with the destruction or disturbance of an existing ecosystem.

Primary Succession

Primary succession can begin on a bare rock surface, pure sand, or standing water. Since succession on rock and sand is somewhat different from that which occurs with watery situations, we deal with them separately. We discuss terrestrial succession first.

Terrestrial Primary Succession

Bare rock is a very inhospitable place for organisms to live. The temperature changes drastically, there is little mois- ture, the organisms are exposed to the damaging effect of the wind, few nutrients are available, and few places are available for organisms to attach themselves or hide. Primarily what is lacking is soil; however, a few kinds of organisms can survive in even this inhospitable environment. This collection of organisms is known as the **pioneer community** because it is the first to colonize bare rock. (See figure 5.1.)

The dominant organism in this initial community is something called a lichen. Lichens are actually mutualistic relationships between two kinds of organisms: algae that are green and carry on photosynthesis and fungi that retain moisture and attach to the rock surface. The growth and development of lichens is a slow process. It may take lichens one hundred years to grow as large as a dinner plate. A lichen serves as a producer in this simple ecosystem, and many tiny organisms may be found associated with a lichen. They may feed on it and use it as a place of shelter, since even a drizzle is like a torrential rain for a microscopic animal. Since lichens are firmly attached to rock surfaces, they also tend to accumulate bits of airborne debris and store small amounts of water that would otherwise blow away or run off the rock surface. Acids produced by the lichen tend to cause the breakdown of the rock substrate into smaller particles. This fragmentation of rock, aided by physical and chemical weathering processes, along with the trapping of debris and the contribution of organic matter by the death of lichens and other organisms, ultimately leads to the accumulation of a very thin layer of soil.

This thin layer of soil is the key to the next stage in the successional process. The layer can support some fungi, certain small worms, insects, bacteria, protozoa, and perhaps a few tiny annual plants. As these organisms grow, reproduce, and die, they contribute additional organic material for the soil-building process. This stage eliminates the lichen community and is then replaced by a community of small perennial plants. The perennial grasses and herbs are eventually replaced by shrubs, which are often replaced by trees

Figure 5.1

Pioneer Organism The lichen growing on this rock is able to accumulate bits of debris, carry on photosynthesis, and aid in breaking down the rock. All of these activities contribute to the formation of a thin layer of soil, which is necessary for plant growth in the early stages of succession.

that require lots of sunlight, which are replaced by trees that can tolerate shade. Eventually, a relatively stable, long-lasting, more complex, and interrelated community of plants, animals, fungi, and bacteria is produced, known as the **climax community.** Each step in this process is called a **successional stage,** or **seral stage,** and the entire sequence of stages—from pioneer community to climax community— is called a **sere.** (See figure 5.2.) The specific kind of climax community produced depends on such things as climate and soil type, which are discussed in greater detail later in this chapter. However, certain characteristics are typical of climax communities regardless of the type.

1. Climax communities are able to maintain their mix of species.
2. They are in energy balance, while successional communities show changes in species and tend to accumulate large amounts of new material (gain energy).
3. They tend to have many more kinds of organisms and kinds of interactions among organisms than does a successional community.

The general trend in climax communities is toward increasing complexity and energy

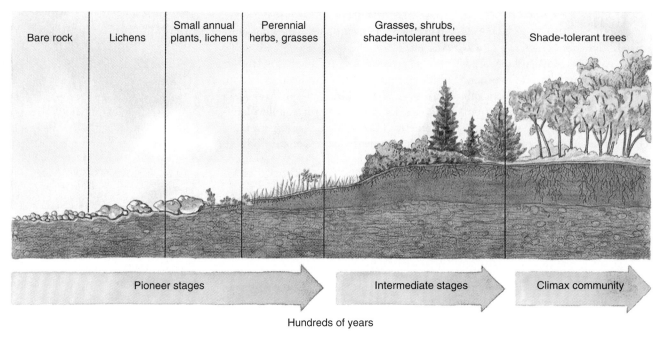

| Bare rock | Lichens | Small annual plants, lichens | Perennial herbs, grasses | Grasses, shrubs, shade-intolerant trees | Shade-tolerant trees |

Pioneer stages → Intermediate stages → Climax community

Hundreds of years

Figure 5.2

Primary Succession on Land   The formation of soil is a major step in primary succession. Until soil is formed, the area is unable to support large amounts of vegetation, which modify the harsh environment. Once soil formation begins, the site proceeds through an orderly series of stages toward a climax community.

efficiency, compared to the successional communities that preceded them.

Aquatic Primary Succession

The principal concepts of land succession can be applied to aquatic ecosystems. Except for the oceans, most aquatic ecosystems are considered temporary. Certainly, some are going to be around for thousands of years, but eventually they will disappear and be replaced by terrestrial ecosystems as a result of normal successional processes. All aquatic ecosystems receive a continuous input of soil particles and organic matter from surrounding land, which results in the gradual filling in of shallow bodies of water like ponds and lakes.

In deep portions of lakes and ponds, only floating plants and algae can exist, but as the amount of sediment accumulates, it becomes possible for submerged plants to become established on the bottom of shallow bodies of water. They carry on photosynthesis, resulting in a further accumulation of organic matter, along with the trapping of sediments

that flow into the pond or lake from streams or rivers. Eventually, as the water becomes shallower, emergent plants become established. They have leaves that float on the surface or project into the air. The network of roots and stems below the surface of the water results in the accumulation of more material, and the water becomes even shallower as material accumulates on the bottom. As the process continues, more sediment accumulates, the wet soil thus formed begins to dry out, and grasses and other plants that can live in wet soil become established. This is often called a wet meadow. Once this occurs, the stage is set for a typical terrestrial successional series of changes, eventually resulting in a climax community. (See figure 5.3.)

Since the shallower portions of most lakes and ponds are at the shore, it is often possible to see the various stages in aquatic succession from the shore. In the central, deeper portions of the lake, there are only floating plants and algae. As we approach the shore, we first find submerged plants like *Elodea* and algal mats, then emergent vegetation like water lilies

and cattails, then grasses and sedges that can tolerate wet soil, and on the shore the beginnings of a typical terrestrial succession resulting in the climax community typical for the area.

In many northern ponds and lakes, sphagnum moss forms thick, floating mats. These mats may allow certain plants that can tolerate wet soil conditions to become established. The roots of the plants bind the mat together and establish a floating bog, which may contain small trees and shrubs as well as many other smaller, flowering plants. (See figure 5.4.) You can recognize that the entire system is floating only when you jump on it and the trees sway or when you step through a weak zone in the mat and sink to your hips in water. Eventually these bogs will become increasingly dry and the normal climax vegetation for the area will succeed the more temporary bog stage.

Secondary Succession

The same processes and activities that drive primary succession result in secondary succession. The major difference

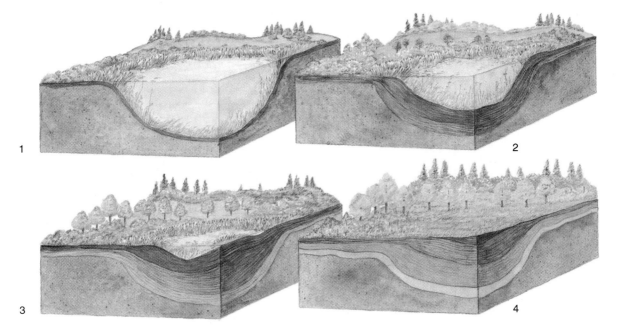

Figure 5.3

Succession from a Pond to a Wet Meadow A shallow pond will fill slowly with organic matter from producers in the pond. Eventually, a floating mat will form over the pond and grasses will become established. In many areas this will be succeeded by a climax forest.

Figure 5.4

Floating Bog In many northern regions, sphagnum moss forms a floating mat that can support the growth of other plants. If a person were to walk on this mat, it would bounce up and down, because it is floating on water.

is that secondary succession occurs when an existing community is destroyed in some way. A forest fire, a flood, or the conversion of a natural ecosystem to agriculture may be the cause. Usually, the destroyed ecosystem is not completely returned to bare rock. Much of the soil may remain, and many of the nutrients necessary for plant growth may be available for the reestab-

lishment of the previously existing ecosystem. Consequently, secondary succession tends to be more rapid than primary succession. Figure 5.5 shows the typical secondary succession found on abandoned farmland in the southeastern United States.

Similarly, when beavers flood an area, the existing trees die and an aquatic ecosystem is established. As the area behind the dam fills in with sediment and organic matter, it goes through a series of stages that eventually return the area to the typical climax community. (See figure 5.6.)

Major Types of Climax Communities: Biomes

Biomes are terrestrial climax communities with wide geographic distribution. (See figure 5.7.) The species involved may not be identical in all parts of the world, but the general structure of the ecosystem and the kinds of niches are similar in broad terms. Two primary nonbiological factors have major impacts on the kind of climax community that develops: precipitation and temperature.

Several aspects of precipitation are important: the total amount of precipitation per year, the form in which it arrives (rain, snow, sleet), and the seasonal distribution of the precipitation. Is it evenly spaced throughout the year? Are there wet and dry seasons?

The temperature can vary considerably. Tropical areas have warm, relatively unchanging temperatures throughout the year. Areas near the poles have long winters with extremely cold temperatures and relatively short, cool summers. Other areas are more evenly divided between cold and warm periods of the year. The interplay between these two major parameters can account for many of the differences seen in climax communities. (See figure 5.8.)

In addition to temperature and precipitation, several other factors may influence the kind of climax community present. Some climax communities seem to rely on periodic fires to prevent the establishment of larger, woody species. Similarly, severe wind may prevent the establishment of trees and cause rapid drying of the soil. Sandy soils tend to dry out quickly and may not allow the establishment of more water-demanding

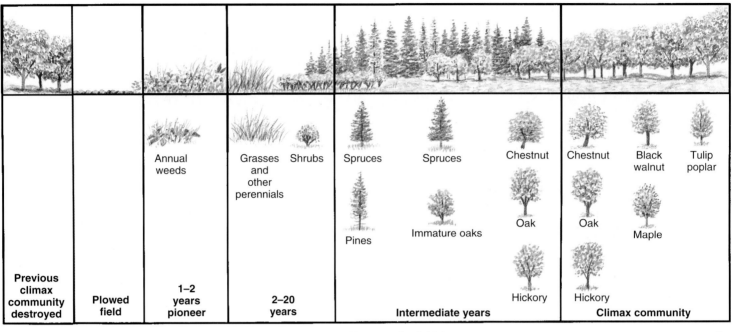

Previous climax community destroyed	Plowed field	1–2 years pioneer	2–20 years	Intermediate years			Climax community		
		Annual weeds	Grasses and other perennials · Shrubs	Spruces · Pines	Spruces · Immature oaks	Chestnut · Oak · Hickory	Chestnut · Oak · Hickory	Black walnut · Maple	Tulip poplar

← 200 years (variable) →

McGee

Figure 5.5

Secondary Succession on Land   A plowed field in the southeastern United States shows a parade of changes over time involving plant and animal associations. The general pattern is for annual weeds to be replaced by grasses and other perennial herbs, which are replaced by shrubs, which are replaced by trees. As the plant species change, so do the animal species.

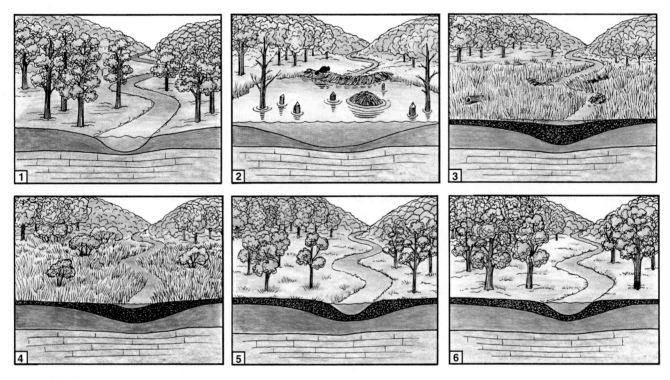

Figure 5.6

Beaver Pond Succession—Secondary Succession from Water   A colony of beavers can dam up streams and kill trees by the flooding that occurs and by using trees for food. Once the site is abandoned, it will slowly return to the original forest community by a process of succession.

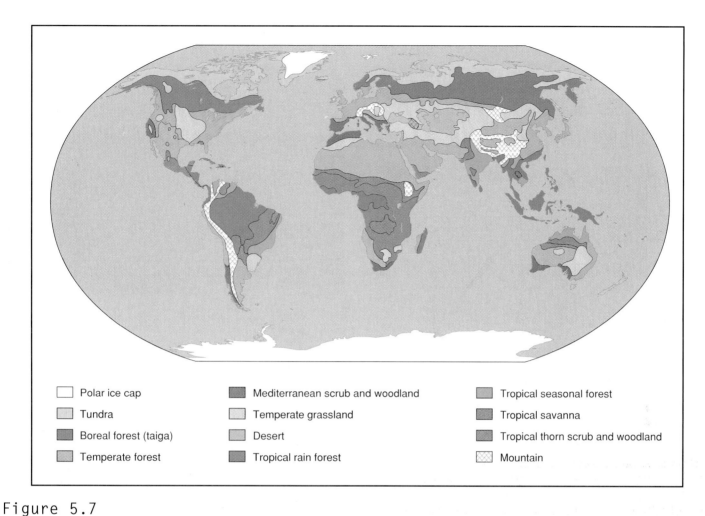

Figure 5.7

Biomes of the World   Although most biomes are named for a major type of vegetation, each includes a specialized group of animals adapted to the plants and the biome's climatic conditions.

Legend:
- Polar ice cap
- Tundra
- Boreal forest (taiga)
- Temperate forest
- Mediterranean scrub and woodland
- Temperate grassland
- Desert
- Tropical rain forest
- Tropical seasonal forest
- Tropical savanna
- Tropical thorn scrub and woodland
- Mountain

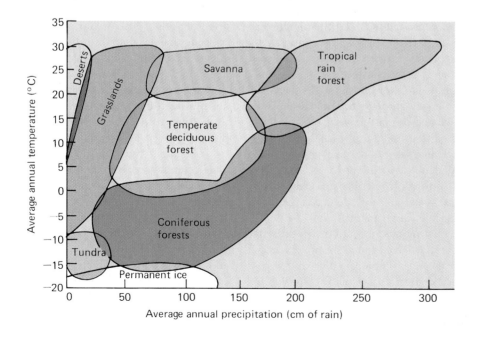

Figure 5.8

Influence of Precipitation and Temperature on Vegetation   Temperature and moisture are two major factors that influence the kind of vegetation that can occur in an area. Areas with low moisture and low temperatures produce tundra; areas with high moisture and freezing temperatures during part of the year produce deciduous or coniferous forests; dry areas produce deserts; moderate amounts of rainfall or seasonal rainfall support grasslands or savannas; and areas with high rainfall and high temperatures support tropical rainforests.

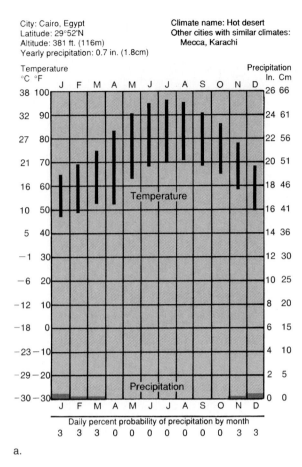

City: Cairo, Egypt
Latitude: 29°52'N
Altitude: 381 ft. (116m)
Yearly precipitation: 0.7 in. (1.8cm)

Climate name: Hot desert
Other cities with similar climates:
Mecca, Karachi

Daily percent probability of precipitation by month
3 3 3 0 0 0 0 0 0 0 3 3

a.

b.

Figure 5.9

Desert (*a*) Climagraph for Cairo, Egypt. (*b*) The desert receives less than 25 centimeters of precipitation per year, yet it teems with life. Cactus, sagebrush, lichens, snakes, small mammals, birds, and insects inhabit the desert. Because daytime temperatures are often high, most animals are active only at night, when the air temperature drops significantly. Cool deserts also exist in many parts of the world, where rainfall is low but temperatures are not high.

species like trees, while extremely wet soils may allow only certain species of trees to grow. Obviously, the kinds of organisms currently living in the area are also important, since their offspring will be the ones available to colonize a new area. Let's look at some of the major biomes and the factors that shape their character.

Desert

Deserts are areas that generally receive fewer than 25 centimeters of precipitation per year. A lack of water is the primary factor that determines that an area will be a desert. (See figure 5.9.) Rain comes in the form of thundershowers at infrequent intervals. Much of the water runs off into gullies and, since the rate of evaporation is high, the rains provide only brief periods of rapid growth. Deserts are also likely to be windy. We often think of deserts as hot, dry wastelands devoid of life. However, many deserts are quite cool during a major part of the year.

Certainly, the Sahara Desert and the deserts of the southwestern United States and Mexico are hot during much of the year, but the desert areas of the northwestern United States and the Gobi Desert in Central Asia can be extremely cold during winter months and have relatively cool summers. Furthermore, the temperature can vary greatly during a 24-hour period. Since deserts receive little rainfall, it is logical that they have infrequent cloud cover. Since no clouds block out the sun, throughout the day the soil surface and the air above it tend to heat up rapidly. After the sun has set, the absence of clouds allows heat energy to be reradiated from the earth, and the area cools off rapidly. Cool to cold nights are typical even in "hot" deserts, especially during the winter months.

Another misconception about deserts is that they have limited species diversity. Many types of organisms live in deserts. They are specially adapted to survive in dry, often hot environments. For example, since water evaporates from the sur-

faces of leaves, many plants have very small leaves that allow them to conserve water. Some even lose their leaves entirely during the driest part of the year. Many desert plants are spiny. Some, like cactus, have the ability to store water in their spongy bodies for use during drier periods. Other plants have parts or seeds that lie dormant until the rains come. Then they grow rapidly, reproduce, and die, or become dormant until the next rains. Even the perennial plants are tied to the infrequent rains. During these times, the plants are most likely to produce flowers and reproduce.

The desert has many kinds of animals, often inconspicuous and overlooked because they are small or are inactive during the hot part of the day. They also aren't seen in large, conspicuous groups. Many lizards, snakes, small mammals, grazing mammals, carnivorous mammals, and birds are common in desert areas. All of the animals that live in deserts are able to survive with a minimal amount of water. Some get nearly all of their water

environmental
close-up

The Changing Nature of the Climax Concept

When European explorers traveled across the North American continent they saw huge expanses of land covered by the same kinds of organisms. Deciduous forests in the East, coniferous forests in the North, grasslands in central North America and deserts in the southwest. These collections came to be considered the steady-state or normal situation for those parts of the world. When ecologists began to explore the way in which ecosystems developed over time they began to think of these ecosystems as the end point or climax of a long journey beginning with the formation of soil and its colonization by a variety of plants and other organisms.

As settlers removed the original forests or grasslands and converted the land to farming, the original "climax" community was destroyed. Eventually, as poor farming practices destroyed the soil, the farms were abandoned and the land was allowed to return to its "original" condition. This secondary succession often resulted in forests that resembled those that had been destroyed. However, in most cases these successional forests contained fewer species and in some cases were entirely different kinds of communities from the originals. These new stable communities were also called climax communities, but they were not the same as the originals.

Ecologists began to recognize that there was not a fixed, predetermined community for each part of the world and began to modify the way they looked at the concept of climax communities. The concept today is a more plastic one. It is still used to talk about a stable stage following a period of change, but ecologists no longer feel that land will eventually return to a "preordained" climax condition. They have also recognized in recent years that the type of climax community that develops depends on many other factors than simply climate. One of these is the availability of seeds to colonize new areas. Two areas with very similar climate and soil characteristics may contain different species because of the seeds available when the lands were released from agriculture. Furthermore, we need to recognize that the only thing that differentiates a "climax" community from a successional one is the time scale over which change occurs. "Climax" communities do not change as rapidly as successional ones. However, all communities are eventually replaced, as were the swamps that produced coal deposits, the preglacial forests of Europe and North America, and the pine forests of the Northeastern U.S.

So what should we do with this concept? Although the climax concept embraces a false notion that there is a specific end point to succession, it is still important to recognize that there is a predictable pattern of change during succession and that later stages in succession are more stable and longer lasting than early stages. Whether we call it a climax community is not really important.

---

from the moisture in the food they eat. They generally have a waterproof outer skin or cuticle, so they lose little water by evaporation. They often limit their activities to the cooler part of the day (the evening) and may spend considerable amounts of time in underground burrows during the day, which allows them to avoid extreme temperatures and to conserve water.

Grassland

Grasslands, also known as **prairies** or **steppes,** are widely distributed over the world. As with deserts, the major factor that contributes to the establishment of a grassland is the amount of available moisture. Grasslands generally receive between 25 and 75 centimeters of precipitation per year. These areas are windy with hot summers and cold to mild winters. Fire is an important force in this biome. Trees, which generally require greater amounts of water, are rare in these areas except along watercourses. (See figure 5.10.) Grasses make up 60 to 90 percent of the vegetation. Many other kinds of flowering plants are interspersed with the grasses.

Characteristic of these areas are large herds of migratory, grazing mammals like bison, wildebeests, wild horses, and various kinds of sheep, cattle, and goats. Most of the grasslands of the world have been converted to agriculture, since the rich, deep soil that developed as a result of the activities of centuries of soil building is useful for growing cultivated grasses like corn (maize) and wheat. The drier grasslands have been converted to the raising of domesticated grazers like cattle, sheep, and horses.

In addition to grazing mammals, many kinds of insects, including grasshoppers and other herbivorous insects, dung beetles (which feed on the dung of grazing animals), and flies (which bite the large mammals), are common. Small herbivorous mammals, such as mice and ground squirrels, are also common. Birds

Grassland Succession

Because there are many kinds of grasslands, it is difficult to generalize about how succession takes place in these areas. Most grasslands in North America have been heavily influenced by agriculture and the grazing of domesticated animals. The grasslands reestablished in these areas may be somewhat different from the original ecosystem. However, there appear to be several stages typically involved in grassland succession.

After land is abandoned from cultivation, a short period of time, perhaps one to three years, elapses in which the field is dominated by annual weeds. In this respect, grassland succession is like deciduous forest succession. The next stage varies in length (10 or more years) and is dominated by annual grasses. Usually, in these early stages, the soil is in poor condition, lacking organic matter and nutrients. After several years, the soil fertility increases as organic material accumulates from the death and decay of annual grasses. This leads to the next stage in development, perennial grasses. Eventually, a mature grassland develops as prairie flowers invade the area and become interspersed with the grasses. In general, throughout this sequence, the soil becomes more fertile and of higher quality.

Because so much of the original North American grassland has been used for agriculture, when the land is allowed to return to a prairie, there may not be seeds of all of the original plants native to the area. Thus the grassland that results from secondary succession may not be exactly like the original; some species may be missing. Consequently, in many managed restorations of prairies, seeds that are no longer available in the local soils are introduced from other sources.

The low amount of rainfall and the fires typical of grasslands generally cause the successional process to stop at this point. However, if more water becomes available or if fire is prevented, woody trees may invade moist sites.

Actively farmed

Recently abandoned

Several years of succession

are often associated with grazing mammals. They eat the insects stirred up by the mammals or feed on the insects that bite them. Other birds feed on seeds and other plant parts. Reptiles (snakes and lizards) often feed on small mammals and insects.

Savanna

Parts of Africa, South America, and Australia have extensive grasslands spotted with occasional trees or patches of trees. (See figure 5.11.) This kind of a biome is often called a **savanna.** These areas of the world are typically tropical, with 50 to 150 centimeters of rain per year. However, the rain is not distributed evenly throughout the year. Typically, a period of heavy rainfall is followed by a prolonged drought. This results in a very seasonally structured ecosystem. The plants and animals time their reproductive activities to coincide with the rainy period, when limiting factors are least confining. The predominant plants are grasses, but many drought-resistant, flat-topped, thorny trees are common.

The trees are also resistant to fire damage. Many of these trees are particularly important because they are involved in nitrogen fixation. They also provide shade and nesting sites for animals. As with grasslands, the predominant mammals are the grazers. Kangaroos in Australia, various species of antelope in Africa, and llamas in South America are examples. Many kinds of rodents, birds, insects, and reptiles are associated with this biome. Among the insects, mound-building termites are particularly common.

City: Tehran, Iran
Latitude: 35° 41′N
Altitude: 4002 ft. (1220m)
Yearly precipitation: 10.1 in. (26cm)

Climate name: Midlatitude dryland
Other cities with similar climates:
Salt Lake City, Ankara

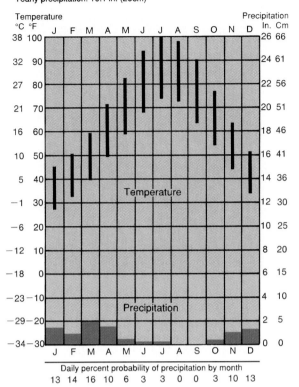

Daily percent probability of precipitation by month

13 14 16 10 6 3 3 0 0 3 10 13

a.

City: Rangoon, Burma
Latitude: 16°46′N
Altitude: 18 ft. (5.5m)
Yearly precipitation: 99.2 in. (250cm)

Climate name: Savanna (monsoon type)
Other cities with similar climates:
Bombay, Calcutta, Miami

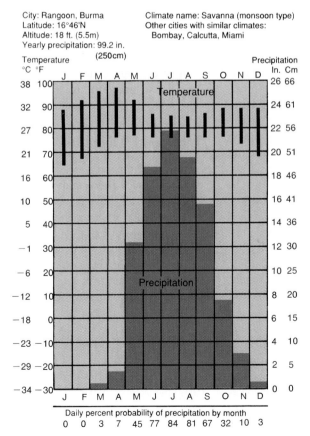

Daily percent probability of precipitation by month

0 0 3 7 45 77 84 81 67 32 10 3

a.

b.

Figure 5.10

Grassland (*a*) Climagraph for Tehran, Iran. (*b*) Grasses are better able to withstand low water levels than are trees. Therefore, in areas that have moderate rainfall, grasses are the dominant plants.

b.

Figure 5.11

Savanna (*a*) Climagraph for Rangoon, Burma. (*b*) Savannas develop in tropical areas that have seasonal rainfall. They typically have grasses as the dominant vegetation with drought- and fire-resistant trees scattered through the area.

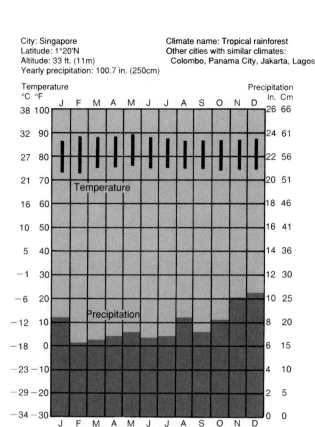

City: Singapore
Latitude: 1°20'N
Altitude: 33 ft. (11m)
Yearly precipitation: 100.7 in. (250cm)

Climate name: Tropical rainforest
Other cities with similar climates:
Colombo, Panama City, Jakarta, Lagos

Daily percent probability of precipitation by month
52 46 42 50 45 43 39 42 47 48 60 58

a.

b.

Figure 5.12

Tropical Rainforest (*a*) Climagraph for Singapore. (*b*) Tropical rainforests develop in areas with high rainfall and warm temperatures. They have an extremely diverse mixture of plants and animals.

Tropical Rainforest

Tropical rainforests are located near the equator in Central and South America, Africa, Southeast Asia, and some islands in the Caribbean Sea and Pacific Ocean. (See figure 5.12.) The temperature is normally warm and relatively constant. There is no frost, and it rains nearly every day. Most areas receive in excess of 200 centimeters of rain per year. Some receive 500 centimeters or more. Because of the warm temperatures and abundant rainfall, most plants grow very rapidly; however, soils are often poor in nutrients because water tends to carry away any nutrients not immediately taken up by plants. Many of the trees have extensive root networks, associated with fungi, near the surface of the soil that allow them to capture nutrients from decaying vegetation before they can be carried away. Most of the nutrients in a tropical rainforest are tied up in the biomass, not in the soil, and, therefore, most of these areas do not make good farmland.

These forests have a very large number of species. A small area of a few square kilometers is likely to have hundreds of species of trees. Balsa, teakwood, and many other ornamental woods are from tropical trees. The trees are usually in two or three layers, with some standing taller than their neighbors. Since most of the sunlight is captured by the trees, many shade-tolerant plants live beneath the trees' canopy. Each tree serves as a surface for the growth of ferns, mosses, orchids, and vines. Associated with this variety of plants is an equally large variety of animals. Insects, such as ants, termites, moths, butterflies, and beetles, are particularly abundant. Birds also are extremely common, as are many climbing mammals, lizards, and tree frogs. Since flowers and fruits are available throughout the year, there are many kinds of nectar- and fruit-feeding birds and mammals. Because of the low light levels, many of the animals communicate by making noise.

More species are found in the tropical rainforests of the world than in the rest of the world combined. Recently, biologists discovered a whole new community of organisms that live in the canopy (leafy tops) of these forests. The dense vegetation reduces the amount of light on the forest floor. This, coupled with the bewildering array of plants and animals, makes study of this kind of ecosystem difficult.

Temperate Deciduous Forest

Forests in temperate areas of the world that have a winter–summer change of seasons typically have trees that lose their leaves during the winter and replace them the following spring. This kind of forest is called a **temperate deciduous forest** and is typical of the eastern half of the United States, parts of south central and southeastern Canada, southern Africa, and many areas of Europe and Asia.

These areas generally receive 100 or more centimeters of relatively evenly distributed precipitation per year. Each area of the world has certain species of trees that are the major producers for the biome. In contrast to tropical rainforests, where individuals of a tree species are

Tropical Rainforests and Farming

Because tropical rainforests are so lush and fast-growing, many people have suggested that they would be excellent for agriculture. In fact, primitive people who live in rainforests have practiced slash-and-burn agriculture for centuries. This method involves cutting and burning the vegetation on a small plot of land. The burning releases the nutrients in the vegetation. If the soil is quickly planted and tilled, the crop covers it and prevents its exposure to the hot sun and erosion caused by the frequent rains. During the first year, farmers can harvest a good crop, but the yield declines each succeeding year unless massive amounts of fertilizer are used. For the native inhabitants of these areas, declining soil fertility was no problem because they simply abandoned the garden and cleared a new site. The old garden was quickly repopulated by the seeds of trees in the surrounding forest and succession occurred, resulting in a return to the original forest community.

One kind of soil found in hot and humid tropical rainforests such as those of South America is called a laterite soil. Abundant rainfall is common in areas where laterite soil is found. Since the soil is porous, it is subjected to a great deal of leaching, which results in the removal of soil nutrients as water flows over and through the exposed soil. Silica is removed, but the soil retains a relatively high concentration of oxides of iron and aluminum. These compounds contribute to the soil's reddish color. Laterite soil gets its name from the Latin word for brick, because whenever it is directly exposed to weathering, it hardens into a bricklike mass. In the past, laterite blocks were used as building materials. A temple in the People's Republic of Kampuchea (Cambodia) is constructed entirely of blocks of laterite soil.

If this is such an infertile soil, why does it support such a lush tropical forest? The high temperature and abundant moisture encourage the growth of plants. The plants rely not on nutrients in the soil, but on a rapid uptake of nutrients from the plants and animals that die. Whenever a plant dies, it decays rapidly, and surrounding plants absorb the nutrients needed for their growth. Tropical forest trees bear mutualistic fungi called mycorrhiza in association with an extensive root network near the surface of the soil. The mycorrhiza quickly penetrate each fallen leaf and take all nutrients back into the roots of their host tree. This rapid uptake prevents the leaching of valuable nutrients from the porous soil due to the abundant rainfall. Nutrients from decaying organisms do not usually go unused in a healthy tropical rainforest.

It is estimated that 10 to 15 percent of the trees in tropical rainforests are legumes, which have symbiotic nitrogen-fixing bacteria associated with their roots. These trees, which grow throughout the forest, provide this much-needed nutrient for the trees and other plants that surround them. Thus, the lush vegetation in tropical rainforests is the result of a very finely tuned ecological system that recycles nutrients as fast as they are released. The nutrients are in the plants, not in the soil.

Because of this, many tropical soils are poor in nutrients and are not suitable for long-term agricultural use. The slash-and-burn agriculture practiced by small tribes did not disrupt the system. The small garden plots were quickly overgrown by the surrounding forest and reverted to the original forest type. The soil was not exposed to extensive erosion and exposure to the sun, and it was not changed to a red-colored, bricklike texture. Most attempts to farm large areas using temperate-region agricultural techniques have failed because of the infertility of the soil and its tendency to harden. The use of this ecosystem for agriculture or forestry must take into account the unique nutrient cycling mechanisms and fragile soil structure, or the tropical rainforest will be permanently destroyed.

scattered throughout the forest, temperate deciduous forests generally have many fewer species, and many forests may consist of two or three dominant tree species. In deciduous forests of North America and Europe, common species are maples, aspen, birch, beech, oaks, and hickories. These tall trees shade the forest floor, where many small flowering plants bloom in the spring. These spring wildflowers store food in underground structures. In the spring, before the leaves come out on the trees, the wildflowers can capture sunlight and reproduce before they are shaded. Many smaller shrubs also may be found in these forests. (See figure 5.13.)

These forests are home to a great variety of insects, many of which use the leaves and wood of trees as food. Beetles, moth larvae, wasps, and ants are examples. The birds that live in these forests are primarily migrants that arrive in the spring of the year, raise their young during the summer, and leave in the fall. Many of these birds rely on the large summer insect population for their food. A few kinds of birds, including woodpeckers, grouse, turkeys, and some of the finches, are year-round residents. Several kinds of small and large mammals inhabit

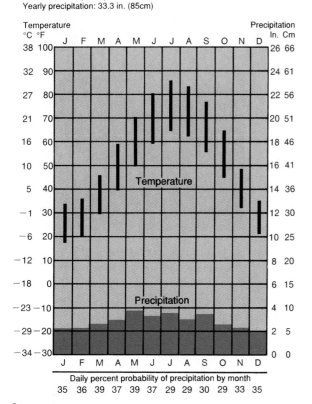

City: Chicago, Illinois
Latitude: 41°52'N
Altitude: 595 ft. (181m)
Yearly precipitation: 33.3 in. (85cm)

Climate name: Humid continental (warm summer)
Other cities with similar climates:
New York, Berlin, Warsaw

Daily percent probability of precipitation by month
35 36 39 37 39 37 29 29 30 29 33 35

a.

b.

Figure 5.13

Temperate Deciduous Forest   (*a*) Climagraph for Chicago, Illinois. (*b*) A temperate deciduous forest develops in areas that have significant amounts of moisture throughout the year, but where the temperature falls below freezing for parts of the year. During this time the trees lose their leaves. This kind of forest once dominated the eastern half of the United States and southeastern Canada.

e n v i r o n m e n t a l
close-up

Forest Canopy Studies

In the past few years, a new frontier of ecological study has developed. Scientists have traditionally looked at forests ecosystems at ground level. Trees were identified, understory species were categorized, and the animals that live on or near the forest floor were studied. Gradually, scientists began to realize that many kinds of plants and animals that are important parts of forest ecosystems rarely decend from the tops of trees to the forest floor. They began to devise methods for studying these canopy-dwelling organisms. Initially, they relied on techniques and gear used by mountain climbers, but

that approach was labor intensive and dangerous. Recently, ecologists have established several sites where large construction cranes have been built in forests. These cranes allow researchers to study the chemistry, climate, and organisms found in the canopy.

Several surprising discoveries have been made, including the discovery of new species of insects. A canopy study in an old-growth temperate rainforest on Vancouver Island, British Columbia, identified more than 60 new species of insects. Researchers estimate that hundreds of new species will even-

Old-Growth Temperate Rainforests of the Pacific Northwest

The coastal areas of northern California, Oregon, Washington, British Columbia, and southern Alaska have an unusual set of environmental conditions that support a special kind of forest, a temperate rainforest. The prevailing winds from the west bring moisture-laden air to the coast. As this air meets the coastal mountains and is forced to rise, it cools and the moisture falls as rain or snow. Most of these areas receive 200 or more centimeters of precipitation per year. This abundance of water, along with fertile soil and mild temperatures, results in a luxuriant growth of plants.

Sitka spruce, Douglas fir, and western hemlock are typical evergreen coniferous trees. Undisturbed (old growth) forests of this region have trees as old as 800 years that are almost as tall as the length of a football field. Deciduous trees of various kinds (red alder, bigleaf maple, black cottonwood) grow in places where they can get enough light. All trees are covered with mosses, ferns, and other plants that grow on their surface. The dominant color is green, since most surfaces have something growing on them.

When a tree dies and falls to the ground, it rots in place and often serves as a site for the establishment of new trees. This is such a common feature of the forest that the fallen, rotting trees are called nurse trees. The fallen trees also serve as a food source for a variety of insects, which are food for a variety of other animals. Some animals, such as the northern spotted owl, the marbled murrelet (a seabird), and the Roosevelt elk seem to be dependent on undisturbed forest.

Because of the rich resource of trees, 90 percent of the original temperate rainforest has been logged. What remains has become a source of controversy. Some maintain that it should be protected as a remnant of the original forest of the region, just as small patches of prairie and eastern woodland have been preserved in other parts of North America. Others maintain that the trees are a resource to be used and that they should be harvested for lumber since they are old and will die in the near future. Both sides in the controversy have taken firm positions so that the issue is not likely to be resolved without legislation or legal action.

tually be identified. Similar studies in tropical forests have identified hundreds more new species of insects. Studies in Panama recognized that vines (lianas) can constitute up to 70 percent of the forest canopy and that these vines compete for sunlight with the trees that support them.

The way in which animals use the canopy is another interesting area of research. Birds, bats, monkeys, squirrels, and insects use specific parts of the forest canopy for food, nesting, and travel routes. Researchers are learning much as they spend time in the forest canopy with the animals.

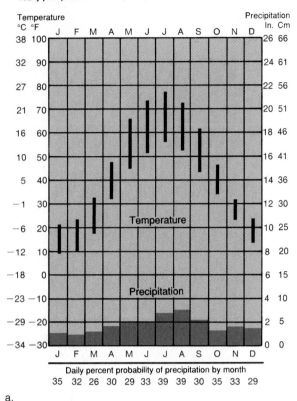

City: Moscow, Russia
Latitude: 55°46'N
Altitude: 505 ft. (154m)
Yearly precipitation: 21.8 in. (55cm)

Climate name: Humid continental (cool summer)
Other cities with similar climates:
Montreal, Winnipeg, Leningrad

Daily percent probability of precipitation by month
35 32 26 30 29 33 39 39 30 35 33 29

a.

b.

Figure 5.14

Taiga, Northern Coniferous Forest, or Boreal Forest   (*a*) Climagraph for Moscow. (*b*) The taiga, northern coniferous forest, or boreal forest occurs in areas with long winters and heavy snowfall. The trees have adapted to these conditions and provide food and shelter for many of the animals that live there.

these areas. Mice, squirrels, deer, moles, and rabbits are common examples. Major predators are foxes, badgers, weasels, coyotes, and birds of prey.

Taiga, Northern Coniferous Forest, or Boreal Forest

Throughout the southern half of Canada, parts of northern Europe, and much of the USSR, there is a coniferous forest known as the **taiga, northern coniferous forest,** or **boreal forest.** Spruces, firs, and larches are the trees most common in these areas. (See figure 5.14.) The climate is one of short, cool summers and long winters with abundant snowfall. The winters are extremely harsh and can last as long as six months. Typically, the soil freezes during the winter. Precipitation

ranges between 25 and 100 centimeters per year, but the climate is humid because the generally low temperatures during all parts of the year reduces evaporation. The landscape is typically dotted with lakes, ponds, and bogs.

The trees are specifically adapted to winter conditions. Winter is relatively dry as far as the trees are concerned because the moisture falls as snow and stays above the soil until it melts in the spring. The needle-shaped leaves are adapted to prevent water loss; the larch loses its needles in the fall. The branches of these trees are flexible, allowing the snow to slide off the pyramid-shaped trees without greatly damaging them. As with the temperate deciduous forest, many of the inhabitants of this biome are temporarily active during the summer. Most birds are migratory and feed on the abundant in-

sect population, which becomes inactive during the long, cold winter. A few woodpeckers, owls, and grouse are permanent residents. Typical mammals are deer, caribou, moose, wolves, weasels, mice, and squirrels. Because of the cold, few reptiles and amphibians live in this biome.

Tundra

North of the taiga is a biome that lacks trees and has a permanently frozen soil layer or **permafrost.** It is known as the **tundra.** (See figure 5.15.) Scattered patches of tundralike communities also are found on mountaintops throughout the world. These are known as **alpine tundra.** Because of the permanently frozen soil and extremely cold, windy climate (up to 10 months of winter), no

City: Fairbanks, Alaska
Latitude: 64°51'N
Altitude: 440 ft. (134m)
Yearly precipitation: 12.4 in. (31.5cm)

Climate name: Subarctic tundra
Other cities with similar climates:
 Yellowknife, Yakutsk

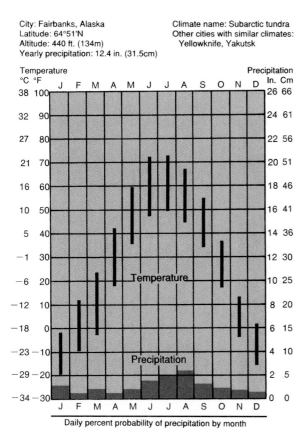

Daily percent probability of precipitation by month
32 21 19 13 29 33 42 48 33 35 33 23

a.

b.

Figure 5.15

Tundra (a) Climagraph for Fairbanks, Alaska. (b) In the northern latitudes and on the tops of some mountains, the growing season is short and plants grow very slowly. Trees are unable to live in these extremely cold areas, in part because there is a permanently frozen layer of soil beneath the surface, known as the permafrost. Because growth is so slow, damage to the tundra can still be seen generations later.

trees can live in the area. During the brief summer, the top layers of the soil thaw, and many plants and lichens (reindeer moss) grow. The plants are short, usually less than 20 centimeters. A few hardy mammals like musk oxen, caribou or reindeer, arctic hare, and lemmings can survive by feeding on the grasses and other plants that grow during the short, cool summer. Although the amount of precipitation is similar to that in some deserts (25 centimeters per year), the short summer is generally moist because of the melting of the snow that falls during the winter. In addition, the permafrost does not let the water sink into the soil, resulting in waterlogged soils and many shallow ponds and pools. Consequently, many waterfowl like ducks and geese migrate to the tundra, where they mate and raise their young. Insects are common during the summer and serve as food for migratory birds. Permanent resident birds are the ptarmigan and snowy owl. No reptiles or amphibians survive in this extreme climate. Because of the very short growing season, damage to this kind of ecosystem is slow to heal, so the land must be handled with a great deal of care.

Altitude and Latitude

The distribution of terrestrial ecosystems is primarily related to precipitation and temperature. The temperature is warmest near the equator and becomes cooler toward the poles. Similarly, as the altitude increases, the average temperature decreases. This means that even at the equator it is possible to have cold temperatures on the peaks of tall mountains. As one proceeds from sea level to the tops of mountains, it is possible to pass through a series of biomes that are similar to what would be encountered as one traveled from the equator to the North Pole. (See figure 5.16.)

Major Aquatic Ecosystems

Terrestrial biomes are determined by the amount and kind of precipitation and by temperatures. Other factors, such as soil type and wind, also play a part. Aquatic ecosystems also are shaped by primary determiners. Four such factors are the ability of the sun's rays to penetrate the water, the nature of the bottom substrate, the water temperature, and the amount of dissolved materials.

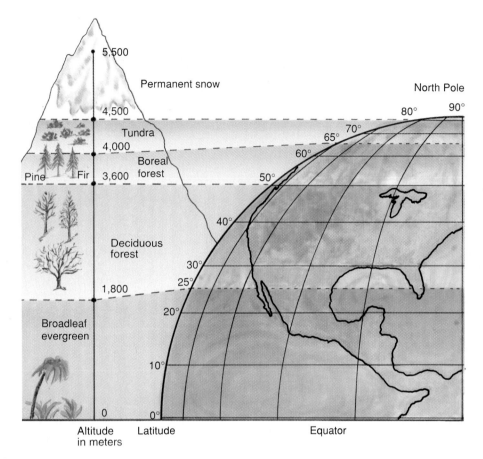

Figure 5.16

Relationship between Altitude, Latitude, and Vegetation As one travels up a mountain, the climate changes. The higher the elevation, the cooler the climate. Even in the tropics tall mountains can have snow on the top. Thus it is possible to experience the same change in vegetation by traveling up a mountain as one would experience traveling from the equator to the North Pole.

Marine Ecosystems

An important determiner of the nature of aquatic ecosystems is the amount of salt dissolved in the water. Those that have little dissolved salt are called **freshwater ecosystems** and those that have a high salt content are called **marine ecosystems.**

Pelagic Marine Ecosystems

In the open ocean, many kinds of crustaceans, fish, and whales swim actively as they pursue food. These kinds of animals are called **pelagic,** and the ecosystem they are a part of is called a **pelagic ecosystem.** As with all ecosystems, the organisms at the bottom of the energy pyramid carry on photosynthesis. In the

ocean, a majority of these organisms are small, microscopic, floating algae called **phytoplankton.** Since sunlight cannot penetrate to great depths in water, these phytoplankton are located in the upper regions. This region where the sun's rays penetrate is known as the **euphotic zone.** Small, weakly swimming animals of many kinds, known as **zooplankton,** feed on the phytoplankton. Zooplankton are often located at a greater depth in the ocean than the phytoplankton but migrate upward at night and feed on the large population of phytoplankton. The zooplankton are in turn eaten by larger animals like fish and larger shrimp, which are eaten by larger fish like salmon, tuna, sharks, and mackerel. (See figure 5.17.)

A major factor that influences the nature of a marine community is the

kind and amount of material dissolved in the water. Probably more important is the amount of nutrients available to the organisms carrying on photosynthesis. Phosphorus, nitrogen, and carbon are all required for the construction of new living material. In water, these are often in short supply. Therefore, the most productive aquatic ecosystems are those in which these essential nutrients are most common. These areas include places in oceans where currents bring up nutrients that have settled to the bottom and areas where rivers deposit their load of suspended and dissolved materials.

Benthic Marine Ecosystems

Organisms that live on the ocean bottom, whether attached or not, are known as **benthic** organisms, and the kind of ecosystem that consists of these organisms is called a **benthic ecosystem.** Some fish, clams, oysters, various crustaceans, sponges, sea anemones, and many other kinds of organisms live on the bottom. In shallow water, sunlight can penetrate to the bottom, and a variety of attached photosynthetic organisms like kelp (commonly called seaweeds) are common. Since they are attached and can grow to very large size, many other bottom-dwelling organisms are associated with them.

The substrate is very important in determining the kind of benthic community that develops. Sand tends to shift and move, making it difficult for large plants or algae to become established, although some clams, burrowing worms, and small crustaceans find sand to be a suitable habitat. Clams filter water for plankton and detritus, or burrow through the sand, feeding on other inhabitants. Muds may provide suitable habitats for some kinds of rooted plants. Muds generally have little oxygen in them but still may be inhabited by a variety of burrowing organisms that feed by filtering the water above them or feed on other animals in the mud. Rocky surfaces in the ocean provide a good substrate for many kinds of large algae. Associated with this profuse growth of algae is a large variety of animals. (See figure 5.18.)

PART TWO Ecological Principles and Their Application

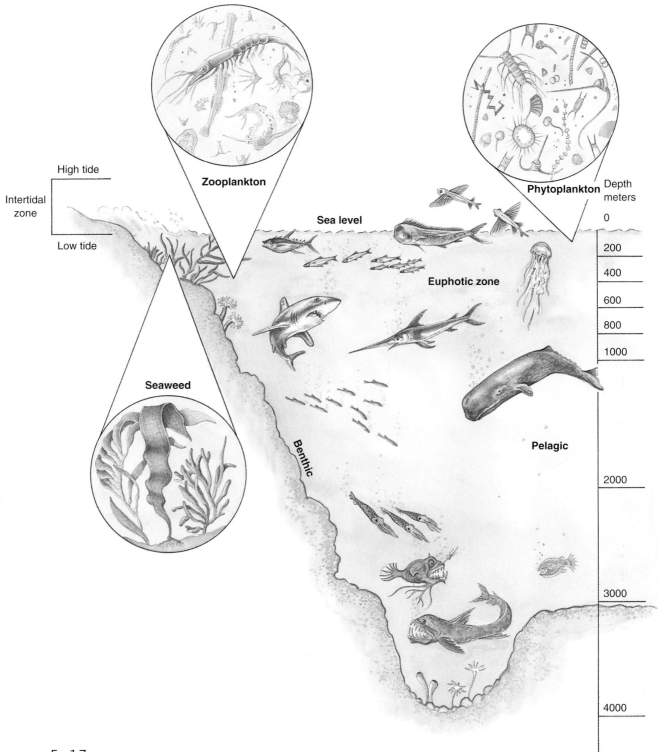

High tide

Intertidal zone

Low tide

Zooplankton

Sea level

Phytoplankton Depth meters

0

200

400

Euphotic zone

600

800

1000

Seaweed

Benthic

Pelagic

2000

3000

4000

Figure 5.17

Marine Ecosystems   All of the photosynthetic activity of the ocean occurs in shallow water called the euphotic zone, either by attached algae near the shore or by minute phytoplankton in the upper levels of the open ocean. Consumers are either free-swimming pelagic organisms or benthic organisms that live on the bottom. Small animals that feed on phytoplankton are known as zooplankton.

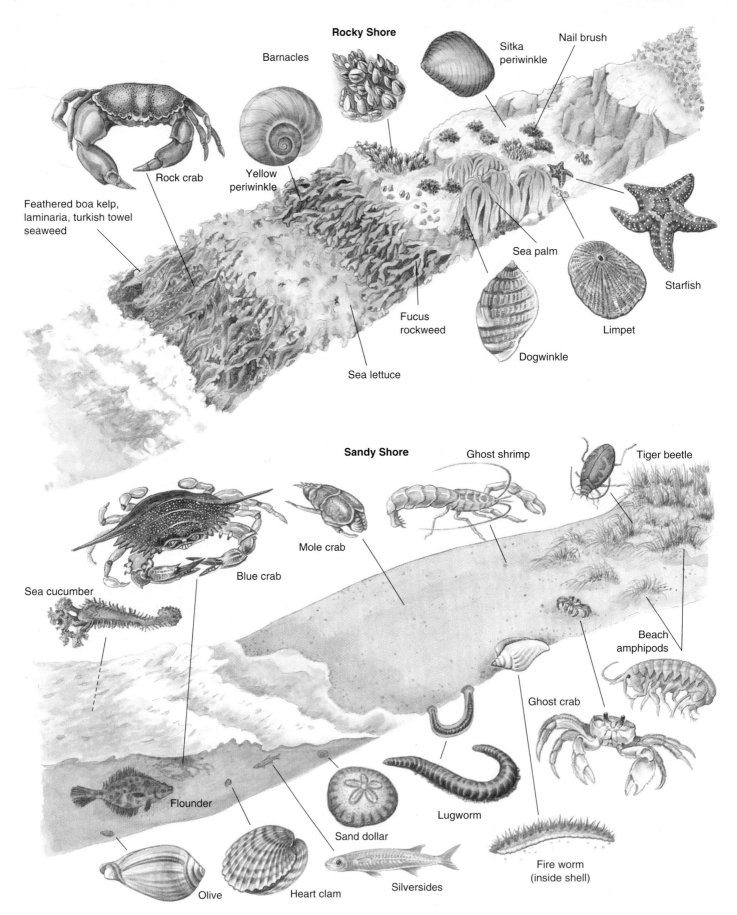

Rocky Shore

Barnacles

Sitka periwinkle

Nail brush

Rock crab

Yellow periwinkle

Feathered boa kelp, laminaria, turkish towel seaweed

Sea palm

Starfish

Fucus rockweed

Limpet

Sea lettuce

Dogwinkle

Sandy Shore

Ghost shrimp

Tiger beetle

Mole crab

Sea cucumber

Blue crab

Beach amphipods

Ghost crab

Flounder

Lugworm

Sand dollar

Fire worm (inside shell)

Olive

Heart clam

Silversides

Figure 5.18

Types of Shores   The kind of substrate determines the kind of organisms that can live near the shore. Rocks provide areas for attachment that sands do not, since sands are constantly shifting. Muds usually have little oxygen in them; therefore, the organisms that live there must be adapted to those kinds of conditions.

Figure 5.19

Coral Reef Corals are small sea animals that secrete external skeletons. They have a mutualistic relationship with certain algae, which allows both kinds of organisms to be very successful. The skeletal material serves as a substrate for many other kinds of organisms to live.

Figure 5.20

Mangrove Swamp Mangroves are tropical trees that are able to live in very wet, salty muds found along the ocean shore. Since they are able to trap additional sediment, they tend to extend farther seaward as they reproduce.

At great depths in the ocean is an ecosystem that must rely on a continuous rain of organic matter from the euphotic zone. These areas are known as abyssal areas, and the ecosystem is known as an **abyssal ecosystem.** Essentially, all of the organisms in this environment are scavengers that feed on whatever drifts their way.

Temperature also has an impact on the kind of benthic community established. Some communities, such as coral reefs or mangrove swamps, are found only in areas where the water is warm. **Coral reef ecosystems** are the result of large numbers of small animals that build cup-shaped external skeletons around themselves. They are able to protrude from their skeletons to capture food and expose themselves to the sun. This is important because corals contain single-celled algae within their bodies. These algae carry on photosynthesis and provide both themselves and the coral with the nutrients necessary for growth. Because they require warm water, coral ecosystems are found only near the equator. Coral ecosystems also require shallow, clear water since the algae must have ample sunlight to carry on photosynthesis. This mutualistic relationship between algae and coral is the basis for a very productive community of organisms. The skeletons of the corals provide a surface upon which many other kinds of animals live. Some of these animals feed on corals directly, while others feed on small plankton and bits of algae that establish themselves among the coral organisms. Many kinds of fish, crustaceans, sponges, clams, and snails are members of coral reef ecosystems. Coral reefs are considered one of the most productive ecosystems on earth. (See figure 5.19.)

Mangrove swamp ecosystems occupy a region near the shore. The dominant organisms are special kinds of trees that are able to tolerate the high salt content of the ocean. In areas where the water is shallow and wave action is not too great, the trees can become established. They have long seeds that float in the water. When the seeds become trapped in mud, they take root. The trees can ex-crete salt from their leaves. They also have extensively developed roots that are above the water, where they can obtain oxygen and prop up the plant. The trees trap sediment and provide places for crabs, jellyfish, sponges, and fish to live. The trapping of sediment and the continual movement of mangroves into shallow areas result in the development of a terrestrial ecosystem in what was once shallow ocean. Mangroves are found in South Florida, the Caribbean, Southeast Asia, Africa, and other parts of the world where conditions are suitable. (See figure 5.20.)

Estuaries

Estuaries, a special category of marine ecosystem, consist of shallow, partially enclosed areas where freshwater enters the ocean. The saltiness of the water in the estuary changes with tides and the flow of water from rivers. The organisms that live here are specially adapted to this set of physical conditions, and the number of species is less than in the ocean or in freshwater. Estuaries are particularly productive ecosystems because of the large amounts of nutrients introduced into the basin from the rivers that run into them. This is further enhanced by the fact that the shallow water allows light to penetrate to most of the water in the basin. Phytoplankton and attached algae and plants are able to use the sunlight and the nutrients for rapid growth. This photosynthetic activity supports many kinds of organisms in the estuary. Estuaries are especially important as nursery sites for fish and crustaceans like flounder and shrimp. The adults enter these productive, sheltered areas to reproduce and then return to the ocean. The young spend their early life in the estuary and eventually leave as they get larger and are more able to survive in the ocean. Estuaries also trap sediment. This activity tends to prevent many kinds of pollutants from reaching the ocean and also results in the gradual filling in of the estuary, which may eventually become a salt marsh and then part of a terrestrial ecosystem.

Freshwater Ecosystems

Freshwater ecosystems differ from marine ecosystems in several ways. The amount of salt present is much less, the temperature of the water can change greatly, the water is in the process of moving downhill, oxygen can often be in short supply, and the organisms that inhabit freshwater systems are different.

Freshwater ecosystems can be divided into two categories: those in which the water is relatively stationary, such as lakes, ponds, and reservoirs, and those in which the water is running downhill, such as streams and rivers.

Lakes and Ponds

Large lakes have many of the same characteristics as the ocean. If the lake is deep, there is a euphotic zone at the top, with many kinds of phytoplankton, and zooplankton that feed on the phytoplankton. Small fish feed on the zooplankton, which are in turn eaten by larger fish. The species of organisms found in freshwater lakes are different from those found in the ocean, but the roles played are similar, so the same terminology is used.

Along the shore and in the shallower parts of lakes, the euphotic zone contains many kinds of flowering plants. They are rooted in the bottom and some have leaves that float on the surface or protrude above the water. These are called **emergent plants.** Cattails, bulrushes, arrowhead plants, and water lilies are examples. Rooted plants that stay submerged below the surface of the water are called **submerged plants.** *Elodea* and *Chara* are examples.

Many kinds of freshwater algae also grow in the shallow water, where they may appear as mats on the bottom or attached to vegetation and other objects in the water. Associated with the plants and algae are a large number of different kinds of animals. Fish, crayfish, clams, and many kinds of aquatic insects are common inhabitants of this mixture of plants and algae. This region, with rooted vegetation, is known as the **littoral zone,** and the portion of the lake that does not have rooted vegetation is called the **limnetic zone.** (See figure 5.21.)

The productivity of the lake is determined by several factors. Temperature is important, since cold temperatures tend to reduce the amount of photosynthesis. Shallow lakes tend to be more productive because most of the water has sunlight penetrating it and, therefore, photosynthesis can occur to the bottom of the lake. Shallow lakes also tend to be warmer as a result of the warming effects of the sun's rays. A third factor that influences the productivity of lakes is the amount of nutrients present. This is primarily determined by the rivers and streams that carry nutrients to the lake. River systems that run through areas that donate many nutrients will carry the nutrients to the lakes. Exposed soil and farmland tend to release nutrients, as do other human activities like depositing sewage into streams and lakes. Deep, cold, nutrient-poor lakes are low in productivity and are called **oligotrophic** lakes. Shallow, warm, nutrient-rich lakes are called **eutrophic lakes.**

Although water (H_2O) has oxygen as part of its molecular structure, this oxygen is not available to organisms. The oxygen that they need is molecular oxygen (O_2), which enters water from the air or as a result of photosynthesis by aquatic plants. Wave action as wind blows across the surface tends to mix air and water and allows more oxygen to dissolve in the water.

Dissolved oxygen content in the water is an important factor since the quantity determines the kinds of organisms that can inhabit the lake. Organic materials in the lake result from the metabolic wastes of organisms that live in and around the lake, plants and animals that die, and parts of organisms, such as leaves from trees, that fall into the lake. The breakdown of these organic materials by bacteria and fungi uses oxygen from the water. This is called the **biochemical oxygen demand or BOD.** The amount and kinds of organic matter determine, in part, how much oxygen is left to be used by other organisms, such as fish, crustaceans, and snails. Many lakes may experience periods when oxygen is low, resulting in the death of fish and other organisms. These topics are discussed in greater depth in chapter 15.

Streams and Rivers

Streams and rivers are a second category of freshwater ecosystem. Since the water is moving, plankton organisms are less important than are attached organisms. Most algae grow attached to rocks and other objects on the bottom. This collection of algae, animals, and fungi is called the **periphyton.** Since the water is shallow, light can penetrate easily to the bottom (except for large or extremely muddy rivers). Even so, it is difficult for photosynthetic organisms to accumulate the nutrients necessary for growth, and most clear-water streams are not very productive. As a matter of fact, the major input of nutrients is from organic matter that falls into the stream from terrestrial sources. These are primarily the leaves from trees and other vegetation, as well as the bodies of living and dead insects. Within the stream is a community of organisms that are specifically adapted to use the debris as a source of food. Bacteria and fungi colonize the organic matter, and many kinds of insects shred the material as they eat it and the fungi and bacteria living on it. The feces (intestinal wastes) of these insects and the tiny particles produced during the eating process become food for other insects that build nets to capture the tiny bits of organic matter that

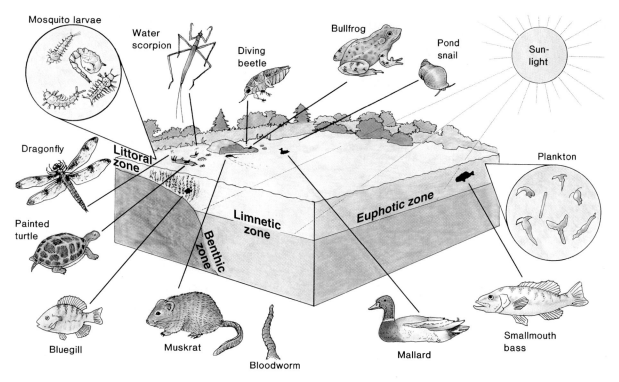

Figure 5.21

Lake Ecosystem   Lakes are similar in structure to oceans except that the species are different because most marine organisms cannot live in fresh water. Insects are common organisms in freshwater lakes, as are many kinds of fish, zooplankton, and phytoplankton.

drift their way. These insects are in turn eaten by carnivorous insects and fish.

Organisms in larger rivers and muddy streams, which have less light penetration, rely on the accumulation of food that drifts their way from the many streams that empty into the river. These larger rivers tend to be warmer and to have slower-moving water. Consequently, the amount of oxygen is usually less, and the species of plants and animals change. Any additional organic matter added to the river system adds to the BOD, further reducing the oxygen in the water. Plants may become established along the shore and contribute to the ecosystem by carrying on photosynthesis and providing hiding places for animals.

Just as estuaries are a bridge between freshwater and marine ecosystems, swamps and marshes are a transition between aquatic and terrestrial ecosystems. **Swamps** contain trees that are able to live in places that are either permanently flooded or flooded for a major part of the year. **Marshes** are dominated by grasses and reeds. Many swamps and marshes are successional stages that eventually become totally terrestrial communities.

Restoring Ecosystems

Wherever humans have had major influence on ecosystems, they have changed them substantially. In most of the world, all the land that can be used for agriculture has been converted to that use. In the process, entire ecosystems have been destroyed and the original complex mixtures of organisms have been replaced by more simple, managed agricultural ecosystems. In North America, much of the original, temperate deciduous forest has been cleared of trees to provide farmland. Similarly, most of the prairie ecosystems have been plowed to produce corn, wheat, and other grains. In other cases, human activity has accidentally altered ecosystems. For example, rivers have been dammed or controlled to provide protection from floods or to facilitate river navigation.

Today, there is interest in restoring some of these altered ecosystems. This is not as easy as it sounds. To restore the Everglades, for example, would require that water currently being diverted for municipal and agricultural use be returned to the Everglades. Restoring a prairie often involves planting seeds of native plants, since farmers would have systematically removed native prairie plants as weeds, thus eliminating the seeds. These seeds must be returned to the soil to reestablish a prairie. Furthermore, since many prairie species are adapted to periodic fires, it may be necessary to burn the prairie as a normal management practice. The absence of fires allows fire-adapted species to be crowded out of the ecosystem.

All of these restorations require an investment of time and money to preserve some small part of the original ecosystem.

- What values are there in protecting special ecosystems?
- Should society bear the cost of removing land from cultivation and replanting it to prairie or temperate forest?
- Who would be inconvenienced if water were returned to the Everglades?

SUMMARY

Ecosystems change as one kind of organism replaces another in a process called succession. Ultimately, a relatively stable stage is reached, called the climax community. Succession may begin with bare rock or water, in which case it is called primary succession, or may occur when the original ecosystem is destroyed, in which case it is called secondary succession. The stages that lead to the climax are called successional stages.

Major regional terrestrial climax communities are called biomes. The primary determiners of the kinds of biomes that develop are the amount and yearly distribution of rainfall and the yearly temperature cycle. Major biomes are desert, grassland, savanna, tropical rainforest, temperate deciduous forest, taiga, and tundra. Each has a particular set of organisms that is adapted to the climatic conditions typical for the area. As one proceeds up a mountainside, it is possible to witness the same kind of change in biomes that occurs if one were to travel from the equator to the North Pole.

Aquatic ecosystems can be divided into marine (saltwater) and freshwater ecosystems. In the ocean, some organisms live in open water and are called pelagic organisms. Light penetrates only the upper few meters of water; therefore, this region is called the euphotic zone. Tiny photosynthetic organisms that float near the surface are called phytoplankton. They are eaten by small animals known as zooplankton, which in turn are eaten by fish and other larger organisms.

The kind of material that makes up the shore determines the mixture of organisms that lives there. Rocky shores provide surfaces for organisms to attach; sandy shores do not. Muddy shores are often poor in oxygen, but marshes and swamps may develop in these areas. Coral reefs are tropical marine ecosystems dominated by coral animals. Mangrove swamps are tropical marine shoreline ecosystems dominated by trees. Estuaries occur where freshwater streams enter the ocean. They are usually shallow, very productive areas. Many marine organisms use estuaries for reproduction.

Insects are common in freshwater and are absent in marine systems. Lakes show similar structure to the ocean, but the species are different. Deep, cold-water lakes with poor productivity are called oligotrophic, while shallow, warm-water, highly productive lakes are called eutrophic. Streams differ from lakes in that most of the organic matter present in streams falls into it from the surrounding land. Thus, organisms in streams are highly sensitive to the land uses that occur near the streams.

REVIEW QUESTIONS

1. Describe the process of succession. How does primary succession differ from secondary succession?
2. How does a climax community differ from a successional community?
3. List three characteristics typical of each of the following biomes: tropical rainforest, desert, tundra, taiga, savanna, grassland, and temperate deciduous forest.
4. What two primary factors determine the kind of terrestrial biome that will develop in an area?
5. How does altitude affect the kind of biome present?
6. What areas of the ocean are the most productive?
7. How does the nature of the substrate affect the kinds of organisms found at the shore?
8. What is the role of each of the following organisms in a marine ecosystem: phytoplankton, zooplankton, algae, coral organisms, and fish?
9. List three differences between freshwater and marine ecosystems.
10. What is an estuary?

KEY TERMS

abyssal ecosystem   89
alpine tundra   84
benthic   86
benthic ecosystem   86
biochemical oxygen demand (BOD)   90
biome   73
boreal forest   84
climax community   71
coral reef ecosystem   89
desert   76
emergent plants   90

estuary   89
euphotic zone   86
eutrophic lake   90
freshwater ecosystem   86
grassland   77
limnetic zone   90
littoral zone   90
mangrove swamp ecosystem   89
marine ecosystem   86
marsh   91
northern coniferous forest   84

oligotrophic lake   90
pelagic   86
pelagic ecosystem   86
periphyton   90
permafrost   84
phytoplankton   86
pioneer community   71
prairie   77
primary succession   71
savanna   78
secondary succession   71
seral stage   71

sere   71
steppe   77
submerged plants   90
succession   71
successional stage   71
swamp   91
taiga   84
temperate deciduous forest   80
tropical rainforest   80
tundra   84
zooplankton   86

RELATED INTERNET SITES

Rainforest Action Network site includes up-to-date information on the state of the world's rainforests.
http://www.ran.org/
Great graphics about rainforest preservation.
http://www.flash.net/~rpf/
Sustainable forests directory with lots of links to other forest topics.
http://homepages.together.net/~wow/Index.htm
Forest Research Institute, a virtual library of forestry information.
http://www.metla.fi/
American forests and policies surrounding them. Global relief, marketing, and more.
http://www.amfor.org/

Southwest Center for Biodiversity, covers dams, logging, and other environmental issues in the southwest.
http://www.envirolink.org/orgs/sw-center/
Coral forest. Maps of reef regions around the world and how to save the reefs. Great links to other coral reef sites.
http://www.blacktop.com/coralforest
Coastal management and resource center.
http://brooktrout.gso.uri.edu/
Great site on the Great Lakes region of North America.
http://www.great-lakes.net/

6

Population Principles

Objectives

After reading this chapter, you should be able to:

- Understand that birthrate and death rate together determine population growth rate.
- Define the following characteristics of a population: natality, mortality, sex ratio, age distribution, reproductive potential, and spatial distribution.
- Explain why the biotic potential for most populations is much greater than needed.
- Describe how a growing population goes through lag, exponential growth, and stable equilibrium phases.
- Describe how the carrying capacity for a population in a region is determined by environmental resistance.
- List the four categories under which limiting factors can be classified.
- Describe how some populations enter a death phase after the stable equilibrium phase.
- Explain why humans are subject to the same forces of environmental resistance as are other organisms.
- Understand the implications of overreproduction.
- Understand that the human population is still growing rapidly.
- Explain how human population growth is influenced by social, theological, philosophical, and political thinking.

Chapter Outline

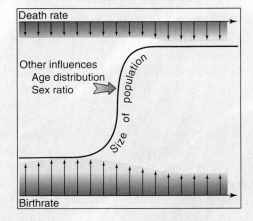

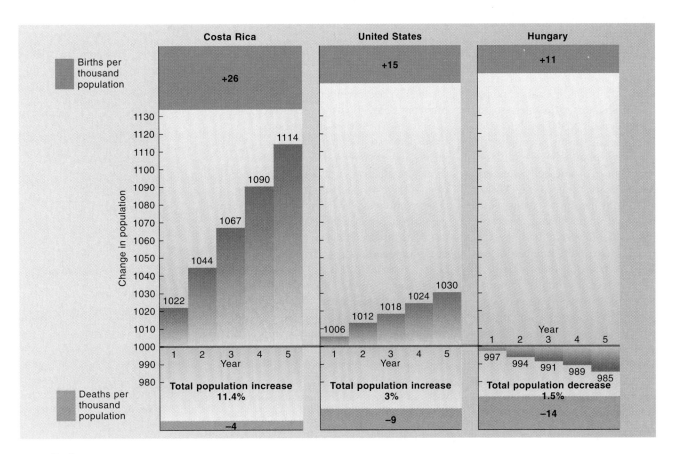

Figure 6.1

Effect of Birthrate and Death Rate on Population Size   For a population to grow, the birthrate must exceed the death rate for a period of time. These three human populations illustrate how the combined effects of births and deaths would change population size if birthrates and death rates were maintained for a five-year period.

Source: Data from World Population Data Sheet 1996, Population Reference Bureau, Inc., Washington, DC.

Population Characteristics

A population can be defined as a group of individuals of the same species inhabiting an area. Just as individuals within a population are recognizable, populations have specific characteristics that distinguish them from one another. Some of these characteristics are natality (birthrate), mortality (death rate), sex ratio, age distribution, growth rates, density, and spatial distribution.

Natality and Mortality

Natality refers to the number of individuals added to the population through reproduction. Bacteria and other tiny organisms are able to divide to form new individuals that are identical to the origi-

nal organism. In plant populations, reproduction results in the production of numerous seeds, but the seeds must land in appropriate soil conditions before they will germinate to produce a plant. Animal species also typically produce large numbers of offspring as a result of sexual reproduction. In human populations, natality is usually described in terms of the **birthrate,** the number of individuals born per one thousand individuals per year. For example, if a population of 2,000 individuals produced 20 offspring during one year, the birthrate would be 10 per thousand per year. The natality for most species is typically quite high. Most organisms produce many more offspring than are needed to replace the parents.

It is important to recognize that the growth of a population is not determined by the birthrate (natality) alone. **Mortality,** the number of deaths per year, is also im-

portant. For most organisms, mortality rates are very high. Of all the seeds that plants produce, very few will result in a mature plant that itself will produce offspring. Many seeds are eaten by animals, some never find proper soil conditions, and those that germinate must compete with other organisms for nutrients and sunlight. In animals, most immature organisms die before they have an opportunity to reproduce. In human population studies, mortality is usually discussed in terms of the **death rate,** the number of people who die per one thousand individuals per year. Compared to the high mortality of the young of most species, the infant death rate of long-lived animals like humans is relatively low. In order for the size of a population to grow, the number of individuals added by reproduction must be greater than the number leaving it by dying. (See figure 6.1.)

Sex Ratio and Age Distribution

A population's ability to grow by reproduction depends on its sex ratio and age distribution. The **sex ratio** refers to the relative numbers of males and females. (Many kinds of organisms, such as earthworms and most plants, have both kinds of sex organs in the same body; sex ratio has no meaning for these species.) Many kinds of organisms have about equal numbers of males and females; however, even minor differences can be important since in all species the number of offspring produced is tied more closely to the number of females than to the number of males. In many game animal populations, where the males are shot and the females are not, there may be an uneven sex ratio in which the females outnumber the males. In many social insect populations (bees, ants, and wasps), the number of females greatly exceeds the number of males at all times, though most of the females are sterile. In humans, about 106 males are born for every 100 females. However, in the United States, by the time people reach their midtwenties, a higher death rate for males has equalized the sex ratio. The higher male death rate continues into old age, when women outnumber men.

As you can see, populations can differ in **age distribution,** the number of individuals of each age in the population. Some are prereproductive juveniles, some are reproducing adults, and some are postreproductive adults. Many kinds of organisms, particularly those that have short life spans, have age distributions that change significantly during the course of a year.

If the majority of a population is made up of reproducing adults, a rapid increase in the number of prereproductive individuals will probably result. Most species of plants and animals typically produce a large number of prereproductive individuals at a specific time of the year (spring, following rain, or when food is plentiful) but have a very high mortality rate among the young. Thus the reproduction of the young declines sharply as time passes. A good example of this type of population is the cottontail rabbit. Many young are produced in the spring, but most of them die, a few reach sexual maturity, and very few live to old age. (See figure 6.2.)

In species that live a long time, it is possible for a stable population to have a balanced age distribution in which the number of individuals in these three categories is relatively constant. For most long-lived plant and animal populations, this means more prereproductive individuals than reproductive individuals, and more reproductive individuals than postreproductive individuals. If the majority of a population is postreproductive, the population declines. This age structure develops in many insect populations in the fall after eggs have been laid.

Human populations exhibit several types of age distribution. (See figure 6.3.) Kenya's population has a large prereproductive and reproductive component. This means that it will continue to increase rapidly for some time. Denmark has a relatively even age distribution and will maintain a nearly constant population size, although it will probably have a declining population in the future. The United States has a very large reproductive component with a declining number of prereproductive individuals. Eventually, the U.S. population will begin to decline if current trends in birthrates and death rates continue.

Population Density and Spacial Distribution

Because of such factors as soil type, quality of habitat, and availability of water, organisms normally are distributed unevenly. Some populations have many individuals clustered into a small space while other populations of the same species may be widely dispersed. **Population density** is the number of organisms per unit area. For example, fruitfly populations are very dense around a source of rotting fruit, while they are rare in other places. Similarly, humans are often clustered into dense concentrations we call cities, with lower densities in rural areas. When the population density is too great, all organisms are injured because they compete severely with each other for necessary resources. Plant populations may compete for water, soil nutrients, or sunlight.

In animal populations, overcrowding might cause exploration and migration

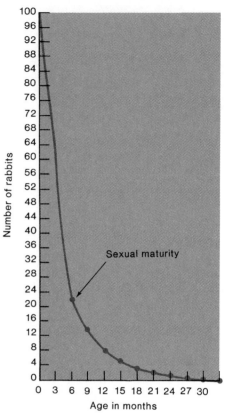

Figure 6.2

Survivorship Curve of Cottontail Rabbit
In most natural populations, mortality rates are so high that very few individuals reach sexual maturity, and even fewer reach old age.

into new areas. This **dispersal** of organisms relieves the overcrowded conditions in the home area and, at the same time, leads to the establishment of new populations. Often, juvenile individuals relieve the overcrowding by leaving. The pressure for out-migration (**emigration**) may be a result of seasonal reproduction leading to a rapid increase in population size, or environmental changes that intensify competition among members of the same species. For example, as water holes dry up, competition for water increases, and many desert birds emigrate to areas where water is still available.

The organisms that leave one population often become members of a different population. This in-migration (**immigration**) may introduce characteristics that were not in the population originally. When Europeans immigrated to North America, they brought genetic and cultural characteristics that

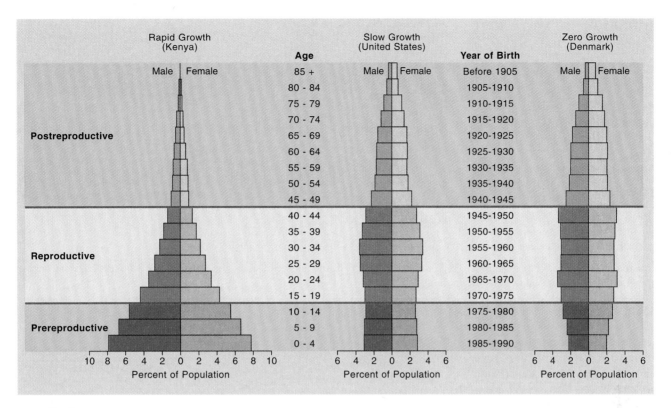

Figure 6.3

Age Distribution in Human Populations The relative numbers of individuals in each of the three categories (prereproductive, reproductive, and postreproductive) are good clues to the future of the population. Kenya has a large number of young individuals who will become reproducing adults. Therefore, this population is likely to grow rapidly. The United States has a declining proportion of prereproductive individuals, and Denmark has a relatively balanced population. Therefore, at present birthrates and death rates, they will probably have declining populations sometime in the future, with Denmark's population declining first.

had a tremendous impact on the existing Native American population. Among other things, Europeans brought diseases that were foreign to the Native Americans. These diseases increased their death rate and lowered the birthrate.

Droughts, wars, and political persecution have caused people to emigrate from their native lands to other countries. In 1995, the United Nations High Commissioner for Refugees estimated that there were 27 million refugees worldwide. Often, the receiving countries are unable to cope with the large influx of new inhabitants.

Summary of Factors that Influence Population Growth Rates

Populations have an inherent tendency to increase in size. However, as we have

just seen, many factors influence the rate at which a population can grow. At the simplest level, the rate of increase is determined by subtracting the number of individuals leaving the population from the number entering. Individuals leave the population either by death or emigration. They enter the population by birth or immigration. Birthrates and death rates are influenced by several factors, including the number of females in the population and their age. In addition, the density of a population may encourage individuals to leave because of intense competition for a limited supply of resources.

A Population Growth Curve

Sex ratios and age distributions directly influence the rate of reproduction within

a population. Each species has an inherent reproductive capacity, or **biotic potential,** which is its ability to produce offspring. Generally, this biotic potential is many times greater than the number of offspring needed to replace the parents when they die. (See figure 6.4.) Consequently, most of the young die; only a few survive to become reproductive adults.

This high reproductive potential results in a natural tendency for populations to increase. For example, two mice produce four offspring, which, if they live, will produce offspring while their parents are also reproducing. Therefore, the population will grow exponentially (2, 4, 8, 16, 32, etc.).

Population growth tends to follow a particular pattern, consisting of a lag phase, an exponential growth phase, and a stable equilibrium phase. Figure 6.5 shows a typical population growth curve. During the first portion of the curve, known as the **lag phase,** the population grows very

Apples

Ducks

Pigs

Figure 6.4

Biotic Potential   The ability of a population to reproduce greatly exceeds the number necessary to replace those who die. Here are some examples of the prodigious reproductive abilities of some species.

slowly because the process of reproduction and growth of offspring takes time. Organisms must mature into adults before they can reproduce. This period is followed by mating and the development of the young into independent organisms. By the time the first batch of young have reached sexual maturity, the parents may be producing a second set of offspring. Since more organisms now are reproducing, the population begins to increase at an exponential rate. This stage is known as the **exponential growth phase.** This growth will continue for as long as the birthrate exceeds the death rate. Eventually, however, the death rate and the birthrate will come to equal one another, and the population will stop growing and reach a relatively stable population size. This stage is known as the **stable equilibrium phase.**

It is important to recognize that although the size of the population may not be changing, the individuals are changing. As new individuals enter by birth or immigration, others leave by death or emigration. For most organisms, the first indication that a population is entering a stable equilibrium phase is an increase in the death rate. A decline in the birthrate may also contribute to the stabilizing of population size. Usually, this occurs after an increase in the death rate. To understand why populations cannot grow continuously, it is necessary to discuss the concept of carrying capacity.

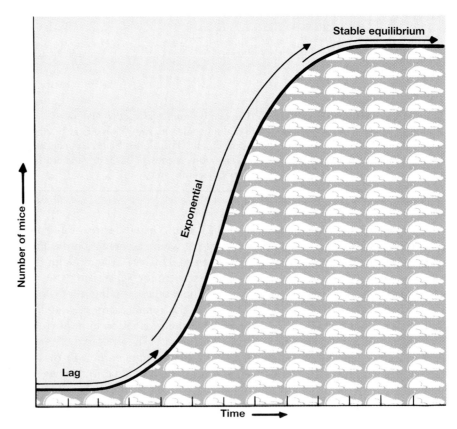

Figure 6.5

Typical Population Growth Curve   In this mouse population, there is little growth during the lag phase. During the exponential growth phase, the population increases rapidly as the offspring reach reproductive age. Eventually the population reaches a stable equilibrium phase, during which the birthrate equals the death rate.

Carrying Capacity

The **carrying capacity** of an area is the number of individuals of a species that can survive in that area over time. For most populations, four factors interact to set the carrying capacity. These are: (1) the availability of raw materials, (2) the availability of energy, (3) the accumulation of waste products and their means of disposal, and (4) interactions among organisms. All of these forces acting together to limit population size is known as **environmental resistance.**

Certain **limiting factors** have a primary role in limiting the size of a population. In some cases, these limiting factors are easy to identify. Lack of food, lack of oxygen, competition with other species, or disease are examples. In other cases the limiting factors may be less obvious. (See figure 6.6.) For example, in grass plants, nitrogen and magnesium in the soil are necessary raw materials for the manufacture of chlorophyll. If these minerals are not present in sufficient quantities, the grass population cannot increase. The application of fertilizers containing these minerals removes this limiting factor, and the individual grass plants grow and reproduce, resulting in a larger population. In effect, the carrying capacity has been increased because this limiting factor has been removed. The carrying capacity has not been eliminated because now some new limiting factor will predominate. Perhaps it will be the amount of water, the number of insects that feed on the grass, or competition for sunlight. Because plants require energy in the form of sunlight for photosynthesis, the amount of light can be a limiting factor for many plants. When small plants are in the shade of trees, they often do not grow well and have small populations.

Accumulation of waste products is not normally a limiting factor for plants, since they produce few wastes, but it can be for other kinds of organisms. Bacteria, other tiny organisms, and many kinds of aquatic organisms that live in small ecosystems like puddles, pools, or aquariums may be limited by wastes. When a small number of a species of bacterium are placed on a petri plate with nutrient

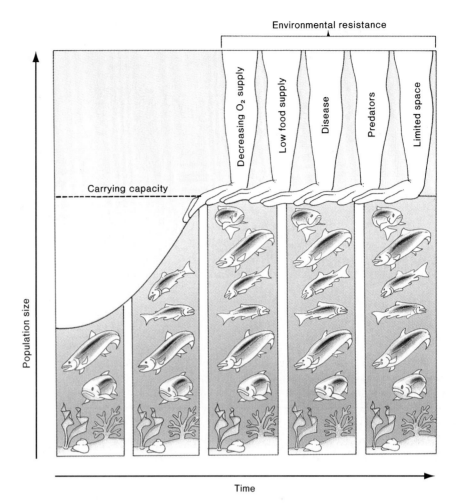

Figure 6.6

Carrying Capacity A number of factors in the environment, such as food, oxygen supply, diseases, predators, and space, determine the number of organisms that can survive in a given area—the carrying capacity of that area. The environmental factors that limit populations are known collectively as environmental resistance.

agar (a jellylike material containing food substances), the population growth follows a curve shown in figure 6.7. As expected, it begins with a lag phase, continues through an exponential growth phase, and eventually levels off in a stable equilibrium phase. However, in this small, enclosed space, there is no way to get rid of the toxic waste products, which accumulate, eventually killing the bacteria. This decline in population size is known as the **death phase.** When a population decreases rapidly, it is said to crash.

Interactions among organisms are also important in controlling population size. Since many birds rely on grass seeds for food, birds have a limiting effect on the size of grass populations. Decomposer organisms may allow for increased populations because they prevent the buildup of toxic wastes. Parasites and predators may cause the premature death of individuals, thus limiting the size of the population. A good example of predator–prey interaction is the relationship between the cat known as the Canada lynx and a member of the rabbit family known as the varying hare. (See figure 6.8.) The varying hare has a high biotic potential that the lynx helps to control by using the hare as food. In turn, the size of the varying hare population determines the size of the lynx population.

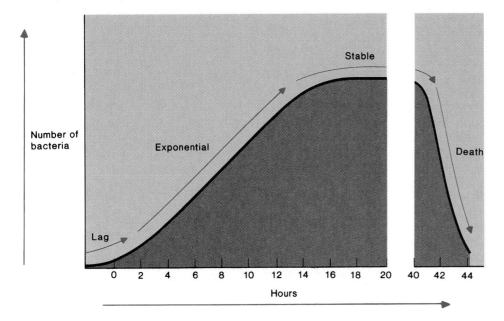

Figure 6.7

Bacterial Growth Curve The change in population size follows a typical population growth curve until waste products become lethal. The buildup of waste products lowers the carrying capacity. When a population begins to decline, it enters the death phase.

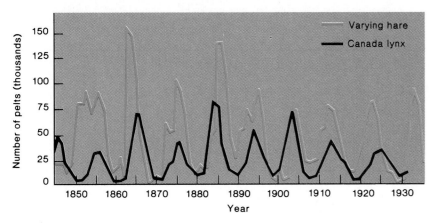

Figure 6.8

Predator/Prey Interaction Interaction between predator and prey species is complex and often difficult to interpret. These data were collected from the records of the number of pelts purchased by the Hudson Bay Company. They show that the two populations fluctuate, with changes in the lynx population usually following changes in the varying hare population.

Source: Data from D.A. MacLulich, "Fluctuations in the Numbers of the Varying Hare (Lepus americanus)," 1937 (reprinted 1974), University of Toronto Press, Toronto.

Reproductive Strategies and Population Fluctuations

So far, we have talked about population growth as if all organisms reach a stable population when they reach the carrying capacity. That is an appropriate way to begin to understand population changes, but the real world is much more complicated. Species can be divided into two broad categories based on their reproductive strategies. **K-strategists** are usually large organisms that have relatively long lives, produce few offspring, and provide care for their offspring. Their populations typically stabilize at the carrying capacity. Their reproductive strategy is to invest a great deal of energy in producing a few offspring that have a good chance of living to reproduce. Deer, lions, and swans are examples of this kind of organism. Humans generally produce single offspring, and even in countries with high infant mortality, 80 percent of the children survive. Generally, populations of K-strategists are limited by density-dependent limiting factors. **Density-dependent limiting factors** are those that become more severe as the size of the population increases. For example, as a population of hawks increases, the hawks begin to compete for available food, such as mice, snakes, and small birds. When food is in short supply, many of the young in the nests die, and population growth slows. They have reached the carrying capacity.

The **r-strategist** is typically a small organism that has a short life, produces many offspring, and does not reach a carrying capacity. Examples are grasshoppers, gypsy moths, and some mice. The reproductive strategy of r-strategists is to expend large amounts of energy producing many offspring but to provide limited care (often none) for them. Consequently, there is high mortality among the young. For example, one female oyster may produce a million eggs, but few of them ever find suitable places to attach themselves and grow. Typically, these populations are limited by **density-independent limiting factors** in which the size of the population has nothing to do with the limiting factor.

Some studies indicate that populations can be controlled by interaction among individuals within the population. A study of laboratory rats shows that crowding causes a breakdown in normal social behavior, which leads to fewer births and increased deaths. The changes observed include abnormal mating behavior, decreased litter size, fewer litters per year, lack of maternal care, and increased aggression in some rats or withdrawal in others. Thus, limiting factors can reduce birthrates as well as increase death rates.

Elephant Harvesting

During the 1980s, uncontrolled hunting of elephants had reduced the African elephant population from 1.3 million to 650,000. In 1989, the Convention on International Trade in Endangered Species of Wild Fauna and Flora (CITES) banned all international trade in elephant products to protect elephant populations that were being decimated by poachers. The ban worked. The price of ivory fell and poaching became unprofitable. Since then, some southern African countries have experienced a population explosion of elephants.

The African elephant, *Loxodonta africana*, requires huge amounts of food (about 180 kg/day) to sustain its large body. The animals strip bark from trees, uproot the trees, and eat large quantities of grass. The elephants can do great damage to crops. Because they have been increasingly confined to national parks and nature preserves, neither of the two original options for relieving population pressure (migration or starvation) is available. Migration results in increased agricultural damage, and increased populations may cause irreparable damage to the protected habitat they currently occupy as they seek food.

Several southern African countries (Botswana, Malawi, Namibia, Zambia, and Zimbabwe) have requested that the ban on the sale of ivory and other elephant products be lifted. These countries along with South Africa, regularly cull their elephant herds to keep them from becoming too numerous. The meat is sold in local markets. Because they do not see elephants as being endangered, they would like to manage their herds in the same way other wildlife populations are managed and achieve the economic benefits of selling ivory and other elephant products such as leather. The income could be used to provide additional funding for park management. At the CITES meeting in Koyoto, Japan, in March 1992, their request was not received favorably and they withdrew it. The international ban on the sale of ivory continues.

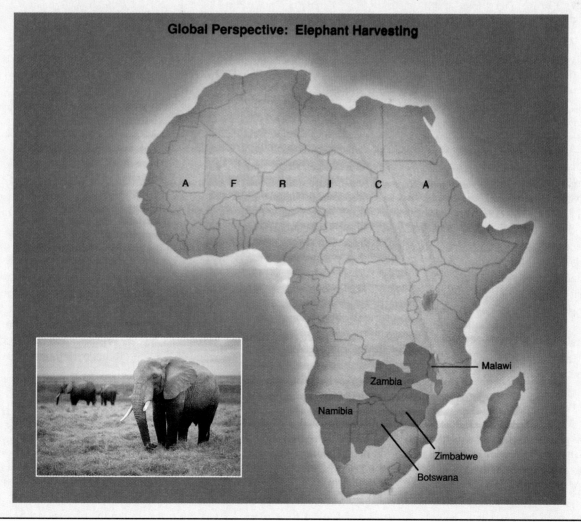

Global Perspective: Elephant Harvesting

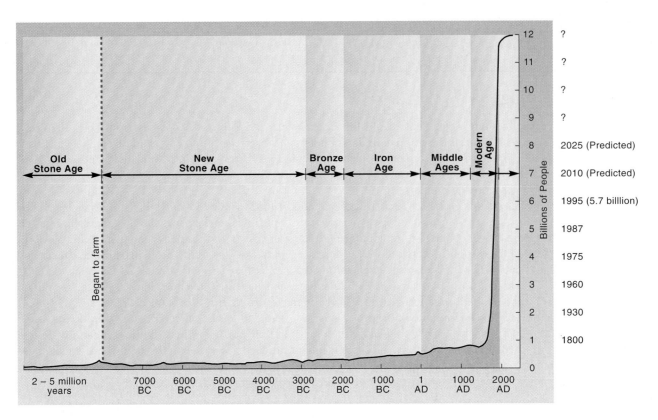

Figure 6.9

Human Population Curve   From A.D. 1800 to A.D. 1930, the number of humans doubled (from one billion to two billion) and then doubled again by 1975 (four billion) and could double again (eight billion) by the year 2025. How long can this pattern continue before the earth's ultimate carrying capacity is reached?

Source: Data from Jean Van Der Tak, et al., "Our Population Predicament: A New Look" in Population Bulletin, *Vol. 34, No. 5, December 1979, Population Reference Bureau, Washington, DC; and more recent data taken from Population Reference Bureau.*

Typical density-independent limiting factors are changing weather conditions that kill large numbers of organisms, the drying up of a small pond, or the death of entire populations due to the destruction of their food source. The population size of r-strategists is likely to fluctuate wildly. They reproduce rapidly, population size increases until some factor causes the population to crash, and then they begin the whole process all over again.

Since humans are K-strategists, it may be difficult for us to appreciate that the r-strategy can be viable from an evolutionary point of view. There is no carrying capacity for temporary resources. Resources that are present only for a short time can be exploited most effectively if many individuals of one species monopolize the resource, while denying other species access to it. Rapid reproduction can place a species in a position to compete against other species that are not

able to increase numbers as rapidly. Obviously, most of the individuals will die, but not before they have left some offspring or resistant stage that will be capable of exploiting the resource should it become available again.

Even K-strategists, however, have population fluctuations for a variety of reasons. One reason is that the environment is not constant from year to year. Floods, droughts, fires, extreme cold, and similar events may affect the carrying capacity of an area, thus causing fluctuations in population size. Epidemic disease or increased predation may also lead to populations that vary from year to year. Figure 6.8 shows rather substantial changes in the populations of lynx and varying hare. The size of the lynx population seems to be tied to the size of the hare population, which is logical since lynx eat hares. However, the causes of the fluctuations in the varying hare popu-

lation are unclear. One possibility is that periodic epidemic disease may cause dense populations of organisms like hares to crash, leading to the crash of their predators' populations.

Although local human populations often show fluctuations, the worldwide human population has increased continually for the past several hundred years. Humans have been able to reduce environmental resistance by eliminating competing organisms, increasing food production, and controlling disease organisms.

Human Population Growth

The human population growth curve has a long lag phase followed by a sharply rising exponential growth phase that is still rapidly increasing. (See figure 6.9.) A major reason for the increase is that the human species has lowered its death rate. When various countries reduce environmental resistance by increasing food production or controlling disease, they share this technology throughout the world. Developed countries send health care personnel to all parts of the globe to improve the quality of life for people in less-developed countries. Physicians offer advice on nutrition, and engineers develop wastewater treatment systems. Improved sanitary facilities in India and Indonesia, for example, decreased deaths caused by cholera. These advancements tend to reduce mortality and directly increase the *quantity* of life, because birthrates remain high.

Let us examine this situation from a different perspective. The world population is currently increasing at an annual rate of 1.5 percent. That may not seem like much, but even at 1.5 percent the population is growing rapidly. It can be difficult to comprehend the impact of a 1 or 2 percent annual increase. Remember that a growth rate in any population compounds itself, since the additional individuals eventually reproduce, thus adding more individuals. One way to look at this growth is to determine how much time is needed to double the population. This is a valuable method because most of us can appreciate what life would be like if the number of people in our locality were doubled, particularly if the number were to occur within our lifetime.

Figure 6.10 shows the relationship between the rate of annual increase for the human population and the number of years it would take to double the population if that rate were to continue. At a 1 percent rate of annual increase, the population will double in approximately 70 years. At a 2 percent rate of annual increase, the population will double in about 35 years. The current worldwide rate of annual increase of about 1.5 percent will double the world population in about 46 years.

What does this very rapid rate of growth mean to the human species? As a species, humans are subject to the same limiting factors as all other species. We cannot increase beyond our ability to acquire raw materials and energy and safely dispose of our wastes. We also must

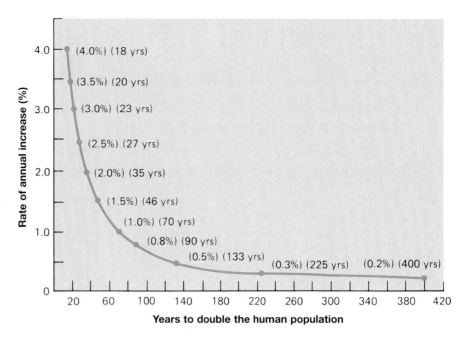

Figure 6.10

Doubling Time for the Human Population   This graph shows the relationship between the rate of annual increase in percent and doubling time. A population growth rate of 1 percent per year would result in the doubling of the population in about seventy years. A population growth rate of 3 percent per year would result in a population doubling in about twenty-three years.

remember that interactions with other species and with other humans will help determine our carrying capacity.

Let us look at these four factors in more detail. Many of us think of raw materials simply as the amount of food available. However, we have become increasingly dependent on technology, and our lifestyles are directly tied to our use of other kinds of resources, such as irrigation water, genetic research, and antibiotics. Food production is becoming a limiting factor for some segments of the world's human population. Malnutrition is a serious problem in many parts of the world because sufficient food is not available. Chapter 7 deals with the problems of food production and distribution and their relationship to human population growth.

The second factor, available energy, involves problems similar to those of raw materials. Essentially all species on earth are ultimately dependent on sunlight for their energy. New, less disruptive methods of harnessing this energy must be developed to support an increasing population. Currently, the world population depends on fossil fuels to raise food, modify the en-

vironment, and move from place to place. When energy prices increase, much of the world's population is placed in jeopardy because incomes are not sufficient to pay the increased costs for food and other essentials.

Waste disposal is the third factor determining the carrying capacity for humans. Most pollution is, in reality, the waste product of human activity. Some people are convinced that disregard for the quality of our environment will be a major limiting factor. In any case, it makes good sense to control pollution and to work toward cleaning our environment.

The fourth factor that determines the carrying capacity of a species is interaction with other organisms. We need to become aware that we are not the only species of importance. When we convert land to meet our needs, we displace other species from their habitats. Many of these displaced organisms are not able to compete with us successfully and must migrate or become extinct. Unfortunately, as humans expand their domain, the areas available to these displaced organisms become more rare. Parks and natural areas

Population Growth of Invading Species

When a new species is introduced into an area suitable for its survival, it has great potential to increase its population size, since it will not have natural enemies to keep mortality high. New species may also be able to compete favorably for available resources and lower the populations of native species. A typical population growth curve starts with a few individuals being released. Once established, the population will increase in number and expand its range. The zebra mussel, for example, is thought to have entered the Great Lakes about 1985. Today, it is found throughout all the Great Lakes and has been introduced into the Mississippi River and its tributaries, where it has been discovered as far south as New Orleans. Similarly, the gypsy moth has spread from its place of original release to much of the forestland of the Midwest, and is still expanding its range. Another invading species is the kudzu vine, which has become a pest in many areas of the southern United States.

Eventually, the invading species occupies all the habitat suitable to it and the population stabilizes. Dandelions and starlings are introduced species that are no longer expanding their range. They have simply become a normal part of the biology of North America.

Dandelion

Zebra mussels

have become tiny refuges for the plants and animals that once occupied vast expanses of land. If these refuges fall to the developer's bulldozer or are converted to agricultural use, many organisms will become extinct. What today seems like an unimportant organism, one that we could easily do without, may someday be seen as an important link to our very survival.

Humans Are Social Animals

Human survival depends upon interaction and cooperation with other humans. Current technology and medical knowledge are available to control human population growth and to improve the health of the people of the world. Why then does the population continue to increase, and why does poverty become worse every year? Humans are social animals who have freedom of choice and frequently do not do what is considered "best" from an unemotional, uninvolved, biological point of view. People make decisions based on history, social situations, ethical and religious considerations, and personal desires.

The biggest obstacle to controlling human population is not biological but falls under the province of philosophers, theologians, politicians, and sociologists. People in all fields need to understand the cause of the population problem if they are to successfully deal with every aspect of it.

Ultimate Size Limitation

The human population cannot increase indefinitely, because, eventually, the weight of the human tissue would equal

Starlings

Kudzu vine

Gypsy moth

the weight of the earth. We can say with certainty that the population will ultimately reach its carrying capacity and level off long before this absurd condition would occur. Many people speculate about the limits of human population growth. In its 1981 report, the United Nations Fund for Population Activities suggested that a moderate estimate of future human population is 10.5 billion people by the year 2110. However, if world human population continues to grow at its current rate of 1.5 percent per year, the population will exceed 11 billion by the year 2040. Some people suggest that a lack of food, a lack of water, or increased waste heat will ultimately control the size of the human population. Still others suggest that, in the future, social controls will limit population growth. It is also important to consider the age structure of the world population. In most of the world, there are many reproductive and prereproductive individuals. Since most of these individuals are currently reproducing or will reproduce in the near future, even if they reduce their rate of reproduction, there will be sharp increase in the number of people in the world in the next few years.

Others are concerned that politically related destruction of humanity will occur. No one knows what the ultimate human population size will be or what the most potent limiting factors will be, but most agree that we are approaching the maximum sustainable human population. If the human population continues to increase, the amount of agricultural land available will not be enough to satisfy the demand for food.

Wolves and Moose on Isle Royale

Isle Royale in Lake Superior has been the site of a long-term study of the relationship between moose and wolf populations. Wolves (probably a single pair) reached Isle Royale by crossing the ice from Canada. By 1958, the population had increased to about 20. About that time, studies began on the relationship between the wolves and moose. A succession of researchers learned a great deal about the relationship between the two animals, the food habits of wolves, and the relationships among various wolf packs. By 1980, the population of wolves was about 50, but then it declined sharply, reaching a low of 13 individuals in 1993. It is thought that the rapid decline in the early 1980s was at least partly the result of a viral disease introduced by domesticated dogs. By 1996, good reproduction allowed the population to rise to 22 wolves. The increase in young animals suggests that they will have high survivorship and that the population is likely to grow in the future.

There is concern about the possible negative effects of inbreeding among the wolves on Isle Royale. Since the entire population is thought to be descended from a mated pair, it is possible that limited genetic variety might ultimately become a problem for the wolves' survival.

The moose population is affected by several factors: food availability, disease, weather, accidents, and wolf predation. Since the reduction in the wolf population in the early 1980s, the moose population has steadily increased. However, during the winter of 1995–96, the moose population suffered a 50 percent reduction. The growth in previous years has reduced the amount of food available and resulted in malnutrition, slow growth, and poor reproduction. A very severe winter and a tick infestation probably contributed to the die-off as well. The wolf population was able to kill many moose and also fed on moose that died of other causes. The high moose death rate provided an ample food resource for wolves. It will be interesting to see if the decline in moose numbers will impact the wolf population in future years.

- When populations of animals become endangered, efforts are often made to preserve them. Should special efforts be made to preserve this wolf population?
- In what ways is this case different from that of populations of California condors or whooping cranes?

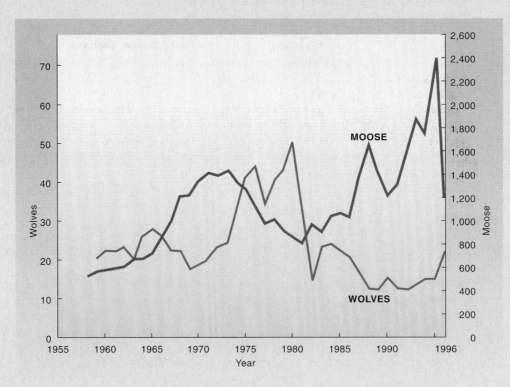

Source: Data from Rolf O. Peterson, "Ecological Studies of Wolves on Isle Royale," Annual Report 1995–96.

PART TWO   Ecological Principles and Their Application

SUMMARY

A population is a group of organisms of the same species that inhabits an area. The birthrate (natality) is the number of individuals entering the population by reproduction during a certain period. The death rate (mortality) measures the number of individuals that die in a population during a certain period. Population growth is determined by the combined effects of the birthrate and death rate.

The sex ratio of a population is a way of stating the relative number of males and females. Age distribution and the sex ratio have a profound impact on population growth. Most organisms have a biotic potential much greater than that needed to replace dying organisms.

Interactions among individuals in a population, such as competition, predation, and parasitism, are also important in determining population size. Organisms may migrate into (immigrate) or migrate out of (emigrate) an area as a result of competitive pressure.

A typical population growth curve shows a lag phase followed by an exponential growth phase and a stable equilibrium phase at the carrying capacity. The carrying capacity is determined by many limiting factors that are collectively known as environmental resistance. The four major categories of environmental resistance are available raw materials, available energy, disposal of wastes, and interactions among organisms. Some populations experience a death phase following the stable equilibrium phase.

K-strategists typically are large, long-living organisms that reach a stable carrying capacity. Their population size is usually controlled by density-dependent limiting factors. r-strategists are generally small, short-living organisms that reproduce very quickly. Their populations do not generally reach a carrying capacity but crash because of some density-independent limiting factor.

The human population is increasing at a rapid rate. The earth's ultimate carrying capacity for humans is not known. The causes for human population growth are not just biological but also social, political, philosophical, and theological.

REVIEW QUESTIONS

1. How are biotic potential and age distribution interrelated?
2. List three characteristics populations might have.
3. Why do some populations grow? What factors help to determine the rate of this growth?
4. Under what conditions might a death phase occur?
5. List four factors that could determine the carrying capacity of an animal species.
6. How do the concepts of birthrate and population growth differ?
7. How does the population growth curve of humans compare with that of bacteria on a petri dish?
8. How do r-strategists and K-strategists differ?
9. As the human population continues to increase, what might happen to other species?
10. All successful organisms overproduce. What advantage does this provide for the species? What disadvantages may occur?

KEY TERMS

age distribution 96
biotic potential 97
birthrate 95
carrying capacity 99
death phase 99
death rate 95

density-dependent limiting factors 100
density-independent limiting factors 100
dispersal 96
emigration 96

environmental resistance 99
exponential growth phase 98
immigration 96
K-strategists 100
lag phase 97
limiting factors 99

mortality 95
natality 95
population density 96
r-strategists 100
sex ratio 96
stable equilibrium phase 98

Slowing down the population of zebra mollusks in the western part of North America.
> http://www.usbr.gov/zebra/wzmtf.html

On-line newspaper dealing with current issues about the world, humans, and the environment.
> ***http://www.usatoday.com

United Nations Population Information Network. Comprehensive access to population data.
> http://www.undp.org/popin/popin.htm

U.S. Census Bureau's official statistics. Shows today's population, estimates, and projections.
> http://www.census.gov/population/www/

Oregon Department of Fish and Wildlife Population Lab. Population analysis of wildlife in the northwestern region of the United States.
> http://www.peak.org/~ursus/lab.html

Population Politics: The carrying capacity of regions in the United States.
> http://csf.colorado.edu/authors/hanson/page58.htm

Animal populations. Examines animal life to determine why some become endangered while others overpopulate.
> http://www.bergen.org/AAST/Projects/ES/AP/

c h a p t e r

7

Human Population Issues

Objectives

After reading this chapter, you should be able to:

- Apply some of the principles discussed in chapter 6 to the human population.
- Differentiate between birthrate and population growth rate.
- Explain the current population situation in the United States.
- Explain how age distribution and women's roles affect population projections.
- Recognize that U.S. society is adjusting as the average age increases.
- Recognize that most of the world still has a rapidly growing population.
- Describe the implications of the hypothesis that a demographic transition occurs.
- Understand how an increasing world population will alter the worldwide ecosystem.
- Understand the implications of the fact that most of the countries of the world are not able to produce enough food to feed their people.
- Explain why less-developed nations have high birthrates and why they will continue to have a low standard of living.
- Recognize that the developed nations of the world will be under greater pressure to share their abundance.

Chapter Outline

Current Population Trends

Population and Standard of Living

The Human Population Issue

Causes of Population Growth

    Biological Reasons for Population Growth

    Social Reasons for Population Growth

Global Perspective: *The AIDS Pandemic*

Environmental Close-Up: *Control of Births*

Global Perspective: *Governmental Policy and Population Control*

    Political Factors That Affect Population Growth

The Demographic Transition Concept

The U.S. Population Picture

Global Perspective: *The Urbanization of the World's Population*

Hunger, Food Production, and Environmental Degradation

Global Perspective: *Population and Poverty: A Vicious Cycle*

Global Perspective: *Canadian Population Overview*

Issues and Analysis: *Population Growth in Mexico*

Anticipated Changes with Continued Population Growth

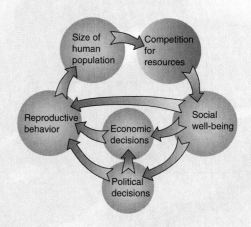

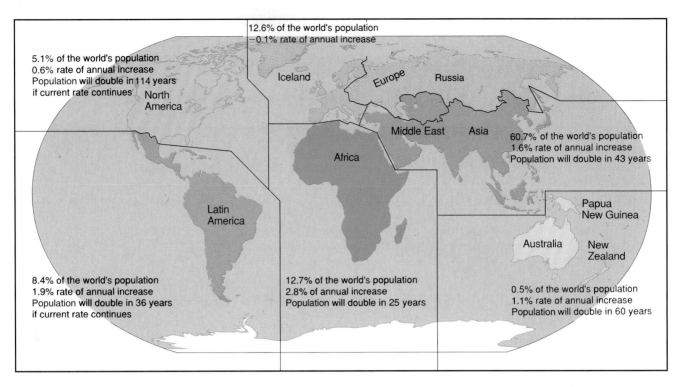

Figure 7.1

Population Growth in the World (1996) The population of the world is not evenly distributed. Currently nearly 82 percent of the world's population is in Latin America, Africa, and Asia. These areas also have the highest rates of increase and are generally considered less developed. Because of the high birthrates, they are likely to remain less developed and will constitute nearly 84 percent of the world's population by the year 2010.

Current Population Trends

The human population predicament is very complex. In order to appreciate it, we must understand current population trends and how they are related to economic conditions. Currently, the world population is over 5.8 billion people. By the year 2025, this is expected to increase to just over 8 billion. Much of this increase is expected to occur in Africa, Asia, and Latin America, which already have over 82 percent of the world population. (See figure 7.1.) If these trends continue, the total population of Africa, Asia, and Latin America will increase from the current 4.72 billion to more than 7 billion by 2025, when these continents will contain 86 percent of the world's people. These regions have not only the highest population growth rates but also the lowest per capita **gross national product (GNP).** The GNP is an index that measures the total goods and

services generated within a country. As you can see from figure 7.2, a wide economic gap exists between less-developed countries with low GNPs and the developed countries with high GNPs. This large difference in economic well-being is reflected in a dissimilarity in the **standard of living,** an abstract measure of the degree to which necessities and comforts of daily life are met. Yet the people of less-developed countries aspire to the same standard of living enjoyed by people in the developed world.

Population and Standard of Living

Standard of living is a difficult concept to quantify since various cultures have different attitudes and feelings about what is desirable. Here, we compare averages of several aspects of the cultures in three countries: (1) the United States, which is an example of a highly devel-

oped industrialized country; (2) Argentina, which is a moderately developed country, and (3) Zimbabwe, which is less developed. Figure 7.3 shows several statistics that relate to the standard of living in these three countries. The United States produces approximately three times more goods and services per person than does Argentina and about fifty times more goods and services than Zimbabwe. U.S. citizens consume more than five times more energy than the average Argentinean and about seventeen times more than the average Zimbabwean. Infant death is high in Zimbabwe, intermediate in Argentina, and low in the United States. In addition, people live longer and go to school longer in the United States than in the other two countries. Zimbabwe is more densely populated than Argentina. The average Zimbabwean is less well fed, a factor related to high mortality and low life expectancy, whereas the average citizen of the United States or Argentina has more food than needed.

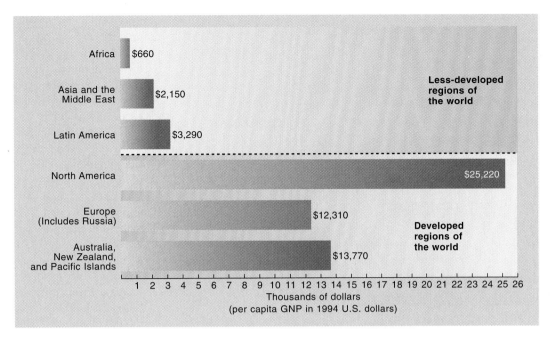

Figure 7.2

Per Capita Gross National Product of Nations   The world can be divided into developed and less-developed regions. One measure of the degree to which a country is developed is its per capita gross national product. A less-developed country produces very little for each person in the country, while a developed country has high productivity and, therefore, also has goods and services available for consumption by its people.

Source: Data from 1996 World Population Data Sheet.

Obviously, tremendous differences exist in the standard of living among these three countries. What the average U.S. citizen would consider poverty level would be considered a luxurious life for the average person in a poorly developed country. Standard of living seems to be closely tied to energy consumption. In general, the higher the per capita energy consumption, the higher the standard of living of the people. This is related to the degree of industrialization or development. However, with industrialization comes pollution, as huge amounts of fossil fuels are consumed to provide the goods and services that contribute to the standard of living.

The Human Population Issue

There are many ways to show that world population growth is a contributing factor in nearly all world problems. Current population growth has led to famine in areas where food production cannot keep pace with increasing numbers of people; political unrest in areas with great disparities in availability of resources (jobs, goods, food); environmental degradation by poor agricultural practices (erosion, desertification); water pollution by human and industrial waste; air pollution caused by the human need to use energy for personal use and for industrial applications; extinctions caused by people converting natural ecosystems to managed agricultural ecosystems; and destructive effects of exploitation of natural resources (strip mining, oil spills, groundwater mining).

Several factors interact to determine the impact of a society on the resources of the country.

$$\begin{array}{c} \text{Impact} \\ \text{of a} \\ \text{Population} \end{array} = \dfrac{\begin{array}{c}\text{Population} \\ \text{Size}\end{array}}{\begin{array}{c}\text{Area} \\ \text{Occupied}\end{array}} \times \begin{array}{c}\text{Degree of} \\ \text{Technological} \\ \text{Development}\end{array}$$

The larger the size of a population, the greater the demand on the resources of the country. However, population size is not the only important factor. **Population density,** the number of people per unit of land area, is also important. A million people spread out over the huge area of the Amazon have much less impact on resources than that same million people in a small island country, because the impact is distributed over a greater land surface.

The degree of technological development is also important. The environmental impact of the developed world is often underestimated because the population in these countries is relatively stable and local environmental conditions are good. However, highly developed countries consume huge amounts of resources. Citizens of these countries eat more food, particularly animal protein, which requires larger agricultural inputs than does a vegetarian diet. They have more material possessions, and consume vast amounts of energy. These developed countries purchase goods and services from other parts of the world, often degrading environmental conditions in less-developed countries. The environmental impact of highly developed regions like

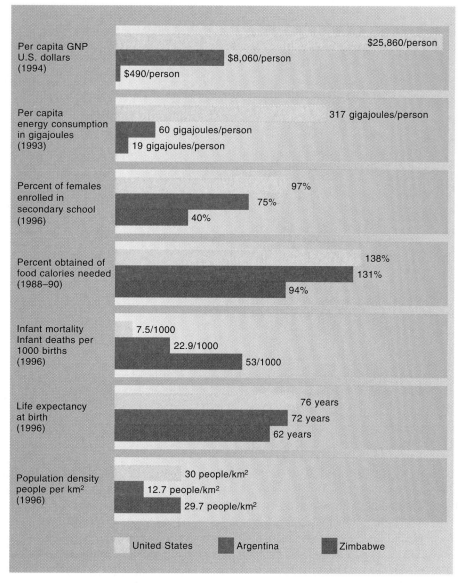

Figure 7.3

Standard of Living in Three Countries Standard of living is a measure of how well one lives. It is not possible to get a precise definition, but when we compare the United States, Argentina, and Zimbabwe, it is obvious that there are great differences in how the people in these countries live. Zimbabwe has a low life expectancy, a high death rate, and low productivity, and the people are often starving. The United States has a high life expectancy, a low death rate, and high productivity, and many people eat too much. Argentina is intermediate in all of these characteristics.

North America, Japan, and Europe is often felt in distant places, while the impact on resources in the home country may be minimal.

While controlling world population would not eliminate the problems, it could reduce the rate at which environmental degradation is occurring. The quality of life for many people in the world would improve. Why, then, does the human population continue to grow at such a rapid rate?

Causes of Population Growth

In chapter 6, we examined populations from a biological point of view. We looked at their characteristics, the causes of growth, and the forces that cause populations to stabilize. All of these biological factors apply to human as well as nonhuman populations. There is an ultimate carrying capacity for the human population. Eventually, limiting factors will cause human populations to stabilize. However, unlike other kinds of organisms, humans are also influenced by social, political, economic, and ethical factors. We have accumulated knowledge that allows us to predict the future. We can make conscious decisions based on the likely course of events and adjust our lives accordingly. Part of our knowledge is the certainty that as populations continue to increase, death rates and birthrates will become equal. This can happen by allowing the death rate to rise or by *choosing* to limit the birthrate. Controlling human population would seem to be a simple process. Once everyone understands that lowering the birthrate is more humane than allowing the death rate to rise, most people should make the "correct" decision; however, it is not quite that simple.

Biological Reasons for Population Growth

The study of human populations, their characteristics, and what happens to them is known as **demography.** Demographers can predict the future growth of a population by looking at several biological indicators. Currently, in almost all countries of the world, the birthrate exceeds the death rate. Therefore, the size of the population must increase. (See table 7.1.) Some countries that have high birthrates and high death rates, with birthrates greatly exceeding the death rates, will grow rapidly (Afghanistan and Ethiopia). Such countries usually have an extremely high mortality rate among children because of disease and malnutrition.

Some countries have high birthrates and low death rates and will grow extremely rapidly (Guatemala and Syria). Infant mortality rates are moderately high in these countries. Other countries have low birthrates, and death rates that closely match the birthrates; they will grow slowly (Sweden and the United Kingdom). These

PART TWO Ecological Principles and Their Application

Table 7.1

Population Growth Rates in Selected Countries (1996)

Country	Births per 1,000 individuals	Deaths per 1,000 individuals	Infant mortality rate (deaths per 1,000 live births)	Rate of natural increase (annual %)	Time needed to double population (years)
Russia	9	15	18	− 0.5	—
Germany	9	11	5.5	− 0.1	—
Belgium	12	10	7.6	0.1	630
Sweden	12	11	4.4	0.1	630
United Kingdom	13	11	6.2	0.2	385
Japan	10	7	4.2	0.2	315
Canada	13	7	6.2	0.6	116
United States	15	9	7.5	0.6	114
Argentina	20	8	22.9	1.2	58
Turkey	23	7	47	1.6	43
India	29	10	79	1.9	37
Uzbekistan	29	7	28	2.3	30
Zimbabwe	35	9	53	2.5	28
Paraguay	34	6	38	2.8	25
Afghanistan	50	22	163	2.8	24
Guatemala	36	7	51	2.9	24
Ethiopia	46	16	120	3.1	23
Syria	44	6	44	3.7	19

Source: Data from Carl Haub and Machiko Yanagishita, World Population Data Sheet 1996, Population Reference Bureau, Washington, DC.

and other more-developed countries typically have very low infant mortality rates. The disruption caused by the political upheaval in the former Soviet Union and Eastern Europe has resulted in several countries (e.g., Russia and Germany) with death rates that exceed birthrates, causing their populations to decline. Because of these countries and the generally low rates of growth in the rest of Europe, the European region as a whole has a declining population.

The most important determinant of the rate at which human populations grow is related to how many women in the population are having children and the number of children each woman will have. The **total fertility rate** of a population is the number of children born per woman per lifetime. A total fertility rate of 2.1 is known as **replacement fertility,** since in the long run, if the total fertility rate is 2.1, population growth will stabilize. A rate of 2.1 is used rather than 2.0

because some children do not live very long after birth and therefore will not contribute to the population for very long. When population is not growing, and the number of births equals the number of deaths, it is said to exhibit **zero population growth.**

A total fertility rate of 2.1 will not necessarily immediately result in a stable population with zero growth, for several reasons. First, the death rate may fall as living conditions improve and people live longer. If the death rate falls faster than the birthrate, there will still be an increase in the population even though it is reproducing at the replacement rate.

The age structure of a population also has a great deal to do with the rate of population growth. If a population has many young people who are raising families or who will be raising families in the near future, the population will continue to increase even if the families limit themselves to two children. Depending

on the number of young people in a population, it may take 20 years to a century for the population of a country to stabilize so that there is no net growth.

Social Reasons for Population Growth

It is clear that the economically developed countries of the world have low fertility rates and low rates of population growth and that the less-developed countries have high fertility rates and high population growth rates. It also appears obvious that reducing fertility rates would be to everyone's advantage; however, not everyone in the world feels that way. Several factors influence a person's desired family size. Some are religious, some are traditional, some are social, and some are economic.

The major social factors that determine family size are the role and desires of

The AIDS Pandemic

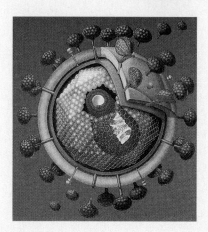

The AIDS (acquired immunodeficiency syndrome) virus has caused a worldwide epidemic, which can be called a pandemic because it continues to spread throughout the world. Millions of people have been infected. The virus was first identified as the cause of AIDS in the late 1970s. Since then, individuals with the infection have been reported in nearly every country in the world. According to the U.S. Centers for Disease Control and Prevention, over 60 percent of all people diagnosed with AIDS in the United States have died. Of those diagnosed with AIDS during 1990, approximately 80 percent had died by the end of 1995.

The disease is spread through direct transfer of body fluids containing the virus into the bloodstream of another person. Sharing of contaminated needles among intravenous drug users and sexual contact are the most likely methods of passage. In the United States, the disease was once considered a problem only for the homosexual community and those who use intravenous drugs. This perception is rapidly changing. Many of the new cases of AIDS are being found in women infected by male sex partners and in children born to infected mothers.

Sub-Saharan Africa has been hit hardest by this disease. The World Health Organization estimates that approximately 65 percent of all persons infected with the AIDS virus (13 million) live in this region of the world. In some large urban hospitals, over 50 percent of the hospital beds are taken up by AIDS patients. In Africa, AIDS has always been primarily a sexually transmitted disease spread through heterosexual contact. Many people believe that in the poor countries of central Africa, permissive sexual behavior and prostitution have created conditions for a rapid spread of the disease. This is particularly evident along major transportation routes.

The economic burden on these countries is tremendous. Those with AIDS symptoms are unable to work and need medical care and medication. Because of poverty there is often little medical care available. Many people have already died from the disease. Others who are currently infected will die in the near future. Over 9 million children have been orphaned, resulting in an economic burden on relatives. Some villages are already beginning to notice a change in the structure of their populations. With the death of young infected adults, villages are composed primarily of older people and children.

women in the culture. In many male-dominated cultures, the traditional role of women is to marry and raise children. Often this role is coupled with strong religious input as well. Typically, little value is placed on educating women, and early marriage is encouraged. In these cultures, women are totally dependent on their husbands and children in old age. Because early marriage is encouraged, fertility rates are high, since women are exposed to the probability of pregnancy for more of their fertile years. Lack of education reduces options for women in these cultures. They do not have the option to not marry or to delay marriage and thus reduce the number of children they will bear. By contrast, in much of the developed world women are educated, delay marriage, and have fewer children. It has been said that the single most important

activity needed to reduce the world population growth rate is to educate women. Whenever the educational level of women increases, fertility rates fall.

Data from the United Nations indicates that 55 percent of African women, 37 percent of Asian women, and 42 percent of Latin American women are married by age 20. The definition of marriage is broad and includes both legally defined marriages and those recognized by local customs. Their fertility rates are 5.7, 2.9, and 3.1 children per woman per lifetime, respectively. In the developed world, the average age of first marriage is much higher, between age 25 and 27; early marriages are rare; and the total fertility rate is about 1.6 children per woman per lifetime.

As women become better educated and obtain higher paying jobs, they be-

come financially independent and can afford to marry later and consequently have fewer children. Better-educated women are also more likely to have access to and use birth control. In economically advanced countries, over 70 percent of women typically use contraception. In the less-developed countries, contraceptive use is much lower—less than 20 percent in Africa, about 60 percent in Latin America, and about 45 percent in Asia (except China). Changing attitudes toward divorce may also be tied to reduced fertility.

It is important to recognize that access to birth control will not solve the population problem. What is most important is the desire of women to limit the size of their families. In developed countries, use of birth control is extremely important in regulating the birthrate. This

e n v i r o n m e n t a l
close-up

Control of Births

The use of technology to control disease and famine has greatly reduced the death rate of the human population. Technological developments can also be used to control the birthrate. A variety of contraceptive methods are available to help people regulate their fertility. Research is continuing to develop more effective, more acceptable, and less expensive methods of controlling conception. Because of cultural and religious differences, some forms of contraception may be more acceptable to one segment of the world's population than another.

The most common methods of contraception are oral contraceptive pills, diaphragms and spermicidal jelly, intrauterine contraceptive devices, spermicidal vaginal foam, condoms, vasectomy, and tubal ligation. The range of effectiveness of these methods, shown in the table, is the result of individual fertility differences and the degree of care employed in the use of each method.

In addition to various methods of conception control, abortion can terminate unwanted conceptions early in pregnancy. Most countries with low birthrates, such as the United States, Japan, and many European countries, allow abortions.

Effectiveness of Fertility Control Methods

Method	Effectiveness*
None	10%
Coitus interruptus (withdrawal)	82%
Natural family planning (determining the time of ovulation)	70–97%
Foams, jellies, and creams alone	80–97%
Cervical cap	82–94%
Diaphragm with jelly	82–98%
Condoms alone	88–90%
Condoms with foam	99%
Birth control pill	98–99%
Intrauterine device (IUD)	99%
Sterilization (tubal ligation or vasectomy)	Nearly 100%
Contraceptive implants (Norplant)	Nearly 100%
Contraceptive injections (Depo-Provera)	Nearly 100%

*Based on 100 couples using the same method for one year. If 10 out of 100 women using a particular method became pregnant, the method is considered 90% effective. The ranges of effectiveness are related to the care with which each method is applied.

---

is true regardless of religion and previous historical birthrates. Italy and France, both traditionally Catholic countries, have birthrates of 9/1,000 and 12/1,000, respectively. The average for the developed countries of the world is 12/1,000. Obviously, women in these countries make use of birth control to help them regulate the size of their families. (See Environmental Close-Up: Control of Births.) By contrast, Mexico, which is also a traditionally Catholic country, has a birthrate of 27/1,000, which is typical of birthrates in the less-developed world regardless of religious tradition.

When women in the less-developed world are asked what family size they desire, they typically suggest four or more

children. However, the desired number is usually fewer than the number they actually have, and both are much higher than the replacement fertility rate of 2.1 children. Access to birth control will allow them to limit the number of children they actually have to their desired number and will allow them to space their children at more convenient intervals, but they still desire more children than the 2.1 needed for replacement. Why do they desire large families? There are several reasons. In areas where infant mortality is high, it is traditional to have large families since several of a woman's children may die before they reach adulthood. This is particularly important in the less-developed world where there is

no government program of social security. Parents are more secure in old age if they have several children to contribute to their needs when they can no longer work.

In less-developed countries, the economic benefits of children are extremely important. Even young children can be given jobs that contribute to the family economy. They can protect livestock from predators, gather firewood, cook, or carry water. In the developed world, large numbers of children are an economic drain. They are prevented from working by law, they must be sent to school at great expense, and they consume large amounts of the family income. Parents in the developed world make an economic

Governmental Policy and Population Control

Governments can have a significant impact on the population growth patterns of nations. Some countries have policies that encourage couples to have children. Canada pays a bonus to couples on the birth of a child. The U.S. tax code indirectly encourages more children by providing exemptions for them. Some countries in Europe are concerned about a future lack of working-age people and are considering programs to encourage births. Many countries have no official population policy, but the position of most of those that do is to reduce the birthrate.

China has long been the most populated nation of the world. It now contains over 1.2 billion people, more than 21 percent of the world's people. Its history of population control is an interesting study of how government policy affects reproductive activity among its citizens. When the People's Republic of China was established in 1949, the population was about 540 million. The official policy of the government was to encourage births because more Chinese would be able to produce more goods and services, and production was the key to economic prosperity.

Because of China's high birthrate and falling death rate, its population increased to 614 million in 1955. This rapid increase and lack of economic growth led government officials to make changes that would lead to population control. Abortions became legal in 1953, and the first family planning program began in 1955, as a means of improving maternal and child health. Birthrates fell. (See graph.) The failure of government programs during the Great Leap Forward, and the establishment of communes to produce food and other products, resulted in widespread famine and increased death rates and low birthrates in the late 1950s and early 1960s. A second family-planning program began in 1962 but was not particularly effective.

The present family-planning policy began in 1971 with the launching of the *wan xi shao* campaign. Translated, the phrase means "later" (marriages), "longer" (intervals between births), and "fewer" (children). This program raised the legal ages for marriage. For women and men in rural areas, the ages were raised to 23 and 25, respectively; for women and men in urban areas, the ages were raised to 25 and 28, respectively. These policies resulted in a reduction of birthrates by nearly 50 percent between 1970 and 1979.

An even more restrictive, one-child campaign was begun in 1978–79. The program offered incentives for couples to restrict their family size to one child. Couples enrolled in the program would receive free medical care, cash bonuses for their work, special housing treatment, and extra old-age benefits. Those who broke their pledge were penalized by the loss of these benefits as well as other economic penalties. By the mid-1980s, less than 20 percent of the eligible couples were signing up for the program. Rural couples particularly desired more than one child. In fact, in a country where nearly 75 percent of the population is rural, the rural total fertility rate was 2.5 children per woman. In 1988, a second child was sanctioned for rural couples if their first child was a girl, which legalized what had been happening anyway.

The current total fertility rate is 1.8 children per woman. Over 80 percent of couples use contraception; the most commonly used forms are male and female sterilization and the intrauterine device. Abortion is also an important aspect of this program, with a ratio of over 600 abortions per 1,000 live births. However, because a large proportion of the population is under 30 years of age, the population is still expected to double in about 66 years.

---

decision about having children in the same way they buy a house or car: "We are not having children right away. We are going to wait until we are better off financially."

Political Factors That Affect Population Growth

Two other factors that influence population growth are political pressures and immigration. Governments often encourage or discourage population growth in several ways. (See Global Perspective: Governmental Policy and Population Control.) They can have stated policies that describe population goals and pass laws that penalize those who fail to meet the goals or reward those who meet them. Governments can also subtly encourage people to regulate their reproductive activities by providing tax advantages or low-cost access to birth control.

The immigration policies of a country also have a significant impact on the rate at which the population grows. Birthrates are currently so low in several European countries, Japan, and China that they will likely have a shortage of those of working age in the near future. One way to solve this problem is to encourage immigration from other parts of the world.

The developed countries are under tremendous pressure to accept immigrants. Their standard of living is a tremendous magnet for refugees or people who seek a better life than is possible where they currently live. The continuing economic and political reorganization in central Europe is causing significant increases in the number of immigrants and placing considerable strain on the social systems of Germany,

It appears that China has undergone a demographic transition of sorts. Birthrates and death rates were high in 1949, death rates fell steadily (except during the famine years), and birthrates have fallen to an all-time low. However, this demographic transition occurred as a result of government family-planning policies. Living standards are still low, with a per capita gross national product of US$530, and the population has not yet become stabile. Compare the graph here with figure 7.4 on the demographic transition.

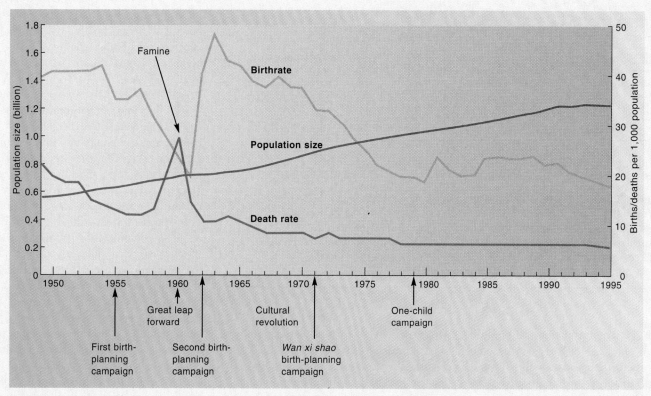

Source: Data from H. Yuan Tien, "China's Demographic Dilemmas" in Population Bulletin 1992, Population Reference Bureau, Inc., Washington, DC and National Family Planning Commission of China; and more recent data taken from Population Reference Bureau.

Austria, and other countries of the region. In the United States, approximately one-third of the population increase experienced each year is the result of immigration.

The Demographic Transition Concept

The relationship between the standard of living and the population growth rate seems to be that countries with the highest standard of living have the lowest population growth rate, and those with the lowest standard of living have the highest population growth rate. This has led many people to suggest that countries naturally go through a series of stages called a **demographic transition.** This model is based on the historical, social, and economic development of Europe and North America. In a demographic transition, the following four stages occur (see figure 7.4):

1. Initially, countries have a stable population with a high birthrate and a high death rate. Death rates often vary because of famine and epidemic disease.
2. Improved economic and social conditions (control of disease and increased food availability) bring about a period of rapid population growth as death rates fall. Birthrates remain high.
3. As countries become industrialized, the birthrates begin to drop because people desire smaller families and make use of contraceptives.
4. Eventually, birthrates and death rates again become balanced, with low birthrates and low death rates.

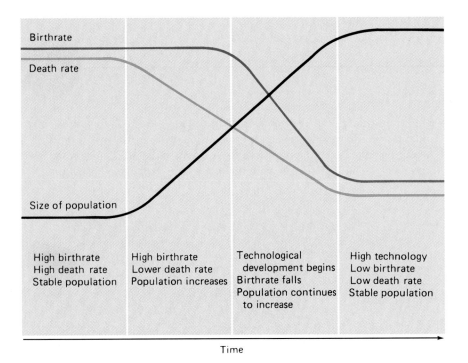

Birthrate

Death rate

Size of population

High birthrate	High birthrate	Technological	High technology
High death rate	Lower death rate	development begins	Low birthrate
Stable population	Population increases	Birthrate falls	Low death rate
		Population continues	Stable population
		to increase	

Time

Figure 7.4

Demographic Transition The demographic transition model suggests that, as a country develops technologically, it automatically experiences a drop in the birthrate. This certainly has been the experience of the developed countries of the world. However, the developed countries make up about 20 percent of the world's population. It is doubtful whether the less-developed countries can achieve the kind of technological advances experienced in the developed world.

This is a very comfortable model because it suggests that, if a country can become industrialized, then social, political, and economic processes will naturally cause its population to stabilize.

However, the model leads to some serious questions. Can the historical pattern exhibited by Europe and North America be repeated in the less-developed countries of today? Europe, North America, Japan, and Australia passed through this transition period when world population was lower and when energy and natural resources were still abundant. It is doubtful whether these supplies are adequate to allow for the industrialization of the major portion of the world currently classified as less-developed.

A second concern is the time element. With the world population increasing as rapidly as it is, industrialization probably cannot occur fast enough to have a significant impact on population growth. As long as people in less-developed countries are poor, there is a

strong incentive to have large numbers of children. Children are a form of social security because they take care of their elderly parents. Only people in developed countries can save money for their old age. They can choose to have children, who are expensive to raise, or to invest money in some other way.

When the countries of Europe and North America passed through the demographic transition, they had access to large expanses of unexploited lands, either within their boundaries or in their colonies. This provided a safety valve for expanding populations during the early stages of the transition. Without this safety valve, it would have been impossible to deal adequately with the population while simultaneously encouraging economic development. Today, less-developed countries may be unable to accumulate the necessary capital to develop economically, since an ever-increasing population is a severe economic drain.

The U.S. Population Picture

As a result of the 1990 census in the United States, several changes have occurred in how demographers view the structure of and future trends in the U.S. population. In many ways, the U.S. population is similar to those of other developed countries of the world with low birthrates and slow population growth. However, the U.S. population includes a **postwar baby boom** component, which has significantly affected population trends. These baby boomers were born during an approximately 20-year period following World War II, when birthrates were much higher than today, and constitute a bulge in the age distribution profile. (See figure 7.5.) As members of this group have raised families, they have had a significant influence on how the U.S. population has grown. Some of the older persons in this group are beginning to retire, and as more of them do, the population will gradually age. By 2030, about 20 percent of the population will be 65 years of age or older.

A changing age structure will lead to social changes as well. The baby boom of the late 1940s and the 1950s encouraged growth in service industries needed by young families. Maternity wards had to be expanded, schools could not be built fast enough, baby-care companies saw unprecedented sales, and the toy industry flourished. Today, these "babies" are in their thirties and forties. They are buying homes, cars, and appliances, but are raising fewer children than did their parents. However, because baby boomers are such a large segment of the population, they have contributed a large number of children to the population. The children of the baby boomers now constitutes a "baby boomlet." In many parts of the country schools are seeing increased numbers of children in the early grades. At the same time, there are more elderly, because people are living longer. That creates a need for additional services for the elderly. This trend toward an aging population will be accentuated as the baby boomers retire. What will the social needs be in the year 2020 when many of these baby boomers will have retired?

Even with the current lower total fertility rate, the population is still

The Urbanization of the World's Population

The United Nations estimates that currently about 45 percent of the world's population lives in cities. This is expected to increase to over 50 percent by 2005 and reach 60 percent by 2025. The economically developed world is currently about 75 percent urban, and the less-developed world is about 37 percent urban. Therefore, the increase in the urban population will occur primarily in the less-developed world where resources to deal with urban problems are least available.

Urbanization is not necessarily bad. The economic activity of the developed world takes place primarily in cities. Cities can be planned and managed to be healthy, interesting places to live. Cities offer jobs, health care, schools, and other services that are usually lacking in rural areas. Often the rural economy of a country is considered to be of low status, and young people wish to move to cities where higher status jobs and desirable cultural amenities are available.

Dense populations of people have a concentrated impact on local resources. Often water must be transported long distances, wastes are difficult to get rid of, and air quality drops as each additional person places a burden on the local environment. Unfortunately, in the less-developed world it is impossible to plan for growth and provide basic resources such as drinking water, sewers, transportation, and housing as fast as the population is growing. New migrants to the cities often lack education or training for the high wage jobs available, have lost the opportunity to raise their own food, and lack the support of their family units. Consequently, when these poor people migrate to the cities, they often live in shanty towns on the edge of the city or form squatter settlements on hillsides and other unused land. These centers of poverty lack basic services since they are often outside the bureaucratic structure of the city. The rich in such cities live a short distance from the poor and may be relatively isolated, but the negative environmental impact of the entire urban region (air and water pollution, transportation problems, etc.) affects all the residents. Even the long-established urban centers of the developed world have major problems. Many contain pockets of poverty that have led to problems of social unrest and crime.

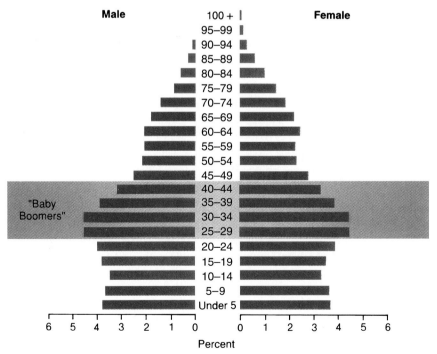

Figure 7.5

Age Distribution of U.S. Population (1990) This graph shows the number of people at each age level. Notice that at age 40 to 44, a bulge begins that ends at about age 25 to 29. These people represent the "baby boom" that followed World War II. In recent years, births have increased.

Source: Data from the U.S. Department of Commerce, Bureau of the Census.

growing by about 1.1 percent per year. About 0.6 percent is the result of natural increases owing to the difference between birthrates and death rates. The remainder is the result of immigration into the United States. The U.S. Census Bureau projects that immigration will increase significantly and account for 50 percent of population growth by the year 2050.

Current immigration policy in the United States is difficult to characterize. Strong measures are being taken to reduce illegal immigration across the southern border. This is in part due to pressures placed on Congress by states that receive large numbers of illegal immigrants. Illegal immigrants add to the education and health care that states must fund. At the same time, some segments of the U.S. economy maintain that they are unable to find workers to do certain kinds of work. Consequently, special guest workers are allowed to enter the country for limited periods to serve the needs of these segments of the economy. There is also a consistent policy of allowing immigration that reunites families of U.S. residents. Obviously, the families that fall into this category are likely to be U.S. citizens who were recent immigrants themselves. It is obvious that most immigration policy is the result of political decisions rather than decisions that relate to population policy or a concern about the rate at which the U.S. population is growing.

Projections based on the 1990 census indicate that the population will grow more rapidly than previously projected, and it does not seem to be moving toward zero growth. Indeed, it appears that the total fertility rate, which is currently about 2 children per woman per lifetime, will increase to 2.12 by the year 2050. This would result in an increase from 248.7 million people in 1990 to about 275 million by 2000 and a further increase to about 383 million by 2050. (See figure 7.6.)

These increases are partly the result of higher fertility rates among some minority populations. The Hispanic and Asian American portions of the population will grow rapidly, and the Caucasian portion will decline. Thus, the future U.S. population will be much more ethnically diverse.

Hunger, Food Production, and Environmental Degradation

As the human population increases, there is an increased demand for food. People must either grow food themselves or purchase it. Most people in the developed world purchase what they need and have more than enough food to eat. Most people in the less-developed world must grow their own food and have very little money to purchase additional food. Typically, these farmers have very little surplus. If crops fail, people starve. Even in countries with the highest population (China and India), the majority of the people live on the land and farm.

The human population can increase only if the population of other kinds of plants and animals decrease. Each ecosystem has a maximum biomass that can ex-

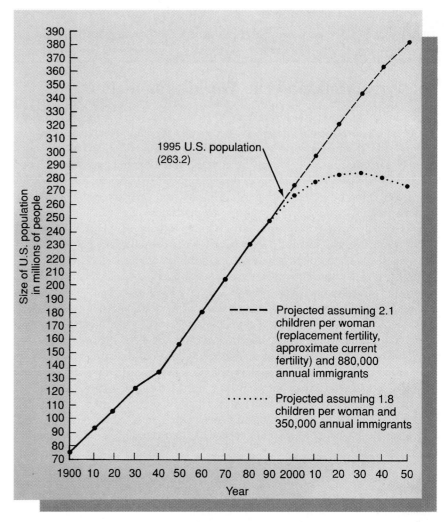

Figure 7.6

Population Growth   The population of the United States has grown continuously until the present. The graph indicates that the size of the population was about 230 million people in 1980. The U.S. Census Bureau has made projections based on birthrate and presumed immigration. The ultimate size of the U.S. population differs considerably, depending on the estimates used.

Source: Data from the U.S. Department of Commerce, Bureau of the Census.

ist within it. There can be shifts within ecosystems to allow an increase in the population of one species, but this always adversely affects certain other populations because they are competing for the same basic resources.

When humans need food, they convert natural ecosystems to artifically maintained agricultural ecosystems. The natural mix of plants and animals is destroyed and replaced with species useful to humans. If these agricultural ecosystems are mismanaged, the region's total productivity may fall below that of the original ecosystem. The dust bowl of North Amer-

ica, desertification in Africa, and destruction of tropical rainforests are well-known examples. In countries where food is in short supply and the population is growing, there is intense pressure to convert remaining natural ecosystems to agriculture. Typically, these areas are the least desirable for agriculture and will not be productive. However, to a starving population, the short-term gain is all that matters. The long-term health of the environment is sacrificed for the immediate needs of the population.

A consequence of the basic need for food is that people in less-developed coun-

PART TWO   Ecological Principles and Their Application

GLOBAL PERSPECTIVE

Population and Poverty: A Vicious Cycle

It is clear that the areas of the world where the human population is growing most rapidly are those that have the lowest standard of living. The developed countries of Europe and North America have abundant wealth and have relatively slowly growing populations. Countries in the southern hemisphere are generally poor and have high population growth rates. What forces are involved? Although not all cases are the same, the following factors seem to be interrelated.

	Countries with Low Population Growth Rates	Countries with High Population Growth Rates
Wealth	Rich	Poor
Educational Level	Nearly all people can read and write	Many people cannot read or write
Status of Women	Educated	Uneducated
	Financially independent	Financially dependent on husband or family unit
Status of Children	Laws prevent children from working	Many children required to work

Let us look at why some of these factors might be interrelated.

1. Poor people cannot afford birth control and, since they are often poorly educated, may not be able to read the directions on how to use various birth control mechanisms. Therefore, they have more children than they may wish to have.
2. Poor people need to obtain income in many ways. Often this includes taking children out of school so that they are able to work on the farm or in other jobs to provide income for the family. Poorly educated people cannot get high-paying jobs and so remain in poverty.
3. Women in poor countries are usually poorly educated and do not have disposable income. Therefore, they are dependent on their husbands or the family unit for their livelihood. Women who do not have independent income are more likely to have children they do not want because they cannot afford birth control.

At the United Nations International Conference on Population and Development held in Cairo, Egypt, in September 1994, attention was focused on several issues.

1. Improving the educational status of women was promoted. This would lead to improved financial standing for women, which could allow them to have fewer children.
2. Access to birth control was advanced as a goal, but was attacked by fundamentalist Muslims and Catholics. The final document was approved by both groups.
3. There was recognition that economic well-being is tied to solving the population problem. However, the huge, rapidly growing population of the poor countries of the world cannot hope to consume at the rate the rich countries do. Furthermore, the rich countries of the world need to reduce their rate of consumption.

---

tries generally feed at lower trophic levels than do those in the developed world. (See figure 7.7.) Converting the less concentrated carbohydrates of plants into more nutritionally valuable animal protein and fat is an expensive process. During the process of feeding plants to animals and harvesting animal products, approximately 90 percent of the energy in the original plants is lost. Therefore, in terms of economics and energy, people in less-developed countries must consume the plants themselves rather than feed the plants to animals and then consume the animals. In

most cases, if the plants were fed to animals, many people would starve to death. On the other hand, a lack of protein in diets that consist primarily of plants can lead to malnutrition. Many people in the less-developed world suffer from a lack of adequate protein, which stunts their physical and mental development.

In contrast, in most of the developed world, meat and other animal protein sources are important parts of the diet. Many people suffer from overnutrition (they eat too much); they are "malnourished" in a different sense. The ecological

impact of one person eating at the carnivore level is about ten times that of a person eating at the herbivore level. If people in the developed world were to reduce their animal protein intake, they would significantly reduce their demands on world resources. Almost all of the corn and soybeans grown in the United States is used as animal feed. If these grains were used to feed people rather than animals, less grains would have to be grown and the impact on farmland would be less.

In countries where food is in short supply, agricultural land is already being

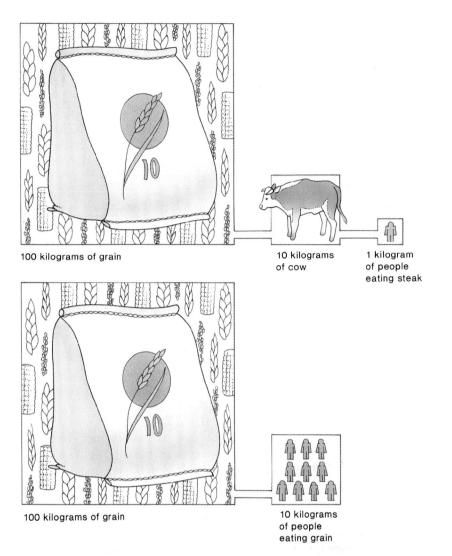

100 kilograms of grain

10 kilograms of cow

1 kilogram of people eating steak

100 kilograms of grain

10 kilograms of people eating grain

Figure 7.7

Populations and Trophic Levels   The larger a population, the more energy it takes to sustain the population. Every time one organism is eaten by another organism, approximately 90 percent of the energy is lost. Therefore, when populations are very large, they usually feed at the herbivore trophic level because they cannot afford the 90 percent energy loss that occurs when plants are fed to animals. The same amount of grain can support ten times more people at the herbivore level than at the carnivore level.

exploited to its limit, and there is still a need for more food. This makes the United States, Canada, Australia, Argentina, New Zealand, and the European Economic Community net food exporters. Many countries, like India and China, are able to grow enough food for their people but do not have any left for export. Others, including many nations of the former Soviet Union, are not able to grow enough to meet their own needs and, therefore, must import food.

A country that is a net food importer is not necessarily destitute. Japan and some European countries are net food importers but have enough economic assets to purchase what they need. Hunger occurs when countries do not produce enough food to feed their people *and* cannot obtain food through purchase or humanitarian aid.

The current situation with respect to world food production and hunger is very complicated. It involves the resources needed to produce food, such as arable land, labor, and machines; appropriate crop selection; and economic incentives. It also involves the maldistribution of food within countries. This is often an economic problem, since the poorest in most countries have difficulty finding the basic necessities of life, while the rich have an excess of food and other resources. In addition, political activities often determine food availability. War, payment of foreign debt, and poor management often contribute to hunger and malnutrition.

Improved plant varieties, irrigation, and improved agricultural methods have dramatically increased food production in some parts of the world. In recent years, India, China, and much of southern Asia have moved from being food importers to being self-sufficient, and in some cases, food exporters.

The areas of greatest need are in sub-Saharan Africa. Africa is the only major region of the world where per capita grain production has decreased over the past few decades. People in these regions are trying to use marginal lands for food production, as forests, scrubland, and grasslands are converted to agriculture. Often, this land is not able to support continued agricultural production. This leads to erosion and desertification.

What should be done about countries that are unable to raise enough food for their people and are unable to buy the food they need? This is not an easy question. A simple humanitarian solution to the problem is for the developed countries to supply food. Many religious and humanitarian organizations do an excellent service by taking food to those who need it, and save many lives. However, the aim should always be to provide temporary help and insist that the people of the country develop mechanisms for solving their own problem. Often, emergency food programs result in large numbers of people migrating from their rural (agricultural) areas to cities, where they are unable to support themselves. They become dependent on the food aid and stop working to raise their own food, not because they do not want to work, but because they need to leave their fields to go to the food distribution centers. Many humanitarian organizations now recognize the futility of trying to feed people with gifts from the developed world. The emphasis must be on self-sufficiency.

PART TWO   Ecological Principles and Their Application

Canadian Population Overview

Although the total Canadian population is slightly more than 11 percent of the U.S. population (30 million versus 265.2 million), the structure of the Canadian population is very similar to that of the United States. The chart shows several comparisons between the two populations for 1996.

	Population size 1996	Birthrate	Death rate	Rate of natural increase	Doubling time	Total fertility rate	Life expectancy	Birth control use	Per capita GNP 1994 US$	% under 15 years old	% over 65 years old
Canada	30 million	13 per 1,000	7 per 1,000	0.6%	116 years	1.6 children/ woman	78 years	73% of married women	19,570	21	12
United States	265.2 million	15 per 1,000	9 per 1,000	0.6%	114 years	2.0 children/ woman	76 years	71% of married women	25,860	22	13

These data are typical for populations in developed nations where birthrates and death rates are low and a relatively large proportion of the population is made up of older members. The relatively small, young proportion of the population, coupled with low total fertility rates, leads to relatively slow population growth. In both countries, a significant amount of the population growth is attributable to immigration. The high standard of living in both countries attracts people who desire a better life.

issues and analysis

Population Growth in Mexico

Mexico currently has a population growth rate of 2.2 percent per year, which will result in a doubling of its population in 32 years. This is an improvement over 1970, when the population was growing at a rate of 3.5 percent. The Mexican government's goal is to reduce the population growth rate to 1 percent. The program includes family planning, sex education, improving the status of women, and educational reform.

The Mexican population is very young; 36 percent are under 15 years of age. As these children mature, they will have children and will need to find jobs. There may not be enough jobs to go around, since current unemployment is high. The government is seeking to create labor-intensive industries to absorb this growing labor force.

The United States has made it more difficult for Mexican nationals to find work north of the border. The U.S. government has set penalties for employers who use illegal immigrant labor and has reduced the number of documented Mexicans who can legally work in the United States. This could have a major impact on the Mexican economy, since 21 percent of Mexicans rely on income from the United States for their support.

Many U.S. and European businesses have built assembly plants in Mexico to make use of the abundant inexpensive labor. The North American Free Trade Agreement (NAFTA) will allow more fluid exchange of goods and services between Canada, the United States, and Mexico. It is unclear what the eventual outcome of free trade will be, but labor leaders in Canada and the United States are concerned about the effect access to low-cost labor will have on their membership:

- Should the U.S. government be involved in setting Mexico's population policy?
- Why can't the Mexican population be limited as easily as European or North American populations have been?
- Should the United States enact policies that make the Mexican population problem worse?

Anticipated Changes with Continued Population Growth

As the world human population continues to increase, pressure for the necessities of life will become greater. Differences in the standard of living between developed and less-developed countries will remain significant because population will increase most in less-developed countries. The supply of fuel and other resources is dwindling. Pressure for these resources will intensify as industrialized countries seek to maintain their current standard of living. People in less-developed countries will seek more land to raise crops and feed themselves unless major increases in food production per hectare occur. Since most of these people live in tropical areas, tropical forests will be cleared for farmland. The resulting erosion or alteration of the soil will make it no longer suitable for forest or crops. This conversion of natural ecosystems to agricultural ecosystems could cause profound changes in the world ecosystem.

Developed countries may have to choose between two alternatives: helping the less-developed countries, thus maintaining their friendship, or isolating themselves from the problems of the less-developed nations. Neither of these policies will be able to prevent a change in lifestyle as world population increases. The resources of the world are finite. Even if industrialized countries continue to use a disproportionate share of the world's resources, the amount available per person will decline as population rises. As world population increases, the less-developed areas will probably maintain their low standard of living. It is difficult to see how that standard of living could get much lower, since some people in these countries are already starving to death. Lifestyles in developed nations will probably become less consumption-oriented. What some people currently view as necessities (meals in restaurants, vacations to remote sites, and two cars per family) will probably become luxuries. Many people who enjoy the freedom of mobility associated with the automobile will have their travel limited as public transportation replaces private transportation. Recreation may cease to involve expensive, energy-demanding machines (powerboats, motorcycles, and electric-powered toys) and emphasize such activities as hiking, bicycling, and reading instead. These changes will not come quickly, unless some catastrophic political or economic force causes major worldwide adjustments. Most likely, changes will occur only as economic pressures affect families. Many economists and political thinkers feel that as the economies of the world become more linked, and jobs flow more freely from country to country, wealth will be redistributed from the former rich countries to emerging economies in developing countries. The possibility of such a redistribution has become a major political issue in many of the developed countries of the world.

SUMMARY

Many of the problems of the world are caused by or made worse by an increasing human population. Currently, the world's population is growing very rapidly. Most of the growth is occurring in the less-developed nations of the world (Africa, Asia, and Latin America) where people have a low standard of living. The developed regions of the world, with their high standard of living, have relatively slow population growth and in some instances have declining populations.

Demography is the study of human populations and the things that affect them. Demographers study the sex ratio and age distribution within a population to predict future growth. Population growth rates are determined by biological factors such as birthrate, which is determined by the number of women in the population and the age of the women, and death rate. Sociological and economic conditions are also important, since they affect the number of children desired by women, which helps set the population growth rate. In developed countries, women usually have access to jobs. Couples marry later, and they make decisions about the number of children they will have based on the economic cost of raising children. In the less-developed world, women marry earlier, and children have economic value as additional workers, future caregivers for the parents, and as status for either or both parents.

The current U.S. population will continue to grow. The average age of the population will increase and the racial mix of the population will change significantly.

The demographic transition model suggests that, as a country becomes industrialized, its population begins to stabilize, but there is little hope that the earth can support the entire world in the style of the industrialized nations. It is doubtful whether there are enough energy and other natural resources to develop the less-developed countries or enough time to change the trends of population growth. Highly developed nations should anticipate increased pressure in the future to share their wealth with less-developed countries.

REVIEW QUESTIONS

1. What is demography?
2. What is demographic transition? What is it based upon?
3. What is a baby boom?
4. What does age distribution of a population mean?
5. List 10 differences between your standard of living and that of someone in a less-developed country?
6. Why do people who live in overpopulated countries use plants as their main source of food?
7. Although predicting the future is difficult, what do you think your life will be like in 10 years? Why?
8. List five changes you might anticipate if world population were to double in the next 50 years.
9. Which three areas of the world have the highest population growth rate? Which three areas of the world have the lowest standard of living?
10. How many children per woman would lead to a stable U.S. population?
11. What role does the status of women play in determining population growth rates?
12. Describe three reasons why women in the less-developed world might desire more than two children.

KEY TERMS

demographic transition 117	population density 111	replacement fertility 113	total fertility rate 113
demography 112	postwar baby boom 118	standard of living 110	zero population growth 113
gross national product (GNP) 110			

RELATED INTERNET SITES

Comprehensive site on population programs and activities in Africa.
   http://www.africa2000.com/
Human dimensions of global environmental change.
   http://www.ciesin.org/
Searchable database of the quarterly bibliography "Population" index.
   http://opr.princeton.edu/popindex/
Zero Population Growth, a grassroots organization concerned with the impact of rapid population growth.
   http://www.zpg.org/zpg/

Center for Demography and Ecology.
   ***http://www.ssc.wisc.edu/cde
World Population Clock, an award-winning site with links to many great population sites.
   ***http://sunsite.unc.edu/lunar/pop.html
Carolina Population Center
   ***http://www.cpc.unc.edu
Population policy briefing on population trends in the United States.
   ***http://www.ucsusa.org/global/popbackpol.html

PART three

All living systems can be described by the flow of energy through them: Energy enables simple forms of matter to be changed into more complex forms and energy is needed to maintain this complexity. Energy sustains the complex technical and social units typical of human populations.

All living things (including humans) rely on the sun as a source of energy. Coal, oil, and natural gas are energy sources available today because organisms in the past captured sunlight energy and stored it in the complex organic molecules that made up their bodies, which were then compressed and concentrated.

Technological development and fossil-fuel exploitations are directly related to one another and have allowed us an increasingly higher standard of living. Chapter 8 traces the development of energy consumption, its interrelationships with economic development and lifestyles, and current uses and demands. Chapter 9 discusses the sources of energy currently being used, as well as the significance of each source and its impact. Chapter 10 discusses nuclear energy, other applications of radioactive substances, and the concerns their use generates.

chapter 8

Energy and Civilization: Patterns of Consumption

Objectives

After reading this chapter, you should be able to:

- Explain why all organisms require a constant input of energy.
- Describe how per capita energy consumption increased as civilization developed from hunting and gathering to primitive agriculture to advanced cultures.
- Describe how advanced modern civilizations developed as new fuels were used to run machines.
- Recognize that coal deposits are not uniformly distributed throughout the world.
- Correlate the Industrial Revolution with social and economic changes.
- Explain how cheap oil and natural gas led to a consumption-oriented society.
- Explain how the automobile changed people's lifestyles.
- Explain why overall energy use in the United States declined during the 1970s and 1980s.

Chapter Outline

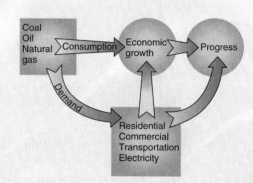

History of Energy Consumption

Every form of life and all societies require a constant input of energy. If the flow of energy through organisms or societies ceases, they stop functioning and begin to disintegrate. Some organisms and societies are more energy efficient than others. In general, history shows that complex industrial societies use the most energy. If societies are to survive, they must continue to expend energy. However, they may need to change their pattern of energy consumption as traditional sources become limited.

Biological Energy Sources

Energy is essential to maintain life. In every ecosystem, the sun provides that energy. (See chapter 4.) The first transfer of energy occurs during photosynthesis, when plants convert light energy into chemical energy in the production of food. Herbivorous animals utilize the food energy in the plants. The herbivores, in turn, are a source of energy for carnivores. Because nearly all of their energy requirements were supplied by food, primitive humans were no different from other animals in their ecosystems. In such hunter-gatherer cultures, nearly all human energy needs were met by using plants and animals as food, tools, and fuel. (See figure 8.1.)

Early in human history, people began to use additional sources of energy to make their lives more comfortable. They domesticated plants and animals to provide a more dependable supply of food. They no longer needed to depend solely upon gathering wild plants and hunting wild animals for sustenance. Domesticated animals also furnished a source of energy for transportation, farming, and other tasks. (See figure 8.2.) Wood provided a source of fuel for heating and cooking. Eventually this biomass energy was used in simple technologies, such as shaping tools and extracting metals.

Increased Use of Wood

Early civilizations, such as the Aztecs, Greeks, Egyptians, Romans, and Chinese,

Figure 8.1

Hunter-Gatherer Society   In this type of society, people obtain nearly all of their energy from the collection of wild plants and the hunting of animals. These societies do not make large demands on fossil fuels.

Figure 8.2

Animal Power   This bas-relief panel from an Egyptian tomb depicts an important accomplishment in the development of human civilization. With the use of domesticated animals, people had a source of power other than their own muscles.

were culturally advanced, but their societies used human muscle, animal muscle, and fire as sources of energy. Except for limited use of some wind-powered and water-powered devices such as ships and canoes, the controlled use of fire was the first use of energy in a form other than food. Wood was the primary fuel. (Wood was also used for building materials and other cultural uses.) The energy provided by wood enabled people to cook their food, heat their dwellings, and develop a primitive form of metallurgy. Such advances separated humans from other animals. When dense populations of humans made heavy use of wood for fuel and building materials, they eventually used up the readily available sources and had to import wood or seek alternative forms of fuel.

Because of a long history of high population density, India and some other parts of the world experienced a wood shortage hundreds of years before Europe and North America did. In many of these areas, animal dung replaced wood as a fuel source. It is still used today in some parts of the world.

Western Europe and North America were able to use wood as a fuel for a longer period of time. The forests of Europe supplied sufficient fuel until the thirteenth century. In North America, vast expanses of virgin forests supplied adequate fuel until the late nineteenth century. Fortunately, when local supplies of wood declined in Europe and North America, coal, formed from fossilized plant remains, was available as an alternative energy source. (See figure 8.3.)

Fossil Fuels and the Industrial Revolution

Fossil fuels are the remains of plants, animals, and microorganisms that lived millions of years ago. (The energy in these fuels is stored sunlight, just as the biomass of wood represents stored sunlight.) During the Carboniferous period, 275 to 350 million years ago, conditions in the world were conducive to the formation of large deposits of fossil fuels. (See figure 8.4.) Ever since machines replaced muscle power, the major energy sources for the world have been fossil remains from the distant past.

Historically, the first fossil fuel to be used extensively was coal. In the early eighteenth century, regions of the world that had readily available coal deposits were able to switch to this new fuel and participate in a major cultural change known as the **Industrial Revolution.** The Industrial Revolution began in England and spread to much of Europe and North America. It involved the invention of machines that replaced human and animal labor in manu-

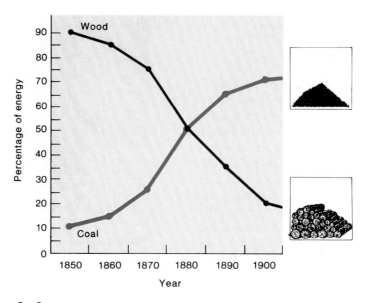

Figure 8.3

Coal Replaces Wood   In 1850, wood furnished 90 percent of U.S. energy, and coal most of the remaining 10 percent. Fifty years later, wood supplied only 20 percent of the energy and coal supplied 70 percent. The remainder was furnished by oil and natural gas.

Source: Data from U.S. Bureau of Mines.

Figure 8.4

Carboniferous Period   Approximately 300 million years ago, this kind of ecosystem was common throughout the world. Plant material accumulated in these swamps and was ultimately converted to coal.

facturing and transporting goods. Central to this change was the invention of the steam engine, which could convert heat energy into the energy of motion. The source of energy for steam engines was either wood or coal; wood was quickly replaced by coal in most cases. Nations without a source of coal, or those possessing coal reserves that were not easily exploited, did not participate in the Industrial Revolution.

Prior to the Industrial Revolution, Europe and North America were predominately rural. Goods were manufactured on a small scale in the home. As machines and the coal to power them became increasingly available, the factory system of manufacturing products replaced the small home-based operation. Because expanding factories required a constantly increasing labor supply, people left the farms and congregated in areas surrounding the factories. Villages became towns, and towns became cities. Widespread use of coal in cities resulted in increased air pollution. In spite of these changes, the Industrial Revolution was viewed as progress. Energy consumption increased, economies grew, and people prospered. Within a span of two hundred years, the daily per capita energy consumption of industrialized nations increased eightfold. This energy was furnished primarily by coal, but a new source of energy was about to be discovered: oil.

The Chinese used some gas and oil as early as 1000 B.C., yet these resources remained virtually untapped until fairly recently. The oil well that Edwin L. Drake, an early oil prospector, drilled in Pennsylvania in 1859 was not the world's first oil well, but it was the beginning of the modern petroleum era. By 1870, oil production in the United States had reached over four million barrels a year and supplied 1 percent of the nation's energy requirements. It grew to nearly 50 percent by 1970 and currently contributes just over 40 percent. (See figure 8.5.)

For the first 60 years of production, the principal use of oil was to make kerosene, a fuel for lamps. The gasoline produced was discarded as a waste product. During this time, the supply of oil exceeded the demand. However, the automobile dramatically increased the demand. In 1900, the United States had

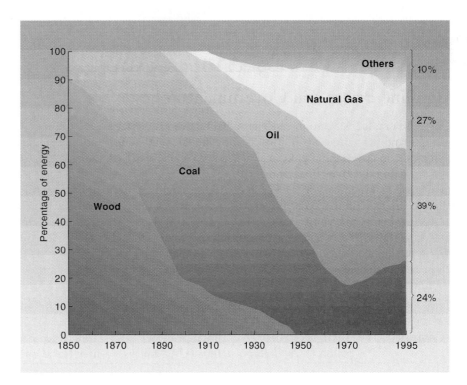

Figure 8.5

Oil Replaces Coal   Just as wood was replaced by coal, coal was later replaced by oil. This graph represents the production of energy by various sources in the United States. Oil has remained a dominant energy source for the past 40 years, coal has decreased slightly, natural gas has increased, and other sources such as nuclear, hydroelectric, wind, and solar power have been increasing.

Source: Data from Annual Energy Review, 1986, Energy Information Administration and more recent data from BP Statistical Review of World Energy 1996 and previous editions.

only 8,000 automobiles. By 1920, it had 8 million cars, and by 1992, 150 million. More oil was needed to make automobile fuel and lubricants.

The use of natural gas did not increase as rapidly as the use of oil. It primarily was used for home heating. In the early 1900s, 90 percent of the natural gas was "flared," that is, burned as a waste product at oil wells.

A series of events involving the U.S. government and the need for more oil was ultimately responsible for increased use of natural gas in the United States. World War II greatly increased the energy demand for manufacturing and transportation. In 1943, a federally financed pipeline was constructed to transport oil within the United States. This 2,000-kilometer pipeline transported oil more efficiently from wells in Texas, Louisiana, and Oklahoma to refineries and factories in the eastern section of the country. In 1944, a longer (2,400 kilometers) federally financed pipeline was

built to increase the flow of oil to the country's eastern and midwestern regions.

After the war, the federal government sold these pipelines to private corporations. The corporations converted the pipelines to transport natural gas. Thus, a direct link was established between the natural gas fields in the Southwest and the markets in the Midwest and East. By 1971, there were 400,000 kilometers of long-range transmission to pipelines and 986,000 kilometers of distribution pipelines in the United States. Approximately 1,613,000 kilometers of natural gas distribution pipelines are used today. Natural gas is used in many parts of the developed world both for home heating and for industrial purposes.

Energy and Economics

Most industrial societies want to ensure a continuous supply of inexpensive energy.

Gasoline Prices and Government Policy

The price of a liter of gasoline is determined by two major factors: (1) the cost of purchasing and processing crude oil into gasoline and (2) various taxes. Most of the differences in gasoline prices among countries are a result of taxes and reflect differences in government policy toward motor vehicle transportation.

A major objective of governments is to collect money to build and repair roads. Governments often charge road users by taxing the fuel their cars or trucks run on. Governments can also discourage the use of automobiles by increasing the cost of fuel. An increase in fuel costs also creates a demand for increased fuel efficiency in all forms of motor transport.

Many European countries raise more money from fuel taxes than they spend on building and repairing roads. The United States, on the other hand, raises approximately 60 percent of the monies needed for roads from fuel taxes. The relatively low cost of fuel in the United States encourages more travel, which increases road repair costs. The cost of taxes to the U.S. consumer is about 28 percent of the retail gasoline price, while in Japan and many European countries, the cost is 60 to 75 percent. (See table.)

World Gasoline Prices and Taxation

Country	Total price per gallon (U.S. dollars)	Total price per liter (U.S. dollars)	Total price per liter (local currency)	Tax per liter (local currency)	Tax % of price
France	$3.85	$1.02	5.3 francs	4.1 francs	77%
Italy	$5.00	$1.32	1,535 liras	1,156 liras	75%
Germany	$3.85	$1.02	1.57 marks	1.1 marks	71%
Spain	$3.75	$0.99	96 pesetas	66.5 pesetas	69%
Britain	$3.69	$0.97	0.51 pound	0.35 pound	69%
Japan	$3.75	$0.99	124 yen	58 yen	47%
United States	$1.19	$0.31	$0.31	$0.084	28%

The higher the price of energy the more expensive goods and services become. To keep costs down, many countries have subsidized their energy industries and maintained energy prices artificially low. International trade in fossil fuels has a major influence on the world economy and politics. The emphasis on low-priced fuels has encouraged high rates of consumption.

Economic Growth and Energy Consumption

There is a direct link between economic growth and the availability of inexpensive energy. The replacement of human and animal energy with fossil fuels began with the Industrial Revolution and was greatly accelerated by the supply of cheap, easy-to-handle, and highly efficient fuels. Because the use of inexpensive fossil fuels allows each worker to produce more goods and services, productivity increased. The result

was unprecedented economic growth in Europe, North America, and the rest of the industrialized world.

In North America, World War II was a prime factor in ending the economic depression of the 1930s. Military activities created millions of defense jobs. Almost everyone was employed, but there was a scarcity of consumer goods. After World War II, consumer goods that had been unavailable during the war were in great demand. Industries set up to produce military goods turned to the production of consumer goods. High employment, a rapidly expanding population, and a supply of inexpensive energy encouraged a period of rapid economic growth.

In Europe and Japan, where much of the industrial base was destroyed by the war, the recovery was slow. However, foreign aid and an intense rebuilding effort on the part of the people resulted in a reindustrialization of these countries. Be-

cause they had to build new facilities, often their technology was better than that of North America where few factories were affected by the war.

The Role of the Automobile

The cheap, abundant energy that fueled industries produced an ever-increasing amount and array of consumer goods. One product was the automobile. The growth of the automobile industry, first in the United States and then in other industrialized countries, led to roadway construction, which required energy. Thus, the energy costs of driving a car were greater than just the fuel consumed in travel. As roads improved, higher speeds were possible. People demanded faster cars, and automobile companies were quick to build them. Bigger and faster cars required more fuel and even better roads. So roads were continually being improved, and better cars were being produced. A cycle of *more chasing more* had begun. In North America and much of Europe, the convenience of the automobile encouraged two-car families, which created a demand for more energy. It requires energy to mine ore, process it into metals, form the metals into automobile components, and transport all the materials. As the economy grew, so did energy needs.

More cars meant more jobs in the automobile industry, the steel industry, the glass industry, and hundreds of other industries. Constructing thousands of kilometers of roads created additional jobs. From their beginnings as suppliers of lamp oil, the oil companies grew into one of the largest industries in the world. Thus, the automobile industry played a major role in the economic development of the industrialized world. All this wealth gave people more money for cars and other necessities of life. The car, originally a luxury, was now considered a necessity.

The car not only created new jobs but also altered people's lifestyles. Vaca-

Figure 8.6

Energy-Demanding Lifestyle Building private homes on large individual lots some distance from shopping areas and places of employment is directly related to the heavy use of the automobile as a mode of transportation. Heating and cooling a large enclosed shopping mall, along with the gasoline consumed in driving to the shopping center, increase the demand for energy.

tioners could travel greater distances. New resorts and chains of motels, restaurants, and other businesses developed to serve the motoring public, creating thousands of new jobs. Because people could live farther from work, they began to move to the suburbs. (See figure 8.6.)

Large shopping centers in suburban areas hastened the decline of central business districts. Today, fewer than 50 percent of retail sales are made in central business districts of North American cities, which has resulted in a loss of jobs in these areas. In Philadelphia, 79 percent of all retail jobs were located in the city in 1930; by 1970, they had declined to 43 percent.

As people moved to the suburbs, they also changed their buying habits. Labor-saving, energy-consuming devices became essential in the home. The vacuum cleaner, dishwasher, garbage disposal, and automatic garage door opener are only a few of the ways human power has been replaced with electrical power. Eleven percent of the electrical energy in North America is used to operate home appliances. Other aspects of our lifestyles illustrate our energy dependence. The small, horse-powered farm of yesterday has grown into the huge, diesel-powered farm of today. Regardless of where we live, we expect Central American bananas, Florida oranges, California lettuce, Texas beef, Hawaiian pineapples, Ontario fruit, and Nova Scotia lobsters to be readily available at all seasons. What we often fail to consider is the amount of energy required to process, refrigerate, and transport these items. The car, the modern home, the farm, and the variety of items on our grocery shelves are only a few indications of how our lifestyles are based on cheap, abundant energy.

How Energy is Used

The amount of energy consumed by countries of the world varies widely. (See table 8.1) The highly industrialized countries use most of the world's energy; less-developed countries use much less. Even countries with the same level of development vary in the amount of energy they use as well as how they use it. To maintain their style of living, individuals in Canada and the United States use about twice as much energy as people in France or Japan and about 25 times as much energy as people in Africa.

Countries also use energy in different ways. Industrialized nations use energy about equally for three purposes: (1) residential and commercial uses, (2) industrial uses, and (3) transportation. Less-developed nations with little industry use most of their energy for residential purposes (cooking and heating). Developing countries use much of their energy to develop their industrial base.

Table 8.1

Energy Consumption 1995	
Region	Energy consumption per capita per year (in tonnes of oil equivalent)
Africa	0.33
Latin America	0.67
Japan	3.72
France	4.05
Germany	4.11
Canada	7.61
United States	7.86

Sources: Data from BP Statistical Review of World Energy, June 1996 and Population Reference Bureau 1995 World Population Data Sheet.

Figure 8.7

Open Fire Cooking   About half of the energy demand in Africa is for cooking. Using a stove instead of an open fire could reduce this energy need by nearly 50 percent.

Residential and Commercial Energy Use

The amount of energy required for residential and commercial use varies greatly throughout the world. Although a country with a high gross domestic product (GDP) uses a large amount of energy, it uses a lower percentage of its energy per capita for residential and commercial needs than does a less-developed country. For example, about 30 percent of the energy used in North America is for residential and commercial purposes, while

in India, 90 percent of the energy is for residential uses. The ways residential and commercial energy is used also vary widely. In North America, 75 percent is used for air conditioning, refrigeration, water heating, and space heating. In India, almost all of the energy is used in the home for cooking since the scarcity and high cost of fuel precludes other uses.

The current pattern of residential and commercial energy use in each region of the world determines what conservation methods will be effective. In Canada, which has a cold climate, 40 percent of the residential energy is used for heating. Proper conservation practices could reduce this by 50 percent. In Africa, almost half of the energy used in the home is for cooking. (See figure 8.7.) Using fuel-efficient stoves instead of open fires could reduce these energy requirements by 50 percent.

Industrial Energy Use

The amount of energy countries use for industrial processes varies considerably. Nonindustrial countries use little energy for industry. Countries that are developing new industries dedicate a high percentage of their energy use to them. They divert energy to the developing industries at the expense of other sectors of their economy. Highly industrialized countries use a significant amount of their energy in industry, but their energy use is high in other sectors as well. In the United States, industry claims about 30 percent of the energy used.

The amount of energy required in a country's industrial sector depends on the types of industrial processes used. Many countries use inefficient processes and could reduce their energy consumption by converting to more energy-efficient ones. However, they need capital investment to upgrade their industries and reduce energy consumption. Some countries cannot afford the upgrade. For example, India, a nation with few coal deposits, still uses outdated open-hearth furnaces to produce steel. These furnaces require nearly double the worldwide energy average to produce a metric ton of steel. The high cost of converting forces India to continue to use this energy-expensive method rather than convert to the more efficient electric-hearth method.

Transporation Energy Use

As with residential, commercial, and industrial uses, the amount of energy used for transportation varies widely throughout the world. In some of the less-developed nations, transportation uses are very small. Per capita energy use for transportation is larger in developing countries, and highest in highly developed countries. (See table 8.2.)

Once a country's state of development has been taken into account, the mix of

Table 8.2

Per Capita Energy Use for Transportation, 1989	
Country	Energy use in gigajoules/capita
India	2
Zimbabwe	4
Mexico	17
Argentina	18
Japan	28
USSR	29
Netherlands	39
Denmark	42
Australia	82
United States	103

Figure 8.8

Public Transportation In regions of the world where energy is expensive, people make maximum use of cheaper public transportation.

bus, rail, water, and private automobiles is the main factor in determining a country's energy use for transportation. In Europe, Latin America, and many other parts of the world, rail and bus transport are widely used because they are more efficient than private automobile travel, governments support these transportation methods, or a large part of the populace is unable to afford an automobile. In countries with high population densities, rail and bus transport is particularly efficient. (See figure 8.8.) In these countries, automobiles require about four times more energy per passenger kilometer than bus or rail transport require. In addition, most of these countries have

high taxes on fuel, which raise the cost to the consumer and encourage the use of public transport. In North America, the situation is different. Government policy has kept the cost of energy low and supported the automobile industry while removing support for bus and rail transport. Consequently, the automobile plays a dominant role, and public transport is primarily used only in metropolitan areas. Rail and bus transport are about twice as energy efficient as private automobiles. Private automobiles in North America consume over 15 percent of the world's oil production, while the rest of the automobiles in the world consume 7 percent. Air

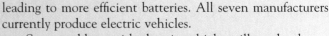
Electric Car Development

Clean air laws and regulations in countries around the world are providing a significant new push for the commercial development of electric vehicles. The Air Resources Board of the California Environmental Protection Agency voted in early 1996 to reaffirm a requirement of zero emission vehicles. Although they voted to reduce the number of electric vehicles required in the late 1990s, they reaffirmed the requirement that 10 percent of all cars and light duty trucks sold in California in 2003 and thereafter be zero emission vehicles. Electric cars are the only zero emission vehicles currently available. Based on the number of automobiles sold in California, the following companies must comply with the requirement: General Motors, Ford, Chrysler, Mazda, Toyota, Honda, and Nissan. They must provide cars with a 125-mile range between charges, and they must participate in research leading to more efficient batteries. All seven manufacturers currently produce electric vehicles.

Some problems with electric vehicles still need to be resolved. The principal technical problem is the development of lower-cost batteries with faster recharge times, longer durability, and less weight. Another problem is that disposing of the batteries, especially lead-acid batteries, is difficult. There is also concern that electric cars simply shift the pollution to a different place—the power plant.

In fact, the total amount of pollution caused by electric cars is less than that caused by gas-powered cars. Also, it is easier to control pollution from a few stationary sources than from millions of mobile sources. Consequently, electric vehicles are expected to improve air quality. In the future, drivers will need to remember to "plug in" their cars rather than "fill it up."

Power pack

Electric car

travel is relatively expensive in terms of energy, although it is slightly more efficient than private automobiles. Passengers, however, are paying for the convenience of rapid travel over long distances.

Electrical Energy

Electrical energy is such a large proportion of energy consumed in most countries that it deserves special comment. Electricity is both a way that energy is consumed and a way that it is supplied. Almost all electric energy is produced as a result of burning fossil fuels. Thus we can look at electrical energy as a use to which fossil fuel energy is put. In the same way we use natural gas to heat homes, we can use natural gas to produce electricity. Because the transportation of electrical energy is so simple and the uses to which it can be put are so varied, electricity is a major energy source for many people of the world.

As with other forms of energy use, electrical consumption in different regions of the world varies widely. The amount of electricity used by all of the less-developed nations of the world, which have about 80 percent of the world's population, is only 66 percent of that used by the United States alone. The per capita use of electricity in North America is 25 times greater than average per capita use in the less-developed countries. In Nepal, the annual per capita use of electricity is 23 kilowatt hours, which is enough to light a 100-watt lightbulb for one week. The per capita consumption of electricity in North America is 270 times greater than in Nepal. There is also a wide variation in electrical consumption among developed countries. For example, the per capita use in Europe is about half that in North America.

The production and distribution of electricity is a major step in the economic development of a country. In developed nations, about a quarter of the electricity is used by industry. The remainder is used primarily for residential and commercial purposes. In nations that are developing their industrial

Alternative-Fuel Vehicles

Electricity is the only zero-emission option now available for vehicles, but many experts believe that other fuels will prove viable in *reduced*-emission vehicles. Here are some of the options.

COMPRESSED NATURAL GAS (CNG)

Status:

Fuels thousands of vehicles already on U.S. highways. Mass production is still a couple of years away, at the earliest.

Advantages:

- Substantial U.S. supply of natural gas.
- Easy—though expensive—to convert conventional gasoline cars to CNG.
- Cheaper: the equivalent of buying gasoline at 70 cents a gallon.

Disadvantages:

- Requires heavy, bulky fuel tanks.
- Public refueling stations expensive to build.
- Lower performance and shorter range than gasoline vehicles.

Emissions Benefits:

- Burns 80% cleaner than gasoline.

METHANOL/FLEX-FUEL

Status:

Some heavy vehicles—including 354 buses operated by Southern California's Metropolitan Transportation Authority—run on pure methanol. More than 10,000 flex-fuel vehicles—running on methanol-gasoline combinations—now on U.S. roads.

Advantages:

- Can be made from natural gas (98% of current production) and a range of renewable sources.
- Requires only modest changes in gasoline engines and fueling infrastructure.
- Higher octane than gasoline, giving 5% better performance.

Disadvantages

- Highly corrosive
- Lower energy content than gasoline, so requires bigger fuel tanks.
- Fuel cost slightly higher than gasoline.

Emissions Benefits:

- An 85% methanol mix burns 30% to 50% cleaner than pure gasoline; pure methanol vehicles potentially could be much cleaner.

HYDROGEN

Status:

Still largely experimental. Vehicles could be powered by hydrogen fuel cells or by engines that burn the gas. Practical, wide-scale use is at least a decade away.

Advantages:

- A high-energy fuel, boosting vehicle range.
- Potentially unlimited fuel supply if produced from water; currently made from natural gas.
- Produces no carbon dioxide, the gas blamed for possible global warming.

Disadvantages:

- Highly explosive, though ultimately could be less dangerous than gasoline.
- Costly and difficult to produce.
- Costly and difficult to store.

Emissions Benefits:

- If produced from water using solar energy—and then used in a fuel cell—would be a zero-emissions technology.

PROPANE

Status:

Broadly used in transportation worldwide. Environmentalists, regulators, and automakers contend that propane makes the most sense for heavy vehicles as a replacement for diesel.

Advantages:

- A proven, low-cost fuel.
- Public refueling infrastructure in place.
- Nontoxic, so storage tanks are exempt from environmental regulations.

Disadvantages:

- Supplies are limited, compared with other alternatives. At most could replace 10% of gasoline.
- Highly flammable.

Emissions Benefits:

- At least 50% cleaner than conventional gasoline.

Sources: Data from California Air Resources Board, California Energy Commission, South Coast Air Quality Management District, Ward's Communications, Ford Motor Co., General Motors Corp., Chrysler Corp., and LP Gas Coalition.

OPEC

In 1973, headlines such as "OPEC Increases Oil Prices" appeared in newspapers throughout the world. Most readers asked: "What is OPEC?" In fact, many people today still are not certain what OPEC is or how it influences their lives.

OPEC, the Organization of the Petroleum Exporting Countries, began in September 1960 when the governments of five of the world's leading oil exporting countries agreed to form a cartel. Three of the original members—Saudi Arabia, Iraq, and Kuwait—were Arab countries, while Venezuela and Iran were not. Today, twelve countries belong to OPEC. These include seven Arab states—Saudi Arabia, Kuwait, Libya, Algeria, Iraq, Qatar, and United Arab Emirates—and five non-Arab members—Iran, Indonesia, Nigeria, Gabon, and Venezuela. (Ecuador was a member but withdrew from OPEC in 1992.) (See the map.) OPEC nations control over 75 percent of the world's estimated oil reserves of 1,017 billion barrels of oil. Middle Eastern OPEC countries control over 60 percent of this total, which makes OPEC and the Middle East important world influences.

OPEC had its origin in political and social events of the 1940s. During that decade, western nations experienced a drastic increase in the demand for oil, first to meet the military needs of World War II and then to meet the increased civilian demand following the war. Domestic oil production could not keep up, so western nations needed new sources of oil.

British and American companies began to develop the vast oil reserves in the Middle East and South America. These companies entered into agreements with local governments whereby the companies supplied capital and developed the oil fields and paid a "royalty," based on production, to the governments. In essence, the oil belonged to the companies, not to the countries where the oil was found. These countries did not have the money or the technology to produce the oil, so they welcomed outside investors.

Very soon, however, many of the countries began to pressure the oil companies for more direct control. Through continuous renegotiation of their agreements, the oil-producing countries gradually gained control of their oil fields. Eventually, the oil industries were nationalized as governments assumed complete ownership and management. This nationalization was complete by the time of the Arab-Israeli War of 1973. By that time, OPEC was supplying 54.6 percent of the world's oil. To protest the war, seven Arab members of OPEC reduced their production. This resulted in a worldwide oil shortage, which caused all oil companies, in OPEC and non-OPEC countries, to increase their prices. In the United States, the price of oil skyrocketed from $3.39 per barrel to $13.93 per barrel.

Oil shortages developed throughout the world. Long lines of cars waited at gas stations. People were urged to lower their thermostats to conserve oil. Congress enacted tax breaks to encourage energy conservation and provided more money for research into other forms of energy, such as solar, wind, and biomass conversion. Auto manufacturers produced smaller, more fuel-efficient cars. With the increase in oil prices came an increase in domestic oil exploration and production.

Energy conservation and increased domestic oil production brought oil prices down to eight dollars a barrel by 1975. Then a second round of OPEC price increases in 1979–1980 increased the price to twenty dollars a barrel. As with the first round of increases in 1973–1974, there was a demand for conservation, alternate forms of energy, and increased oil production. The combination of conservation and increased domestic production again erased the oil shortages and actually led to an oil surplus in the 1980s. That led to a decline in oil prices and convinced many people that there was no oil shortage.

---

base, over half of the electricity is used by industry. For example, industries use 55 percent of the electricity used in Mexico and 70 percent of that used in South Korea.

Energy Consumption Trends

From a historical point of view, it is possible to plot changes in energy consumption. Economics, politics, public attitudes, and many other factors must be incorporated into an analysis of energy use trends.

In 1995, world energy consumption was around 22.3 million metric tons of oil equivalent per day, an increase of 17 percent since 1985. Of this total, conventional fossil fuels—oil, natural gas, and coal—accounted for about 90 percent.

Over half of world energy is consumed by the 25 countries that are members of the Organization for Economic Co-operation and Development (OECD). These countries (Australia, New Zealand, Japan, Canada, Mexico, the United States, and the countries of Europe) are the developed nations of the world. In the last decade, per capita energy consumption in OECD countries has risen moderately while economic growth has continued. There has also been a shift toward service-based economies, with energy-intensive industries moving to non-OECD countries. In contrast, in countries that are becoming more economically advanced energy consumption is increasing at a faster rate.

Oil remains the world's major source of energy, accounting for about 39 percent of primary energy demand. Coal accounts for 24 percent, and nat-

OPEC's share of the world's oil market has declined to about 40 percent. This has important economic implications because many of the OPEC nations are less developed, and during the period of high oil prices, they borrowed large sums of money from the industrial nations. Now they are unable to repay this money. If they do not pay it, many banks in the western nations could experience large financial losses.

During the 1980s, OPEC members had important differences concerning oil pricing and production rules. These differences weakened the cartel. Then the 1990 invasion of Kuwait by Iraq deeply divided many of the OPEC countries.

Owing in part to the friction caused by the Kuwait conflict and the decline in world oil prices during the early 1990s, OPEC's power has continued to slide. This situation could change rapidly, however, in the event of renewed hostilities in the Middle East or a dramatic rise in the cost of oil worldwide.

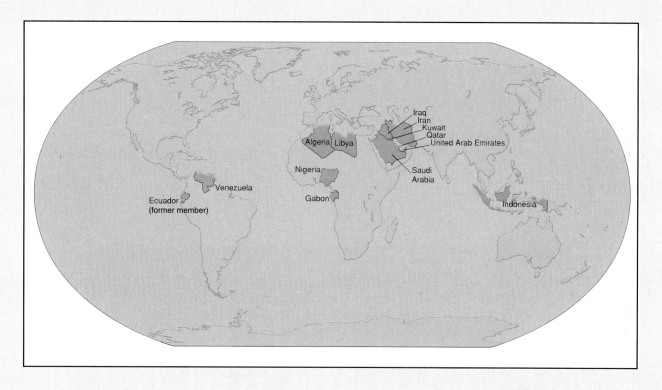

ural gas for 27 percent; the remainder is supplied mainly by nuclear energy and hydropower.

Since 1973, the year of the first "oil shock," demand for natural gas has increased faster than the demand for other fossil fuels. From 1973 to 1995, consumption of natural gas increased by about 80 percent, coal consumption increased by about 40 percent, and oil consumption by about 15 percent.

Since 1950, North American energy consumption has steadily increased. However, there were two periods of decline, one beginning in 1973 and the other in 1979. Both of these episodes were the result of political turmoil in the Middle East and the increasing influence of OPEC. (See Global Perspective: OPEC.) (See figure 8.9.) The same trend occurred in Western Europe, Japan, and Australia, resulting in a slowing of the world's growth in energy consumption. (See figure 8.10.) Several factors contributed to these declines in energy consumption. Increased prices for oil and all other forms of energy forced businesses and individuals to become more energy conscious and to expand efforts to conserve energy. From 1970 to 1983, the amount of energy used for heat per dwelling in the United States declined by 20 percent. Comparable reductions were made in Denmark, West Germany, and Sweden.

However, since 1980, energy consumption has increased in North America and Europe. Like the decreases in energy use in 1973 and 1979, the increase since 1980 is due to the price of oil. In 1979, oil was selling for about forty dollars a barrel. Beginning in 1980, the price began to drop, and in 1995, oil was

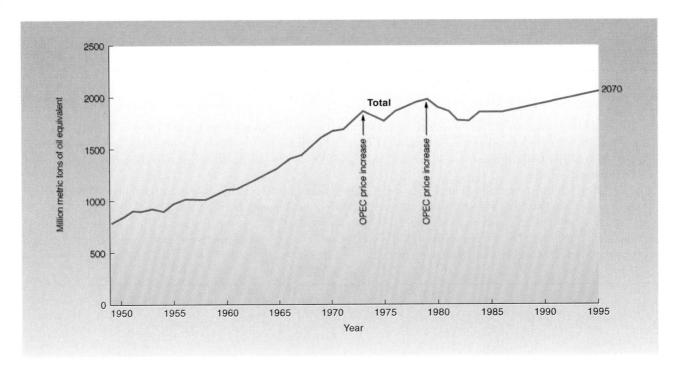

Figure 8.9

Changes in U.S. Energy Consumption   Energy consumption in the United States experienced a gradual increase until 1973 when the OPEC countries increased the price of oil. The increase in price reduced consumption. A similar action in 1979 resulted in another decrease. However, since 1983 consumption has steadily increased.

Source: Data from Annual Energy Review 1990 and BP Statistical Review of World Energy 1996.

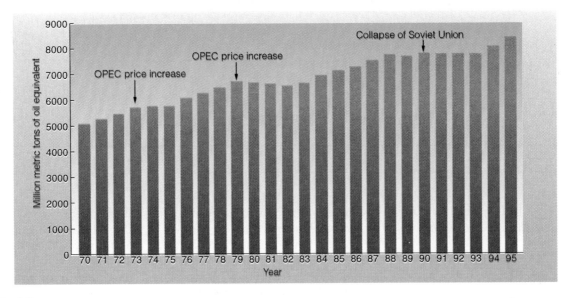

Figure 8.10

World Energy Consumption   Energy consumption has increased steadily for the past 25 years. The slowdown in growth beginning in 1973 and the decline in the early 1980s is the result of price increases instituted by OPEC. The slight decline noted in 1991 and 1992 is the result of decreased energy consumption in the former Soviet Union and Eastern Europe, and is related to economic problems in those countries.

Source: Data from BP Statistical Review of World Energy, June 1996.

Oil, War, and the Persian Gulf

In August, 1990, Iraq invaded Kuwait. While many factors led to the invasion, it was oil that attracted the world's attention. The United States rushed to the aid of Saudi Arabia to ensure that Saddam Hussein, the ruler of Iraq, would not also invade that country and further threaten the world's oil supply. Prior to the Kuwait invasion, Iraq had supplied roughly 35 million barrels of oil per month to the United States, while Saudi Arabia had supplied 42 million barrels. The United States imports about 275 million barrels monthly. Nearly one-third of U.S. oil imports were jeopardized by the war.

Oil prices throughout the world leaped after the August invasion, hitting a high of $41 per barrel in September 1990. This was more than double the preinvasion price and much higher than the $9 per barrel price in 1986. On January 17, 1991, the United States, Great Britian, and several other countries attacked Iraq from the air. As the air strikes showed initial success, the price of oil dropped quickly to the mid-$20s per barrel. The price per barrel rose and fell as the war to liberate Kuwait changed from week to week. In late February 1991, the U.S. discovered evidence that Iraq had ignited many of the oil wells in Kuwait. By 1993, oil production in Kuwait was back to prewar levels.

As long as the oil-rich Middle East is politically unstable, world prices for oil will continue to fluctuate. The degree of fluctuation will depend on how dependent the United States and other oil-consuming nations remain on Persian Gulf oil.

issues and analysis

Energy Development in China

China has approximately 20 percent of the world's population yet uses about 10 percent of the energy consumed in the world. About 80 percent of China's population live in rural areas, where most of the energy is produced from biomass, including fuelwood and animal wastes, which generate methane gas. China has significant oil reserves and huge supplies of coal, but much of its energy technology is out of date. Many industries are fueled by coal burning, which contributes to very poor air quality, particularly in the winter months when coal is also used for space heating.

China has plans to quadruple its industrial production by exploiting its vast coal reserves. Without improved air-pollution technology, however, the quality of the air will suffer. China needs industrialized countries to provide technology to develop its industry and control pollution.

- Should western industrialized nations supply technology and thereby encourage China's industrialization?
- The burning of coal increases carbon dioxide in the air and contributes to greenhouse warming. Should China be discouraged from using its coal resources?
- Do other countries have an obligation to supply advanced technology so that China will use its energy efficiently and reduce the carbon dioxide it produces?

selling for less than twenty dollars a barrel. When Iraq invaded Kuwait in August 1990, prices rose dramatically, but dropped back at the end of hostilities. (See Global Perspective: Oil, War, and the Persian Gulf.)

During the 1980s, with energy costs declining, people in North America and Europe became less concerned about their energy consumption. They used more energy to heat and cool their homes and buildings, used more home appliances, and bought bigger cars. Obviously, the two primary factors that determine energy use are political stability in parts of the world that supply oil, and the price of that oil. Insecurity leads to a price rise, but governments can also manipulate prices by changing taxes, granting subsidies, and using other means. The energy consumption behavior of most people is motivated by economics rather than a desire to wisely use energy resources.

SUMMARY

A constant supply of energy is required by all living things. Energy has a major influence on society. A direct correlation exists between the amount of energy used and the complexity of civilizations.

Wood furnished most of the energy and construction materials for early civilizations. Heavy use of wood in densely populated areas eventually resulted in shortages, so fossil fuels replaced wood as a prime source of energy. Fossil fuels were formed from the remains of plants, animals, and microorganisms that lived about 300 million years ago. Fossil-fuel consumption

in conjunction with the invention of labor-saving machines resulted in the Industrial Revolution, which led to the development of technology-oriented societies today in the developed world.

Throughout the world, residential and commercial uses, industries, transportation, and electrical utilities require energy. Because of financial, political, and other factors, nations vary in the amount of energy they use as well as in how they use it. Analysts expect the worldwide demand for energy to increase steadily.

REVIEW QUESTIONS

1. Why was the sun able to provide all energy requirements for human needs before the Industrial Revolution?
2. In addition to food, what energy requirements does a civilization have?
3. Why were some countries unable to use the technologies developed during the Industrial Revolution?
4. What factors caused a shift from wood to coal as a source of energy?
5. How were energy needs in World War II responsible for the subsequent increased consumption of natural gas?
6. What part does government regulation play in changing the consumption of natural gas and oil?
7. Why was much of the natural gas that was first produced wasted?
8. What was the initial use of oil? What single factor was responsible for a rapid increase in oil consumption?
9. List the three purposes for which a civilization uses energy.
10. Why is OPEC important in the world's economy?

KEY TERMS

fossil fuels   130
Industrial Revolution   130

RELATED INTERNET SITES

Bay Area Green Web and Urban Ecology, focusing on urban ecology, energy, and planning.

   ***http://www.sci.tamucc.edu/stjs/

Energy analysis program focusing on energy use, economics, and the environmental implications of energy use.

   ***http://eande.lbl.gov/EAP/EAP.html

Award-winning site from the Arizona State University photosynthesis center explaining the processes of photosynthesis.

   ***http://photoscience.la.asu.edu/photosyn/default.html

Office of Mobile Sources, controls air pollution emissions from motor vehicles.

   http://www.epa.gov/OMSWWW/

E and P Environment, a biweekly newsletter reporting on environmental issues affecting the oil and gas industry.

   http://www.epeonline.com/

Energy Sources

Objectives

After reading this chapter, you should be able to:

- Differentiate between resources and reserves.
- Identify peat, lignite, bituminous coal, and anthracite coal as steps in the process of coal formation.
- Recognize that natural gas and oil are formed from ancient marine deposits.
- Explain how various methods of coal mining can have negative environmental impacts.
- Explain why surface mining of coal is used in some areas and underground mining in other areas.
- Explain why it is more expensive to find and produce oil today than it was in the past.
- Recognize that secondary recovery methods have been developed to increase the proportion of oil and natural gas obtained from deposits.
- Recognize that transport of natural gas is still a problem in some areas of the world.
- Explain why the amount of energy supplied by hydroelectric power is limited.
- Describe how wind, geothermal, and tidal energy are used to produce electricity.
- Recognize that wind, geothermal, and tidal energy can be developed only on areas with the proper geologic or geographic features.

- Describe the use of solar energy in passive heating systems, active heating systems, and the generation of electricity.
- Recognize that fuelwood is a major source of energy in many parts of the less-developed world and that fuelwood shortages are common.
- Describe the potential and limitations of biomass conversion and waste incineration as a source of energy.
- Recognize that energy conservation can significantly reduce our need for additional energy sources.

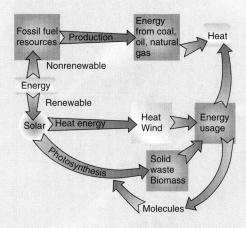

Chapter Outline

Energy Sources

Chapter 8 outlined the historical development of energy consumption and how advances in civilizations were closely linked to the availability and exploitation of energy. New manufacturing processes relied on dependable sources of energy. Technology acclerated in the twentieth century. Between 1900 and 1990, world energy consumption increased by a factor of twelve, the quantity of products manufactured increased by about thirtyfold, but population increased only about threefold. As societies used more energy, they produced goods and services more efficiently. (See table 9.1.)

The energy sources most commonly used by industrialized nations are the fossil fuels: oil, coal, and natural gas, which supply about 90 percent of the world's commercially traded energy. Fossil fuels were formed hundreds of millions of years ago. They are the accumulation of energy-rich organic molecules produced by organisms as a result of photosynthesis over millions of years. We can think of fossil fuels as concentrated, stored solar energy. The rate of formation of fossil fuels is so slow that no significant amount of fossil fuels will be formed over the course of human history. Since we are using these resources much faster than they can be produced and the amount of these materials is finite, they are known as **nonrenewable energy sources.**

Eventually, human demands will exhaust the supplies of coal, oil, and natural gas. Coal reserves are estimated to be about 1 trillion metric tons, and the world consumes approximately 3.3 billion metric tons per year. At this rate of consumption, coal reserves will last about 300 years. The world's oil reserves are estimated to be about 140 billion metric tons. With current world usage of about 3.2 billion metric tons per year, this is a 43-year supply of oil. The approximately 140 trillion cubic meters of natural gas reserves will last 66 years at current rates of use.

In addition to nonrenewable fossil fuels, there are several renewable energy sources. **Renewable energy sources** replace themselves or are continuously present as a feature of the solar system. For example, in plants, photosynthesis converts light energy into chemical energy.

Table 9.1

World Population, Economic Output, and Fossil-Fuel Consumption			
	Population (billions)	Gross world product (trillion 1980 dollars)	Fossil-fuel consumption (billion tons coal equivalent)
1900	1.6	0.6	1
1950	2.5	2.9	3
1990	5.3	18.4	12

Sources: Population statistics from United Nations; gross world products in 1900, author's estimate, and in 1950 from Herbert R. Block, The Planetary Product in 1980: A Creative Pause? (Washington, D.C.: U.S. Department of State, 1981), with updates from International Monetary Fund; fossil fuel consumption in 1900 from M. King Hubbert, "Energy Resources" in Resources and Man (Washington, D.C.: National Academy of Sciences, 1969); for remaining years, Worldwatch estimates based on data from American Petroleum Institute and U.S. Department of Energy.

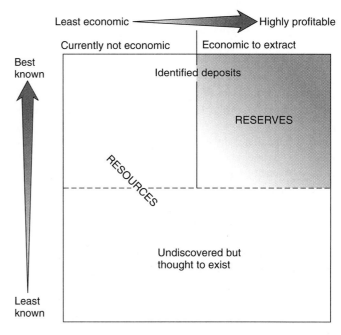

Figure 9.1

Resources and Reserves Each term describes the amount of a natural resource present. Reserves are those known deposits that can be profitably obtained using current technology under current economic conditions. Reserves are shown in the box in the upper-left-hand corner in this diagram. The darker the color, the more valuable the reserve. Resources are a much larger quantity that includes undiscovered deposits and deposits that currently cannot be profitably used, although it might be feasible to do so if technology or market conditions change.

Source: Adapted from the U.S. Bureau of Mines.

$$\text{Carbon Dioxide} + \text{water} + \frac{\text{Sunlight}}{\text{Energy}} \rightarrow \frac{\text{Biomass}}{(\text{Chemical} + \text{Oxygen Energy})}$$

This energy is stored in the organic molecules of the plant as wood, starch, oils, or other compounds. Any form of biomass—plant, animal, alga, or fungus—can be traced back to the energy of the sun. Since biomass is constantly being produced, it is a form of renewable energy. Solar, geothermal, and tidal energy are renewable energy sources because they are continuously available. Anyone who has ever lain in the sun, seen a geyser or hot

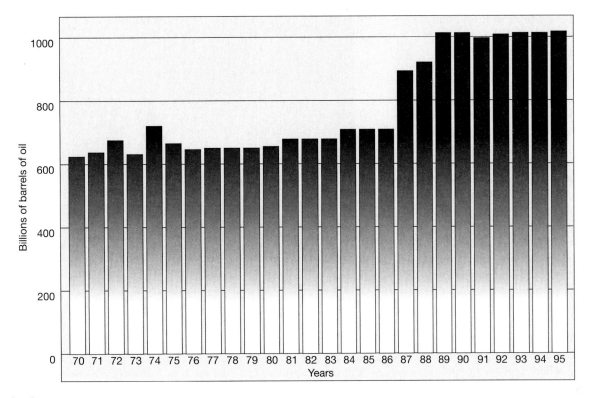

Figure 9.2

Changes in Oil Reserves   The figure shows the changes in oil reserves over a 25-year period. The changes that occurred in 1987 and 1989 are the result of changes in reporting, not the result of new discoveries. Since 1989 new discoveries and revisions have matched consumption. Thus, reserves have remained nearly constant.

Source: Data from BP Statistical Review of World Energy, June 1996.

Resources and Reserves

When discussing deposits of nonrenewable resources, such as fossil fuels, we must differentiate between deposits that can be exploited and those that cannot. From a technical point of view, a **resource** is a naturally occurring substance of use to humans that can *potentially* be extracted using current technology. **Reserves** are known deposits from which materials *can* be extracted profitably with existing technology *under present economic conditions.* It is important to recognize that *reserves* is an economic concept and is only loosely tied to the total quantity of a material present in the world. Therefore, reserves are smaller than resources. (See figure 9.1.) Both terms are used when discussing the amount of mineral or fossil-fuel deposits a country has at its disposal. This can cause considerable confusion if the difference between these concepts is not understood. The total amount of a resource such as coal or oil changes only by the amount used each year. The amount of a reserve changes as technology advances, new deposits are discovered, and economic conditions vary. There can be large changes in the amount of reserves, while the resource remains almost constant. (See figure 9.2.)

When we read about the availability of fossil fuels, we must remember that energy is needed to extract the wanted material from the earth. If the material is dispersed or hard to reach, it is expensive to extract. If the cost of removing and processing a fuel is greater than the fuel's market value, no one is going to produce it. Also, if the amount of energy used to produce, refine, and transport a fuel is greater than its potential energy, the fuel will not be produced. A net useful energy yield is necessary to exploit the resource. However, in the future, new technology or changing prices may permit the profitable removal of some fossil fuels that currently are not profitable. If so, those resources will be reclassified as reserves.

To further illustrate the concept of reserves and how technology and economics influence their magnitude, let us look at the history of oil. The ancient Chinese are said to have been the first to use oil as a fuel. The only oil available to them was the small amount that naturally seeped out of the ground. These seepages represented the known oil reserves as well as the known oil resources at that time.

Nearly two thousand years passed before oil reserves increased significantly. When the first oil well in North America was drilled in Pennsylvania in 1859, it greatly expanded the estimate of the amount of oil in the earth. There was a sudden increase in the known oil reserves. In the years that followed, new deposits were discovered. Better drilling techniques led to the discovery of deeper oil deposits, and offshore drilling established the location of oil under the ocean floor. At the time of their initial discovery, these deep deposits and the offshore deposits added to the estimated size of the world's oil resources. But they did not necessarily add to the reserves because it was not always profitable to extract the oil. With advances in drilling and pumping methods and increases in oil prices, it eventually became profitable to obtain oil from many of these deposits. As it became economical to extract them, they were reclassified as reserves.

The amount of fossil fuel reserves is in a constant state of flux. For example, prior to 1973, many oil wells were capped because it was not profitable to remove the oil. The oil still in the ground at these wells was not economic to produce and was not included in the reserves category. During the oil embargos of 1973–1974 and 1979–1980, when some of the OPEC countries reduced oil production, the price of oil increased. With the increase in oil prices, these wells became profitable, and the oil was reclassified as reserves. Then, in the 1980s, an oil surplus developed because the high oil prices resulting from the embargo caused increased production worldwide. Prices fell. Many of these same wells were capped again, and the oil was no longer classified as part of the reserve. With the exception of 1990 and 1991 when Iraq invaded Kuwait, the price of oil in the 1990s has been relatively stable.

Fossil-Fuel Formation

Fossil fuels are the remains of once living organisms that were preserved and altered as a result of geologic forces. There are significant differences in the formation of coal from that of oil and natural gas.

Coal Formation

Tropical freshwater swamps covered many regions of the earth 300 million years ago. Conditions in these swamps favored extremely rapid plant growth, resulting in large accumulations of plant material. Because this plant material collected under water, decay was inhibited, and a spongy mass of organic material formed, called

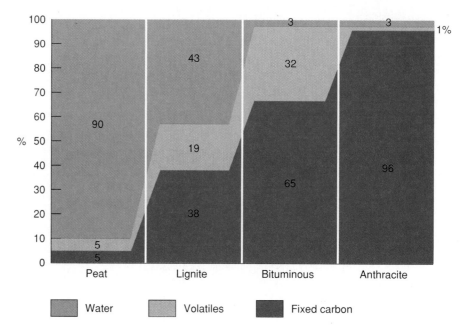

Figure 9.3

Composition of Coal   The high percentage of water in peat and lignite make them low-energy-yielding fuels. Increased pressure and heat decreases the moisture content and increases the percentage of carbon during the formation of bituminous or anthracite coal. Bituminous and anthracite coal, therefore, are better sources of energy.

peat. Peat deposits are 90 percent water, 5 percent carbon, and 5 percent volatile materials. (See figure 9.3.) In some parts of the world, peat is cut, dried, and used as fuel. However, because of its high water content, it is regarded as a low grade of fuel.

Due to geological changes in the earth, some of the swamps containing peat were submerged by seas. The plant material that had collected in the swamps was now covered by sediment. The weight of the plant material plus the weight of the sediment on top of it compressed it into a harder form of low-grade coal known as lignite, which contains less water and a higher proportion of burnable materials.

If the weight of the sediment was great enough, the heat from the earth high enough, and the length of time long enough, the lignite was changed into bituminous (soft) coal. The major change from lignite to bituminous coal is a reduction in the water content from over 40 percent to about 3 percent. If the heat and pressure continued over time, some of the bituminous coal was changed to anthracite (hard) coal, which is about 96 percent carbon. Through this combination of events, which occurred over hundreds of millions

of years, present-day coal deposits were created. Most parts of the world have significant coal deposits. (See figure 9.4.)

Oil and Natural Gas Formation

Oil and natural gas, like coal, are products from the past. They probably originated from microscopic marine organisms. When these organisms died and accumulated on the ocean bottom and were buried by sediments, their breakdown released oil droplets. Gradually, the muddy sediment formed rock called shale, which contained dispersed oil droplets. Although shale is common and contains a great deal of oil, extraction from shale is difficult because the oil is not concentrated. If a layer of sandstone formed on top of the oil-containing rock, and an impermeable layer of rock formed on top of the sandstone, conditions were suitable for oil pools to form. Usually the trapped oil does not exist as a liquid mass but rather as a concentration of oil within sandstone pores, where it accumulates because water and gas pressure force it out of the shale. (See figure 9.5.) These accumulations of oil

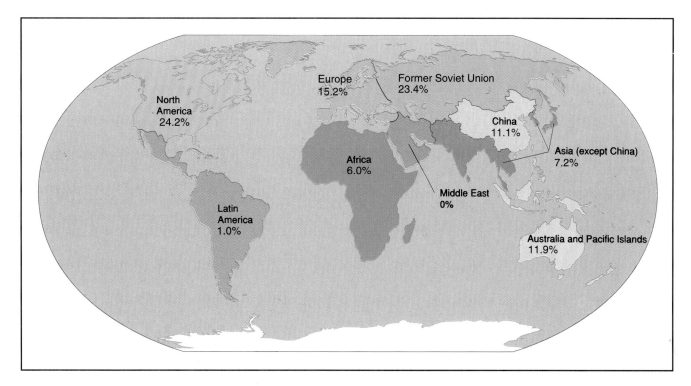

Figure 9.4

Recoverable Coal Reserves of the World 1995   The percentage indicates the coal reserves in different parts of the world. This coal can be recovered under present local economic conditions using available technology.

Source: Data from BP Statistical Review of World Energy, June 1996.

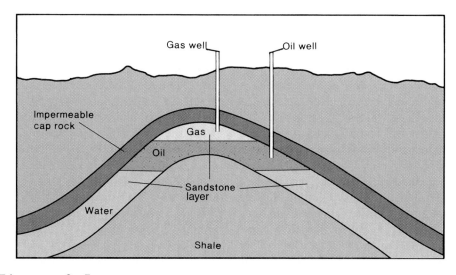

Figure 9.5

Crude Oil and Natural Gas Pool   Water and gas pressure force oil and gas out of the shale and into sandstone beneath the impermeable rock.

the organic material changed to lighter, more volatile (easily evaporated) hydrocarbons than those found in oil. The most common hydrocarbon in natural gas is the gas methane (CH_4). Water, liquid hydrocarbons, and other gases may be present in natural gas as it is pumped from a well.

The conditions that led to the formation of oil and gas deposits were not evenly distributed throughout the world. Figure 9.6 illustrates the geographic distribution of oil reserves and figure 9.7 illustrates the geographic distribution of natural gas reserves. Some of these deposits are easy to exploit, while others are not.

Issues Related to the Use of Fossil Fuels

Of the world's commercial energy, over 90 percent is furnished by the three nonrenewable fossil-fuel resources: coal, oil, and natural gas. Coal supplies about 27 percent, oil supplies about 40 percent,

are more likely to occur if the rock layers were folded by geological forces.

Natural gas, like coal and oil, was formed from fossil remains. In fact, the geological conditions favorable for oil formation are the same as those for natural gas, and the two fuels are often found together. However, in the formation of natural gas,

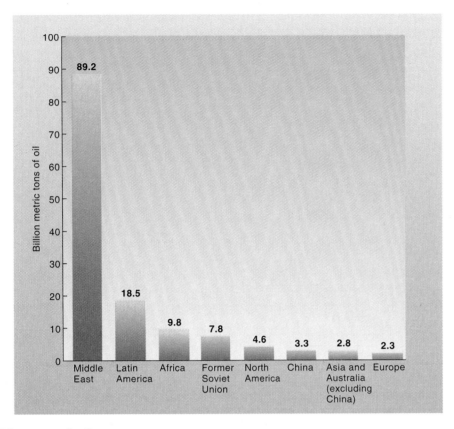

Figure 9.6

World's Oil Reserves 1995 The world's supply of oil is not distributed equally. Certain areas of the world enjoy an economic advantage because they control vast amounts of oil. Reserves are given in billions of metric tons of oil.

Source: Data from BP Statistical Review of World Energy, June 1996.

and natural gas supplies about 23 percent. Each fuel has advantages and disadvantages and requires special techniques for its production and use.

Coal Use Issues

Coal is the world's most abundant fossil fuel, but it supplies only about 27 percent of the energy used in the world. It varies in quality and is generally classified in three categories: lignite, bituminous, and anthracite. Lignite coal has a high moisture content and is crumbly in nature, which makes it the least desirable form. Bituminous coal is the most widely used because it is the easiest to mine and the most abundant. It supplies about 20 percent of the world's energy requirements. Coal is primarily used for electric power generation and other industrial uses. For most uses, anthracite coal is the most desirable because it furnishes more energy than the

other grades and is the cleanest burning. But anthracite is not as common and is usually more expensive because it is found at great depths and is difficult to obtain.

Because coal was formed as a result of plant material being buried under layers of sediment, it must be mined. There are two methods of extracting coal: surface mining and underground mining. **Surface mining** (strip mining) involves removing the material on top of a vein of coal, called, **overburden,** to get at the coal beneath. (See figure 9.8.) Coal is usually surface mined when the overburden is less than 100 meters thick. This type of mining operation is efficient because it removes 100 percent of the coal in a vein and can be profitably used for a seam of coal as thin as half a meter. For these reasons, surface mining results in the best utilization of coal reserves. Advances in the methods of surface mining and the development of better equipment have increased the

amount of surface mining in the United States from 30 percent of the coal production in 1970 to more than 60 percent today. This trend toward increased surface mining has also occurred in Canada, Australia, and the former Soviet Union.

If the overburden is thick, surface mining becomes too expensive, and the coal is extracted through **underground mining.** The deeply buried coal seam can be reached in two ways: In flat country, where the vein of coal lies buried beneath a thick overburden, the coal is reached by a vertical shaft. (See figure 9.9a.) In hilly areas, where the coal seam often comes to the surface along the side of a hill, the coal is reached from a drift-mine opening (See figure 9.9b.)

The mining, transportation, and use of coal as an energy source presents several significant problems. Surface mining disrupts the landscape, as the topsoil and overburden are moved to get at the coal. It is possible to minimize this disturbance by reclaiming the area after mining operations are completed. (See figure 9.10.) However, reclamation rarely, if ever, returns the land to its previous level of productivity. The cost of reclamation is passed on to the consumer in the form of higher coal prices. Underground mining methods do not disrupt the surface environment as much as surface mining does, but subsidence (sinking of the land) occurs if the mine collapses. In addition, large waste heaps are produced around the mine entrance from the debris that must be removed and separated from the coal.

Health and safety are important concerns related to coal mining, which is one of the most dangerous jobs in the world. This is particularly true with underground mining. Many miners suffer from **black lung disease,** a respiratory condition that results from the accumulation of fine coal-dust particles in the miners' lungs. The coal particles inhibit the exchange of gases between the lungs and the blood. The health care costs and death benefits related to black lung disease are an indirect cost of coal mining. Since these costs are partially paid by the federal government, their full price is not reflected in the price of coal but is paid by taxpayers in the form of federal taxes.

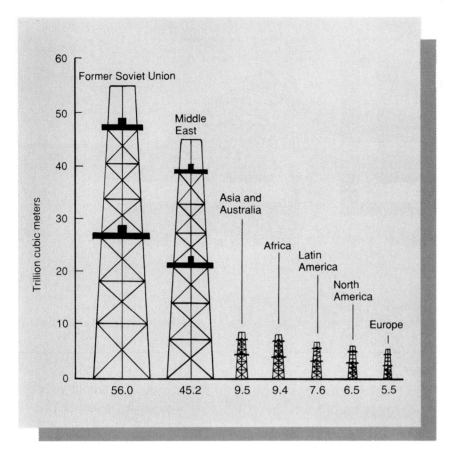

Figure 9.7

Natural Gas Reserves 1995 Natural gas reserves, like oil and coal, are concentrated in certain regions of the world. Figures are trillion cubic meters.

Source: Data from BP Statistical Review of World Energy, June 1996.

Figure 9.8

Surface Mining Large power draglines are used to remove the overburden, which is piled to the side. The coal can then be loaded into trucks. When the coal has been removed, the overburden is placed back in the trench.

Because coal is bulky, shipping presents a problem. Generally, the coal can be used most economically near where it is produced. Rail shipment is usually the most economical way of transporting it from the mine. Rail shipment costs include the expense of constructing and maintaining the tracks, as well as the cost of the energy required to move the long strings of railroad cars. In some areas, the coal is transferred from trains to ships.

Coal is dirty and its mining and transport generate a great deal of dust. The large amounts of coal dust released into the atmosphere at the loading and unloading sites can cause local air pollution problems. If a boat or railroad car is used to transport coal, there is the expense of cleaning it before other types of goods can be shipped. In some cases, the coal can be ground and mixed with water to form a slurry that can be pumped through pipelines. This helps to alleviate some of the air pollution problems without causing significant water pollution problems.

Air pollution from coal burning releases millions of metric tons of material into the atmosphere and is responsible for millions of dollars of damage to the environment. The burning of coal for electric generation is the prime source of this type of pollution.

Since coal is a fossil fuel formed from plant remains, it contains sulfur, which was present in the proteins of the original plants. Sulfur is associated with **acid mine drainage** and air pollution. Acid mine drainage occurs when the combined action of oxygen, water, and certain bacteria causes the sulfur in coal to form sulfuric acid. Sulfuric acid can seep out of a vein of coal even before the coal is mined. However, the problem becomes most acute when the coal is mined and the overburden is disturbed, allowing rains to wash the sulfuric acid into streams. Streams may become so acidic that they can support only certain species of bacteria and algae. Today, many countries regulate the amount of runoff allowed from mines, but underground and surface mines abandoned before these regulations were enacted continue to pollute the water.

Currently, a form of acid pollution called acid deposition is becoming serious.

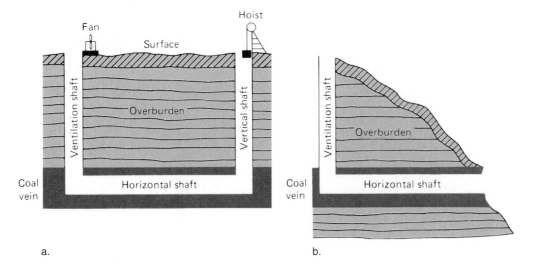

a.

b.

Figure 9.9

Underground Mining If the overburden is too thick to allow surface mining, underground mining must be used. (*a*) If the coal vein is not exposed, a vertical shaft is sunk to reach the coal. (*b*) In hilly areas, if the vein is exposed, a drift mine is used in which miners enter from the side of the hill.

a.

b.

Figure 9.10

Surface-Mine Reclamation (*a*) This photograph shows a large area that has been surface mined with little effort to reclaim the land. The windrows created by past mining activity are clearly evident, and little effort has been made to reforest the land. By contrast, (*b*) is an example of proper surface-mining reclamation. The sides of the cut have been graded and planted with trees. The topsoil has been returned, and the level land is now productive farmland.

Acid deposition occurs when coal is burned and sulfur oxides are released into the atmosphere, causing acid-forming particles to accumulate on a surface. Each year, over 150 million metric tons of sulfur dioxide are released into the atmosphere worldwide. This problem is discussed in greater detail in chapter 17.

The release of carbon dioxide from the burning of coal has become a major issue in recent years. Increasing amounts of carbon dioxide in the atmosphere are said to contribute to global warming. Environmentalists have suggested that the amount of coal used be decreased, since the other fossil fuels (oil and natural gas) produce less carbon dioxide for an equivalent amount of energy.

Because coal is bulky and dusty and often has a high sulfur content resulting in air pollution, people seek alternative sources of fuel. The most common alternatives for coal are oil and natural gas.

Oil Use Issues

Oil has several characteristics that make it superior to coal as a source of energy. Its extraction causes less environmental damage than does coal mining. It is a more concentrated source of energy than coal, it burns with less pollution, and it can be moved easily through pipes. These characteristics make it an ideal fuel for automo-

a.

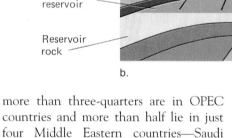

b.

Figure 9.11

Offshore Drilling Once the drilling platform is secured to the ocean floor, a number of wells can be sunk to obtain the gas or oil.

Source: (b) American Petroleum Institute.

biles. However, it is often difficult to find. Today, geologists use a series of tests to locate underground formations that may contain oil. When a likely area is identified, a test well is drilled to determine if oil is actually present. The many easy-to-reach oil fields have already been tapped. Drilling now focuses on smaller amounts of oil in less accessible sites, which means that the cost of oil from most recent discoveries is higher than that of the large, easy-to-locate sources of the past. As oil deposits on land have become more difficult to find, geologists have widened the search to include the ocean floor. Building an offshore drilling platform can cost millions of dollars. To reduce the cost, as many as 70 wells may be sunk from a single platform. (See figure 9.11.) Total world proven reserves of oil in 1995 were estimated at 1,000 billion barrels. Of this,

more than three-quarters are in OPEC countries and more than half lie in just four Middle Eastern countries—Saudi Arabia, Iraq, Kuwait, and Iran.

One of the problems of extracting oil is removing it from the ground. If the water or gas pressure associated with a pool of oil is great enough, the oil is forced to the surface when a well is drilled. When the natural pressure is not great enough, the oil must be pumped to the surface. Present technology allows only about one-third of the oil in the ground to be removed. This means that two barrels of oil are left in the ground for every barrel produced. In most oil fields, **secondary recovery** is used to recover more of the oil. Secondary recovery methods include pumping water or gas into the well to drive the oil out or even starting a fire in the oil-soaked rock to liquefy thick oil.

As oil prices increase, more expensive secondary recovery methods will be used. More will be obtained from expensive deep wells and offshore wells.

Processing crude oil to provide useful products generates a variety of problems. Oil as it comes from the ground is not in a form suitable for use. It must be refined. The various components of crude oil can be separated and collected by heating the oil in a distillation tower. (See figure 9.12.) After distillation, the products may be further refined by "cracking." In this process, heat, pressure, and catalysts are used to produce a higher percentage of volatile chemicals, such as gasoline, from less volatile liquids such as diesel fuel and furnace oils. It is possible, within limits, to obtain many products from one barrel of oil. In addition, petrochemicals from oil serve as raw materials for a variety of

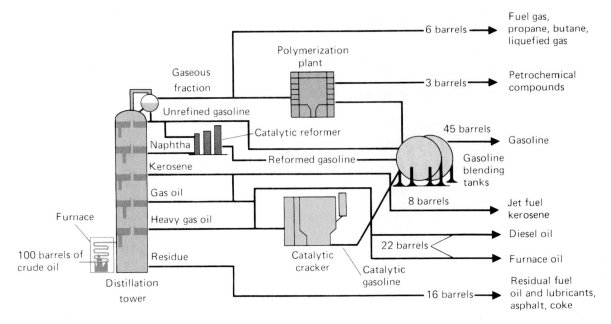

Figure 9.12

Uses of Crude Oil A great variety of products can be obtained from the distilling and refining of crude oil. A barrel of crude oil produces slightly less than half a barrel of gasoline. This figure shows the many steps in the refining process and the variety of products that can be obtained from crude oil.

synthetic compounds. (See figure 9.13.) All of these processing activities are opportunities for accidental or routine releases that cause air or water pollution. The petrochemical industry is the major contributor to air pollution.

The environmental impacts of producing, transporting, and using oil are somewhat different from those of coal. Oil spills in the oceans have been widely reported by the news media. (See figure 9.14.) However, these accidental spills are responsible only for about a third of the oil pollution resulting from shipping. Almost 60 percent of the oil pollution in the oceans is the result of normal, routine shipping operations. Oil spills on land can contaminate soil and underground water. The evaporation of oil products and the incomplete burning of oil fuels cause air pollution. These problems are discussed in chapter 17.

Natural Gas Use Issues

Natural gas, the third major source of fossil-fuel energy, supplies 23 percent of the world's energy. The drilling operations to obtain natural gas are similar to those used for oil. In fact, a well may yield both oil and natural gas. As with oil, secondary recovery methods that pump air or

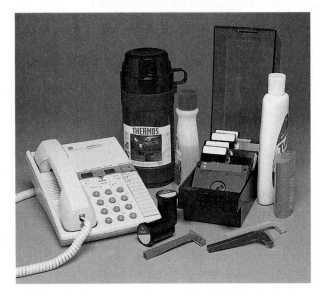

Figure 9.13

Oil-Based Synthetic Materials These common household items are produced from chemicals derived from oil. Although petrochemicals represent only about 3 percent of each barrel of oil, they are extremely profitable for the oil companies.

water into a well are used to obtain the maximum amount of natural gas from a deposit. After processing, the gas is piped to the consumer for use.

Transport of natural gas still presents a problem in some parts of the world. In the Middle East, Mexico, Venezuela, and

Nigeria, wells are too far from consumers to make pipelines practical, so much of the natural gas is burned as a waste product at the wells. However, new methods of transporting natural gas and converting it into other products are being explored. At −162°C, natural gas becomes a liquid and

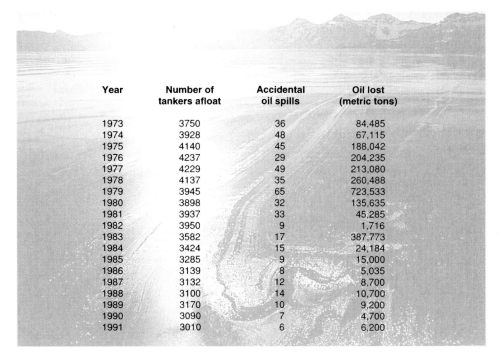

Year	Number of tankers afloat	Accidental oil spills	Oil lost (metric tons)
1973	3750	36	84,485
1974	3928	48	67,115
1975	4140	45	188,042
1976	4237	29	204,235
1977	4229	49	213,080
1978	4137	35	260,488
1979	3945	65	723,533
1980	3898	32	135,635
1981	3937	33	45,285
1982	3950	9	1,716
1983	3582	17	387,773
1984	3424	15	24,184
1985	3285	9	15,000
1986	3139	8	5,035
1987	3132	12	8,700
1988	3100	14	10,700
1989	3170	10	9,200
1990	3090	7	4,700
1991	3010	6	6,200

Figure 9.14

Oil Spills Accidents involving oil tankers are sources of water pollution.

Source: Data from Tanker Advisory Center Reference data for 1973–1987 cited in World Resources 1986, Basic Books, Inc.

has only 1/600 of the volume of its gaseous form. Tankers have been designed to transport **liquefied natural gas** from the area of production to the area of demand. In 1995, over 90 billion cubic meters of natural gas were shipped between countries as liquefied natural gas. This is nearly 4.5 percent of the natural gas consumed in the world. Of that amount, Japan imported 58 billion cubic meters.

Many people are concerned about accidents that might cause the tankers to explode. Another, safer process converts natural gas to methanol, a liquid alcohol, and transports it in that form. As the demand for natural gas increases, the amount of it wasted will decrease and new methods of transportation will be employed. Higher prices will make it profitable to transport natural gas greater distances from the wells to the consumers.

Of the three fossil fuels, natural gas is the least disruptive to the environment. A natural gas well does not produce any unsightly waste, although there may be local odor problems. Except for the danger of an explosion or fire, it poses no harm to the environment during transport. Since it is

clean burning, it causes almost no air pollution. The products of its combustion are carbon dioxide and water.

Although natural gas is used primarily for heat energy, it does have other uses, such as the manufacture of petrochemicals and fertilizer. Methane contains hydrogen atoms that are combined with nitrogen from the air to form ammonia, which can be used as fertilizer.

More than two-thirds of natural gas reserves are located in Russia and the Middle East. In fact, more than a third of total reserves are located in just ten giant fields; six are located in Russia and the remainder are in Qatar, Iran, Algeria, and the Netherlands.

From 1985 to 1995, world production of natural gas rose by more than 25 percent. In Russia, which has the largest natural gas reserves, production increased by 40 percent from 1985 through 1991 but has since fallen as a result of the economic disruption associated with the collapse of the Soviet Union.

Because of oil development from beneath the North Sea, several European countries (the United Kingdom, Norway,

and the Netherlands) have sizable reserves of natural gas. During oil shortages in the 1970s, they increased their use of natural gas to compensate for the increased cost of oil. This trend has continued; natural gas consumption increased by 28 percent from 1985 to 1995.

Renewable Sources of Energy Currently Being Used

The three nonrenewable fossil-fuel sources of energy—coal, oil, and natural gas—furnish about 90 percent of the world's commercially traded energy. Nuclear power provides over 7 percent of the world's energy. The remainder is supplied by renewable energy sources, including hydroelectric power, tidal power, geothermal power, wind power, solar energy, fuelwood, biomass conversion, and solid waste. These data do not include the use of fuelwood, animal dung, and other locally produced energy sources typically used in the less-developed world.

Hydroelectric Power

People have long used water to power a variety of machines. Some early uses of water power were to mill grain, saw wood, and run machinery for the textile industry. Today, water power is used almost exclusively to generate electricity. As the water flows from higher to lower levels, it supplies the energy to turn a generator and produce electricity. Hydroelectric power plants are commonly located on artificial reservoirs. (See figure 9.15.) The impounded water represents a potential energy source. In some areas of the world where the streams have steep gradients and a constant flow of water, hydroelectricity may be generated without a reservoir. Such sites are usually found in mountainous regions and can support only small power-generating stations. At present, hydroelectricity produces about 2.5 percent of the world's commercially traded energy.

It is important to recognize that the construction of a reservoir for a hydroelectric plant causes environmental and

Hydroelectric Sites

Slightly over 17 percent of the potential hydroelectric sites of the world have been developed. (See the table.) The World Energy Conference estimates that the electricity produced by hydropower will increase six times by the year 2020. The less-developed countries, which have developed about 10 percent of their hydropower, will experience most of this growth.

The projected increase will come mainly from the development of plants on large reservoirs. However, construction of "mini-hydro" (less than 10 megawatts) and "micro-hydro" (less than 1 megawatt) plants is also increasing. Such plants can be built in remote places and supply electricity to small areas. China has built over 80,000 such small stations, and the United States has nearly 1,500.

About 50 percent of the U.S. hydroelectric capacity has been developed. However, this statistic is somewhat misleading. The Wild and Scenic Rivers Act (1968) prevents the construction of dams on designated streams. Presently, 37 potential hydroelectric sites are on streams protected by this act. The hydroelectric generating potential often quoted for the United States includes these areas, even though, at present, construction of plants at these sites is not possible. The U.S. political climate, which now favors the protection of certain rivers, might change if the demand for more energy becomes acute.

Developed Hydroelectric Sites, 1990

Region	Percent of hydropower developed
Asia	14
South America	17
Africa	10
North America	55
Russia	9
Europe	90
Oceania	34

Source: Data from Survey of Energy Resources, World Energy Conference.

Figure 9.15

Hydroelectric Power Plant   The water impounded in this reservoir is used to produce electricity. In addition, this reservoir serves as a means of flood control and provides an area for recreation.

social problems, including loss of fertile farmland, destruction of the natural aquatic ecosystem, relocation of entire communities (including people and buildings), and a reduction in the amount of nutrient-rich silt deposited on downriver agricultural lands. The building of the Tellico Dam in Tennessee was delayed for several years because it might have caused the extinction of the snail darter, a fish that lived only in streams that would be flooded. The construction of the Aswan Dam in Egypt resulted in the displacement of eighty thousand people. It created an environment that increased schistosomiasis, a waterborne disease caused by flatworm parasites that spend part of their life cycle in snails that live in slowly moving water. The irrigation canals built to distribute the water from the Aswan Dam provide ideal conditions for the snails. Now many of the people who use the canal water for cooking, drinking, bathing, and as a sewer are infected with the flatworm parasites.

Even though hydroelectric projects cause problems, new sites continue to be developed. From 1984 to 1995, the energy furnished by hydroelectricity for world use increased by about 25 percent. The most significant change took place in South America, where hydroelectric use increased about 40 percent over that time.

China is currently constructing a huge hydroelectric dam, known as the Three Gorges Dam, on the Yangtze River. If it is completed, it would be the largest hydroelectric dam in the world. It would displace

Figure 9.16

Tidal Generating Station The Rance River Estuary Power Plant in France is the world's largest tidal electrical generating station.

1.2 million people. Experts have expressed concern about the rate at which the reservoir will fill with sediment and about the structural soundness of the design. The project is expected to take 20 years to complete, but financing is a problem and there has been a lack of international support.

Tidal Power

Another source of energy related to local geological conditions is tidal flow. The gravitational pull of the sun and the moon, along with the earth's rotation, causes tides. The tidal movement of water represents a great deal of energy. For years, engineers have suggested that this moving water could be used to produce electricity. The principle is the same as that employed in a hydroelectric plant. As water flows from a higher level to a lower one, it can be used to generate electricity. The greater the difference between high and low tides the more energy can be extracted.

Since tidal changes are greatest near the poles and are accentuated in narrow bays and estuaries, suitable sites of constructing power plants are limited. In the 1930s, the United States explored the possibility of constructing a tidal electrical generating facility at Passamaquoddy, Maine, on the Bay of Fundy. After spending considerable time and money on a feasibility study, it abandoned the idea.

In 1966, France constructed a commercial tidal generating station. (See figure 9.16.) Located on the Rance Estuary on the Brittany coast, it is the world's

Figure 9.17

Geothermal Power Plant In this plant, the steam obtained from geothermal wells is used in the production of electricity.

only large tidal generating station. It was built to generate 240 megawatts, but because of the tides, it usually produces only 62 megawatts. This satisfies the electric power needs of about 100,000 people (less than 0.2 percent of the population of France).

The British government studied the feasibility of constructing a 7,200 megawatt tidal station in the Severn Estuary in southwestern England. It would have thirty times the capacity of the French station. The facility could be built for a price comparable to that of a coal-fired plant of similar generating capacity, but it would disrupt the normal estuary flow and would concentrate pollutants in the area. The plant has not been built.

Geothermal Power

The earth's core, with temperatures as high as 4,400°C, is a molten mass of material possessing vast amounts of energy. In some regions, this material sometimes breaks through the earth's crust and produces volcanoes. In other regions, the hot material is close enough to the surface to heat underground water and form steam. Geysers and hot springs are natural areas where this steam and hot water come to the surface. In areas where the steam is

trapped underground, **geothermal energy** is tapped by drilling wells to obtain the steam. The steam is then used to power electrical generators. At present, geothermal energy is practical only in areas where this hot mass is near the surface. (See figure 9.17.)

The United States has about half of the world's geothermal electrical generating capacity, but more than 130 generating plants are operating in twelve other countries. In addition to the Philippines, which have half the generating power that the United States does, Italy, Mexico, Japan, New Zealand, and Iceland produce sizable amounts of electricity by geothermal methods.

California produces 40 percent of the world's geothermal electricity, about 1,900 megawatts. One megawatt provides energy for about one thousand households. The Pacific Gas and Electric Company (PG&E) has been producing electricity from geothermal energy since 1960. PG&E's complex of generating units located north of San Francisco is the largest in the world and provides 700 megawatts of power, enough for 700,000 households or about 2.9 million people.

The state of Hawaii has considered building two 25-megawatt geothermal

Electricity from the Ground Up

The naturally occurring heat beneath the planet's surface powers volcanoes, hot springs, and geysers such as Yellowstone National Park's Old Faithful. In a few places in the world, such as Iceland and parts of California, natural heat and natural water come together in sufficient quantities to provide energy. But in most places where there is abundant subsurface water, useful heat is five kilometers or more below the surface, beyond the reach of economical drilling technology. In places where heat is within reach, there is often no subsurface water.

In the search for cheap, nonpolluting energy, scientists at the Los Alamos National Laboratory in New Mexico have begun a process to "mine heat." Mining heat is actually a new effort to use geothermal energy. In this process, water is pumped at high pressure three kilometers down into the earth through a well. When the water comes back up through a parallel well, it has been heated beyond the boiling point by the earth's natural heat. The hot dry rock project is an attempt to bring heat and water together as a geothermal energy source.

On average, temperatures below the earth's surface increase about 27°C per kilometer. In much of the western United States, however, residual heat from ancient volcanoes increases temperatures by more than 69°C/km. At the Los Alamos site, the temperature at the base of the well is about 240°C.

According to researchers at Los Alamos, a fairly conservative estimate is that there are at least 500,000 quads (quadrillion British Thermal Units, or BTUs) of useful heat in hot dry rock at accessible drilling depths beneath the United States. This is about 6,000 times the total amount of energy used in the country in one year.

The fundamental question now being asked is if the energy can be developed economically for commercial use. At this point, the answer is no; however, if new technology is developed, the answer could change. Other questions also must be addressed: Will the heat in a particular well last long enough to justify the cost? Can the system be sealed to minimize water leakage? Can a constant flow of water be sustained?

Even if it proves commercially competitive, hot dry rock mining might encounter obstacles involving site access, hookups to distant power grids, and legal disputes over water rights. If such problems can be overcome, maybe one day the energy used to run your computer will come from water heated kilometers beneath the surface of the Earth.

plants on the island of Hawaii. Most of the electricity currently generated for the state comes from oil-fired power plants; geothermal-generated electricity would be less costly and reduce the risk of accidental oil spills. However, the plants would be built in a rainforest area that environmentalists believe would be destroyed by the construction.

In addition to producing electricity, geothermal energy is used directly for space heating. In Iceland, half of the geothermal energy is used to produce electricity and half is used for heating. In the capital, Reykjavík, all of the buildings are heated with geothermal energy at a cost that is less than 25 percent of what it would be if oil were used.

Geothermal energy does present some environmental problems. The steam contains hydrogen sulfide gas, which has the odor of rotten eggs and is an unpleasant form of air pollution. The minerals in the steam corrode pipes and equipment, causing maintenance problems. The minerals are also toxic to fish.

Other ways to use the heat in the earth are less dependent on special geologic features. All objects contain heat energy, which can be extracted from and transferred to other locations. A refrigerator extracts heat from the interior and exports it to the coils on the back of the

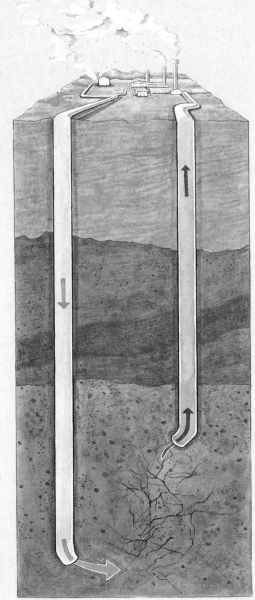

① Recirculated water is pumped at high pressure 4 to 6 miles below the earth's surface where the temperature reaches up to 400° F hotter than the earth's surface.

② As the water is forced into the rock, the extreme pressure causes the rocks to break.

⑤ Steam from the hot water generates electricity.

④ Pipes recapture the hot water and carry it back to the earth's surface.

③ "Hot rocks" heat the water as it percolates through the rock fissures.

unit. Similarly, pipes placed within the earth can extract heat and transfer it to a home. It is important to recognize that such an application of geothermal energy requires the expenditure of energy through an electric-powered device, so the amount of heat energy extracted costs something. In some instances, obtaining heat in this manner is less expensive than other traditional sources, such as oil or natural gas, but it is not a "free" source of heat.

Wind Power

As the sun's radiant energy strikes the earth, it is converted into heat, which warms the atmosphere. The earth is unequally heated because various portions receive different amounts of sunlight. Since warm air is less dense and rises, cooler, denser air flows in to take its place. This flow of air is wind. For centuries, wind has been used to move ships, grind grains,

pump water, and do other forms of work. In more recent times, wind has been used to generate electricity. (See figure 9.18.)

In 1994, California was using about 14,600 wind turbines to produce about 300 megawatts of electricity at a cost competitive with newly constructed coal or nuclear plants. This is about 40 percent of the world's electricity produced by wind. The other major countries involved in wind power are Germany, Denmark,

India, and the Netherlands. The California Energy Commission plans to produce 8 percent of the state's electricity by wind by the year 2000, and many other regions of the country plan wind power installations. India, China, Germany, and Spain have plans for major increases in wind-generated electricity through the years 2005. In Inner Mongolia, nomadic herders carry with them two thousand small, portable, wind-driven generators that provide electricity for light, television, and movies in their tents as well as for electric fences to contain their animals.

A steady and dependable source of wind makes the use of wind power more productive in some regions than others. Wide, open areas, such as the Great Plains in North America, are better suited for wind power than are heavily wooded areas. Wind-generated electricity is usually used in conjunction with other sources of electricity that take over when the wind does not blow. Wind generators do have some negative effects. The moving blades are a hazard to birds and produce a noise that some find annoying. In addition, some people consider the sight of a large number of wind generators, such as are found in California, to be visual pollution.

Solar Energy

The sun is often mentioned as the ultimate answer to the world's energy problems. It provides a continuous supply of energy that far exceeds the world's demands. In fact, the amount of energy received from the sun each day is six hundred times greater than the amount of energy produced each day by all other energy sources combined. The major problem with solar energy is its intermittent nature. It is only available during the day and in cloudy weather is less available than when it is sunny. All systems that use solar energy must store or use supplementary sources of energy when sunlight is not available. Because of differences in the availability of sunlight, some parts of the world are more suited to the use of solar energy than others.

Solar energy is utilized in three ways:

1. In a passive heating system, the sun's energy is converted directly

Figure 9.18

Wind Energy Fields of wind-powered generators such as these can produce large amounts of electricity.

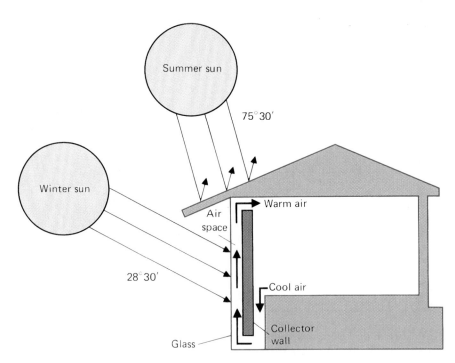

Figure 9.19

Passive Solar Heating The length of overhang in this home is designed for solar heating at the latitude of St. Louis, Missouri (38°N). In this design, a wall 30–40 centimeters thick is used to collect and store heat. The collector wall is located behind a glass wall and faces south. During a midwinter day, when the sun's angle is 28 degrees, light energy is collected by the wall and stored as heat. At night, the heat stored in the wall is used to warm the house. Natural convection causes the air to circulate past the wall, and the house is heated. During a midsummer day, when the sun's angle is 75 degrees, the overhang shades the collector wall from the sun.

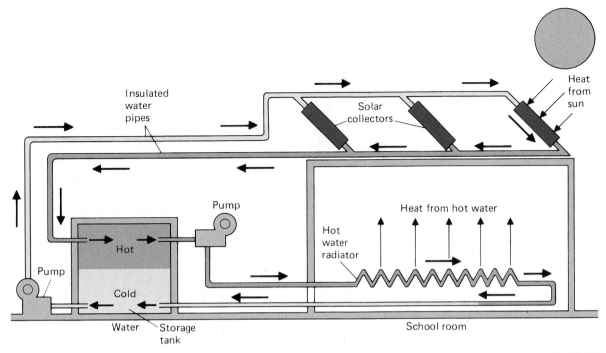

Insulated water pipes

Solar collectors

Heat from sun

Pump

Hot

Cold

Pump

Water   Storage tank

Heat from hot water

Hot water radiator

School room

Figure 9.20

Active Solar System Collectors provide 65 percent of the hot water in this dental school building in California.

into heat for use at the site where it is collected.

2. In an active heating system, the sun's energy is converted into heat, but the heat must be transferred from the collection area to the place of use.

3. The sun's energy also can be used to generate electricity, which may be used to operate solar batteries or may be transmitted along normal transmission lines.

Passive Solar Systems

Anyone who has walked barefoot on a sidewalk or a blacktopped surface on a sunny day has experienced the effects of passive solar heating. In a **passive solar system,** light energy is transformed to heat energy when it is absorbed by a surface. Some of the earliest uses of passive solar energy were to dry food and clothes and to evaporate seawater to produce salt. In fact, solar energy is still used for these purposes. Homes and buildings can also be designed to use passive solar energy for heating. (See figure 9.19.) Such systems require a large window through which sunlight enters and a large mass that collects and stores the heat. This large mass may be a thick wall or floor that would be a normal part of the room.

Since there are no moving parts, a passive solar system is maintenance free. None of the energy is used to transfer heat within the system, and there are no operating costs. However, passive solar design is usually only practical in new construction.

Active Solar Systems

In addition to a solar collector, an **active solar system** requires a pump and a system of pipes to transfer the heat from the site of its production to the area to be heated. (See figure 9.20.) Active solar systems are most easily installed in new buildings, but in some cases can be installed in existing structures. A major consideration in the use of an active solar system is the initial cost of installation. An active system requires a specially designed collector, consisting of a series of liquid-filled tubes; a pump; and pipes to transfer the warm liquid from the collector to the space to be heated. Because an active system has moving parts, it also has operation and maintenance costs like any other heating or cooling system.

Rock, water, or specially produced products are used to store heat. The hot liquid in the pipes heats the storage medium, which releases its heat when the sun is not shining.

Solar-Generated Electricity

When the first **photovoltaic cell,** a bimetallic unit that allows the direct conversion of sunlight to electricity, was developed by Bell Laboratories in 1954, it was regarded as an expensive novelty. However, as more efficient batteries were developed and production costs were reduced, practical uses were found for photovoltaic cells. By the mid 1980s, more than 60 million solar calculators were being produced annually. These calculators used over 10 percent of the photovoltaic cells manufactured.

Photovoltaic cells appear to be emerging as sources of small amounts of electricity for special uses like calculators and for running equipment in remote regions. The normal system of generating electricity in large, centrally located plants and distributing it by high tension lines is costly, so it is only practical in highly populated areas. Photovoltaic cells are more practical in remote regions of the world. Over ten thousand solar-electric homes have been built in parts of Alaska and the Australian outback. (See figure 9.21.) The French government has subsidized the installation of over two thousand solar electric units on eighteen islands in the Pacific. These units provide electricity for a thousand homes and five hospitals. Many of the developing countries of the world will introduce electricity to villages through the use of photovoltaics rather than the use of generators that require fuel and

Figure 9.21

Solar Energy In some remote areas, solar energy is an economical method of electricity production.

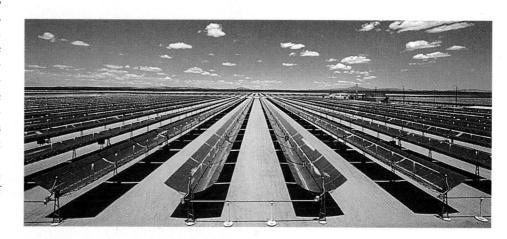

Figure 9.22

Solar Generation of Electricity This solar-powered electricity generating plant is capable of generating electricity at a cost that is competitive with other methods of generating electricity.

distribution lines. Two-thirds of the current U.S. production of photovoltaic cells is being exported.

The price of photovoltaic cells has been falling as better technology is developed. Eventually, the cells may become competitive with other energy sources, particularly as the cost of fossil fuels rises.

Solar energy is also being used to generate electricity in a more conventional way. A company named Luz International in California has built solar collector troughs that can heat oil in pipes to 390°C.

(See figure 9.22.) This heat can be transferred to water, which is turned into steam that is used to run conventional electricity-generating turbines. As with photovoltaic cells, the cost of producing electricity in this manner is falling and is becoming competitive with conventional sources.

Limitations of Solar Energy

Solar energy provides less than 1 percent of the world's energy for several reasons. The most obvious is that it only works

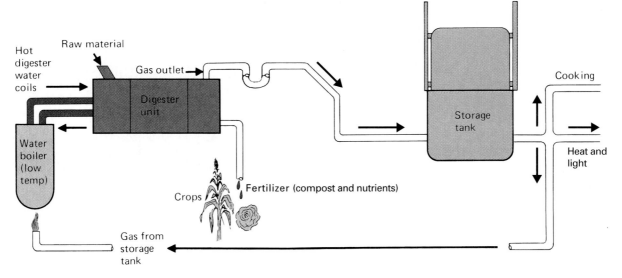

Figure 9.23

Methane Digester   In the digester unit, anaerobic bacteria convert animal waste into methane gas. This gas is then used as a source of fuel. The sludge from this process serves as a fertilizer. In many less-developed countries, this type of digester has the advantages of providing a source of energy and a supply of fertilizer and managing animal wastes, which helps reduce disease.

during the day, which means that some type of heat or electrical storage mechanism is needed for night use, which adds to the expense of relying on solar energy. The fact that solar heating is most practical in new construction also limits its use. In colder climates, solar heat is inadequate as the sole source of heat, and some type of a conventional heating system is required for backup. Climate is also a problem, since many areas have extensive cloudy periods, which reduce the amount of energy that can be collected. Although the price of collectors and related equipment has decreased in recent years, many collector systems are still expensive. For example, electricity from photovoltaic cells is still more expensive than conventional electric generating systems. Prices of photovoltaic cells continue to fall, however, and the cost of generating electricity from conventional fuels is rising.

Biomass Conversion

Biomass is any accumulation of organic material produced by living things. The most commonly used biomass sources are fuelwood, agricultural residue from the harvesting of crops, crops grown for their energy content, and animal waste. (Because of its major impact on world energy

resources, fuelwood will be discussed separately in the next section.) These traditional, often noncommercial sources of fuel provide more than 10 percent of the world's energy, but are not reported in most statistics about global energy. In many developing countries these sources of fuel are a large proportion of the energy available.

Biomass conversion is the process of obtaining energy from the chemical energy stored in biomass. It is not a new idea; burning wood is a form of biomass conversion that has been used for thousands of years. Biomass can be burned directly as a source of heat for cooking, burned to produce electricity, converted to alcohol, or used to generate methane. (See figure 9.23.) The People's Republic of China has 500,000 small methane digesters in homes and on farms, India has 100,000, and Korea has 50,000. Brazil is the largest producer of alcohol from biomass. The low price of sugar coupled with the high price of oil have prompted Brazil to use its large crop of sugar cane as a source of energy. Alcohol provides 50 percent of Brazil's automobile fuel.

Biomass conversion raises some environmental and economic concerns. Countries that use large amounts of biomass for energy are usually those that have food

shortages. Biomass conversion means that fewer nutrients are being returned to the soil, and this compounds the food shortage. If the price of food rises or the price of oil falls, there could be less biomass conversion.

The energy required to produce usable energy stocks from biomass must be taken into account. Growing corn to produce alcohol requires large energy inputs. The amount of energy present in the alcohol produced from the corn is actually less than the amount of energy that went into producing the alcohol. Obviously, this makes no sense from an energy point of view. However, the convenience of a liquid fuel may be worth paying for in economic terms.

Fuelwood

In less-developed countries, fuelwood has been the major source of fuel for centuries. In fact, fuelwood is the primary source of energy for nearly half of the world's population. In these regions, the primary use of fuelwood is for cooking.

The use of fuelwood as a prime energy source, a rapid population increase, and the high cost of other types of fuel have combined to create some serious environmental problems in many areas of the world. It is estimated that 1.3 billion people are not able to obtain enough fuelwood or must harvest fuelwood at a rate that exceeds its growth. This has resulted in the destruction of much forestland in Asia and Africa and has hastened the rate of desertification in these regions. (See figure 9.24.)

Because of its bulk and low level of energy compared to equal amounts of coal or oil, fuelwood is not practical to transport over a long distance, and most of it is used locally. In the United States, Norway, and Sweden, fuelwood furnishes 10 percent of the energy for home heating. Canada obtains 3 percent of its total energy, not just home-eating energy, from fuelwood. Most of this energy is used in forest product industries, such as lumbering and paper mills.

Burning wood is also a source of air pollution. Studies indicate that more than 75 organic compounds are released when wood is burned, 22 of which are hydrocarbons known or suspected to be carcinogens. However, these are usually released in small amounts. No studies have been conducted to determine if wood burning produces them in large enough quantities to be harmful. Often, woodstoves are not operated in the most efficient manner, and high amounts of particulate matter and other products of incomplete combustion, such as carbon monoxide, are released, contributing to ill health and death.

In areas with a high population density, the heavy use of fuelwood releases large amounts of fly ash into the air. In Missoula, Montana, in recent years, 55 percent of the particles in the air during the summer were from burning wood. In the winter, wood was responsible for 75 percent of the particles. A number of steps have been taken to reduce air pollution resulting from burning wood. Some cities, such as London, have a total ban on burning wood. Vail, Colorado, permits only one wood-burning stove per dwelling. Many areas require woodstoves to have special pollution controls that reduce the amount of particulates and other pollutants released.

Solid Waste

Residents of New York City discard 24,000 metric tons of waste each day, or 1.8 kilograms per person. In fact, New Yorkers

Figure 9.24

Desertification The demand for fuelwood in many regions has resulted in the destruction of forests. This is a major cause of desertification.

Table 9.2

Amount of Solid Waste Produced per Capita	
High-income cities	*Kilograms per day*
New York, United States	1.8
Singapore	1.6
Tokyo, Japan	0.94
Rome, Italy	0.72
Low-income cities	
Tunis, Tunisia	0.56
Medellín, Colombia	0.54
Calcutta, India	0.51
Kano, Nigeria	0.46

lead the world in the production of municipal trash. High-income cities such as New York usually produce more waste than low-income cities. (See table 9.2.) About 80 percent of this waste is combustible and, therefore, represents a potential energy source. (See figure 9.25.)

"Trash power," the use of municipal waste as a source of energy, requires several steps. First, the waste must be sorted so that the burnable organic material is separated from the inorganic material. The sorting is accomplished most economically by the person who produces the waste. That means the producer must separate trash before putting it out for pickup. It must be separated into garbage, burnable materials, glass, and metals, and picked up by compartmentalized collect-

ing trucks. Second, the efficient use of waste as a source of fuel requires a large volume and a dependable supply. If a community constructs a facility to burn 200 metric tons of waste a day, the community must generate and collect 200 metric tons of waste each day.

Communities have been burning trash as a means of reducing the volume of waste for a number of years. The first consolidated incineration of waste was done in Nottingham, England, in 1874. Burning the municipal waste of Munich, Germany, not only reduces the volume of the waste but supplies the energy for 12 percent of the city's electricity. Rotterdam, the Netherlands, operates a 55-megawatt power plant from its garbage. In the United States, only 3 percent of

Figure 9.25

Waste to Energy   Municipal trash can be burned to produce heat and electricity. This refuse pit is used to feed hoppers of high-temperature furnaces.

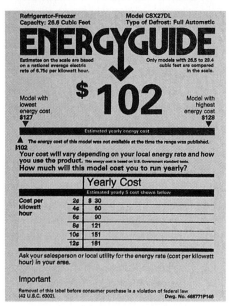

household waste is burned. This is low in comparison to the 26 percent burned in Japan, 51 percent in Sweden, and 75 percent in Switzerland.

Burning trash is not a way to make a profit. It is a way to decrease the cost of trash disposal because it reduces the need for landfill sites. Baltimore produces gas from 1,000 metric tons of waste per day. This gas generates steam for the Baltimore Gas and Electric Company. The daily waste also yields 80 metric tons of a charcoal-like material (char), 70 tons of ferrous metals, and 170 metric tons of glass. These products all have potential uses. The char can be burned as an additional source of energy. The ferrous metals can be sold, reducing the cost of operating the plant, as well as conserving mineral resources. The glass can be sold for recycling, further reducing the economic costs of the plant.

Although the burning of trash reduces the trash volume and furnishes energy, it poses environmental concerns, one of which is air pollution. Many of the older incinerating plants do not comply with today's air-quality standards. Also, much of the waste material, such as bleached paper and plastics, contains chlorine-containing organic compounds.

When burned, these compounds can form dioxins, which are highly toxic and suspected carcinogens. Another problem associated with waste to energy systems is the popularity of recycling. Many of the items that are now recycled such as plastics and wood, have high heat content. The reduction in the amount of these items in the waste stream reduces its value as an energy source.

Energy Conservation

Conservation is not a way of generating energy, but it is a way of reducing the need for additional energy consumption. Some conservation technologies are sophisticated, while others are quite simple. For example, if a small, inexpensive wood-burning stove were developed and used to replace open fires in the less-developed world, energy consumption in these regions could be reduced by 50 percent.

Several technologies that reduce energy consumption are now available. (See figure 9.26.) Highly efficient fluorescent light bulbs that can be used in regular incandescent fixtures give the same amount of light for 25 percent of the energy, and they produce less heat. Since lighting and

Figure 9.26

Energy Conservation   The use of fluorescent light bulbs, energy-efficient appliances, and low-emissive glass could reduce energy consumption significantly.

The James Bay Power Project

The James Bay power project is a development that would harness the energy of almost every drop of water in the rivers flowing through 350,000 square kilometers of northeastern Quebec—more that one-fifth of Canada's largest province. The water would be collected in vast reservoirs behind powerhouses on the main rivers. While some water would be released year-round to spin turbines and generate electricity, the system would be geared more to winter months, when demand for power is at its peak, primarily in the heavily populated northeastern United States. The reservoir levels would drop as much as 20 meters as water would be released and generating stations would be pushed to capacity. Cascading rivers would be dammed and diverted to create the reservoirs, flooding a combined area larger than the surface of Lake Ontario. Some rivers would be reduced to a trickle; others would simply be submerged.

The scene of all this proposed activity is a wilderness of lakes, rivers, lichens, and peat bogs along the east side of James Bay and the southeast coast of Hudson Bay. The region is home to many animal species, including moose, caribou, beaver, and lynx. The cold lakes and fast-flowing rivers are filled with fish, and the coastline is a rich habitat for fish, birds, whales, and seals. These resources are a crucial source of food and income for the Cree and Inuit natives. The coastal waters are internationally renowned resting and breeding grounds for millions of migratory birds.

According to Hydro-Quebec, the company undertaking the development, the $40 billion project of dikes, powerhouses, roads, and transmission lines could be inserted into an unspoiled northern wilderness with minimal environmental harm. Supporters of the project extol the jobs and income that would result and the subsequent exports of electricity it would bring for Quebec.

Hydro-Quebec has awarded billions of dollars worth of engineering and supply contracts to Quebec firms, enabling the firms to develop high-technology products and become international competitors. In addition, Hydro-Quebec says that every kilowatt of power from James Bay would cut the amount that power plants fueled by coal, oil, or nuclear energy would have to generate at greater risk to the environment.

Critics have argued that the project would create a few long-term jobs while taking a devastating toll on the environment. They also worry that the 9,000 local Cree and Inuit would lose their source of food, their livelihood, and their identity. Native people have battled the project from the outset. In 1975, after winning an injunction in Quebec Superior Court and then losing on appeal, the Grand Council of the Cree agreed to let the project proceed in return for some control over use of the land and in the environmental review process. In 1990, The Cree asked the Federal Environmental Assessment Review Office for a public review of the project. In 1992, the New York Power Authority, intended to be the Quebec power project's largest customer, put off final approval of a $15 billion contract to buy electricity, citing, in part, fears raised by the Crees that the project would flood a vast ecosystem and ruin their way of life.

Environmental concerns of the project include erosion, siltation, destruction of critical habitat, nutrient loading and algae blooms from decaying vegetation, water temperature increase, and the mixing of salt water and fresh water. In addition, a small clam that is the main source of food for millions of birds would be severely impacted. The loss of this clam from changes in salinity and temperature could threaten many species of migratory birds.

In 1994 the Quebec government halted the second major portion of the project when it cancelled a major hydroelectric development on the Great Whale River, which empties into Hudson Bay. Critics consider this to be a victory for the natives of the area and their environment. Supporters decry the loss of jobs and the economic development potential. While many questions remain unanswered, the overriding question is: Can humans limit their appetite for power so that megaprojects such as the one at James Bay do not need to be considered?

---

air conditioning (which removes the heat from inefficient incandescent lighting) account for 25 percent of U.S. electricity consumption, widespread use of these lights could significantly reduce energy consumption. Low-emissive glass for windows can reduce the amount of heat entering a building while allowing light to enter. The use of this glass in new construction and replacement windows could have a major impact on the energy picture. Many other technologies, such as automatic dimming devices or automatic light-shutoff devices, are being used in new construction.

The shift to more efficient use of energy needs encouragement. Often, poorly designed, energy-inefficient buildings and

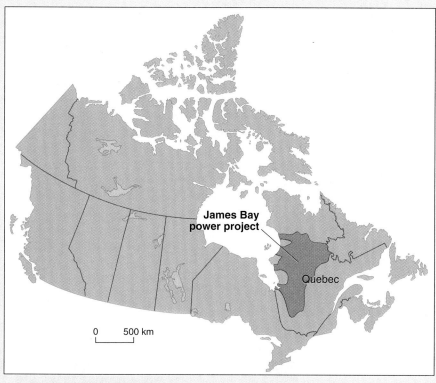

Location of the James Bay power project.

Native Cree fishing for trout on a free-flowing river in the area scheduled for development.

Dams like these have created vast, often sediment-laden reservoirs in the Northern Quebec wilderness.

machines can be produced inexpensively. The short-term cost is low, but the long-term cost is high. The public needs to be educated to look at the long-term economic and energy costs of purchasing poorly designed buildings and appliances.

Electric utilities have recently become part of the energy conservation picture. In some states, they have been allowed to make money on conservation efforts; previously they could only make money by building more power plants. This encourages them to become involved in energy conservation education, because teaching their customers how to use energy more efficiently allows them to serve more people without building new power plants.

The Arctic National Wildlife Refuge and Oil

The Artic National Wildlife Refuge has been a source of controversy for many years. The major players are environmentalists who seek to preserve this region as wilderness; the state of Alaska, which funds a major portion of its activities with dividends from oil production; Alaska residents, who receive a dividend payment from oil revenues; oil companies that want to drill in the refuge; and members of Congress who see the oil reserves in the region as important economic and political issues.

In 1960, 8.9 million acres were set aside as the Arctic National Wildlife Range. Passage of the Alaskan National Interest Lands Conservation Act in 1980 expanded the range to 19.8 million acres and established 8.6 million acres as wilderness. The act also renamed the area the Artic National Wildlife Refuge. There are international implications to this act. The refuge borders the Northern Yukon National Park. Many animals, particularly members of the Porcupine caribou herd, travel across the border on a regular yearly migration. The United States is obligated by treaty to protect these migration routes.

The act requires specific authorization from Congress before oil drilling or other development activities can take place on the coastal plain in the refuge. The coastal plain has the greatest concentration of wildlife, is the calving ground for the Porcupine caribou, and has the greatest potential for oil production. Several attempts have been made to authorize oil drilling in the refuge, most recently in 1996. In each case the potential authorization caused a collision of three forces: environmental protection, economic development, and political benefit. Furthermore, there are great differences of opinion within each of the competing interest groups. Some Alaskan citizens support drilling; others oppose it. Members of Congress are similarly split. Even members of the Department of the Interior have provided conflicting testimony about the risks and benefits of drilling for oil in the refuge.

These forces will continue to operate. As demand for energy increases, the pressures to drill will increase. Unfortunately, it is likely that the ultimate decision will be made for narrow political reasons rather than as a result of assessing the situation from a broader perspective.

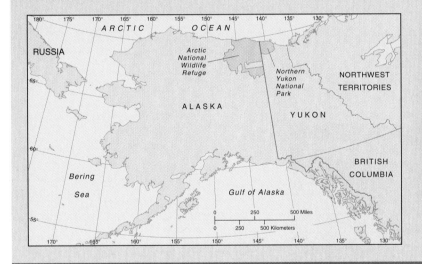

SUMMARY

A resource is a naturally occurring substance of use to humans that can potentially be extracted using current technology. Reserves are known deposits from which materials can be extracted profitably with existing technology under present economic conditions.

Coal is the world's most abundant fossil fuel. Coal is obtained by either surface mining or underground mining. Problems associated with coal extractions are disruption of the landscape due to surface mining and subsidence due to underground mining. Black lung disease, waste heaps, water and air pollution, and acid mine drainage are additional problems. Oil was originally chosen as an alternative to coal because it was more convenient and less expensive. However, the supply of oil is limited. As oil becomes less readily available, multiple offshore wells, secondary recovery methods, and increased oil exploration will become more common. Natural gas is another major source of fossil-fuel energy. The primary problem associated with natural gas is transport of the gas to consumers.

Fossil fuels are nonrenewable: The amounts of these fuels are finite. When the fossil fuels are exhausted, they will have to be replaced with other forms of energy, probably renewable forms. Hydroelectric power can be increased significantly, but its development must flood areas and in so doing may require the displacement of people. The use of geothermal and tidal energy is limited by geographic locations. Wind power may be used to generate electricity but may require wide, open areas and a large number of wind generators. Solar energy can be collected and used in either passive or active systems and can also be used to generate electricity. Lack of a constant supply of sunlight is solar energy's primary limitation. Fuelwood is a minor source of energy in industrialized countries but is the major source of fuel in many less-developed nations. Biomass can be burned to provide heat for cooking or to produce electricity, or it can be converted to alcohol or used to generate methane. In some communities, solid waste is burned to reduce the volume of the waste and also to supply energy.

Energy conservation can reduce energy demands without noticeably changing standards of living.

REVIEW QUESTIONS

1. Why are fossil fuels important?
2. Distinguish between reserves and resources.
3. What are the advantages of surface mining of coal compared to underground mining? What are the disadvantages of surface mining?
4. Compare the environmental impacts of the use of coal and the use of oil.
5. What are some limiting factors in the development of new hydroelectric generating sites?
6. What factors limit the development of tidal power as a source of electricity?
7. In what parts of the world is geothermal energy available?
8. Why can wind be considered a form of solar energy?
9. Compare a passive solar heating system with an active solar heating system.
10. What problems are associated with the use of solid waste as a source of energy?
11. List three energy conservation techniques.

KEY TERMS

acid mine drainage   149
active solar system   159
biomass   161
black lung disease   148
geothermal energy   155
liquefied natural gas   153
nonrenewable energy
   sources   144
overburden   148
passive solar system   159
photovoltaic cell   160
renewable energy sources   144
reserves   145
resources   145
secondary recovery   151
surface mining   148
underground mining   148

RELATED INTERNET SITES

Studies in earth science, fuel sciences, and mineral economics.
   http://www.ems.psu.edu/
Working solar house in Maine.
   http://solstice.crest.org/renewables/wlord/index.html
Award-winning site with a lot of information on solar energy uses.
   http://www.netins.net/showcase/solarcatalog/

Interesting site on wind energy. Lots of links to alternative energy sites.
   http://www.econet.org/awea/

chapter 10

Nuclear Energy: Benefits and Risks

Objectives

After reading this chapter, you should be able to:

- Explain how nuclear fission has the potential to provide large amounts of energy.
- Describe how a nuclear reactor produces electricity.
- Describe the basic types of nuclear reactors.
- Explain the steps involved in the nuclear fuel cycle.
- List concerns regarding the use of nuclear power.
- Explain the problem of decommissioning a nuclear plant.
- Describe how high-level radiation waste is stored.
- Describe the accident at Chernobyl.
- Explain how a breeder reactor differs from other nuclear reactors.
- List the technical problems associated with the design and operation of a liquid metal fast-breeder reactor.
- Explain the process of fusion.

Chapter Outline

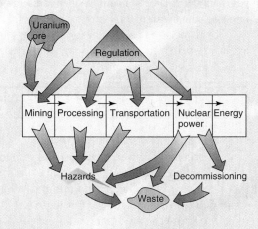

The Nature of Nuclear Energy

Energy from disintegrating atomic nuclei has a tremendous potential to do good for the people of the world. We routinely use X-rays to examine bones for fractures, treat cancer with radiation, and diagnose disease with the use of radioactive isotopes. We get a significant amount of our electrical energy from nuclear power plants. Engineers in the former Soviet Union used nuclear explosions to move large amounts of earth and rock to construct dams, canals, and underground storage facilities. The U.S. government briefly considered the possibility of using nuclear explosions to build a new Panama Canal. (See figure 10.1.)

On the other hand, nuclear energy has the potential to do great harm. The Japanese cities of Hiroshima and Nagasaki were destroyed by nuclear bombs. Military uses of nuclear energy have left a legacy of radioactive wastes. In many cases these wastes have been mismanaged or carelessly disposed of.

In order to understand where nuclear energy comes from, it is necessary to review some of the aspects of atomic structure presented in chapter 3. All atoms are composed of a central region called the nucleus, which contains positively charged protons, and neutrons that have no charge. Moving around the nucleus are smaller, negatively charged electrons.

Since the positively charged particles in the nucleus repel one another, energy is needed to hold the protons and neutrons together. However, some isotopes of atoms are **radioactive;** that is, the nuclei of these atoms are unstable and spontaneously decompose. Neutrons, electrons, protons, and other larger particles are released during nuclear disintegration, and a great deal of energy is released as well. The rate of decomposition is consistent for any given isotope. It is measured and expressed as **radioactive half-life,** which is the time it takes for one-half of the radioactive material to spontaneously decompose. Table 10.1 lists the half-lives of several radioactive isotopes.

Nuclear disintegration releases energy from the nucleus as **radiation,** of which there are three major types: **Alpha radiation** consists of a moving particle composed of two neutrons and two protons. Alpha radiation usually travels through air for only about a meter and can be stopped by a sheet of paper. **Beta radiation** consists of electrons released from nuclei. The electrons also travel only a short distance through air and are stopped by a layer of clothing. **Gamma radiation** is a type of electromagnetic radiation, like X-rays, light, and radio waves. It can pass through several centimeters of concrete. If the radiation reaches living tissue, equivalent doses of beta and gamma radiation cause equal amounts of biological damage. Alpha radiation can cause up to twenty times more damage than beta or gamma radiation because it is able to cause more changes in tissue as it passes through.

The release of neutrons is particularly important in obtaining energy from nuclear disintegration. In addition to releasing alpha, beta, and gamma radiation when they disintegrate, the nuclei of atoms release neutrons. When moving neutrons hit the nuclei of certain other atoms, they cause those nuclei to split. This process is known as **nuclear fission.** If these splitting nuclei also release neutrons, they can strike the nuclei of other atoms, which also disintegrate, resulting in a continuous process called a **nuclear chain reaction.** Only certain kinds of atoms are suitable for the development of a nuclear chain reaction. The two materials commonly used in nuclear reactions are uranium 235 and plutonium 239. In addition, there must be a certain quantity of nuclear fuel (a critical mass) in order for a nuclear chain reaction to occur. It is this process that results in the large amounts of energy released from bombs or nuclear reactors.

The History of Nuclear Energy Development

The first controlled fission of an atom occurred in Germany in 1938, but the United States was the first country to develop an atomic bomb. In 1945, the U.S. military dropped atomic bombs on the Japanese cities of Hiroshima and Nagasaki. The incredible devastation of

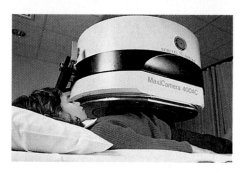

Figure 10.1

Uses of Nuclear Energy Each of these photographs illustrates the uses of the energy available from the splitting of atoms.

these two cities demonstrated the potential of nuclear energy for destruction and for peaceful uses. For many years, most atomic research involved military applications of nuclear energy as bombs and as power sources for ships. During the 50 years following World War II, the two major military powers of the world—the United States and the Soviet Union—conducted secret nuclear research projects related to the building and testing of bombs. This continued to be a primary focus of nuclear research until the recent changes in the former Soviet Union, which led to a world in which nuclear war is much less of a concern. A legacy of this military research is a great deal of soil, water, and air contaminated with

Table 10.1

The Half-life of Some Radioactive Isotopes	
Radioactive isotope	**Half-life**
Iodine 132	2.4 hours
Technetium 99	6.0 hours
Rhodium 105	36.0 hours
Xenon 133	5.3 days
Barium 140	12.8 days
Cerium 144	284.0 days
Cesium 137	30 years
Carbon 14	5,730 years
Uranium 234	250,000 years
Chlorine 36	300,000 years
Beryllium 10	4.5 million years
Potassium 40	1.3 billion years
Helium 4	12.5 billion years

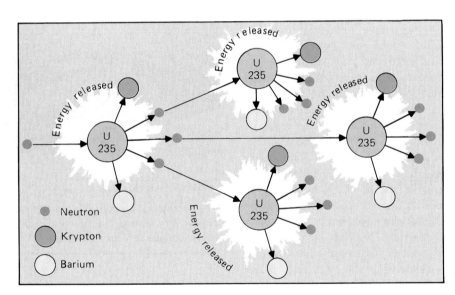

Figure 10.2

Nuclear Fission Chain Reaction   When a neutron strikes a nucleus of U-235, energy is released, krypton and barium are produced, and several neutrons are released. These new neutrons may strike other atoms of U-235, causing their nuclei to split. This series of events is called a chain reaction.

radioactive material. Many of these contaminated sites have come to light recently and require major cleanup efforts. The U.S. Department of Energy has begun to clean up the pollution created by its weapons production activities.

After World War II, people began to see the potential for using nuclear energy for peaceful purposes rather than as weapons. The world's first electricity-generating reactor was constructed in the United States in 1951, and the Soviet Union built its first reactor in 1954. In December 1953, President Dwight D. Eisenhower, in his "Atoms for Peace" speech, made the following prediction:

Nuclear reactors will produce electricity so cheaply that it will not be necessary to meter it. The users will pay an annual fee and use as much electricity as they want.

Atoms will provide a safe, clean, and dependable source of electricity.

More than 40 years have passed since Eisenhower's predictions. Although nuclear power currently is being used throughout the world as a reliable source of electricity, it has not fulfilled such overly optimistic promises. Several serious accidents have caused worldwide concern about safety, and construction of most new nuclear power projects has stopped. At the same time, many energy experts predict a rebirth of the nuclear power industry as energy demands increase and a new generation of safer nuclear power plants is designed. These experts believe the public will favor nuclear power plants because they do not generate carbon dioxide, which is thought to contribute to global warming.

Nuclear Reactors

A **nuclear reactor** is a device that permits a controlled nuclear fission chain reaction. In the reactor, neutrons are used to cause a controlled fission of heavy atoms, such as uranium. **Uranium-235 (U-235)** is a radioactive isotope used to fuel nuclear fission reactors. When the nucleus of a U-235 atom is struck by a slowly moving neutron from another atom, the nucleus of the atom of U-235 is split. When this occurs, two to three rapidly moving neutrons are released, along with large amounts of energy. This energy is an important product of nuclear fission reactions. The neutrons released strike the nuclei of other atoms of U-235 and also cause them to undergo fission, which, in turn, releases more energy and more neutrons, thus resulting in a chain reaction. (See figure 10.2.) Once begun, this chain reaction continues to release energy until the fuel is spent or the neutrons are prevented from striking other nuclei.

In addition to fuel rods containing uranium, reactors contain control rods of cadmium, boron, graphite, or other non-fissionable materials used to control the rate of fission by absorbing neutrons. When control rods are lowered into a reactor, they absorb the neutrons produced by fissioning uranium. There are fewer neutrons to continue the chain reaction, and the rate of fission decreases. If the control rods are withdrawn, more fission occurs, and more particles, radiation, and heat are produced.

The fuel rods housed in a reactor are surrounded by water or some other type of moderator. A **moderator** absorbs energy, which slows neutrons, enabling them to split the nuclei of other atoms more effectively. Fast-moving neutrons are less effective at splitting atoms than slow-moving neutrons. As U-235 undergoes fission, the energy of the fast-moving neutrons is transferred to water; the neutrons slow down, and the water is heated.

In the production of electricity, a nuclear-powered reactor serves the same function as any fossil-fueled boiler. It produces heat, which converts water to steam to operate a turbine that generates electricity. After passing through the turbine, the steam must be cooled, and the water is returned to the reactor to be heated again.

Various types of reactors have been constructed to furnish heat for the production of steam. They differ in the moderator used, in how the reactor core is cooled, and in how the heat from the core is used to generate steam. Water is the most commonly used reactor-core coolant and also serves as a neutron moderator. **Light-water reactors (LWR),** which make up 75 percent of reactors operating today, use ordinary water, which contains the lightest, most common isotope of hydrogen, having an atomic mass of one. The two types of LWRs are **boiling-water reactors (BWR)** and **pressurized-water reactors (PWR).**

In a BWR (the type of construction used in about 50 percent of the nuclear reactors in the world), the water functions as both a moderator and reactor-core coolant. (See figure 10.3.) Steam is formed within the reactor and transferred directly to the turbine, which generates electricity. A disadvantage of the BWR is that the steam passing to the turbine must be treated to remove any radiation. Even then, some radioactive material is left in the steam; therefore, the generating building must be shielded.

In a PWR (the type of construction used in over 20 percent of the nuclear reactors in the world), the water is kept under high pressure so that steam is not allowed to form in the reactor. (See figure 10.4.) A secondary loop transfers the heat from the pressurized water in

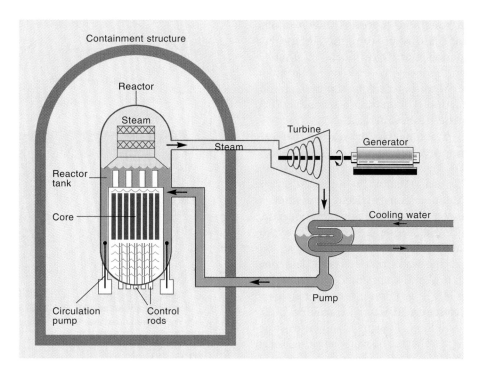

Figure 10.3

Boiling-Water Reactor (BWR)   A boiling-water reactor is a type of light-water reactor that produces steam to directly power the turbine and produce electricity. Water is used as a moderator and as a reactor-core coolant.

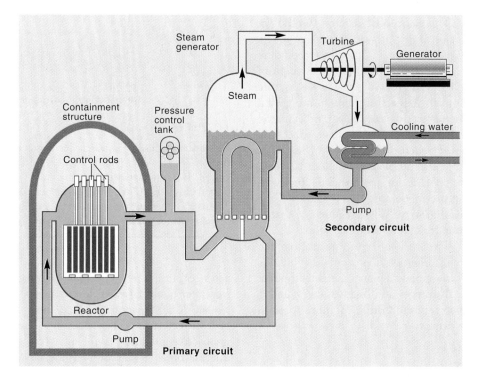

Figure 10.4

Pressurized-Water Reactor (PWR)   A pressurized-water reactor is a type of light-water reactor that uses a steam generator to form steam and a secondary loop to transfer this steam to the turbine.

the reactor to a steam generator. The steam is used to turn the turbine and generate electricity. Such an arrangement reduces the risk of radiation in the steam but adds to the cost of construction by requiring a secondary loop for the steam generator.

A second type of reactor that uses water as a coolant is a **heavy-water reactor (HWR),** developed by Canadians. It uses water that contains the hydrogen isotope deuterium in its molecular structure as the reactor-core coolant and moderator. Since the deuterium atom is twice as heavy as the more common hydrogen isotope, the water that contains deuterium weighs slightly more than ordinary water. An HWR also uses a steam generator to convert regular water to steam in a secondary loop. Thus, it is similar in structure to a PWR. The major advantage of an HWR is that naturally occurring uranium isotopic mixtures serve as a suitable fuel. This is possible because heavy water is a better neutron moderator than is regular water, while other reactors require that the amount of U-235 be enriched to get a suitable fuel. Since it does not require enriched fuel, the operating costs of an HWR are less than that of an LWR.

The **gas-cooled reactor (GCR)** was developed by atomic scientists in the United Kingdom. Originally, carbon dioxide served as a coolant for a graphite-moderated core. As in the HWR, natural isotopic mixtures of uranium are used as a fuel. Presently, the GCR has been modified into the **high-temperature, gas-cooled reactor (HTGCR).** (See figure 10.5.) In this reactor, helium serves as the coolant for an enriched fuel in a graphite-moderated core. The HTGCR is similar to a PWR in the use of a steam generator and a secondary loop.

The various types of reactors represent differing approaches to building safe, economical plants. Each method has its advantages and disadvantages, and its supporters and critics. Modifications will continue to be made to these basic types.

Breeder Reactors

During the early stages of the development of nuclear power plants, breeder re-

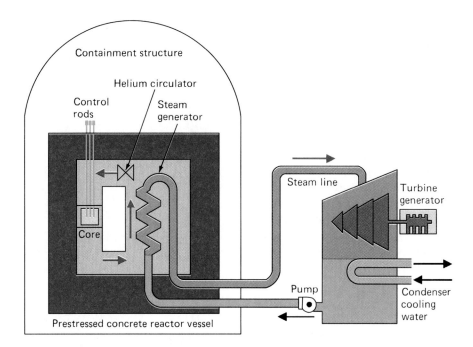

Figure 10.5

High-Temperature, Gas-Cooled Reactor (HTGCR) This type of reactor uses graphite as a moderator and the inert gas helium as the reactor-core coolant. A steam generator forms steam and a secondary loop is used to transfer steam to the turbine.

Source: Atomic Industrial Forum, U.S. Council for Energy Awareness.

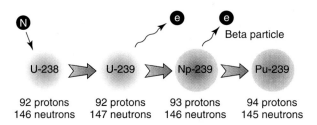

Figure 10.6

Formation of Pu-239 in a Breeder Reactor When a fast-moving neutron (N) is absorbed by the nucleus of a U-238 atom, a series of reactions results in the formation of Pu-239 from U-238. Two intermediate atoms are U-239 and Np-239, which release beta particles (electrons) from their nuclei. (A neutron in the nucleus can release a beta particle and become a proton.) This is an important reaction because, while the U-238 does not disintegrate readily and therefore is not a nuclear fuel, the Pu-239 is fissionable and can serve as a nuclear fuel.

actor construction was seen as the logical step after nuclear fission development. A regular fission reactor produces heat to generate electricity but does not form radioactive products useful as a fuel. A **nuclear breeder reactor** is a nuclear fission reactor that produces heat to be converted to steam to generate electricity and also forms a new supply of radioactive isotopes. If a fast-moving neutron hits a non-

radioactive uranium-238 (U-238) nucleus and is absorbed, an atom of radioactive **plutonium-239 (Pu-239)** is produced. (See figure 10.6.) In a breeder reactor, water is not used as a moderator because water slows the neutrons too much and Pu-239 is not produced. Breeder reactors need a moderator that allows the neutrons to move more rapidly and that also has good heat transfer properties.

The **liquid metal fast-breeder reactor (LMFBR)** appears to be the most promising model. In this type of reactor, the fuel rods in the core are surrounded by rods of U-238 and liquid sodium. The energy of the neutrons released from U-235 in the fuel rods heats the sodium to a temperature of 620°C. In addition to furnishing heat, these fast-moving neutrons are absorbed by the rods containing U-238, and some of these atoms are converted into Pu-239. After approximately 10 years of operation, during which electricity is produced, the LMFBR will have also produced enough radioactive material to operate a second reactor.

There are some serious drawbacks to the LMFBR. Sodium reacts violently if it comes into contact with water or air. Therefore, costly, highly specialized equipment is required to contain and pump the sodium. If these systems fail, the sodium boils, which allows the chain reaction to proceed at a faster rate and could damage the reactor, leading to a nuclear accident.

There are also problems in the startup of an LMFBR. When a breeder reactor is starting up or going on-line after a shutdown, the solid sodium cannot be moved by the pumps until it becomes a liquid. This presents a technical problem in developing LMFBRs.

Another problem is that reaction rates are extremely rapid and very difficult to regulate. The instrumentation needed to monitor an LMFBR must be of extremely high quality because control of the reaction involves precise adjustments over very short periods of time.

Finally, the product of a breeder reactor, plutonium-239, is extremely hazardous to humans who come in contact with it. And, because plutonium-239 can be made into nuclear weapons, it must be transported, processed, or produced under very close security. The more breeder reactors in use, the more difficult the security problems and the more likely the chance that the small amount of plutonium needed to manufacture a bomb could be stolen.

Because of these problems, no breeder reactors are scheduled for commercial use in the United States. In fact, the only experimental U.S. breeder reactor is sched-

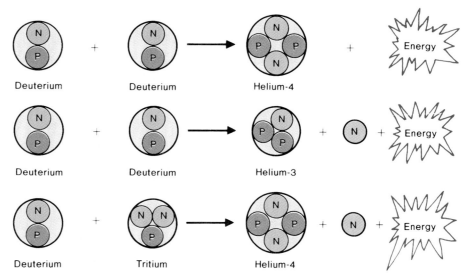

Figure 10.7

Nuclear Fusion In nuclear fusion, small atomic nuclei are combined to form heavier nuclei. Large amounts of energy are released when this occurs. Different isotopes of hydrogen can be used in the process of fusion. There are three possible types of fusion: (1) Two deuterium isotopes can combine to form helium-4 and energy; (2) two deuterium isotopes can combine to form helium-3, a free neutron, and energy; and (3) a deuterium and a tritium isotope can combine to form helium-4, a free neutron, and energy.

uled to be shut down. The Clinch River Fast-Breeder Reactor Plant near Knoxville, Tennessee, which was to be operational in the 1970s, never was completed.

In Europe, as in the United States, the development of breeder reactors has slowed. In 1981, after running smoothly for eight years, a 250-megawatt French prototype plant developed a leak in the cooling system that resulted in a sodium fire. This raised some serious questions about the safety of breeder reactors. In 1982, the French government announced that it was scaling down the planned construction of breeder reactors from five to one. Today, it is doubtful that even that one will be built.

Nuclear Fusion

Another aspect of nuclear power that may have promise for the future involves the even more advanced technology of nuclear fusion. When two lightweight atomic nuclei combine to form a heavier nucleus, a large amount of energy is released. This process is known as **nuclear fusion.** The energy produced by the sun is the result of fusion. Most studies of fusion have in-

volved small atoms like hydrogen. Most hydrogen atoms have one proton and no neutrons in the nucleus. The hydrogen isotope deuterium (H^2) has a neutron and a proton. Tritium (H^3) has a proton and two neutrons. When deuterium and tritium isotopes combine to form heavier atoms, large amounts of energy are released. (See figure 10.7.) The energy that would be released by combining the deuterium in 1 cubic kilometer of ocean water would be greater than that contained in the world's entire supply of fossil fuels.

Although fusion could solve the world's energy problems, technology must answer several questions before we can actually use fusion power. Three conditions must be met simultaneously if fusion is to occur: high temperature, adequate density, and confinement. If heat is used to provide the energy necessary for fusion, the temperature must approach that of the center of the sun. At the same time, the walls of the vessel confining the atoms must be protected from the heat, or they will vaporize. However, the main problem is containment of the nuclei. Because they have a positive electric charge, the nuclei repel one another.

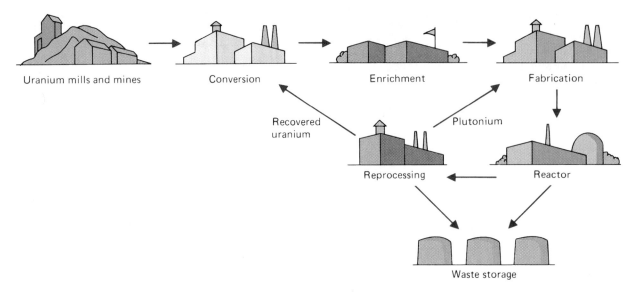

Figure 10.8

Steps in the Nuclear Fuel Cycle The process of obtaining nuclear fuel involves mining, extracting the uranium from the ore, concentrating the U-235, fabricating the fuel rods, installing and using the fuel in a reactor, and disposing of the waste. Some countries reprocess the spent fuel as a way of reducing the amount of waste they must deal with.

Source: Data from U.S. Department of Energy.

Even though, in theory, fusion promises to furnish a large amount of energy, technical difficulties appear to prevent its commercial use in the near future. Even the governments of nuclear nations are budgeting only modest amounts of money for fusion research. And, as with nuclear fission and the breeder reactor, economic costs and fear of accidents may continue to delay the development of fusion reactors.

The Nuclear Fuel Cycle

In order to appreciate the consequences of using nuclear fuels to generate energy, it is important to understand how the fuel is processed. The nuclear fuel cycle begins with the mining operation. (See figure 10.8.) Low-grade uranium ore is obtained by underground or surface mining. The ore contains about 0.2 percent uranium by weight. After it is mined, the ore goes through a milling process. It is crushed and treated with a solvent to concentrate the uranium. Milling produces yellow-cake, a material containing 70 to 90 percent uranium oxide.

Naturally occurring uranium contains about 99.3 percent U-238 and only 0.7 per-cent U-235. This concentration of U-235 is not high enough for most types of reactors, so the amount of U-235 must be increased by enrichment. Since the masses of the isotopes U-235 and U-238 vary only slightly, and there is no chemical difference, enrichment is a difficult and expensive process. However, it increases the U-235 content from 0.7 percent to 3 percent.

Fuel fabrication converts the enriched material into a powder, which is then compacted into pellets about the size of a pencil eraser. These pellets are sealed in metal fuel rods about 4 meters in length, which are then loaded into the reactor.

As fission occurs, the concentration of U-235 atoms decreases. After about three years, a fuel rod does not have enough radioactive material to sustain a chain reaction, and the spent fuel rods must be replaced by new ones. The spent rods are still very radioactive, containing about 1 percent U-235 and 1 percent plutonium. These rods are a major source of the radioactive waste material produced by a nuclear reactor.

When nuclear reactors were first being built, scientists proposed that spent fuel rods could be reprocessed. The re-maining U-235 could be enriched and used to manufacture new fuel rods. Since plutonium is radioactive, it could also be fabricated into fuel rods. Besides providing new fuel, reprocessing would reduce the amount of nuclear waste. However, the cost of producing fuel rods by reprocessing was found to be greater than the cost of producing fuel rods from ore, and the United States closed its reprocessing facilities. At present, India, Japan, Russia, France, and the United Kingdom operate reprocessing plants that reprocess spent fuel rods as an alternative to storing them as nuclear waste.

Each step in the nuclear fuel cycle involves the transport of radioactive materials. The uranium mines are some distance from the processing plants. The fuel rods must be transported to the power plants, and the spent rods must be moved to a reprocessing plant or storage area. Each of these links in the fuel cycle presents the possibility of an accident or mishandling that could release radioactive material. Therefore, the methods of transport are extremely carefully designed and tested before they are used. Many people are convinced that the transport of radioactive materials is hazardous, while others are satisfied that the utmost care is being taken and that the risks are extremely small.

Ultimately, both high-level and low-level radioactive wastes must be stored. Thus, each step in the nuclear fuel cycle, from the mining of uranium to the storage of nuclear waste, poses health and environmental concerns.

Nuclear Material and Weapons Production

Producing nuclear materials for weapons and other military uses involves many of the same steps used to produce nuclear fuel for power reactors. In fact, the nuclear power industry is an outgrowth of the weapons industry. In the United States, the Department of Energy currently is responsible for nuclear research for both weapons and peaceful uses and stewardship of the facilities used for research and weapons production. Some facilities are used for both processing nuclear fuel and providing materials for weapons. In both cases, uranium must be mined, concentrated, and transported to sites of use. Furthermore, military uses involve the production and concentration of plutonium. The production and storage facilities invariably become contaminated, as does the surrounding land.

Research and production facilities have typically dealt with hazardous chemicals and low-level radioactive wastes by burying them, pumping them into the ground, storing them in ponds, or releasing them into rivers. Despite environmental regulations that prevented such activities, the Department of Energy (formerly the Atomic Energy Commission) maintained that it was exempt from such federal environmental legislation. As a result, the DOE has become the steward of a large number of sites that are contaminated with both hazardous chemicals and radioactive materials. The magnitude of the problem is huge. There are 3,365 square miles of DOE properties that are or have been involved in weapons development or production. These include

- 3,700 contaminated sites,
- 330 underground storage tanks with high-level radioactive waste,
- more than a million 55 gallon drums of radioactive, hazardous, or mixed waste in storage,
- 5,700 sites where wastes are moving through the soil, and
- millions of cubic meters of low-level and high-level radioactive wastes.

The Department of Energy has pledged to clean up these sites by 2019. Environmental cleanup is now the largest single item in its budget. Several U.S. sites are currently being cleaned up, but things are not going smoothly. Local residents and the states that are hosts to these facilities distrust the DOE and are insisting that the sites be cleaned completely. This may not be technologically or economically possible. Furthermore, environmental cleanup is a new mission for the DOE and the department is having difficulty adjusting. The cleanup process will take many years and require the expenditure of tens of billions of dollars.

An additional problem has arisen as a result of the reduced importance of nuclear weapons. The political disintegration of the Soviet Union and Eastern Europe has made large numbers of nuclear weapons, both in the East and West, unnecessary. This has probably made the world a safer place, but how are nations of the world to dispose of their nuclear weapons? Some nuclear material can be diverted to fuel use in nuclear reactors, and some reactors can be modified to accept enriched uranium or plutonium. The security of these materials, particularly in the former Soviet Union, is a concern. Investigators have uncovered several incidents in which plutonium has been sold to other countries.

Nuclear Power Concerns

Nuclear power has provided a significant amount of electricity for the people of the world. Currently, over 7 percent of the energy consumed worldwide comes from nuclear power. However, several accidents have raised questions about safety, radiation has been released into the air

and water, the production and transportation of fuel have caused contamination, and the disposal of wastes is a continuing problem.

Reactor Safety: The Effects of Three Mile Island and Chernobyl

Although there have been accidents at nuclear power plants since they were first used, two relatively recent accidents have had the greatest effect on people's attitudes toward nuclear power plant safety: Three Mile Island in 1979 and Chernobyl in 1986. (See Environmental Close-up: Three Mile Island).

Chernobyl is a small city in the Ukraine near the border with Belarus, north of Kiev. As is true of many small cities in the world, most people had never heard of it. (See figure 10.9.) However, in the spring of 1986, the world's largest nuclear accident catapulted Chernobyl into the news.

At 1:00 A.M. on April 25, 1986, at Chernobyl Nuclear Power Station-4, a test was begun to measure the amount of electricity that the still-spinning turbine would produce if the steam were shut off. This was important information since the emergency core cooling system required energy for its operation and the coasting turbine could provide some of that energy until another source became available. The amount of steam to the turbine was reduced by lowering the control rods into the reactor. But the test was delayed because of a demand for electricity. This delay resulted in a shift change of workers as the test was in progress.

Several serious safety violations and mistakes contributed to the explosion. At 11:10 P.M. on April 25, the monitoring units were set for low-level operation, but the operators failed to program the computer to maintain power at 700 megawatts, and output dropped to 30 megawatts. This presented a need to rapidly increase the power, and many of the control rods were withdrawn. Meanwhile, an inert gas (xenon) had accumulated on the fuel rods. The gas absorbed the neutrons and slowed the rate of power increase. In an attempt to obtain more power, operators withdrew all the control rods. This was a second serious safety violation.

environmental *close-up*

Three Mile Island

On March 14, 1979, the valves to three auxiliary pumps at the unit two reactor at the Three Mile Island nuclear plant closed. At 4:00 a.m. on March 28, 1979, the main pump at unit two broke down. The auxiliary pumps failed to operate because of the closed valves, and the electrical-generating turbine stopped. At this point, an emergency coolant should have flooded the reactor and stabilized the temperature. The coolant did start to flow, but a faulty gauge indicated that the reactor was already flooded. Relying on this faulty reading, an operator overrode the automatic emergency cooling system and stopped it.

Without the emergency coolant, the reactor temperature rose rapidly. Two hours after the main pump failed, the temperature and pressure were still rising, and an alarm sounded. An hour later, the control rods stopped fission, but a partial core meltdown had occurred. Later that day, as pressure in the primary loop increased, radioactive steam was vented into the atmosphere. Also, part of the plant's ventilating system was contaminated with radioactive gases.

Problems continued to develop, and on March 29, a bubble of hydrogen formed at the top of the reactor. The next day, children and pregnant women were advised to evacuate the area. Five days after the accident, it was announced that the immediate danger had passed. But by this time, unit two

had been severely damaged. More than a year passed before anyone entered the plant.

Following the accident at Three Mile Island, Metropolitan Edison, the operators of the plant, and the Nuclear Regulatory Commission were criticized for their actions at the time of the accident and their failure to clean up afterward. The crippled reactor was eventually defueled in 1990 at a cost of about $1 billion. It has been placed in monitored storage until the unit one reactor, which is still operating reaches the end of its useful life. At that time both reactors will be decommissioned.

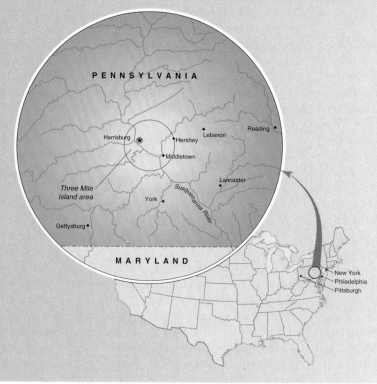

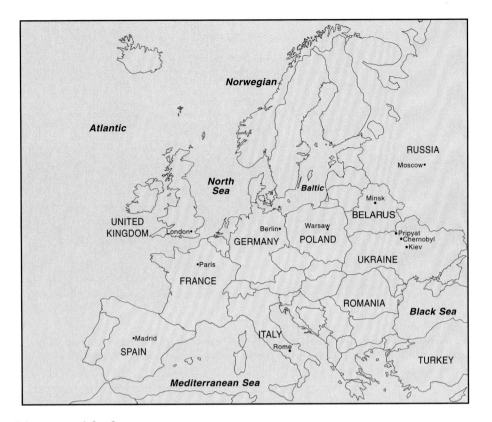

Figure 10.9

Chernobyl The town of Chernobyl became infamous as the location of the world's worst nuclear power plant accident.

Figure 10.10

The Accident at Chernobyl An uncontrolled chain of reactions in the reactor of unit four resulted in a series of explosions and fires. See the circled area in the photograph.

At 1:00 A.M. on April 26, the operators shut off most emergency warning signals and turned on all eight pumps to provide adequate cooling for the reactor following the completion of the test. Just as final stages of the test were beginning, a signal indicated excessive reaction in the reactor. In spite of the warning, the operators manually blocked the automatic reactor shutdown and began the test.

As the test continued, the power output of the reactor rose beyond its normal level and continued to rise. The operators activated the emergency system designed to put the control rods back into the reactor and stop the fission. But it was too late. The core had already been deformed, and the rods would not fit properly; the reaction could not be stopped. In 4.5 seconds, the energy level of the reactor increased two thousand times. The fuel rods ruptured, the cooling water turned into steam, and a steam explosion occurred. The lack of cooling water allowed the reactor to explode. The explosion blew the 1,000 metric ton concrete roof from the reactor and the reactor caught fire. In less than 10 seconds, Chernobyl became the scene of the world's worst nuclear accident. A core meltdown had occurred. (See figure 10.10.) It took 10 days to bring the runaway reaction under control. By November, the damaged reactor was entombed in a concrete covering, but the hastily built structure, known as the sarcophagus, may have structural flaws. The Ukrainian government is planning to have a second containment structure built around the current structure.

The immediate consequences were 31 fatalities; 500 persons hospitalized, including 237 with acute radiation sickness; and 116,000 people evacuated. Of the evacuees, 24,000 received high amounts of radiation. The delayed effects are more difficult to assess. Many people suffer from illnesses they feel are related to their exposure to the fallout from Chernobyl. In 1996 it became clear that at least one delayed effect was strongly correlated with exposure. Children or fetuses exposed to fallout are showing increased frequency of thyroid cancer. The thyroid gland accumulates iodine, and radioactive iodine 131 was released from Chernobyl. It is still too early to tell if there are other delayed health effects.

More than a year after the disaster at Chernobyl, the decontamination of 27 cities and villages within 40 kilometers of Chernobyl was considered finished. This does not mean that the area was cleaned up, only that all practical measures were completed. Some areas were simply abandoned. The largest city to be affected was Pripyat, which had a population of 50,000 and was only 4 kilometers from the reactor. A new town was built to accommodate those displaced by the accident, and Pripyat remains a ghost town. Sixteen other communities within the 40-kilometer radius have been cleaned up, and the original inhabitants have returned, although there is still controversy about the safety of these towns.

One important impact of Chernobyl is that it deepened public concern about the safety of nuclear reactors. Even before Chernobyl, between 1980 and 1986, the governments of Australia, Denmark, Greece, Luxembourg, and New Zealand had officially adopted a "no nuclear" policy. Since 1980, 10 countries have canceled nuclear plant orders or mothballed plants under construction. Argentina canceled four plants; Brazil, eight; China, eight; Mexico, 18; and the United States, 54. There have been no orders for new plants in the United States since 1974. Sweden, Austria, and the Philippines have decided to phase out and dismantle their nuclear power plants.

Before Chernobyl, 65 percent of the people in the United Kingdom were opposed to nuclear power plants; after Chernobyl, 83 percent were against them. In Germany, opposition increased from 46 percent to 83 percent. Opposition in the United States rose from 67 to 78 percent. (See figure 10.11.) Even the French, who have the greatest commitment to nuclear power, were against it by 52 percent. The number of new nuclear plants being constructed has been reduced. In 1996 there were 34 nuclear plants under construction—before Chernobyl there were 160. Many of these plants were cancelled. In addition, many plants have been shut down prematurely, resulting in the prediction that the amount of energy furnished by nuclear fission reactors will actually decline in the future.

Figure 10.11

Anti-nuclear Demonstration   This anti-nuclear demonstration held in 1991 at Rocky Flats, Colorado, is an expression of the concern people have about the safety of using nuclear reactions for power or weapons.

At the same time some people are predicting the death of nuclear power as a viable energy source, others are examining ways to make nuclear power facilities safer. They expect nuclear power to be an important energy source in the future. At both Three Mile Island and Chernobyl, operator error caused or contributed to the accidents. Operators manually stopped normal safety actions from taking place. However, a contributing factor was the design: active mechanical processes had to work properly to shut the reactors down. Many of the new designs for reactors include passive mechanisms that will shut down the reactor, special catching basins for the reactor core should it melt down, and better containment buildings.

Nuclear power will continue to be a part of the energy mix, particularly in countries that lack fossil fuel reserves. Furthermore, as fossil fuels are used up, the pressure for energy will probably create a market for nuclear power technology.

Exposure to Radiation

Although nuclear accidents are spectacular, frightening, and can cause immediate death because of the incredible amount of energy involved, the long-term problems resulting from exposure to radiation are even more worrisome. Radiation is converted to other forms of energy when it is absorbed by matter. When organisms are irradiated, this energy conversion causes damage at a cellular, tissue, organ, or organism level. The degree and kind of damage vary with the kind of radiation, the amount of radiation, the duration of the exposure, and the type of cells irradiated.

Radiation can also cause mutations, which are changes in the genetic messages within cells. Mutations can cause two quite different kinds of problems. Mutations that occur in the ovaries or testes can form mutated eggs or sperm, which can lead to abnormal offspring. Care is usually taken to shield these organs from unnecessary radiation. Mutations that occur in other tissues of the body may manifest themselves as abnormal tissue growths known as cancer. Two common cancers that are strongly linked to increased radiation exposure are leukemia and breast cancer. Because mutations are essentially permanent, they may accumulate over time. Therefore, the accumulated effects of radiation over many years may result in the development of cancer later in life.

Table 10.2

Radiation Effects		
Source	**Dose**	**Biological effect**
Nuclear bomb blast or exposure in a nuclear facility	100,000 rems/incident	Immediate death
X-rays for cancer patients	10,000 rems/incident	Coma, death within one to two days
	1,000 rems/incident	Nausea, lining of intestine damaged, death in one to two weeks
	100 rems/incident	Increased probability of leukemia
	10 rems/incident	Early embryos may show abnormalities
Upper limit for occupationally exposed people	5 rems/year	Effects difficult to demonstrate
X-ray of the intestine	1 rem/procedure	Effects difficult to demonstrate
Upper limit for release from nuclear installations (except nuclear power plants)	0.5 rem/year	Effects difficult to demonstrate
Natural background radiation	0.1–0.2 rem/year	Effects difficult to demonstrate
Upper limit for release by nuclear power plants	0.005 rem/year	Effects difficult to demonstrate

Human exposure to radiation is usually measured in **rems** (*Roentgen Equivalent Man*), a measure of the biological damage to tissue. The effects of large doses (1,000 to 1,000,000 rems) are easily seen and can be quantified, because there is a high incidence of death at these levels, but demonstrating known harmful biological effects from smaller doses is much more difficult. (See table 10.2.) Moderate doses (10 to 1,000 rems) are known to increase the likelihood of cancer and birth defects. The higher the dose, the higher the incidence of abnormality. Lower doses may cause temporary cellular changes, but it is difficult to demonstrate long-term effects. Thus, the effects of low-level, chronic radiation generate much controversy. Some people feel that all radiation is harmful, that there is no safe level, and that special care must be taken to prevent exposure. (See figure 10.12.) Others feel that the increased risk of low-level radiation is extremely small and that current radiation standards are adequate to protect the public, especially in light of the benefits of radiation, such as medical diagnoses and electrical energy. Current research is trying to assess the risks associated with repeated exposure to low-level radiation.

Each step in the nuclear fuel cycle poses a radiation exposure problem, beginning with the mining of uranium. Al-

Figure 10.12

Protective Equipment Persons working in an area subjected to radiation must take steps to protect themselves. These workers are wearing protective clothing and filtering the air they breathe.

though the radiation level in the ore is low, miners' prolonged exposure to low-level radiation increases their rates of certain cancers, such as lung cancer. Uranium miners who smoke have an even higher lung cancer rate. After mining, the ore must be crushed in a milling process. This releases radioactive dust into the at-

mosphere, so workers are chronically exposed to low levels of radioactivity. The crushed rock is left on the surface as mine tailings. There are over 150 million metric tons of low-level radioactive mine tailings in the United States, and at least as much in the rest of the world. These tailings constitute a hazard because they can be dispersed into the environment. (See figure 10.13.)

In the enrichment and fabrication processes, the main dangers are exposure to radiation and accidental release of radioactive material into the environment. Transport also involves exposure risks. Most radioactive material is transported by highways or railroads. If a transporting vehicle were involved in an accident, radioactive material could be released into the environment. Even after fuel rods have been loaded into the reactor, people who work in the area of the reactor risk radiation exposure. Countries that reprocess spent fuel rods must be concerned about exposure of workers and the possible theft of rods or reprocessed material by terrorist organizations that would use the radioactive material to construct an atomic bomb.

Thermal Pollution

Thermal pollution is the addition of waste heat to the environment. This is particularly a problem in aquatic environments, since many aquatic organisms are very sensitive to changes in temperature. All industrial processes release waste heat, so the problem of thermal pollution is not unique to nuclear power plants. In both fossil-fuel and nuclear plants, generating steam to produce electricity results in a great deal of waste heat. In a fossil-fuel plant, half of the heat energy produces electricity, and half is lost as waste heat. In a nuclear power plant, only one-third of the heat generates electricity, and two-thirds is waste heat. Therefore, the less efficient nuclear power plant increases the amount of thermal pollution more than does an equivalent fossil-fuel power plant. To reduce the effects of this waste heat, utilities build costly cooling facilities. Cooling processes usually involve water, so nuclear power plants often are constructed next to a water source. In some cases, water is drawn directly from lakes, rivers, or oceans and returned. In other cases, it is supplied by giant cooling towers. (See figure 10.14.)

Decommissioning Costs

All industrial facilities have a life expectancy, that is, the number of years they can

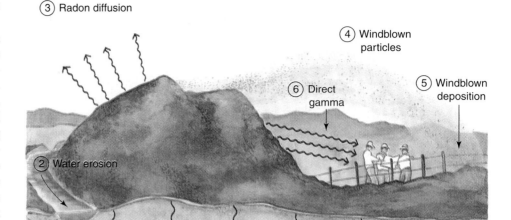

④ Windblown particles

⑤ Windblown deposition

⑥ Direct gamma

② Water erosion

① Leaching

Potable aquifer

Figure 10.13

Uranium Mine Tailings Even though the amount of radiation in the tailings is low, the radiation still represents a threat to human health. Radioactivity may be dispersed throughout the environment and come in contact with humans in several ways: (1) radioactive materials may leach into groundwater; (2) radioactive materials may enter surface water through erosion; (3) radon gas may diffuse from the tailings and enter the air; (4) persons living near the tailings may receive particles in the air, (5) come in contact with objects that are coated with particulate, or (6) receive direct gamma radiation.

Source: Data from U.S. Environmental Protection Agency.

be profitably operated. The life expectancy for an electrical generating plant, whether fossil-fuel or nuclear, is about 30 to 40 years, after which time the plant is demolished. With a fossil-fuel plant, the demolition is relatively simple and quick. A wrecking ball and bulldozers reduce the plant to rubble, which is trucked off to a landfill. The only harm to the environment is usually the dust raised by the demolition.

Demolition of a nuclear plant is not so simple. In fact, nuclear plants are not demolished, they are decommissioned. **Decommissioning** involves removing the fuel, cleaning surfaces, and permanently preventing people from coming into contact with the contaminated buildings or equipment. Today, several nuclear plants in the world are shut down and waiting to be decommissioned, and the number of plants scheduled for decommissioning will grow. By 2000, 62 of the 110 nuclear plants

Figure 10.14

Cooling Towers Cooling towers draw air over wet surfaces. The evaporation of the water cools the surface, removing heat from the process and releasing the heat into the air.

in the United States will be 20 years old or older. The Nuclear Regulatory Commission authorizes their operation for 40 years but is considering extending authorization for

The Nuclear Legacy of the Soviet Union

The tremendous political changes that have occurred in the former Soviet Union and Eastern Europe have brought to light the magnitude of nuclear contamination caused by those nations' unwise use of nuclear energy. The Soviet Union used nuclear energy for several purposes: generating power for electricity and nuclear-powered ships, testing and producing weapons, and blasting to move earth. The accident at Chernobyl was only the most recent and most public of many problems associated with the use of nuclear energy in the Soviet Union. About 13 nuclear reactors of the Chernobyl type are still operating in the former Soviet Union. These are considered unsafe by most experts, including former Soviet scientists.

The production of nuclear fuels and weapons and the reprocessing of nuclear fuels, all of which produced nuclear wastes, occurred at three major sites: Chelyabinsk, Tomsk, and Krasnoyarsk. These facilities were secret until the recent breakup of the Soviet Union. At Chelyabinsk, radioactive waste was dumped into the Techa River and Lake Karachai. Other nuclear wastes were secretly dumped into the ocean or were buried in shallow dumps. It is estimated that there are 600 secret nuclear dump sites in and around Moscow alone.

More than 100 nuclear bombs were exploded to move earth for mining purposes, to create underground storage caverns, or to "dig" canals. In addition, 467 nuclear blasts were conducted to test nuclear weapons at a site near Semipalatinsk in Kazakhstan. As a result, nuclear fallout contaminated farmland to the northeast.

The Barents Sea and the Kara Sea have been polluted by nuclear tests and the dumping of nuclear waste. At least 15 nuclear reactors from obsolete ships were dumped into the Kara Sea. Seals that live in the area have high rates of cancer.

None of these problems is unique to the former Soviet Union. Similar secrecy surrounded the early development of nuclear power in the United States and the rest of the world. What makes the situation in the former Soviet Union different is its magnitude, caused by decades of secrecy and a commitment to nuclear weapons as the primary deterrent against enemies. It can be argued that the threat posed to the Soviet Union by its former enemies contributed to the reckless development of nuclear power.

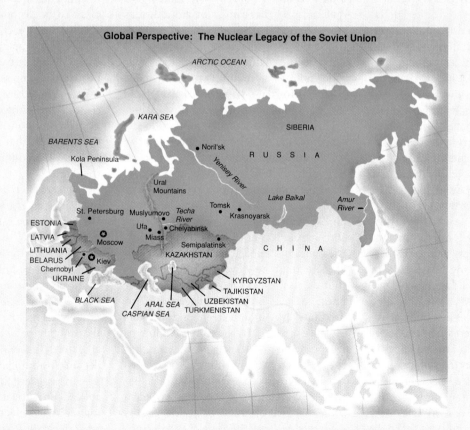

Global Perspective: The Nuclear Legacy of the Soviet Union

an additional 20 years. This would put off the need for decommissioning for a few years, but eventually it will be necessary. Twelve U.S. plants have been removed from service and are already awaiting decommissioning. Canada is in the process of decommissioning three plants.

Two problems are associated with decommissioning a nuclear power plant. First, since few plants have been decommissioned and there are many plant designs, engineers are not sure exactly how it should be done. This leads to the second problem; no one is certain how much it will cost. When a plant must be decommissioned, it is shut down. All the fuel rods must be removed; all of the water used as a reactor-core coolant, as a moderator, or to produce steam must be drained; and the reactor and generator pipes must be cleaned and flushed. The spent fuel rods, the drained water, and the material used to clean the pipes are all radioactive and must be safely stored or disposed of.

When a nuclear plant is constructed, none of it is radioactive. By the time a plant is decommissioned, much of it is radioactive. This radioactivity is the result of contamination and activated material. Radioactivity from contamination could involve the pipes, the containment building, the control building, and certain auxiliary buildings. The amount of such contamination depends on how much fuel leaked when the plant was in operation. Washing decontaminates the surface of these buildings. But even after surface decontamination, most of the equipment and building material is still radioactive. Such material, along with all the solutions used in the decontamination, must be disposed of safely.

Activated material results when nonradioactive atoms in the reactor and the shield are converted into radioactive isotopes as a result of being bombarded by neutrons. The half-life of the activated materials may vary from five years for cobalt to 80,000 years for nickel. These radioactive materials cannot be removed because they are within the structures of the plant themselves, not just sitting on the surface.

Utilities have three decommissioning options: (1) decontaminate and dismantle the plant as soon as it is shut down; (2) put the plant in mothballs for 20 to 100 years to allow radioactive

materials that have a short half-life to disintegrate and then dismantle the plant; (3) entomb the plant by covering the reactor with reinforced concrete and placing a barrier around the plant.

Originally, entombment was thought to be the best method. But because of the long half-life of some of the radioactive material and the danger of groundwater contamination, this method now appears to be the least favorable. Two-thirds of U.S. utilities with nuclear plants favor immediate decontamination and dismantling. Japan plans to mothball its plants for 10 years before dismantling them. France and Canada plan to mothball for a longer period and then dismantle.

Costs associated with decommissioning include dismantling the plant and then packaging, transporting, and burying the wastes. Because of contamination and activated material, ordinary dismantling methods cannot be used to demolish the buildings. Special remote-control machinery must be developed. Techniques that do not release dust into the atmosphere must be used (one suggestion is to do the work underwater), and workers must wear protective clothing. Adding to the problems and the cost is the fact that the protective equipment and the expensive remote-control machinery may become so contaminated that they will have to be disposed of safely. Waste disposal is estimated to be 40 percent of the decommissioning costs. Table 10.3 shows the

volume of material that would be handled from a 1,100-megawatt power plant.

Cost projections for decommissioning a single plant range from a low estimate of $50 million to a high of $3 billion. If the latter figure is accurate, it may cost more to decommission some plants than it cost to construct them. The first nuclear power plant to be decommissioned was a small, 22-megawatt reactor at Elk River, Minnesota, that had operated for only four years. The Department of Energy was responsible for decommissioning it. The process was completed in 1974, three years after the decommissioning was begun, at a cost of $6.15 million. Part of the decommissioning was the burial of 3,360 cubic meters of decontaminated material from the plant.

It has been proposed that, when a nuclear plant is constructed, a decommissioning fund be established. This has not been favorably received, since utilities have difficulty raising money to build the plants in the first place. They are reluctant to add decommissioning costs, which would raise electrical rates even more. However, nine states in the United States require that money be collected and placed into an external account to finance the costs of decommissioning.

In Sweden, the federal government collects an annual tax for decommissioning. A separate account is set up for each reactor. A similar arrangement is used in Switzerland and Germany. However, most

Table 10.3

Low-Level Radioactive Contaminants from Decommissioning a 1,100-megawatt Pressurized-Water Reactor

Material	Volume (cubic meters)
Radioactive	618
Activated	
Metal	484
Concrete	707
Contaminated	
Metal	5,465
Concrete	10,613
Total	17,887

Sources: Data from Worldwatch Paper 69, "Decommissioning: Nuclear Power's Missing Link" and U.S. Nuclear Regulatory Commission.

utilities worldwide ignore the cost of decommissioning, which is why most of the reactors that have been shut down have not been decommissioned. The problem of decommissioning may become a legacy for future generations, which means that the costs of decommissioning nuclear plants will be borne by those who had no voice in their construction and who did not use the electricity these plants produced.

Radioactive Waste Disposal

When the world entered the atomic age, the problem of the disposal of nuclear waste was not fully appreciated. Low-level radioactive waste is generated by nuclear power plants, military facilities, hospitals, and research institutions. High-level radioactive waste results from spent fuel rods and obsolete nuclear weapons.

High-Level Radioactive Waste

In the United States, 380,000 cubic meters of highly radioactive military waste are temporarily stored in Idaho, South Carolina, New York, and Washington. Two million cubic meters of low-level radioactive military and commercial waste are buried at various sites. In addition, about 30,000 metric tons of high-level radioactive waste from spent fuel rods are being stored in special storage ponds at nuclear reactor sites. Many plants are running out of storage space and have been authorized by the Nuclear Regulatory Commission to store radioactive waste in aboveground casks, because the storage ponds will not accommodate additional waste.

In other countries, spent fuel rods are either stored in special ponds or sent to reprocessing plants. Even though reprocessing is more expensive than manufacturing fuel rods from ore, some countries reprocess as an alternative to waste storage. The reprocessing plants in France and the United Kingdom accept domestic and foreign fuel rods, and any fuel rods fabricated by the former Soviet Union can be returned to reprocessing plants in Russia.

At this point in time, no country has a permanent storage solution for the disposal of high-level radioactive waste. Most experts feel the best solution is to bury it in a stable geologic formation. Several countries are investigating storage in salt deposits.

The waste would be placed in borosilicate glass containers. Each container would then be sealed in a thick-walled, stainless steel canister. Due to the high temperature of the radioactive waste, the containers would be stored aboveground for 10 years. At the end of this time, the temperature of the container would be reduced, and the material could be buried in a salt deposit 600 meters below the surface.

Sweden intends to store its waste 500 meters underground in granite. Site construction is not planned until 2010, when Sweden's 12 reactors are scheduled to be decommissioned. (Sweden plans to discontinue nuclear power generation of electricity, but some now doubt that this will occur.) France plans to use its reprocessing plants and "temporary" storage indefinitely.

The politics concerning the disposal of high-level radioactive waste are probably as critical as developing a suitable method. No communities want a radioactive disposal site in their area. In December 1982, the U.S. Congress passed legislation calling for a high-level radioactive disposal site to be selected by March 1987 and to be completed by 1998. In 1984, the Department of Energy stated that it was already three years behind schedule. Final site selection occurred in 1989, nearly three years later than the date set by legislation. The location is Yucca Mountain, Nevada. The site was chosen for several reasons. It is in an unpopulated area near the Nevada Test Site where several nuclear devices were exploded. It is stable geologically. It is very unlikely to experience earthquakes or volcanos. It is a very dry area and the water table is about 600 meters below the mountain, so groundwater will not likely be contaminated. These are important considerations because the site must be stable enough to last thousands of years.

Work has begun on the series of tunnels that would serve as storage places for the waste. (See figure 10.15) However, current work is primarily exploratory and is seeking to characterize the likelihood of earthquake damage and the movement of water through sediments. If completed, the facility would hold about 70,000 metric tons of spent fuel rods and other highly radioactive material. It will not be completed before 2015, and by that time the total amount of waste produced by nuclear

power plants will exceed the storage capacity of the site. Furthermore, the amount of waste from nuclear weapons production greatly exceeds that produced by nuclear power plants. Local protests against the site are continuing, Nevada can refuse the site (its refusal could be overridden by Congress), and Congress has not appropriated adequate funds to proceed rapidly. Therefore, it is still uncertain if or when it will be open to receive waste.

Low-Level Radioactive Waste

Low-level radioactive waste includes the cooling water from nuclear reactors, material from decommissioned reactors, radioactive materials used in the medical field, protective clothing worn by persons working with radioactive materials, and materials from many other modern uses of radioactive isotopes. Disposal of this type of radioactive waste is very difficult to control. Estimates indicate that much of it is not disposed of properly.

In 1970, a U.S. moratorium halted the dumping of radioactive waste in the oceans. Prior to this, the United States placed some 90,000 barrels of radioactive waste on the ocean floor. European countries also have dumped both high-level and low-level radioactive material into the Atlantic Ocean. Before 1983, when ocean dumping was halted, these countries disposed of 90,000 metric tons of radioactive waste in the ocean.

The United States produces about 800,000 cubic meters of low-level radioactive waste per year. This is presently being buried in disposal sites in Nevada, South Carolina, and Washington. However, these states balked at accepting all the country's low-level waste. In 1980, Congress set a deadline of 1986 (later extended to 1993) for each state to provide for its own low-level radioactive waste storage site. Later, a change allowed several states to cooperate and form regional coalitions, called compacts. Under this arrangement, one state would provide a disposal site for the entire compact. Only California and Texas did not enter into a compact arrangement. The extended deadline of 1993 has passed. However, most states still do not have a permanent place to deposit low-level radioactive wastes.

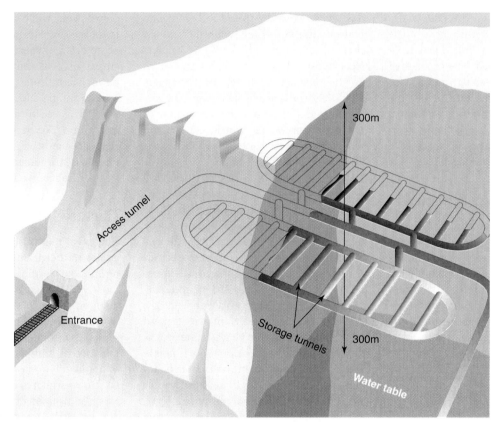

Figure 10.15

High-Level Nuclear Waste Disposal   Current plans for the disposal of high-level nuclear waste involve placing materials in tunnels underground at Yucca Mountain, Nevada. The material could be stored about 300 meters below the surface and about 300 meters above the water table. Because of the dry climate it is considered unlikely that water infiltrating the soil would move nuclear material downward to the water table.

SUMMARY

Nuclear fission is the splitting of the nucleus of an atom. The resulting energy can be used for a variety of purposes. The splitting of U-235 in a nuclear reactor can be used to heat water to produce steam that generates electricity. Various kinds of nuclear reactors have been constructed, including boiling-water reactors, pressurized-water reactors, heavy-water reactors, gas-cooled reactors, and experimental breeder reactors. Scientists are also conducting research on the possibilities of using fusion to generate electricity. All reactors contain a core with fuel, a moderator to control the rate of the reaction, and a cooling mechanism to prevent the reactor from overheating.

The nuclear fuel cycle involves mining and enriching the original uranium ore, fabricating it into fuel rods, using the fuel in reactors, and reprocessing or storing the spent fuel rods. The fuel and wastes must also be transported. At each step in the cycle, there is danger of exposure. During the entire cycle, great care must be taken to prevent accidental releases of nuclear material.

The use of nuclear materials for military purposes has resulted in the same kinds of problems caused by nuclear power generation. The reduced risk of nuclear warfare has resulted in a need to destroy nuclear weapons and clean up the sites contaminated by their construction.

Public acceptance of nuclear power plants has been declining. Initial promises of cheap electricity have not been fulfilled because of expensive construction, cleanup, and decommissioning costs. The accidents at Three Mile Island in the United States and Chernobyl in the Ukraine have accelerated concerns about the safety of nuclear power plants.

Other concerns are the unknown dangers associated with exposure to low-level radiation, the difficulty of agreeing to proper long-term storage of high-level and low-level radioactive waste, and thermal pollution.

REVIEW QUESTIONS

1. How does a nuclear power plant generate electricity?
2. Name the steps in the nuclear fuel cycle.
3. What is a rem?
4. What is a nuclear chain reaction?
5. How will nuclear fuel supplies, the cost of decommissioning facilities, and storage and ultimate disposal of nuclear wastes influence the nuclear power industry?
6. What happened at Chernobyl, and why did it happen?
7. Describe a boiling-water reactor, and explain how it works.
8. How is plutonium-239 produced in a breeder reactor?
9. Why is plutonium-239 considered dangerous?
10. Why is fusion not currently being used as a source of energy?
11. What are the major environmental problems associated with the use of nuclear power?
12. List three environmental problems associated with the construction and subsequent de-emphasis of nuclear weapons.

KEY TERMS

alpha radiation   169
beta radiation   169
boiling-water reactor
   (BWR)   171
decommissioning   180
gamma radiation   169
gas-cooled reactor (GCR)   172

heavy-water reactor
   (HWR)   172
high-temperature, gas-cooled
   reactor (HTGCR)   172
light-water reactor (LWR)   171
liquid metal fast-breeder
   reactor (LMFBR)   173

moderator   171
nuclear breeder reactor   172
nuclear chain reaction   169
nuclear fission   169
nuclear fusion   173
nuclear reactor   170
plutonium-239 (Pu-239)   172

pressurized-water reactor
   (PWR)   171
radiation   169
radioactive   169
radioactive half-life   169
rem   179
thermal pollution   180
uranium-235 (U-235)   170

RELATED INTERNET SITES

Savannah River site. The Department of Energy deploys technology to clean up industrial and nuclear waste.
   http://www.srs.gov/
Department of Energy office of high energy and nuclear physics.
   http://www.acl.lanl.gov/DOE/o_henp.html
Nuclear Energy Institute—electricity for cleaner air.
   http://nei.org/
Good start for links to other nuclear pages.
   http://hawkeye.me.utexas.edu:80/~ans/
Nuclear Base, a newspaper database on the civil nuclear industry and related information.
   http://www.users.zetnet.co.uk/n-base/
Nuclear Power and Space Exploration, discusses a wide variety of nuclear topics.
   http://funnelweb.utcc.utk.edu/~cramsey/

Award-winning, searchable site focusing on the potential locations for storing radioactive wastes and spent nuclear fuel.
   http://www.ymp.gov/
Education web site on fusion energy.
   http://fusioned.gat.com/
Advantages of fusion over other energy sources.
   http://www.pppl.gov/oview/pages/fusion_advantage.html
United Kingdom Atomic Energy Authority offers a thorough education about fusion energy for lay people.
   http://www.fusion.org.uk

PART four

The success of the human species is a result of our ability to change our surroundings so that we have tools, food, water, and shelter. Our use of mineral resources is important to many aspects of modern society. Chapter 11 discusses how humans exploit mineral resources and use natural ecosystems. Often, this means that existing natural ecosystems are altered significantly or replaced by agricultural ecosystems. In some cases, organisms are driven to extinction by human activities.

Chapter 12 emphasizes the use of the land surface, while chapter 13 discusses the nature and wise use of soil. Land and soil are not the same. Land is the part of the world that is not covered by water, while soil is a thin layer on the land's surface that supports plant growth. Land-use decisions must take into account the nature of the soil.

Agriculture allows us to support about 5.8 billion people on the earth. The use of the soil, discussed in chapter 13, and the use of pesticides, discussed in chapter 14, are both important to food production. Misuse of the soil or pesticides can lead to the loss of usable farmland or to the poisoning of people.

Water is necessary for human life. It is important as a nutrient, but also for agriculture, industrial processes, transportation, and many other functions. Chapter 15 deals with a wide range of water-use issues and conservation practices.

chapter 11

Human Impact on Resources and Ecosystems

Objectives

After reading this chapter, you should be able to:
- Recognize that humans have an increasing impact on natural ecosystems.
- Define pollution.
- Differentiate between renewable and nonrenewable resources.
- Recognize that mineral resources are unevenly distributed, which creates international trade in these commodities.
- List three types of costs associated with mineral exploitation.
- Understand that some wilderness areas still have minimal human influence.
- Appreciate the ways humans modify forests.
- Identify causes of desertification.
- Recognize that aquatic systems are modified by terrestrial changes.
- Identify changes that occur to aquatic systems as a result of human activity.
- Recognize that wildlife management focuses on specific species.
- Appreciate that waterfowl management is an international problem.
- Recognize that extinction is a natural process.
- Recognize that humans increase the rate of extinction.
- Identify ways that humans cause extinctions.
- Recognize that many extinctions can be prevented if societies are willing to preserve critical habitats and prevent the hunting of endangered species.

Chapter Outline

The Changing Role of Human Impact

Historical Basis of Pollution

Renewable and Nonrenewable Resources

Mineral Resources

    Costs Associated with Mineral Exploitation

    Steps in Mineral Exploitation

    Recycling As an Alternative

Areas with Minimal Human Impact— Wilderness and Remote Areas

Ecosystems Modified by Human Use

    Forests

    Forest Management Practices

Environmental Close-Up: *The Northern Spotted Owl*

    Rangelands

    Modified Aquatic Ecosystems

Environmental Close-Up: *Exploiting the Oceans' Bounty*

    Modifying Ecosystems to Manage Wildlife

Environmental Close-Up: *Native American Fishing Rights*

    Replacement of Natural Ecosystems with Human-Managed Ecosystems

Natural Selection and Extinction

Human-Accelerated Extinction

Why Worry about Extinction?

What Is Being Done to Prevent Extinction?

Global Perspective: *The History of the Bison*

Global Perspective: *Hope for the Rhinoceros*

Environmental Close-Up: *Whaling*

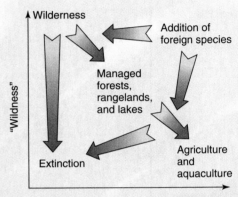

Environmental Close-Up: *The California Condor*

Environmental Close-Up: *Animals That Benefit from Human Activity*

Issues and Analysis: *Costa Rican Forests Yield Tourists and Medicines*

Environmental Close-Up: *Extinction of the Dusky Seaside Sparrow*

The Changing Role of Human Impact

At one time, a human was just another consumer somewhere in the food chain. Humans fell prey to predators and died as a result of disease and accident just like other animals. The tools they used were primitive, so these people did not have a long-term effect on their surroundings. They only minimally exploited mineral and energy resources. They mined chert, obsidian, and raw copper for the manufacture of tools, and used certain other mineral materials, such as salt, clay, and ocher, as nutrients, for making pots, or as pigments. Aside from occasional use of surface seams of coal or surface oil seeps, early humans met most of their energy needs by biomass.

As human populations grew, and as their tools and systems of use became more advanced, the impact that a single human could have on his or her surroundings increased tremendously. The use of fire was one of the first events that marked the capability of humans to change ecosystems. Although in many ecosystems fires were natural events, the use of fire to capture game and to clear land for gardens returned climax communities to earlier successional stages more frequently than normal.

As technology advanced, wood was needed for fuel and building materials, land was cleared for farming, streams were dammed to provide water power, and various mineral resources were exploited to provide energy and build machines. These modifications allowed larger human populations to survive, but always at the expense of previously existing ecosystems.

Today, with about 5.8 billion people on the earth, nearly all of the surface of the earth has been affected in some way by human activity. Even the polar ice caps show the effects of human activity. Various kinds of organic pollutants and lead residues from the burning of leaded gasoline can be identified in the layers of ice that build up from the continuous accumulation of snow in these areas. (The amount of lead is currently decreasing because some countries have been making the transition from buring leaded fuel to unleaded fuel in automobiles.)

Historical Basis of Pollution

Pollution is usually defined as something that people produce in large enough quantities that it interferes with our health or well-being. Two primary factors that affect the damage done by pollution are the size of the population and the development of technology that "invents" new forms of pollution.

When the human population was small and people lived in a primitive manner, the wastes produced were biological and so dilute that they usually did not constitute a pollution problem. People used what was naturally available and did not manufacture many products. Humans, like any other animal, fit into their natural ecosystems. Their waste products were **biodegradable** materials. Biodegradable materials are a source of food for decomposers, the organisms that break the material down into simpler chemicals, such as water and carbon dioxide.

Pollution began when human populations became so concentrated that their waste materials could not be broken down as fast as they were produced. As the population increased, people began to congregate and establish cities. The release of large amounts of smoke and other forms of waste into the air caused an unhealthy condition because the pollutants were released faster than they could be absorbed by the atmosphere.

Throughout history, humans have made numerous attempts to eliminate the misery caused by hunger and disease. In general, we rely on science and technology to improve our quality of life. However, technological progress often offers short-term solutions that in the process can create new pollution problems.

The development of the steam engine allowed machines to replace human labor but increased the amount of smoke and other pollutants in the air. The modern chemical industry has produced many extremely valuable synthetic materials (plastics, pesticides, medicines) but has also produced toxic pollutants.

Even identifying pollution is not always easy. To some the smell of a little wood smoke in the air is pleasant; others do not like the odor. A business may consider advertising signs valuable and necessary; others consider them to be visual pollution. (See figure 11.1.) Even the presence of chemicals in drinking water can be difficult to classify as a clear-cut example of pollution. For example, a small town discovered traces of arsenic in its groundwater supply. Toxic heavy metals such as arsenic certainly should be considered hazardous, but we do not know how much arsenic contamination is allowable before harm occurs. Are the small quantities of arsenic in groundwater great enough that, over a period of time, they will accumulate and cause damage to individuals? Will children, with smaller body masses, be more affected than adults? Is the arsenic a normal part of the groundwater, or is it the result of the heavy spraying of arsenic-containing pesticides on apple trees in the past?

Certainly, if the arsenic is the result of human activity and is causing health problems, it is a pollutant. If it is a natural part of the groundwater, it may still be a health hazard to be dealt with but is technically not a pollutant.

Renewable and Nonrenewable Resources

Modern technologies have allowed us to exploit natural resources to a much greater extent than our ancestors were capable of. **Natural resources** are structures and processes that humans can use for their own purposes but cannot create. If the supply of a resource is very large and the demand for it is low, the resource may be thought of as free. Sunlight, land, and air are often not even thought of as natural resources because their supply is so large. If a resource has been consumed or if it was rare initially, it is expensive. Pearls and gold and other precious metals fall into this category. In the past, land was considered a limitless natural resource, but as the population grew and the demand for food,

Fish kill—an indication of water contamination

Smoke from stack—contains particulate material, which could cause lung problems

Feed lot—odor pollution as well as a source of water contamination from surface runoff

Strip—visual pollution, which is an annoyance but not a health hazard

Smog—an indication that thermal inversion has kept the air contamination in the valley

Traffic—fumes (HC, NO_x, PAN, etc.) from internal combustion engines cause eye irritation

Nuclear power plant cooling towers—possible thermal pollution and radiation hazards

Litter—an indication that we need to become more aware of how we dispose of materials

Health hazard ... **Annoyance**

Figure 11.1

Forms of Pollution Pollution is produced in many forms. Some are major health concerns, whereas others merely annoy.

lodging, and transportation increased, we began to realize that land is a finite, nonrenewable resource. Unwise or unplanned use can result in inappropriate use or severe damage to soil. (See figure 11.2.)

The landscape is also a natural resource, as we see in countries with a combination of mountainous terrain and high rainfall that can be used to generate hydroelectric power. Rivers, forests, scenery,

climates, and wildlife populations are additional examples of natural resources. Resources, especially those like land, air, and water, become more valuable as we begin to exploit them more intensively.

Natural resources are usually categorized as either renewable or nonrenewable. **Renewable resources** can be formed or regenerated by natural processes. Soil, vegetation, animal life, air, and water are re-

newable primarily because they naturally undergo repair and a cleansing process.

Nonrenewable resources are not replaced by natural processes, or rate of replacement is so slow as to be ineffective. Therefore, when nonrenewable resources are used up, they are gone, and a substitute must be found or we must do without.

Mineral Resources

The geological and biological processes that formed mineral resources in certain places on the earth occurred many millions of years ago. Since that time, landmasses have been divided into political entities. Some have a greater wealth of mineral resources than others. Because no country has within its territorial limits all of the mineral resources it needs, an international exchange has developed. In particular, the industrially developed countries import many of the minerals they need from countries that have the resources but no economic ability to develop them. Table 11.1 shows several minerals and the countries that supply them. North America lacks a large number of materials and must import them from other countries, often in less-developed areas of the world.

Not only are mineral resources unevenly distributed, but those that are easiest to use and the least costly to extract have been exploited. If we continue to use mineral resources, they will be harder to find and more costly to develop. As with energy, North America is one of the primary consumers of the world's mineral resources. Reasonable estimates are that North America consumes over 30 percent of the minerals produced in the world each year, which is a disproportionate share, given that the combined population of the United States and Canada is about 5.1 percent of the world's population.

Costs Associated with Mineral Exploitation

Costs are always associated with the exploitation of any natural resource. These costs fall into three categories. First, the **economic costs** are those monetary costs

Figure 11.2

Mismanagement of a Renewable Resource Although soil is a renewable resource, extensive use can permanently damage it. Many of the world's deserts were formed or extended by unwise use of farmland. This photograph shows a once-productive farm, now abandoned to the wind and sand because the soil was mistreated and allowed to erode.

Table 11.1

Distribution of Some Mineral Resources

Mineral	% consumed in U.S. that was imported	Major supplier countries (1982–1994)
Columbium	100	Brazil, Canada, Thailand
Mica (sheet)	100	India, Belgium, France
Strontium	100	Mexico, Spain
Manganese	100	South Africa, Gabon, France
Bauxite	100	Jamaica, Guinea, Australia
Platinum group	88	South Africa, former USSR, United Kingdom
Tantalum	85	Thailand, Brazil, Germany
Cobalt	82	Zaire, Zambia, Canada
Chromium	80	South Africa, Zimbabwe, Turkey
Nickel	74	Canada, Australia, Norway
Tin	73	Malaysia, Thailand, Brazil
Potassium	67	Canada, Israel, former USSR
Cadmium	54	Canada, Australia, Mexico
Zinc	30	Canada, Mexico, Peru

Source: Data from the Statistical Abstract of the United States, 1994.

necessary to exploit the resource. Money is needed to lease or buy land, build equipment, pay for labor, and buy energy to run the equipment.

A second category is the **energy cost** of exploiting the resource. It takes energy to extract, concentrate, and transport mineral materials to manufacturing sites.

Since energy costs money, energy costs are ultimately converted to economic costs. When energy is inexpensive, inefficient processes may be profitable; however, when the cost of energy rises, energy-intensive processes will be eliminated.

A third way to look at costs is in terms of environmental effects. Air pollution, water pollution, animal extinction, and loss of scenic quality are all **environmental costs** of resource exploitation. Environmental costs are often deferred costs. They may not even be recognized as costs at first but become important after several years. Environmental costs may also be lost opportunities or lost values because the resource could not be used for another purpose. Lately, environmental costs are being converted to economic costs as more strict controls on pollution are enforced. It takes money to clean up polluted water and air, or to reclaim land that has been removed from biological production by mining.

These three categories of costs (economic, energy, and environmental) apply to several steps that lead from the mineral source in its undisturbed state to the manufacture of a finished product. These steps are exploration, mining, refining, transportation, and manufacturing.

Steps in Mineral Exploitation

The costs associated with locating new sources of minerals (exploration) are primarily economic because exploration takes time and new technology. There are also some energy costs and some very small environmental costs. As the better sources of mineral resources are used up, it will be necessary to look for minerals in areas that are more difficult to explore, such as under the oceans. Therefore, both economic and energy costs will increase. Some areas, such as national parks and preserves, have been off-limits to mineral exploration. As current reserves of mineral resources are used up, pressures will build to explore in these protected areas, resulting in increased environmental costs.

Once a mineral resource has been located and the decision made to exploit it, the resource must be taken from the earth. Mining involves large expenditures of money to pay for labor and the

Figure 11.3

A Strip-Mining Operation It is easy to see the important impact a mine of this type has on the local environment. Unfortunately, many mining operations are located in areas that are also known for their scenic beauty.

Table 11.2

Metals in Ores		
Metal	% metal needed in the ore for profitable extraction	Price range per kilogram
Iron	30.0	
Chromium	30.0	Less than a dollar
Aluminum	20.0	
Nickel	1.5	
Tin	1.0	Several dollars
Copper	0.5	
Uranium	0.1	Tens of dollars

construction of machines and equipment. Energy must be purchased for operation.

In addition to the economic and energy costs there are significant environmental costs. Mining affects the environment in several ways. All mining operations involve the separation of the valuable mineral from the surrounding rock. The surrounding rock must then be disposed of in some way. These pieces of rock are usually piled on the surface of the earth, where they are known as mine tailings and present an eyesore. It is also difficult to get vegetation to grow on these deposits. Some mine tailings contain materials (such as asbestos, arsenic, lead, and radioactive materials) that can be harmful to humans and other living things.

Many types of mining operations require vast quantities of water for the extraction process. The quality of this water is degraded, so it is unsuitable for drinking, irrigation, or recreation. Since mining disturbs the natural vegetation in an area, water may carry soil particles into streams and cause erosion and siltation. Some mining operations, such as strip mining, rearrange the top layers of the soil, which lessens or eliminates its productivity for a long time. (See figure 11.3.) Strip mining has disturbed approximately 75,000 square kilometers of U.S. land, an area equivalent to the state of Maine.

After ore is mined, it must be processed to remove the desired materials from the rocks in which they are embedded. Table 11.2 shows the percentage of desired material present in several ores. Ore processing poses economic costs: physical facilities and equipment must be built, people must be employed, and fuel must be purchased for the operation. The energy costs are also significant because refining is basically the process of concentrating the desired material. Natural physical processes tend to disperse materials; therefore, energy must be expended to concentrate them. In many cases, this energy expenditure is the major cost associated with obtaining the mineral. As energy costs increase and ores become more diluted, the economics of producing some materials may change so drastically that seeking substitute materials that require less energy expenditure is the best alternative.

Extracting materials from ores also involves environmental costs in the form of air and water pollution. These environmental costs are being converted to economic costs as regulations on industry require less environmental damage than had been previously tolerated.

Transportation is another component of the overall cost of extracting minerals from the earth. Transportation is involved in the actual mining process, in getting the ore to the refinery, and in moving the concentrated mineral to the site where it will be made into a finished product. Transportation costs are primarily in the form of money and energy.

Finally, there is the cost of making the finished product from the concentrated ore. Figure 11.4 reviews this entire process by looking at what happens from the time iron atoms are mined until finished steel is produced.

Recycling As an Alternative

To fully appreciate mineral resource use we should consider the concept of recycling. Many minerals extracted from the earth are not actually consumed or used up; they are just temporarily within a structure or a process. When the material is reclaimed and used again for another structure or process, it is recycled. (See chapter 18.)

Empty aluminum beverage cans are no longer useful. However, the aluminum atoms can be reprocessed into new cans or other aluminum products. For recycling to be economically practical, the obsolete items must be recaptured before they have dispersed into the environment. For example, waste oil is relatively easy to recapture at the gasoline station where engine oil is replaced, but difficult to recapture if the oil is dumped on the ground or into a sewer.

Costs from all categories (economic, energy, and environmental) are involved in reprocessing, but the costs are often less because mining or refining is unnecessary. In many industries, however, the economic cost is lowest if new materials are mined and refined rather than if used materials are reprocessed. For example, disassembling a building and reusing its parts is more costly than using new construction

PART FOUR Human Influences on Ecosystems

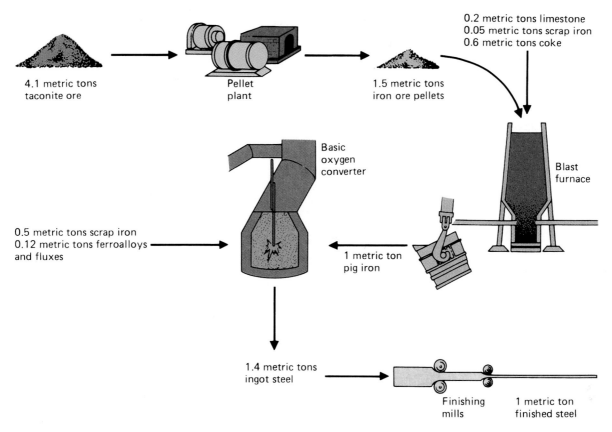

Figure 11.4

Steps in Steel Manufacture This diagram reviews the process from the mining of taconite ore to the production of finished steel. Each step involves the production of pollutants, labor and capital costs, and the expenditure of energy to concentrate the ore.

materials. However, when the energy cost and environmental costs are taken into account, recycling is often a less expensive alternative.

One reason recycling is not more common is that, historically, the monetary costs for energy have been extremely low. As the cost of energy rises, more attention will be given to recycling minerals rather than mining new ones. Manufacturing products in an environmentally harmonious way will become economically advantageous in the future.

Areas with Minimal Human Impact— Wilderness and Remote Areas

There are still many areas of the world that have had minimal human impact. Some of these are remote areas with harsh environmental conditions, such as the continent of Antarctica, northern arctic areas, the tops of some tall mountains, or extremely arid areas, such as desert areas of Africa, central Australia, and central Asia. Most of these have been explored and found to be too harsh to sustain human habitation because growing food would be impossible.

Many other areas of the world, such as tropical rainforests, currently support ecosystems that are little affected by humans but are under threat because they are capable of supporting agriculture or other human uses. The primary factor that will determine the survival of these natural areas is population pressure. As the human population grows, it will need more food and more space. This ultimately leads to the destruction of natural ecosystems and their replacement by human-modified ecosystems. Many countries have established parks and other special designations of land use to protect areas of natural beauty or communities of organisms thought worthy of protection. This has been most noticeable in Africa, Central and South America, North America, and Australia. Until recently, these continents had large amounts of land that were relatively untouched. (See figure 11.5.)

Europe, the British Isles, and the agricultural regions of North America have almost none of their original vegetation remaining. The forests were cut for shipbuilding, fuel, and agriculture, and the prairies were converted to cropland. The forests that remain are small remnants of the original communities and have been modified by the introduction of many exotic species of plants and animals that have replaced the original inhabitants. Historically, the pressures to modify the environment were greatest in Europe because of the large population and the Industrial Revolution.

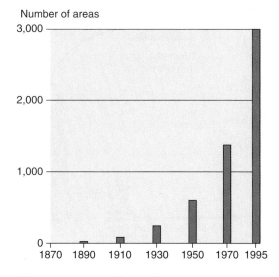

Growth of global network of protected areas, 1890-1995

Number of areas

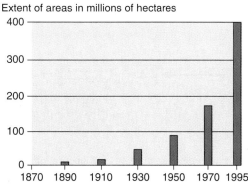

Extent of areas in millions of hectares

Figure 11.5

Growth in Major Park and Wilderness Areas of the World   Although there has been a steady increase in the number of protected areas and the number of hectares protected, many of the areas are in more-developed countries that can afford to set aside land for nonproductive uses.

Source: Data from International Union for Conservation of Nature and Natural Resources (IUCN), the United Nations List of National Parks and Protected Areas, 1985, Gland, Switzerland, and more recent 1995 data.

Parks and protected lands allow a variety of uses, depending on the country and its laws. Some are used primarily as tourist attractions, which generate funds for the country. Others are set up primarily to protect a species or community of organisms, while others simply seek to restrict use so that certain environmental standards are met. For example, harvesting in a forest might be restricted to a certain number of trees each year, and the number of people who visit a particular scenic site might be limited. A particularly sensitive issue is the designation of certain areas as wilderness. Although definitions of wilderness vary from country to country, the U.S. Congress, in the Wilderness Act of 1964, defined **wilderness** as "an area where the earth and its community of life are untrampled by man, where man himself is a visitor who does not remain."

Recently, there has been intense pressure from members of Congress from oil-producing states such as Alaska to allow oil exploration in areas currently designated as wilderness, such as the Arctic National Wildlife Refuge. As population increases and resources become more scarce, this pressure will become greater throughout the world. (See table 11.3)

The oceans of the world represent a major area that has been modified very little by human activities. Some areas, particularly those near the shore, receive intense use, but many areas of the open ocean are poorly understood and little affected by humans.

Ecosystems Modified by Human Use

Humans modify natural ecosystems in several fundamental ways. Removing some individuals or species from the ecosystem is one way. Encouraging the reproduction of certain species and the introduction of exotic species are other common activities. Typically, the most productive natural ecosystems are the first to be modified by human use and the most intensely managed.

Forests

Most forest ecosystems have been extensively modified by human activity. Originally, almost half of the United States, three-fourths of Canada, almost all of Europe, and significant portions of the rest of the world were forested. The forests were removed for fuel, building materials, to clear land for farming, and just because they were in the way. This activity returned the forests to an earlier successional stage, which resulted in the loss of certain animal and plant species that required mature forests as part of their niche.

Today in Australia, a small squirrel-sized marsupial known as a numbat is totally dependent on old forests that have termite- and ant-infested trees. The insects are numbats' major source of food, and the hollow trees and limbs caused by the insects' activities provide numbats with places to hide. Clearly, tree harvesting will have a negative impact on this species, which is already in danger of extinction. (See figure 11.6.)

Because of increasing world population, forested areas are being pressured to provide firewood, lumber, and agricultural land. These same forests are important refuges for many species of plants and animals, and are often essential as protectors of watersheds. If the

Table 11.3

Population Size and Availability of Renewable Resources, Circa 1990, With Projections for 2010

	Circa 1990 (million)	2010 (million)	Total change (percent)	Per capita change (percent)
Population	5,290	7,030	+33	
Fish catch (tons)[1]	85	102	+20	−10
Irrigated land (hectares)	237	277	+17	−12
Cropland (hectares)	1,444	1,516	+5	−21
Rangeland and pasture (hectares)	3,402	3,540	+4	−22
Forests (hectares)[2]	3,413	3,165	−7	−30

[1]Wild catch from fresh and marine waters, excludes aquaculture.
[2]Includes plantations; excludes woodlands and shrublands.

Sources: Population figures from U.S. Bureau of the Census, Department of Commerce, International Data Base, unpublished printout, November 2, 1993; 1990 irrigated land, cropland, and rangeland from U.N. Food and Agriculture Organization (FAO), Production Yearbook 1991; fish catch from M. Perotti, chief, Statistics Branch, Fisheries Department, FAO, private communication, November 3, 1993; forests from FAO, Forest Resources Assessment 1990, 1992 and 1993. For detailed methodology, see State of the World 1994, among other sources.

Figure 11.6

A Specialized Marsupial—the Numbat This small marsupial mammal requires termite- and ant-infested trees for its survival. The insects serve as food and the hollow limbs and logs provide hiding places. Loss of old-growth forests with diseased trees will lead to the numbat's extinction.

trees are removed, flooding is more common, and much valuable water is lost as runoff.

In general, all continents contain significant amounts of forested land. Table 11.4 shows that North America, Africa, Latin America, and the former USSR contain nearly 80 percent of the densely forested land in the world. However, these regions are responsible for less than 60 percent of the wood products produced. Asia and Africa produce about the same amount but with only about 24 percent of the world's forests. (See table 11.5.)

The areas of the world that could increase timber production are the tropical forests and the northern forests of the former USSR and Canada. Both of these areas present some problems. The far north of the former USSR and Canada is largely forested, but exploiting these remote areas would require extensive transportation expansion. In addition, these are slow-growing coniferous forests, so there are questions about how intensively these forests can be exploited before they are not growing as fast as they are being harvested. Good forest management seeks a sustained yield from the forest. This requires that the rate of harvest be approximately the same as the rate of regrowth.

The tropical forests of Asia, South America, and Africa are another underutilized source of timber. However, tropical forests have a diverse mixture of tree species that require different kinds of harvesting techniques than have traditionally been used in northern temperate forests. Also, because of soil deficiency, tropical forests are not as likely to regenerate after logging as are many of the temperate forests. If they do not regenerate, they must be considered nonrenewable resources. If these forests are to be used, a new set of forestry principles will be needed to establish a renewable tropical forest industry. Currently, few of these forests are being managed for long-term productivity; they are being harvested on a short-term economic basis only, as if they were nonrenewable resources. Tropical forests in Latin America are being deforested at a rate of nearly 1 percent per year, those in Africa are being deforested at about 0.8 percent per year, and those in

Table 11.4

Distribution of World Forest Resources, 1990

Region of the world	Millions of hectares of closed forest	% of the world total
Latin America	966	28.1
USSR (former)	755	21.9
Africa	545	15.8
Asia	489	14.2
North America	457	13.3
Europe	140	4.1
Australia, and Pacific area	88	2.6
World total	3440	100.0

Source: Data from the Food and Agricultural Organization of the UN.

Table 11.5

World Wood Production

Region of the world	Production in million cubic meters of round wood—1991–1993	% of world total
Asia	1123.0	32.9
United States	491.0	14.4
Africa	539.7	15.8
Latin America	421.1	12.6
USSR (former)	292.0	8.6
Europe	319.1	9.4
Canada	172.7	5.0
Pacific area	44.1	1.3
World total	3411.5	100.0

Source: Data from the Food and Agricultural Organization of the UN.

Asia are being deforested at 1.2 percent per year.

The deforestation of large tracts of land significantly impacts climate. Forested land very effectively traps rainfall and prevents its rapid runoff. Furthermore, the large amount of water transpired from the leaves of trees tends to increase the humidity of the air in forested areas. The shade provided by trees and the evaporation of water also tend to moderate temperature extremes. Destruction of huge areas of forest can result in regional climate change.

More recently, people have become concerned about preserving the potential of forests to trap carbon dioxide. Trees trap large amounts of carbon dioxide as a result of photosynthesis and may help to prevent increased carbon dioxide levels that contribute to global warming. (See chapter 17 for a discussion of global warming.)

Another complicating factor is that human population pressures are the greatest in tropical regions of the world. More people need more food, which means that some forestland will be converted to agriculture, thus reducing the stocks of timber available and also reducing the land available for forest production.

Forest Management Practices

Whenever a resource is exploited, several interests come into conflict. Two major ones are the economy and the environment. Economic factors are easy to measure. The cost of exploitation and the financial return are the primary issues. The environmental viewpoint is often difficult to put into monetary terms and must rely on ethical or biological arguments. Modern forest management practices in many parts of the world involve a compromise between these two points of view.

The forests of the world are known quantities. The economic worth of the standing timber can be assessed, and the value of forests for wildlife and watershed protection can be given a value. However, beyond that, the harvesting of forests changes the ecosystem from its original "wilderness" character. Logging removes the trees and, therefore, the habitat for many kinds of animals that require mature stands of timber. The pine martin, grizzly bear, and cougar all require forested habitat that is relatively untouched by human activity. An area that has recently been logged is not very scenic, and the roads and other changes often irreversibly alter the area's wilderness character. Obviously, wilderness and logging cannot coexist. Therefore, it becomes necessary to decide whether a forest is going to be used for the production of timber or designated a wilderness area.

Several concerns must be addressed when it is decided that a forested area will be logged. Removing trees exposes the soil to increased erosion. Because the soil's water-holding ability is related to the amount of organic material and roots it has, denuded land allows water to run off rather than sink into it. Soil particles can wash into streams, where they cause siltation. The loss of soil particles reduces the soil's fertility. The particles that enter streams may cover spawning sites and eliminate fish populations. If the trees along the stream are removed, the water will be warmed by the increased sunlight. This may also have a negative effect on the fish population.

The roads necessary for moving equipment and removing logs are another

environmental close-up

The Northern Spotted Owl

In June 1990, the northern spotted owl was listed as a threatened species. Threatened species are those that are likely to become endangered if current conditions do not change. The northern spotted owl lives in old-growth coniferous forests in the Pacific Northwest. As with most carnivores at the top of the food chain, the northern spotted owl requires large areas of forest for hunting. Listing the owl as threatened requires that the U.S. Forest Service develop and implement a plan to protect it. Since the owl is found only in mature old-growth forests, and since it requires large areas for hunting, the plan must set aside large tracts of forested land and protect from logging.

Logging in the Pacific Northwest has already used up most of the trees on private land. Most of the remaining old-growth forest, which is about 10 percent of its original area, is on land managed by the U.S. Forest Service. Many of the trees in these forests are several centuries old, and forest replacement following logging takes an extremely long time. This time scale would not allow for the continued existence of the northern spotted owl if major sections of the remaining old-growth forest were logged.

The timber industry is an important element in the economy of the Pacific Northwest. The U.S. Forest Service had been selling about 300 square kilometers of forested land per year. If the old-growth forest is not available for harvest, forestry workers will lose their jobs. Those who make their living from the forest argue that there are already adequate wilderness areas and areas unfit for logging to supply the northern spotted owl with all the space it needs. They also argue that their livelihood is endangered in a way similar to the owl, and ask who is more important.

problem. Constant travel over these roads removes vegetation and exposes the bare soil to more rapid erosion. When the roads are not properly located and constructed, they eventually become gulleys and serve as channels for the flow of water. Most of these environmental problems can be minimized by using properly engineered roads and appropriate harvesting methods. Logging also disturbs wildlife by increasing human access and has resulted in the loss of large tracts of winter range.

One of the most controversial logging practices is **clear-cutting.** (See figure 11.7.) As the name implies, all of the trees in a large area are removed, which is a very economical method of harvesting. However, clear-cutting exposes the soil to significant erosive forces. In large blocks it may slow reestablishment of forest and have significant effects on wildlife.

On some sites, clear-cutting is still a reasonable method of harvesting trees. Sites with gentle slopes, where regrowth is rapid, have little erosion and environmental damage is limited. This is especially true if any streams in the area are left with a border of mature trees. The roots of the trees help to retain the stream banks and retard siltation. The shade provided by the trees also helps to prevent warming of the water, which might be detrimental to some fish species.

Clear-cutting can be very destructive on sites with steep slopes or where regrowth is slow. Under these circumstances, it may be possible to use **patchwork clear-cutting.** With this method, smaller areas are clear-cut among patches of untouched forest. This reduces many of the problems associated with clear-cutting and can also improve conditions for species of game animals that flourish in successional forests but not in mature forests. For example, deer, grouse, and rabbits benefit from a mixture of mature forest and early-stage successional forest.

Sites where natural reseeding or regrowth is slow may need to be replanted with trees, a process called **reforestation.** Reforestation is especially important in many of the conifer species, which often require bare soil to become established. Many of the deciduous trees will resprout from stumps or grow quickly from the seeds that litter the forest floor, so reforestation is not as important in deciduous forests.

Selective harvesting of some species of trees is also possible but is not as efficient or as economical as other methods, from the point of view of the harvesters. It allows them, however, to take individual,

Figure 11.7

A close clustering of large, clear-cut plots can lead to a loss of species and accelerated soil erosion.

Figure 11.8

Red-Cockaded Woodpecker The red-cockaded woodpecker requires old, living pines that have a disease known as red heart in which to build its nest. Diseased, old trees are rare in intensely managed forests. Therefore, the birds are endangered.

mature, high-value trees without causing much change to the forest ecosystem.

Forestlands can be managed to meet different needs. The forest products industry would prefer even-aged stands of a single species of tree for the most efficient harvest. Wildlife managers would prefer to see a variety of tree species at different stages of maturity to encourage many species of wildlife. A single-species, even-aged forest does not support as wide a variety of wildlife as does a mixed-age forest.

Some species even require diseased trees. For example, in the southern pine forests of the United States, the red-cockaded woodpecker requires old, diseased pine trees in which to build its nests. (See figure 11.8.) A well-managed forest plantation does not provide these sites.

Proper planning and effort can integrate several uses of the forest. Trees can be harvested in small enough patches to encourage wildlife but large enough to be economical. Some older trees, or some patches of old-stand timber, can be left as refuges for species that require them as a part of their niche.

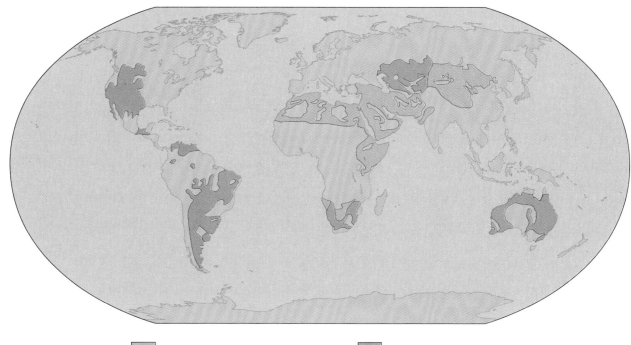

Nomadic herding   Stock raising on ranges

Figure 11.9

Nomadic Herding The arid and semiarid regions of the world will not support farming without irrigation. In many of these areas livestock can be raised. Permanent ranges occur where rainfall is low but regular. Nomadic herders can utilize areas that have irregular, sparse rainfall.

Many lumber companies maintain forests as crops and seek to manage them in the same way farmers manage crops. They plant single species, even-aged forests of fast-growing hybrid trees that have been developed in the same way high-yielding agricultural crops have been. Competing species are controlled by the use of fire in some forests, and insects are controlled by aerial spraying. In these intensively managed forests, some single-species plantations mature to harvestable size in 20 years, rather than in the approximately 100 years typical for naturally re-producing mixed forests. However, the quality of the lumber products is reduced. Such single-species forests also contain a decreased variety of organisms since different species of wildlife have differing food and cover requirements.

The trees planted in many managed forests may be exotic species. *Eucalyptus* forests have been planted in South America, and most of the forests in northern England and Scotland have a mixture of native pines and imported species from the European mainland and North America.

Road construction, fire control, and control of fungus and insect pests modify forests used for timber. For this reason, many people are attempting to protect forests being exploited for the first time. Of particular concern are the tropical rain-forests that are under intense pressure to provide both wood products and land that can temporarily be used for agriculture.

Rangelands

Many of the arid and semiarid lands of the world are being used to raise domesticated or semidomesticated animals in permanent open ranges or nomadic herds. (See figure 11.9.) Most of these areas will not support agriculture but can support sparse populations of native or introduced species of grazing animals. These areas have been modified by the animals' selective eating habits, which tend to reduce certain species of plants and encourage others.

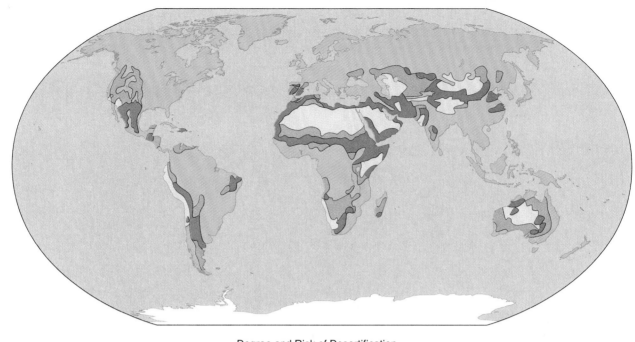

Degree and Risk of Desertification

High   Moderate   Existing Desert

Figure 11.10

Desertification Arid and semiarid areas can be converted to deserts by overgrazing or unsuccessful farming practices. The loss of vegetation increases erosion by wind and water, increases the evaporation rate, and reduces the amount of water that infiltrates the soil. All of these conditions encourage the development of desertlike areas.

(Top) Sources: United Nations Map of World Desertification, *United Nations Food and Agriculture Organization, United Nations Educational, Scientific, and Cultural Organization, and the World Meterological Organization for the United Nations Conference on Desertification, 1977, Nairobi, Kenya.*

In most of these areas, the animals that graze are exotic, non-native species. Sheep, cattle, and goats that are native to Europe and Asia have been introduced into the Americas, Australia, New Zealand, and many areas of Africa. The introduction of exotic species has reduced the numbers of certain native species. For example, sheep and cattle have replaced the bison in North America, and many native antelope species in Africa have been negatively affected by cattle. In an effort to increase the productivity of rangelands, management techniques may specifically eliminate certain species of plants not useful as food for the grazing animals, or specific grasses may be planted that are not native to the area.

In many parts of the world, where human population pressures are great, overgrazing is a severe problem. As populations increase, desperate people attempt to graze too many animals on the land and cut down the trees for firewood. If overgrazed, many plants die, and the loss of plant cover, subjects the soil to wind erosion, resulting in a loss of fertility, which further reduces the land's ability to support vegetation. The cutting of trees for firewood has a similar effect, but it is especially damaging because many of these trees are legumes, which are important in nitrogen fixation. Their removal further reduces soil fertility. All of this results in degradation of the land to a more desertlike ecosystem. This process of converting arid and semiarid land to desert because of improper use by humans is called **desertification.**

Desertification can be found throughout the world but is particularly prevalent in northern Africa, where rainfall is irregular and unpredictable, and where many people are subsistence farmers or nomadic herders who are under considerable pressure to provide food for their families. (See figure 11.10.)

Modified Aquatic Ecosystems

For several reasons, the most productive areas of the ocean are those close to land. In shallow water, the entire depth (water column) is exposed to sunlight. Plants and algae can carry on photosynthesis, and biological productivity is high. The

nutrients washed from the land also tend to make these waters more fertile. Furthermore, land masses modify currents that bring nutrients up from the ocean bottom. Many of the commercially important fish species are bottom-dwellers, but fishing for them at great depths is not practical. Therefore, fishing pressure is concentrated on areas where the water is shallow and relatively nutrient rich.

Since commercial fishing targets certain species, it can change species composition in heavily fished areas. The commercial fishing industry has been attempting to market fish species that previously were regarded as unacceptable to the consumer. These activities are the result of reduced catches of desired species. Examples of "newly discovered" fish in this category are monkfish and orange roughy.

Thirty years ago, anchovy fishing off the coast of Peru was a major industry. From 1971 to 1972, the catch dropped dramatically. Overfishing was believed to be one of the major contributing factors. This was aggravated by an increase in the area's water temperature, which prevented nutrient-rich layers from rising to the euphotic zone. Thus, productivity at all trophic levels decreased. Currently, the Peruvian anchovy fishery is considered to be overexploited and not sustainable.

Because freshwater ecosystems are small and more intimately associated with human activities, few have not been considerably altered. Changes in water quality and the introduction of exotic species are two primary alterations. The warming of water due to thermal pollution and the addition of nutrients have made it impossible for some native species to survive. The effects of water pollution are discussed in chapter 15.

The management of fish populations is similar to that of other wild animals. Fish require cover, such as logs, stumps, rocks, and weed beds, so they can escape from predators. They also need special areas for spawning and raising young. These might be a gravel bed in a stream for salmon, a sandy area in a lake for bluegills or bass, or an estuary for many species of marine fish. A fisheries biologist tries to manipulate some of the features of the habitat to enhance them for the desired

species of game fish. This might take the form of providing artificial spawning areas or cover. Regulation of the fishing season so that the fish have an opportunity to breed is also important.

In addition to these basic concerns, the fisheries biologist pays special attention to water quality. Whenever people use water or disturb land near the water, water quality is affected. For example, toxic substances kill fish directly, and organic matter in the water may reduce the oxygen. The use of water by industry or the removal of trees lining a stream warms the water and makes it unsuitable for certain species. Poor watershed management results in siltation, which covers spawning areas, clogs the gills of young fish, and changes the bottom so food organisms cannot live there. The fisheries biologist is probably as concerned about what happens outside the lake or stream as what happens within it.

The introduction of exotic fish species has greatly affected naturally occurring freshwater ecosystems. The Great Lakes, for example, have been altered considerably by the accidental and purposeful introduction of fish species. The sea lamprey, smelt, carp, alewife, brown trout, and several species of salmon are all new to this ecosystem. (See figure 11.11.) The sea lamprey is parasitic on lake trout and other species and nearly eliminated the native lake trout population. Controlling the lamprey problem requires use of a very specific larvicide that kills the immature lamprey in the streams. This technique works because mature lamprey migrate up streams to spawn, and the larvae spend a year or two in the stream before migrating downstream into the lake. With the partial control of lamprey, lake trout populations have been increasing. (See figure 11.12.) Recovery of the lake trout is particularly desirable, since at one time it was an important commercial species.

Another accidental introduction to the Great Lakes, the alewife, a small fish of little commercial or sport value, became a problem during the 1960s, when alewife populations were so great that they died in large numbers and littered beaches. Various species of salmon were introduced at about this time in an attempt to control

environmental *close-up*

Exploiting the Oceans' Bounty

Throughout history, the world's oceans have been a source of unlimited seafood available to all who fished. This, however, is no longer the case. Today, nearly 40 percent of the world's oceans have been locked up by national territorial claims and fishing zones. These 200-nautical-mile limits began to be established in the late 1940s with the development of long-range fishing fleets.

Expanding the boundaries, however, has not stopped the overexploitation of certain species that are favored for their commercial value. (See map for distribution.) Wealthy nations continue to buy access to the waters of poorer nations. Harvesting technology grows more sophisticated and the annual catch gets scarcer and smaller. Despite a rise in production of the Pacific Ocean (see top graph) in the 1980s, the tonnage of seafood harvested worldwide has reached a plateau after peaking in 1989.

The world demand for seafood continues to grow. With an annual per capita consumption of 67 kilograms (147

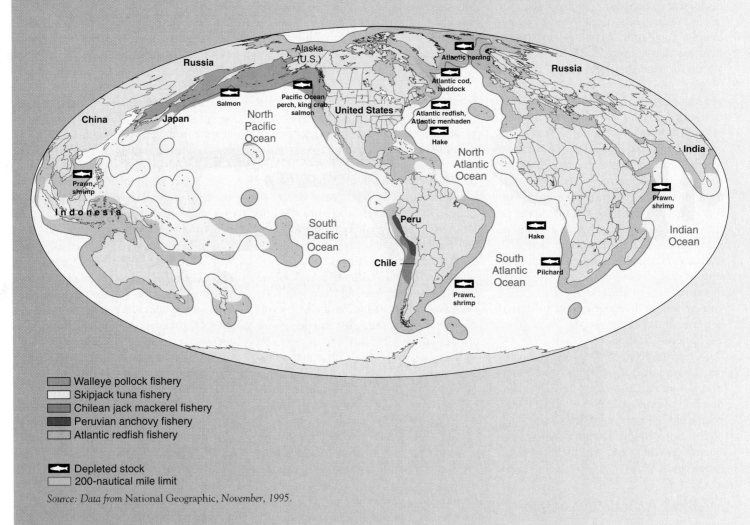

- Walleye pollock fishery
- Skipjack tuna fishery
- Chilean jack mackerel fishery
- Peruvian anchovy fishery
- Atlantic redfish fishery

- Depleted stock
- 200-nautical mile limit

Source: Data from National Geographic, *November, 1995.*

pounds), Japan is the world's biggest consumer of seafood. This world demand and the decline in natural stocks has prompted a dramatic increase in fish farming in several parts of the world. For example, in 1995, salmon farming in British Columbia outstripped the commercial fishery for the first time: 80 salmon farms had a harvest worth $167 million compared with a commercial catch worth about $75 million. While fish farming will continue to provide more and more of the world's seafood, that does not excuse the ignorance, short-sightedness, and greed of many of the world's commercial fishers and the potential irreparable damage their fishing practices have done to the sustainability of species.

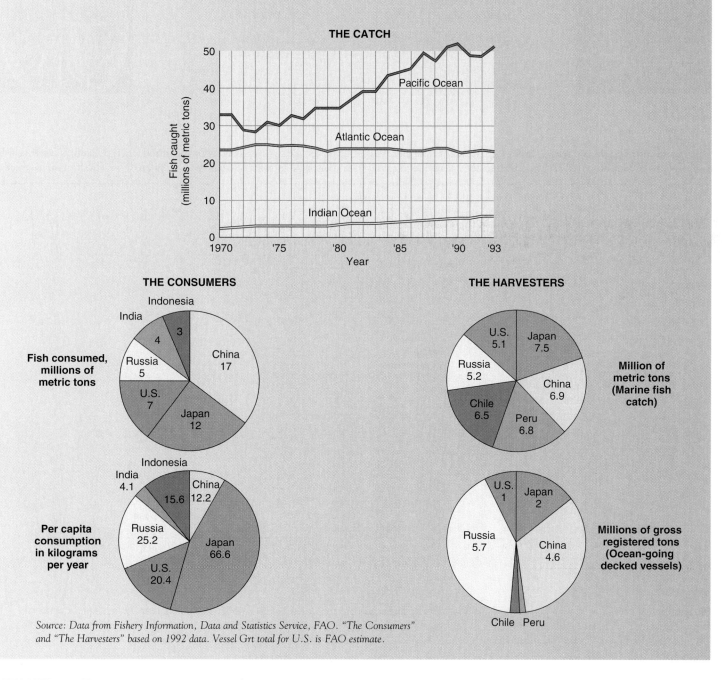

THE CATCH

THE CONSUMERS

THE HARVESTERS

Source: Data from Fishery Information, Data and Statistics Service, FAO. "The Consumers" and "The Harvesters" based on 1992 data. Vessel Grt total for U.S. is FAO estimate.

Lake trout

Native	Introduced
Brook lamprey	Sea lamprey
Lake trout	Brown trout
Brook trout	Rainbow trout
Whitefish	Pink salmon
Herring	Coho salmon
Ciscoes	Chinook salmon
	Atlantic salmon
Suckers	
Chubs	Carp
Shiners	
Catfish	
Bullheads	
	Smelt
	Alewife
White bass	White perch
Smallmouth bass	Crappies
Rock bass	Sunfish
Yellow perch	
Walleye	

Brown trout

Smelt

Yellow perch

Figure 11.11

Native and Introduced Fish Species in the Great Lakes   The Great Lakes have been altered considerably by the introduction of many non-native fish species. Some were introduced accidentally (lamprey, alewife, and carp), and others were introduced on purpose (salmon, brown trout, rainbow trout).

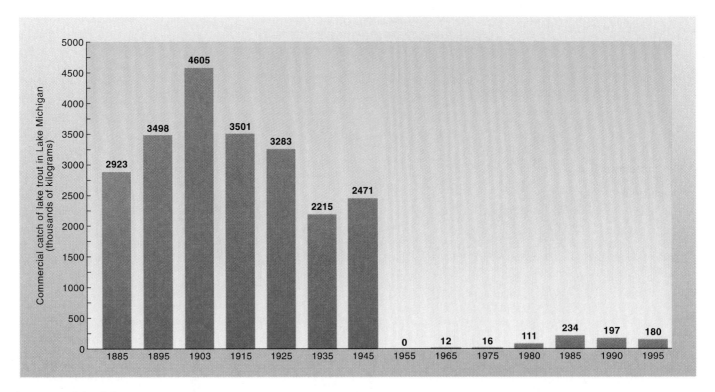

Figure 11.12

The Impact of the Lamprey on Commercial Fishing   The lamprey entered the Great Lakes in 1932. Because it is an external parasite on lake trout, it had a drastic effect on the population of lake trout in the Great Lakes. As a result of programs to prevent the lamprey from reproducing, the number of lamprey has been reduced somewhat, and the lake trout population is recovering with the aid of stocking programs.

PART FOUR   Human Influences on Ecosystems

the alewife and replace the lake trout population, which had been depressed by the lamprey. While this salmon introduction has been an economic success and has generated millions of dollars for the sport fishing industry, it may have had a negative impact on some native fish like the lake trout, with which the salmon probably compete. Salmon also migrate up streams, where they disrupt the spawning of native fish. Most species of salmon die after spawning, which causes a local odor problem.

Fish management in much of the world includes managing for both recreational and commercial food purposes. The fisheries resource manager must try to satisfy two interest groups. Both sports fishers and those who harvest for commercial purposes must adhere to regulations. However, the regulations usually allow the commercial fisher to use different methods, such as nets.

Streams and rivers are modified for navigation, irrigation, flood control, or power production purposes, all of which may alter the natural ecosystem. These topics are discussed in greater detail in chapter 15.

Modifying Ecosystems to Manage Wildlife

Many kinds of terrestrial wild animals are desired as game animals or for other purposes, and efforts are made to improve conditions for these species. Several techniques, some of which result in ecosystem modifications, are used to enhance certain wildlife populations. These techniques include game and habitat analysis, population census methods, stocking areas with game species, predator control, establishing refuges, and habitat management.

Managing a particular species requires understanding the habitat needs of that species. An animal's habitat must provide the following five requirements: food and water, cover for escaping from enemies, cover for protection against the elements, cover for resting and sleeping, and cover for mating and for rearing young. **Cover** refers to any set of physical features that conceals or protects animals from the elements or enemies.

Figure 11.13

Cover Requirements for Quail   Quail need several kinds of cover to be successful in an area. When raising young, they need nesting of tall grass and weeds to provide protection from predators. In the winter, they need thickets to provide cover from weather and predators.

Animals are highly specific in their habitat requirements. For example, bobwhite quail must have a winter supply of food that protrudes above the snow. They require a supply of small rocks called grit, which the birds use in their gizzards to grind food. During most of the year, they need a field of tall grass and weeds as cover from natural predators. However, such protection is not available during the winter, so a thicket of brush is necessary. The thicket serves as cover from predators and shelter from the cold winter weather. Most areas provide suitable cover for resting, but for sleeping, quail prefer an open, elevated location, which allows the birds to quickly take flight if attacked at night. When raising young, they require grassy areas with some patches of bare ground where the young can sun themselves and dry out if they get wet. All of these requirements must be available within a radius of 400 meters, because this is the extent of a quail's normal daily travels. (See figure 11.13.)

Once habitat requirements are known, the desirable population size must be ascer-tained. The population must be checked regularly to see if it is within acceptable limits. Population censuses are used to provide these data. If the population is below the desired number, organisms may be artificially introduced. This is referred to as stocking. Many species that are hunted are actually introduced, and areas are stocked with these introduced species. The ring-necked pheasant was originally from Asia but has been introduced into Europe, Great Britain, and North America. All of the large game animals of New Zealand are introduced species, since there were none in the original biota of these islands. Several species of deer have been introduced into Europe from Asia. Many of these are raised in deer parks, where the animals are similar to free-ranging domestic cattle. They may be hunted for sport or slaughtered to provide food.

Since wildlife management often involves the harvesting of animals by hunting for sport and for meat, seasons are usually regulated to assure adequate reproduction and provide the largest possible health population during the hunting season. Hunting seasons usually occur in the fall because many animals normally die during the winter. Winter taxes the animal's ability to stay warm and is also a time of low food supplies in most temperate regions. A well-managed wildlife resource allows for a large number of animals to be harvested in the fall and still leaves a healthy population to survive the winter and reproduce during the following spring. (See figure 11.14.)

Often, predator or competitor populations have been reduced to allow optimal survival of game species. In intensely managed European systems, gamekeepers are often charged with the responsibility of killing predators or unwanted competitors. With more wild and less intensely managed species, as is typical in North America, very little predator or competitor control is exercised. At one time, it was thought that populations of game species could be increased substantially if predators were controlled. In Alaska, for example, the salmon-canning industry claimed that bald eagles were reducing the salmon population, and a bounty was placed on the bald eagle. From 1917 to 1952, 128,000 eagles were killed for bounty money in Alaska. This theory of predator control to increase populations of game species has not proven to be valid in most cases, however, since the predators do not normally take the prime animals anyway. They are more likely to capture sick or injured individuals not suitable for game hunting.

For some species, such as ground nesting birds, it may make some sense to control predators that eat eggs or capture the young, but in most cases, humans have a greater impact through habitat modification and hunting than do the natural predators. Although at one time they were thought to be helpful, bounties and other forms of predator management have largely been eliminated in North America. In fact, the pendulum has swung the other way. In 1993, a controversial attempt was made to allow hunting of wolves in Alaska because the wolves were thought to be killing moose in large numbers. Since in many parts of the world the major predators of game species are humans, regulation of hunting is a form of predator control.

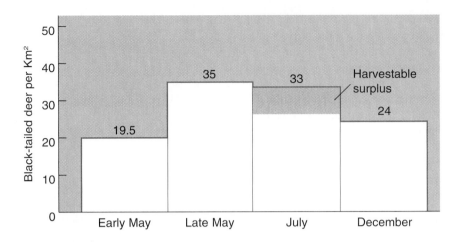

Figure 11.14

Managing a Wildlife Population   The seasonal changes in this population of black-tailed deer are typical of many game species. The hunting season is usually timed to occur in the fall so that surplus animals will be harvested before winter, when the carrying capacity is lower.

(Top) Source: Data from R.D. Taber and R.F. Dasmann, "The Dynamics of Three Natural Populations of the Deer Odocoileus hemionus columbianus," in Ecology 38(2): 233–46, 1957.

Many species may require refuges where they are protected or where the habitat is manipulated to assure their reproduction. Elephants in Africa are often hunted illegally for their ivory. Protected refuges that are patrolled by armed game wardens are necessary to prevent their extinction in much of Africa. Many native prairie species benefit from the exclusion of introduced grazing animals like cattle and sheep. (See figure 11.15.)

The Kirtland warbler is an endangered species that nests in the central part of Michigan, although evidence now indicates that some of these birds may be mating in the neighboring states of Wisconsin and Minnesota. The nesting

Figure 11.15

Habitat Protection   The area on the left side of the fence has been protected from cattle grazing. This area provides a haven for many native species of plants and animals that cannot survive in heavily grazed areas.

environmental
close-up

Native American Fishing Rights

Throughout many parts of the United States, particularly in the Pacific Northwest and the Great Lakes states, there is a continuing controversy over fishing rights. This conflict is unique because it involves treaties that were made 100 to 150 years ago between the U.S. government and Native American nations. It has become a major political, economic, social, and legal issue in some states, such as Washington, Michigan, Wisconsin, and Minnesota, and has involved the entire court system, from local courts to the U.S. Supreme Court. The controversy revolves around the interpretation of treaty language. It has, on several occasions, turned to violence and has divided many communities.

According to the wording of many treaties entered into in the 1800s, the rights of Native Americans to fish would not be infringed upon by the states. Native Americans claim the treaties give them the legal authority to engage in commercial fishing enterprises even when such fishing may be restricted or banned altogether for the general public.

On the other side of the argument are many state officials and sport fishers who believe that Native Americans are seriously endangering populations of such fish as salmon and trout by their uncontrolled harvesting for commercial purposes. They further argue that many species being taken by Native Americans belong to the entire state and not only to a certain group, because the fish are stocked or planted by the state. Another concern is the fishing techniques used by Native Americans. In the 1800s, when the treaties were signed, commercial fishing technology was limited. Today, however, Native American commercial fishers use nylon nets, power boats, depth finders, and other technological aids that enable them to catch much larger quantities of fish than they could with traditional fishing practices.

In the early 1970s, when sport fishers complained that stocks of fish in Lake Michigan were being depleted because of Native Americans' gill-net fishing, the state of Michigan tried to regulate Native American fishing. The issue ended up in court, and in 1978, a U.S. district court judge in Grand Rapids upheld fishing rights granted under 1836 and 1855 treaties between two Chippewa tribes and the U.S. government. The federal judge ruled that the state had no authority

in the matter because the issue in question involved a federal treaty, and the state did not have the right to make regulations contrary to the treaty. Only Congress had such power. Subsequent to the decision, the tribes and the state negotiated changes in the areas in which the tribes could fish in an effort to reduce the potential for conflict between Native American fishers and sport or other commercial fishers.

A similar case was decided by a U.S. district judge in Tacoma, Washington, who interpreted treaties signed in 1854 and 1855 and ruled that Native Americans could catch up to 50 percent of the salmon that passed through their tribal land on the way to other parts of the state. In the face of protests by the non-native commercial fishing industry, the federal judge took over the regulation of salmon fishing in the state. Commercial fishers, who outnumbered Native American fishers by a ratio of approximately eight to one, feared for their livelihood. The case eventually went to the U.S. Supreme Court. In 1979, the Court ruled, in a six-to-three decision, to uphold the federal treaties that entitled Native Americans to half the salmon caught in the area of Puget Sound in Washington state. It further held that the 50 percent figure had to include the salmon that Native Americans caught on their lands for home consumption and religious ceremonies as well as the fish they caught for commercial purposes. The high court also voted to uphold the right of the district court to continue supervising the fishing industry because of the state's resistance to the interpretation of the treaties.

Clearly, the issues surrounding Native American fishing rights are complex and broad in scope. There are purely biological questions involving a resource and its wise use, economic issues involving families' livelihoods, cultural and religious concerns pertaining to Native Americans' use of a resource, legal questions involving federal and state conflicts, and moral questions relating to the unfair treatment of Native Americans in the past. In such conflicts, there is seldom one right answer. While the courts have ruled on the cases in Michigan and Washington, and the states and Native Americans are trying to use the fishing resource wisely, the problem still exists and is likely to continue for years to come.

cover required for this species is young, even-aged stands of jack pine. The areas where these birds nest have been protected, and, in addition, the habitat has been modified to provide for optimal nesting. Portions of their nesting area are periodically burned to assure future nesting sites. This is important because the jack pine cone will only open if subjected to fire. So fire is a very important part of the ecological niche of the Kirtland warbler.

Burning old stands of jack pine to allow for regrowth is an example of **habitat management.** Once the food habits, predators, and cover requirements of a game species are well understood, the habitat can be altered to optimize it for the desired species. Habitat management may take the form of encouraging some species of plants that are the preferred food of the game species. For example, habitat management for deer may involve encouraging the growth of many young trees, saplings, and low-growing shrubs by cutting the timber in the area and allowing the natural regrowth to supply the food and cover the deer need. Both forest management and deer management may have to be integrated in this case because some other species of animals, like squirrels, will be excluded if mature trees are cut, since they rely on the seeds of trees as a major food source.

Given suitable habitats and protection, most wild animals can maintain a sizable population. In general, organisms produce more offspring than can survive. Figure 11.16 illustrates the reproductive potential of both quail and white-tailed deer. High reproductive potential, protection from hunting, and the management and restoration of suitable habitats have reversed the drastic population declines of deer in some parts of the world. In Pennsylvania, where deer were once extinct, the number is now in excess of 900,000. Even in predominately agricultural states, deer have made a comeback. In Indiana, a herd of over 280,000 deer has grown from an initial stocking of 35 individuals.

Other animals have shown comparable changes in population size. In Zimbabwe, unlike in most African countries, the number of elephants increased from about 200 in 1900 to 40,000 in 1987. This is probably not a sustainable population, and the government would like to reduce the number to about 30,000. The wild turkey population in the United States has increased from about 20,000 birds in 1890 to 4 million birds today.

Waterfowl (ducks, geese, swans, rails, etc.) present some special management problems because they are migratory. **Migratory birds** can fly thousands of kilometers and, therefore, can travel north in the spring to reproduce during the summer months and return to the south when cold weather freezes the ponds, lakes, and streams that serve as their summer homes.

Quail

	Adults	Young	Total
1st year	2	14	16
2nd year	16	112	128
3rd year	128	896	1024

Deer

	Adults	Yearlings	Fawns	Total
1st year	2	0	2	4
2nd year	2	2	2	6
3rd year	4	2	4	10
4th year	6	4	6	16
5th year	10	6	10	26

Figure 11.16

Reproductive Potential   If we assume no mortality, animals have a reproductive capacity far above what is required to just keep the population stable. In the real world, mortality generally keeps populations from growing beyond the ability of their habitat to sustain them.

(See figure 11.17.) Because many waterfowl nest in Canada and winter in parts of the United States and Central America, an international agreement is necessary to manage and prevent the destruction of this wildlife resource. Habitat management has taken several forms. In Canada, where much of the breeding occurs, government and private organizations such as Ducks Unlimited have worked to prevent the draining of small ponds and lakes that provide nesting areas for the birds. In addition, new impoundments have been created where it is practical. Because birds migrate southward during the fall hunting season, a series of wildlife refuges provide havens from hunters. In addition, these refuges may be used to raise local populations of birds. During the winter, many of these birds congregate in the southern United States. Refuges in these areas are important overwintering areas where the waterfowl can find food and shelter.

Replacement of Natural Ecosystems with Human-Managed Ecosystems

Over large parts of the world, the natural ecosystems that once existed have been completely destroyed and replaced with intensely managed agricultural ecosystems. This is a natural consequence of human population growth. As the population has increased, more and more land has been converted to agriculture. Today, very little additional land can be converted. All of the best land has already been used, leaving only marginal areas that have relatively low productivity. (See table 11.6.) Temperate forests and grasslands have been most affected by this conversion to agriculture. What today supports wheat, rice, and corn was originally prairie or forest. Chapters 13 and 14 provide a more in-depth look at agricultural ecosystems and how they differ from naturally occurring ecosystems.

Natural Selection and Extinction

Extinction is the death of a species, the elimination of all the individuals of a particular kind. It is a natural and common event in the long history of biological evolution. Of the estimated 500 million species of organisms that are believed to have ever existed on earth since life began,

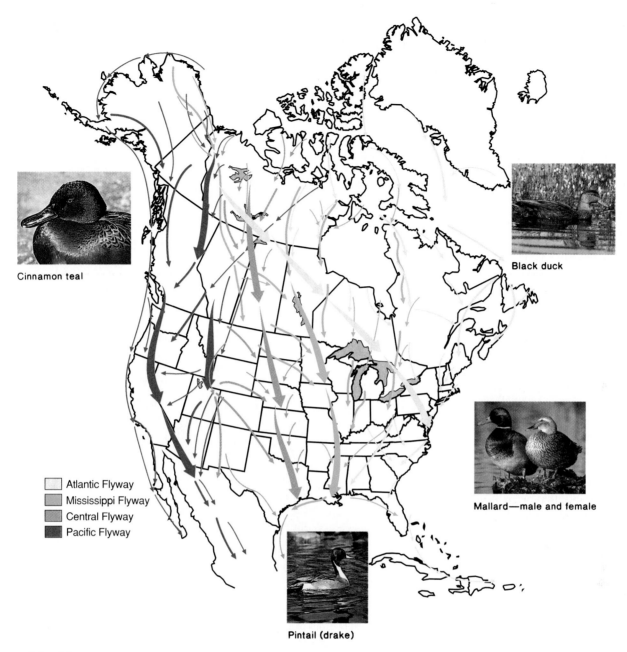

Figure 11.17

Migration Routes for North American Waterfowl   Migratory waterfowl follow traditional routes when they migrate. These have become known as the Atlantic, Mississippi, Central, and Pacific flyways. Many of these waterfowl are hatched in Canada, migrate through the United States, and winter in the southern United States or Mexico.

perhaps 5 million to 10 million are currently active. This represents an extinction rate of 98 to 99 percent. Obviously, these numbers are estimates, but the fact remains that extinction has been the fate of most species of organisms. However, the process of **speciation** produces new species. Thus, new species are continually added while at the same time those that are not

able to adapt to a changing environment are eliminated.

Recall that a species is a population of organisms. Every individual is a member of a species and shares many characteristics that are typical of the species. For example, all dogs have four legs, two eyes, hair, and a tail of some kind. But there is also variation among individuals within a

species. Some are taller, can run faster, have quicker reflexes, or are better able to hear, smell, or see. This individual variation allows some organisms to cope better with their surroundings. Most of these characteristics are at least partly the result of the genetic material, that is, genes and DNA, that the individuals inherited from their parents.

Table 11.6

Human-Induced Land Degradation Worldwide, 1945 to Present

Region	Overgrazing	Deforestation	Agricultural Mismanagement (million hectares)	Other[1]	Total	Degraded area as share of total vegetated land (percent)
Asia	197	298	204	47	746	20
Africa	243	67	121	63	494	22
South America	68	100	64	12	244	14
Europe	50	84	64	22	220	23
North & Central America	38	18	91	11	158	8
Oceania	83	12	8	0	103	13
World	679	579	552	155	1,965	17

[1]Includes exploitation of vegetation for domestic use (133 million hectares) and bioindustrial activities, such as pollution (22 million hecatres).

Source: Data from Worldwatch Institute based on "The Extent of Human-Induced Soil Degradation," Annex 5 in L.R. Oldeman, et al., World Map of the Status of Human-Induced Soil Degradation (Wageningen, Netherlands: United Nations Environment Programme and International Soil Reference and Information Centre.

Individuals that have a combination of characteristics that allow them to be more successful are more likely to live long enough to reproduce, and are also likely to have larger numbers of offspring. Their offspring are likely to have many of the characteristics that made their parents successful and will outnumber the offspring of parents that were not well adapted. Thus the characteristics typical of the species change slightly over time.

If two populations of the same species are isolated from one another for a long time, and the environments of the two locations are very different, the process of natural selection can result in the two populations becoming so different from one another that they do not interbreed and so are considered different species. (See figure 11.18.) Keep in mind that environments change over time, so a set of characteristics will not always be suitable. For example, at one time, much of Canada and the northern parts of the United States were covered with glaciers. The climate was quite different from what exists today and the plants and animals that existed were also different.

Because the environment is continually changing, a species must adapt or become extinct. If this adaptation occurs over a long period, the original species may evolve into a new species. In this

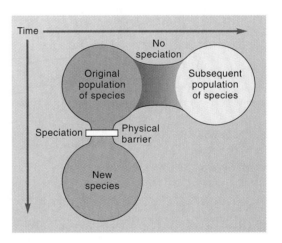

Figure 11.18

Speciation Populations of a species may become isolated from one another for a long time as a result of a barrier. As they encounter new environmental conditions, selection occurs, and the isolated population becomes adapted to local environmental conditions. If isolation is total and lasts for a long time, a new species may develop. If no barrier prevents gene exchange between populations, a new species is not likely to devlop, but the population will still adapt to new environmental conditions.

situation, the original species eventually becomes extinct, but before it does, many of its genes (particularly the favorable ones) are passed on to the newly evolving species. An example of this is the extinction of the dinosaurs. As earth's environment changed, it became less favorable for dinosaurs. However, before their extinction, many of the favorable genes in this reptile line formed the basis for the

newly evolving mammals and birds. This form of gradual extinction and replacement by other, better-adapted groups is common in evolutionary history. The fossil record shows many groups of organisms that flourished for millions of years before more advanced groups replaced them.

Studies of modern local extinctions suggest that certain kinds of species are

Table 11.7

Probability of Becoming Extinct	
Most likely to become extinct	**Least likely to become extinct**
1. Low population density	1. High population density
2. Found in small area	2. Found over large area
3. Specialized niche	3. Generalized niche
4. Low reproductive rates	4. High reproductive rates

more likely than others to become extinct. (See table 11.7.) Species that have small populations of dispersed individuals are more prone to extinction because successful breeding is more difficult than in species that have large populations of relatively high density. Some kinds of organisms, such as carnivores at higher trophic levels in food chains, typically have low populations but also have low rates of reproduction compared to their prey species.

Organisms in small, restricted areas are also prone to extinction because an environmental change in their locale can eliminate the entire species at once. Organisms scattered over large areas are much less likely to be negatively affected by one event.

Specialized organisms are also more likely to become extinct than are generalized ones. Since specialized organisms rely on a few key factors in the environment, anything that negatively affects these factors could result in their extinction, whereas generalists can use alternate resources.

Rabbits and rats are good examples of animals that are not likely to become extinct soon. They have high population density and a wide geographic distribution. In addition, they have high reproductive rates and are generalists that can

live under a variety of conditions and use a variety of items as food. The cheetah is much more likely to become extinct because it has a low population density, is restricted to certain parts of Africa, has low reproductive rates, and has very specialized food habits. It must run down small antelope, in the open, during daylight, by itself. Similarly, the entire wild whooping crane species consists of about 130 individuals that are restricted to small winter and summer ranges that must have isolated marshes. (Captive and experimental populations bring the total number to about 200 individuals.) In addition, their rate of reproduction is low.

Human-Accelerated Extinction

Humans are among the most successful organisms on the face of the earth. We are adaptable, intelligent animals with high reproductive rates and few enemies. As our population increases, we displace other kinds of organisms. This has resulted in an accelerated rate of extinction. Wherever humans become the dominant organism, extinctions occur. Sometimes, we use other animals directly as food. In doing so, we reduce the popu-

lation of our prey species. Since our population is so large and because we have an advanced technology, catching or killing other animals for food is relatively easy. In some cases, this has led to extinctions. The passenger pigeon in North America, the moas (giant birds) of New Zealand, and the bison and wild cattle of Europe were certainly helped on their way to extinction by people who hunted them for food.

We use organisms for a variety of purposes in addition to food. Many plants and animals are used as ornaments. Flowers are picked, animal skins are worn, and animal parts are used for their purported aphrodisiac qualities. In the United States, many species of cactus are being severely reduced because people like to have them in their front yards. In other parts of the world, rhinoceros horn is used to make dagger handles or is powdered and sold as an aphrodisiac. Because some people are willing to pay huge amounts of money for these products, unscrupulous people are willing to take the chance of poaching these animals for the quick profit they can realize. Table 11.8 shows some of the organisms, or their parts, that are highly prized by buyers. The World Wildlife Fund estimates that illegal trade in wild animals globally produces $2 billion to $30.5 billion per year. These activities have already resulted in local extinctions of some plants and animals and may be a contributing factor to the future extinction of some species.

Some organisms are extinct because they were regarded as pests. Many large predators have been locally exterminated because they preyed on the domestic animals that humans use for food. Mountain lions and grizzly bears in North America have been reduced to small, isolated populations, in part because they were hunted to reduce livestock loss. Tigers in Asia and the lion and wolf in Europe were reduced or eliminated for similar reasons. Even though commercial hunters killed thousands of passenger pigeons, their ultimate extinction was caused primarily by the increased agricultural use of the forests. Passenger pigeons ate the acorns of oaks and the beechnuts from beech, and also relied on the forests for communal nesting

Table 11.8

Typical Species Prices (1990 Rates)

International species	Price ($U.S.)	North American species	Price ($U.S.)
Olive python	1,500	Bald eagle	2,500
Rhinoceros horn	12,500/pound	Golden eagle	200
Tiger skin (Siberian)	3,500	Gila monster	200
Tiger meat	130/pound	Peregrine falcon	10,000
Cockatoo	2,000	Grizzly bear	5,000
Leopard	8,500	Grizzly bear claw necklace	2,500
Snow leopard	14,000	Polar bear	6,000
Elephant tusk	250/pound	Black bear paw pad	150
Walrus tusk	50/pound	Reindeer antlers	35/pound
Mountain gorilla	150,000	Mountain lion	500
Giant panda	3,700	Mountain goat	3,500
Ocelot	40,000/coat	Saguaro cactus	15,000
Imperial Amazon macaw	30,000		

Source: Data from U.S. Fish and Wildlife Service.

sites. When the forests were cleared, the pigeons became pests to farmers, who shot them to protect their crops from being eaten by the birds.

Some extinctions of pest species are considered desirable. Most people would not be disturbed by the extinction of black widow spiders, mosquitoes, rats, or fleas. In fact, people work hard to drive some species to extinction. For example, in the *Morbidity and Mortality Weekly Report* (October 26, 1979), the U.S. Centers for Disease Control triumphantly announced that the virus that causes smallpox was extinct in the human population after many years of continuous effort to eliminate it.

The most important cause of extinctions related to human activity is habitat alteration. (See figure 11.19.) Whenever humans populate an area, they change it by converting the original ecosystem into something that supports themselves. Forests and grasslands have been converted to agricultural and grazing lands. In addition, humans have introduced new species to grow or graze on these lands. These new plants and animals compete with the native organisms for nutrients and living space,

and often, the native organisms lose. (See Global Perspective: The History of the Bison.) The African elephant and various species of rhinoceros are in danger because they are hunted for ivory or horn, but it was the original fragmentation of their habitat for agricultural purposes that first placed the animals in conflict with humans. (See Environmental Close-Up: Hope for the Rhinoceros.)

In much of the tropical world today, rainforests are being cleared to provide grazing land or agricultural land for an expanding human population. This activity destroys existing forests and fragments them into small islands. Scientists studying the effects of this activity have noticed that, as a forest is reduced to small patches, many species of birds disappear from the area. The same kinds of activities certainly happened in Europe, where little of the original forest is left. In North America, the eastern deciduous forests were reduced, which probably resulted in the extinctions of some animals and plants. Almost all of the original prairie in the United States has been replaced with agricultural land, resulting in the loss of some species.

Humans also cause extinction in less direct ways. The building of dams changes the character of rivers, making them less suitable for some species. Air and water pollution may kill all life in an area. For example, acid precipitation has lowered the pH of some lakes so much that all life has been eliminated. In other cases, indirect human activity may selectively eliminate species that are less tolerant to the pollutants being released, or the introduction of exotic species may eliminate existing species. The accidental introduction of the chestnut blight fungus into the United States resulted in the loss of the chestnut tree over much of the Appalachian region. Some species that were dependent on these trees probably were also eliminated.

Why Worry about Extinction?

If extinction is a natural event, and if the species that become extinct are those that cannot effectively cope with the activities of humans, why should we worry about them? There are several answers to that question. First of all, strictly from a selfish point of view, many species we know little about may be useful to us. Many plants have chemicals in them that can be used as medicines. If we drive them into extinction, we may be eliminating a potentially useful product. Also, as the human population increases, the need for different kinds of food plants and animals will increase. Once they are eliminated, we have lost the opportunity to use them for our own ends. Most of the wild ancestors of our most important food grains, like maize (corn), wheat, and rice, are thought to be extinct. What would our world be like if these plants had not been domesticated before they went extinct as wild populations? Because of this concern, many agricultural institutes and universities maintain "gene banks" of wild and primitive stocks of crop plants.

Another interesting answer to the question is that certain organisms in some ecosystems appear to play pivotal roles.

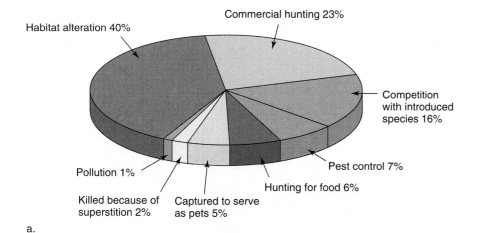

Causes of Extinction   Many organisms have become extinct as a result of human activities. The most important of these is the indirect destruction of habitat by humans modifying the environment for ther own purposes. Part (b) shows the species extinct and at risk in North America (percentage by group).

(b) Source: Data from the Nature Conservancy and the Network of Natural Heritage Programs and Conservation Data Centers 1993.

needlessly eliminated by unthinking activity of the human species. This philosophical statement has nothing to do with economic or social value; it is an ethical position.

What Is Being Done to Prevent Extinction?

Efforts to prevent human-caused extinctions are difficult to assess. Some countries have enacted legislation to protect species that are in danger of becoming extinct. Species that receive special protection usually are given some sort of designation, such as endangered or threatened. **Endangered species** are those that have such small numbers that they are in immediate jeopardy of becoming extinct. **Threatened species** could become extinct if a critical factor in their environment were changed.

Most of the interest in preventing human-caused extinctions comes from developed countries. There, however, the problem is less acute because vulnerable species have already been eliminated. Extinction is a greater potential problem in tropical, less-developed countries. Many biologists estimate that there may be as many species in the tropical rainforests of the world as in the rest of the world combined. Unfortunately, extinction prevention is not a major issue in many less-developed countries. This difference in level of interest is understandable since the developed world has surplus food, higher disposable income, and higher education levels, while people in many of the less-developed countries, where population growth is high, are most concerned with immediate needs for food and shelter, not with long-term issues like extinction.

The forests of Central America are already two-thirds gone. In the Amazon basin, about 100 square kilometers of tropical forestland are cleared each year for farms or by logging companies. Because many tropical countries are less developed, many of the logging companies are owned by foreign companies that are interested only in the short-term economic gain from harvesting the forests,

Accidental extinction of one of these species could be devastating to the ecosystem and the humans that use it. For example, the sardine fishery off the coast of southern California and the anchoveta fishery off the coast of Peru appear to have been fundamentally altered by overfishing. Although the details are not known, it is thought that, once the population was significantly reduced, other organisms filled their niche, making it impossible for their population to return to its original size. The people who depended on these fisheries now must find other ways to make a living.

A third answer, according to many people, is that all species have a fundamental right to exist without being

The History of the Bison

The original ecosystem in the central portions of North America was a prairie dominated by a few species of grasses. The eastern prairie, where moisture was greater, had grasses up to 2 meters tall, while the dryer western grasslands were populated with shorter grasses. Many kinds of animals lived in this area, including prairie dogs, grasshoppers, many kinds of birds, and bison. The bison was the dominant organism. Millions of these animals roamed the prairies of North America with few predators other than the Native Americans, who used the bison for food, their hides for shelter, and their horns for tools and ornaments. The relationship between the bison and Native Americans was a predator-prey relationship in which the humans did not significantly reduce bison numbers.

When Europeans came to North America, they changed this relationship drastically. European-born Americans sought to convert the prairie to agriculture and ranching. However, two things stood in their way: the Native Americans, who resented the intrusion of the "white man" into their territory, and the bison. Since many of the Native American tribes had horses and a history of warlike encounters with other tribes, they attempted to protect their land from this intrusion. The bison was a competitor in the eyes of the settlers, since it was impossible to use the land for agriculture or ranching with the millions of bison occupying so much area. The U.S. government established a policy of controlling the bison and the Native Americans: Since bison were the primary food source of Native Americans in many areas, eliminating them would result in the starvation of many Native Americans, which would eliminate them as a problem for the frontier settler. In 1874, the secretary of the interior stated that "…the civilization of the Indian was impossible while the buffalo remained on the plains." Another example of this kind of thinking was expressed by Colonel Dodge, who was quoted as saying, "Kill every buffalo you can, every buffalo dead is an Indian gone." Bison were killed by the millions. Often, only their hides and tongues were taken; the rest of the animal was left to rot. Years later, the bones from these animals were collected and ground up to provide fertilizer. By 1888, the bison was virtually eliminated.

A few bison were left in the Canadian wild and in remote mountain areas of the United States, while others survived in small captive herds. Eventually, in the early 1900s, the U.S. government established the national Bison Range near Missoula, Montana. The Canadian government established a bison reserve in Alberta. These animals have proven to be useful since many people now desire meat with a lower fat content, which bison have. Crossbreeding cattle with bison has led to a new breed of cattle with reduced fat content known as a beefalo. The animal that was barely saved from extinction by a few thoughtful individuals may contribute to better health for all.

not in the long-term ecological health of the forest and the country.

Nevertheless, many of the governments of less-developed countries have responded to pressures and suggestions from outside sources and have established preserves and parks to protect species in danger of extinction. This does not solve the problem, however, since the areas must be protected from poachers and unauthorized agricultural activity of people who sneak into them. These people are responding to basic biological and economic pressures to provide food for their families. Hiring security forces to patrol such areas is expensive for many of these countries.

Even in countries where interest in extinction prevention is relatively high, there is a cultural bias in favor of protecting certain kinds of organisms. Most endangered or threatened species are birds, mammals, some insects (particularly butterflies), a few mollusks and fish, and certain categories of plants. Bacteria, fungi, reptiles, most insects, and many other inconspicuous organisms rarely show up on endangered species lists, even though they play vital roles in the nitrogen cycle, in the carbon cycle, and as decomposers.

Several international organizations work to prevent the extinction of organisms. The International Union for the Conservation of Nature and Natural Resources (IUCN) estimates that, by the year 2000, at least 500,000 species of plants and animals may be exterminated. The IUCN classifies species in danger of extinction into four categories: endangered, vulnerable, rare, and indeterminate. Endangered species are those whose survival is unlikely if the conditions threatening their extinction continue. These organisms need action by people to preserve them, or they will become extinct. Vulnerable species

Hope for the Rhinoceros

After many years of disappointing news regarding the fate of the black rhino, there may be some good news to report. Prized by poachers for its horn, the animal survives in just a few pockets of southern Africa. Seven years ago there were more than 2,000 in Zimbabwe. By the end of 1994 the number had dropped to under 300. But now that number seems to have stopped falling, and may even have risen somewhat.

Only a third of Zimbabwe's rhinos live in their traditional areas. Security there has been increased, and game wardens have had less trouble with poachers. The other two-thirds have been moved to private ranches, where they are almost as free as their parents were and much safer.

The Imire game ranch is one example. The Imire is home to a herd of seven rhinos brought there in 1987 as tiny calves, after poachers had slaughtered their mothers. Nobody knew at the start how rhinos would react to such a change. In particular, would they breed? The good news is that they do. In 1995 there were 12 births on game ranches in Zimbabwe. While real concern remains for the fate of the black rhino, it is a positive sign to see new births on the game ranches.

are those that have decreasing populations and will become endangered unless causal factors, such as habitat destruction, are stopped. Rare species are primarily those that have small worldwide populations and that could be at risk in the future. Indeterminate species are those that are thought to be extinct, vulerable, or rare, but so little is known about them that they are impossible to classify.

Although the IUCN is a highly visible international conservation organization, it has very little power to effect change. It generally seeks to protect species in danger by encouraging countries to complete inventories of plants and animals within their borders. It also encourages the training of plant and animal biologists within the countries involved. (There is currently a critical shortage of plant and animal biologists who are familiar with the organisms of the tropics.) The IUCN also encourages the establishment of preserves to protect species in danger of extinction.

The U.S. Endangered Species Act was passed in 1973. This legislation gave the

federal government jurisdiction over any species that were designated as endangered. About 300 U.S. species and subspecies have been so designated by the Office of Endangered Species of the Department of the Interior. (See figure 11.20.) The Endangered Species Act directs that no activity by a governmental agency should lead to the extinction of an endangered species and that all governmental agencies must use whatever measures are necessary to preserve these species.

The key to preventing extinctions is preservation of the habitat required by the endangered species. Consequently, many U.S. governmental agencies and private organizations have purchased sensitive habitats or have managed areas to preserve suitable habitats for endangered species. Setting aside certain land areas or bodies of water forces government and private enterprise to confront the issue of endangered species. The question eventually becomes one of assigning a value to the endangered species. This is not an easy task; often, politics become involved, and the endangered species does not always win.

A case in point is the controversy that surrounded Tennessee's Tellico Dam project in 1978. The U.S. Supreme Court declared that completion of the $116 million federal project would result in a violation of the Endangered Species Act because the dam would threaten the survival of an endangered species called the snail darter, a tiny fish about 8 centimeters long that lived in the stream that would be altered by the dam's construction. Developers, however, were not deterred. They lobbied in Congress to have all federally funded projects exempted from the act. Conservationists lobbied for the preservation of the act as it was originally written. Eventually, in 1978, Congress amended the Endangered Species Act so that exemptions to the act could be granted for federally declared major disaster areas or for national defense, or by a seven-member Endangered Species Review Committee. Because this group has the power to sanction the extinction of an organism, it has been nicknamed the "God Squad." If the committee found that the economic benefits of a project outweighed the harmful

Blackfooted ferret

Whooping crane

Mission blue butterfly

Galápagos tortoise

Giant panda

Persistent trillium

Figure 11.20

Endangered Species Many species have been placed on the endangered species list. These are plants and animals that are present in such low numbers that they are in immediate danger of becoming extinct.

environmental
close-up

Whaling

The International Whaling Commission (IWC) was formed in 1946 to regulate the harvesting of whales. Populations of whales had been severely reduced by extensive harvesting by many nations. The development of large factory ships allowed whaling nations to go anywhere in the ocean to seek their prey. The IWC issued recommendations on the numbers and kinds of whales that could be harvested. The purpose was to protect those species of whales whose populations had been severely depleted while allowing whaling nations to protect their whaling industries and the income and jobs they provided. In 1986, amid concerns about the low numbers of many species of whales, the IWC banned all whaling for commercial purposes because the quota system did not appear to be working. The moratorium on whaling resulted in increased populations of some species of whales to the point that the scientific committee of the IWC recommended that the ban be lifted on some of the more plentiful species such

as the minke whale. The IWC, however, voted in May 1993 to maintain the ban.

Norway, a country with a strong environmental record, resumed whaling anyway, insisting that species of whales with large populations are a valuable resource that can be safely exploited in a sustainable manner without damaging the population. Many game species are hunted in this way without driving them to extinction. Iceland withdrew from the IWC and began whaling in 1994. Japan may resume whaling as well.

Further complicating the situation is a strong environmental movement that insists that all species of marine mammals should be protected, whales need not be harvested because whale products have modern substitutes, and the number of jobs involved in whaling is very small. The whaling nations will need to withstand intense international pressure if they are to continue their traditional whaling industries.

ecological effects, it would exempt a project from the Endangered Species Act. At their first meeting, the review board denied the request to exempt the Tellico Dam project on the grounds that the project was economically unsound. This should have stopped the Tellico Dam project. However, nine months later, as a result of several political maneuvers, Congress appropriated money to complete the dam. It is now complete and full of water. The snail darters that once dwelled on the site were transplanted to nearby rivers and by 1981 appeared to be reproducing.

The amendments to the Endangered Species Act also weakened the ability of the U.S. government to add new species to the endangered and threatened lists. Before a species can be listed, it is now necessary to determine the boundaries of its critical habitat, prepare an economic impact study, and hold public hearings—all within two years of the proposal of the listing.

The California Condor

The California Condor (*Gymnogyps californianus*) is thought to have been adapted to feed on the carcasses of large mammals found in North America during the Ice Age. With the extinction of the large, Ice-Age mammals, their major food source disappeared. By the 1940s, their range had shrunk to a small area near Los Angeles, California. Further fragmentation of their habitat caused by human activity, death by shooting, and death by eating animals containing lead shot reduced the total population to about 17 animals by 1986. A low reproductive potential made it difficult for the species to increase in numbers. They do not become sexually mature until six years of age, and females typically lay one egg every two years.

In 1987, all the remaining wild condors were captured to serve as breeding populations. The plan was to raise young condors in captivity, leading to a large population that would ultimately allow for the release of animals back into the wild. Captive condors bred successfully, and females could be induced to lay a second egg if the first egg was removed from the nest. This increased the number of offspring produced per female and resulted in a captive population of 103 individuals by 1996. In 1991, two condors were released into the wild, and six more were released in 1992. In 1996, a second population of condors was introduced in Arizona near the Utah border. There has been unanticipated mortality from contact with power lines and other factors. However, the behavior of the animals seems natural and there is hope that they may eventually reproduce in the wild.

environmental
close-up

Animals That Benefit from Human Activity

Some animals thrive wherever humans are common. Rats and mice have migrated throughout the world, where they feed on the crops that humans raise. In some areas, as much as 50 percent of the crop may be lost to rodents and insects. In some parts of the world, bounties are given for every rat killed. Skunks, foxes, coyotes, rabbits, and many birds thrive where human activity provides a mixture of woodland, farmland, and land devoted to housing.

Pigeons are common animals in many cities, where predation is reduced and they have access to bits of food left by human activity. Storks in Europe build their large nests of sticks in chimneys, and they are protected by the local people. Many other birds, such as barn swallows, chimney swifts, and the European house sparrow, use human structures (as indicated by the birds' common names) as places to build their nests.

Peregrine falcons have been introduced into cities, where they nest on window ledges and rooftops and feed on pigeons, which are often considered pests. Many people place birdhouses and bird-feeding stations near their homes. Bird feeders have extended the range of many birds that would not normally be able to survive the winter months.

Manatees benefit from human activity in a different way. Periods of cold weather in Florida can threaten this tropical aquatic mammal. When the water temperature drops below 18°C (68°F) manatees can suffer hypothermia. During such cold periods it is common to see hundreds of manatees congregating near the warm water discharge points of electrical generation facilities. During the winter of 1996, biologists counted nearly 700 manatees at five power plant sites in Florida. This is more than one-third of all the manatees believed to remain in North America.

Costa Rican Forests Yield Tourists and Medicines

The Central American republic of Costa Rica has been very successful in protecting a significant amount of its remaining forests. It has a rich variety of different kinds of tropical forests and other natural resources. Mangrove swamps, cloud forests, rainforests, dry tropical forests, volcanoes, and beaches on both the Pacific and the Caribbean Sea are all part of the mix found in an area about the size of West Virginia.

Nearly 20 percent of the land in Costa Rica is protected as parks, reserves, and refuges. This is the result of several factors. The government is committed to preserving natural areas. A major part of the parks program involves educating the local people about the values of the parks, including the biological value of the large number of species of plants and animals and the eocnomic value of the parks as a tourist attraction. The job of educating the people is made easier by the fact that 93 percent of the people are literate.

Many jobs in Costa Rica are in the developing ecotourism market. People who wish to visit natural areas require guides, transportation, food, and lodging. The jobs created in these industries encourage local people to preserve their natural resources because their livelihood depends on it. Furthermore, when people are employed, they are less likely to try to convert forested land into farmland.

In addition to using the natural resources of forests for tourists, the government of Costa Rica has an agreement between its Instituto Nacional de Biodiversidad (INBio) and Merck and Company, Inc., to prospect for possible drugs from many tropical plants and animals found in the forests. In return, Merck will pay $1 million to the Costa Rican government. Specially trained local parataxonomists (technicians trained in the identification of plants and animals) are involved in this search, resulting in additional jobs.

- What conditions have contributed to the apparent success of Costa Rica in protecting its forests?
- What are possible sources of failure in other countries?
- Who benefits from ecotourism?

MAP OF COSTA RICA
Not to Scale

NICARAGUA

CARIBBEAN

PACIFIC OCEAN

PANAMA

Elegant Vacations

National Parks and Wildlife Refuges

1. Isla Bolaños Wildlife Refuge
2. Santa Rosa National Park
3. Lomas Barbudal Biological Reserve
4. Palo Verde National Park and the Dr. Rafael Lucas Rodriguez Caballero Wildlife Refuge
5. Barra Honda National Park
6. Peñas Blancas Forest Refuge
7. Guayabo, Negritos, and Pájaros Islands Biological Reserves
8. Ostional National Wildlife Refuge
9. Carara Biological Refuge
10. Curú National Wildlife Reserve
11. Cabo Blanco-Absolute Nature Reserve
12. Manuel Antonio National Park
13. Caño Island Biological Reserve
14. Golfito National Wildlife Reserve
15. Corcovado National Park
16. Isla del Coco National Park
17. Rincón de la Vieja National Park
18. Poás Volcano National Park
19. Braulio Carrillo National Park
20. Irazú Volcano National Park
21. Guayabo National Monument
22. Tapanti National Wildlife Reserve
23. Chirripó National Park and Costa Rica-Panamá International Friendship Park
24. Hitoy-Cerere Biological Reserve
25. Caño Negro National Wildlife Refuge
26. Barra del Colorado National Wildlife Refuge
27. Tortuguero National Park
28. Cahuita National Park
29. Gandoca-Manzanillo National Wildlife Refuge
30. Guanacaste National Park

environmental *close-up*

Extinction of the Dusky Seaside Sparrow

On June 16, 1987, the last dusky seaside sparrow (*Ammodramus maratimus nigrescens*) died in captivity at Walt Disney World's Discovery Island Zoological Park in Orlando, Florida. The bird was a male that was probably about 12 years old. Originally, this subspecies and several other subspecies were found in the coastal salt marshes on the Atlantic coast of Florida. (A subspecies is a distinct population of a species that has several characteristics that distinguish it from other populations.) One other subspecies, the Smyrna seaside sparrow (*Ammodramus maratimus pelonata*), is believed to have become extinct several years ago, and a third subspecies, the Cape Sable seaside sparrow (*Ammodramus maratimus mirabilis*), was listed as an endangered species in 1967. Before the deaths of the last remaining dusky seaside sparrows, a few males were crossed with another subspecies, Scott's seaside sparrow (*Ammodramus maratimus peninsulae*). Thus, the hybrid offspring between these two subspecies contain some of the genes that made the dusky seaside sparrow unique.

The endangerment and extinction of these birds was a direct result of the land development and drainage that destroyed the salt-marsh habitat to which they were adapted. The development of Cape Canaveral as a major center for the U.S. space program also modified much of the birds' original habitat and was a partial cause of their extinction.

SUMMARY

Pollution is the result of technological advancements and increased population density. It is defined as anything produced by humans in a quantity that interferes with the health or well-being of an organism.

Natural resources are structures and processes that can be used by people but cannot be created by them. Renewable resources can be regenerated or repaired; nonrenewable resources are consumed.

Mineral resources must be extracted from ores, a process that requires energy and money, and also changes the environment. Mineral resources are not evenly distributed around the world, so most countries must purchase some of their minerals from foreign sources. The major steps of mineral exploitation are exploration, mining, processing the ore, transportation, and manufacturing the finished product. Ultimately, all costs of mineral exploitation are reduced to monetary costs. Recycling reduces the demand for new sources of mineral deposits but does not necessarily save money.

As the human population increases, we alter more and more ecosystems as we attempt to feed and house ourselves. However, some remote areas have been changed very little by humans. Tropical rainforests represent some of the last large wilderness areas, but they are being rapidly converted to other uses by the constant pressure of growing populations. Many of these wilderness areas should be protected because of the rich diversity of species they contain.

We change natural ecosystems by replacing them with agricultural ecosystems, we alter species mixtures by introducing plants and animals, and we reduce populations by harvesting trees and animals for our use. Forests must be cut in a manner that allows for regrowth so that the soil is not exposed to the erosional effects of wind and water. Cutting small areas and reforestation help to prevent these problems.

Grazing of arid and semiarid lands can be a valuable way to provide food for people. However, the land is often overgrazed and then may be degraded to a desert that is not capable of supporting the animals needed to feed growing populations.

Aquatic ecosystems are modified by pollution and activities that occur on the land adjacent to the water. Land that is devoid of vegetation erodes and fills streams and lakes with sediment. Warming of the water is also likely to occur if trees are removed from the stream side. Many exotic species of fish have been introduced into the freshwater ecosystems of the world. This alters normal food chains and reduces the populations of native species.

Often, areas are managed for certain species of wildlife. This involves careful planning and habitat manipulation to provide the best possible population for hunting. Some areas are intensely managed, as in many European game parks, while in other parts of the world, the game animals lead a more normal wild life. Waterfowl present a unique problem because they migrate across international boundaries.

Extinction is a normal consequence of not being able to adapt to changes in the environment. However, since humans have such a great influence on nearly every ecosystem in the world, they have been the cause of increased rates of extinction. Many people recognize the value of species, both as possible helpers of humans and for their own intrinsic worth, and are trying to preserve sensitive habitats so that species will not be driven to extinction because of the appropriation of their habitats for other uses.

REVIEW QUESTIONS

1. Name three ways humans directly alter ecosystems.
2. Why is the impact of humans greater today than at any time in the past?
3. Define pollution. Has the definition changed with time?
4. What are three kinds of costs associated with resource exploitation?
5. Why is recycling usually more energy efficient than mining new raw materials?
6. List three problems associated with forest exploitation.
7. What is desertification? What causes it?
8. What effects do increased temperature and increased organic matter have on aquatic ecosystems?
9. List six techniques utilized by wildlife managers.
10. What special problems are associated with waterfowl management?
11. What is extinction, and why does it occur?
12. Why should humans worry about extinction? Do you?
13. List three actions that can be taken to prevent extinctions.

KEY TERMS

biodegradable   189
clear-cutting   197
cover   205
desertification   201
economic costs   190
endangered species   213

energy costs   191
environmental costs   191
extinction   208
habitat management   207
migratory birds   208

natural resources   189
nonrenewable resources   190
patchwork clear-cutting   197
pollution   189
reforestation   197

renewable resources   190
selective harvesting   197
speciation   209
threatened species   213
wilderness   194

RELATED INTERNET SITES

Human dimensions of ecosystem management in Oregon.
    http://www.orst.edu/groups/hdnr/
The impact of ecotourism. Facts and articles on ecotourism.
    http://www.oneworld.org/panos/panos_eco2.html
Preservation of endangered species, biodiversity, and habitats.
    http://www.igc.org/igc/issues/habitats/
Nature Conservancy, a searchable index loaded with links.
    http://www.tnc.org/
Earth Sanctuaries—Australia's leading organization advocating conservation of endangered species.
    http://www.esl.com.au/

Award-winning site focusing on whale and marine mammal research.
    http://whale.wheelock.edu/
Oregon Department of Fish and Wildlife, focusing on wildlife and endangered species.
    http://www.dfw.state.or.us/
Texas Department of Range, Wildlife, and Fisheries Management.
    http://www.ttu.edu/~rwfmhp/

chapter 12

Land-Use Planning

Objectives

After reading this chapter, you should be able to:

- List the relative amounts of land used for crops, livestock, urban development, and other uses in the United States.
- Explain the impact that water has on the location and development of cities.
- Explain why farmland surrounding cities was used for housing.
- Explain how taxation may influence land use.
- Explain why floodplains and wetlands are often mismanaged.
- List several exclusionary land uses and several multiple land uses.
- Describe the economic impact of recreation.
- Recognize that people desire outdoor recreation.
- Explain why recreational areas are needed in urban locations.
- Explain why some land must be designated for particular recreational uses, such as wilderness areas, and why that decision sometimes invites disagreement from those who do not desire to use the land in the designated way.
- List the steps in the development and implementation of a land-use plan.
- Describe methods of enforcing compliance with land-use plans.
- List the problems associated with local land-use planning.
- Explain the advantage of regional planning and the problems associated with it.

Chapter Outline

Historical Land Use and the Development of Cities

Global Perspective: *State of the World's Lands*

Global Perspective: *State of the World's Forests*

The Rise of Suburbia

Some Problems Associated with Unplanned Urban Growth

   Loss of Farmland

   Floodplain Problems

   Wetlands Misuse

Environmental Close-Up: *Wetlands Loss in Louisiana*

   Other Land-Use Problems

Multiple Land Use

Recreational Land Use

   Urban Recreation

   Outdoor Recreation

   Conflicts over Recreational Land Use

Land-Use Planning Principles

   Regional Planning

   Urban Transportation Planning

Issues and Analysis: *Decision Making in Land-Use Planning: The Malling of America*

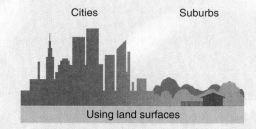

Cities   Suburbs

Using land surfaces

Conflicting desires
- Industry
- Housing
- Transportation
- Commercial development
- Recreation

Planning resolves conflicts

Historical Land Use and the Development of Cities

Land is a nonrenewable resource because it cannot be newly formed by natural processes. Once used for certain activities, it cannot be used for other purposes.

In most regions of the world, natural and planted forestlands have decreased over the past decade. The greatest declines have occurred in Africa (3.1 percent) and South America (4.1 percent). Worldwide, forest and woodland areas have shrunk 3.6 percent. Some of this land has been converted to cropland, which has increased globally by 1.3 percent since 1980. The greatest increases in cropland have occurred in South America and Africa, regions that have simultaneously experienced a net loss of forest and woodland.

Of the land in the United States, 45 percent is used for crops and livestock, about 45 percent is forests and natural areas, and the remaining land is used intensively by people in urban centers. (See figure 12.1.) This urban use includes transportation corridors between the urban centers. Canada is 54 percent forested and wooded and uses only 8 percent of its land for crops, and livestock. A large percentage of its remaining land is wilderness in the north.

This pattern of land use differs greatly from conditions a century ago. Many early settlers immigrated to the New World from crowded European cities. The promise of open space and available land was a primary incentive for those who left their homes. As the North American population increased, the original rural landscape developed villages and cities.

Because the developing countries in North America had no system of railroads or highways, the primary method of transportation was by water. Without a reliable transportation network, cities could not grow and industry could not prosper. Thus, early towns were usually built near rivers, lakes, and oceans. Typically, cities developed as far inland as

Other uses 52.8%   **For crops and livestock 47.2%**

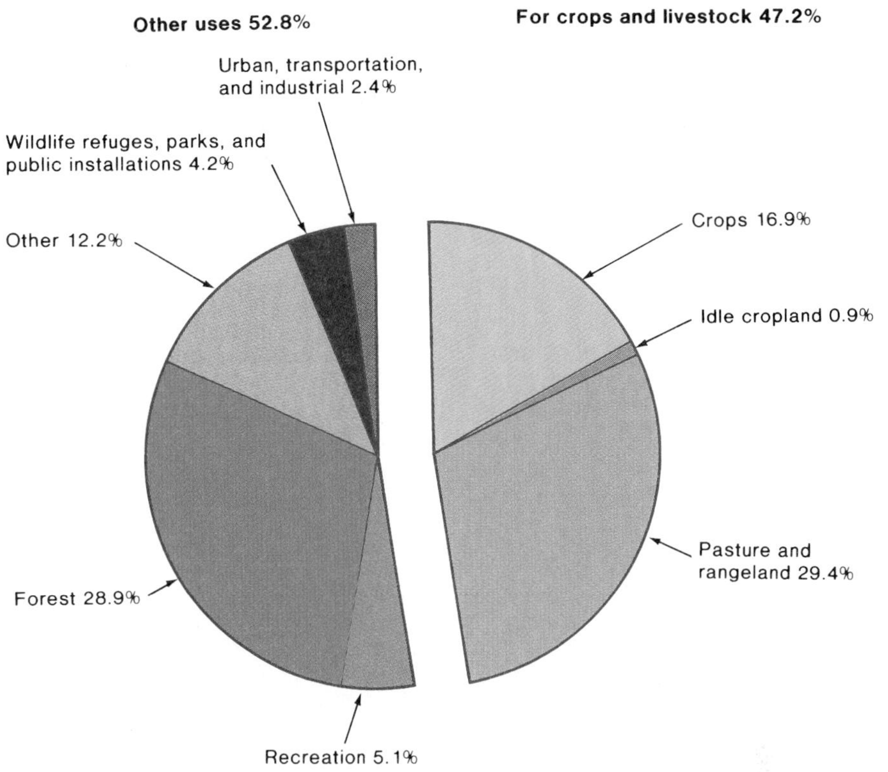

Figure 12.1

Land Use in the United States   Forty-seven percent of the land in the United States is used for crops and livestock, while less than 3 percent is used for urban centers and transportation.

Source: Data from the Statistical Abstract of the United States, *1995.*

rivers were navigable. Abrupt changes in elevation required transshipment of goods, and cities often grew at these points (Buffalo, New York; Sault Sainte Marie, Ontario). Water met many of the needs of villages and small towns, such as drinking water, transportation, power, and waste disposal. Without access to water, St. Louis, Montreal, Chicago, Detroit, Vancouver, and other cities would not have developed. (See figure 12.2.)

Industrial use of water became important as cities grew, and most industrial development took place on the waterfront. Large factories replaced small gristmills, sawmills, and blacksmith shops. Industry eventually took over most of the waterfront. There was little control of industry activities, so the waterfront became polluted, unhealthy, and an undesirable place to live. Anyone who could afford to do so moved away from the original city center.

Land use and water are intimately interrelated. Development of a water resource for industrial use or transportation affects the land bordering the water. The type of land use similarly affects the watercourse. In urban centers, the "decision" to use waterfront land for industrial development was forced by a need for transportation and power. Once this pattern had begun, a series of events eventually resulted in the development of suburban metropolitan regions.

Cities, which are usually located next to water, are often surrounded by farmlands. River valleys provide flat, rich land for crops. Until transportation systems became well developed, farms needed to be close to the city to market their produce. This rich farmland adjacent to the city was the most easily developed space for city expansion. As the population of the city grew, demand for land increased. As the city's land prices rose, people and businesses began to look for cheaper land

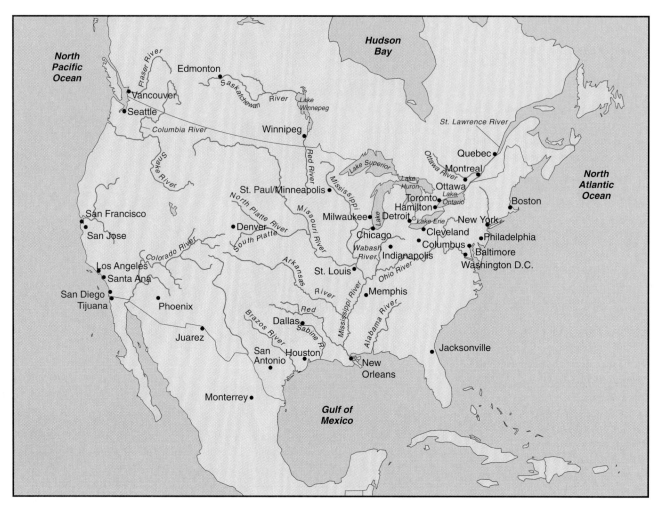

Figure 12.2

Water and Urban Centers   Note that most of the large urban centers are located on water. Water is an important means of transportation and was a major determining factor in the growth of cities. The cities shown have populations of 500,000 or more.

on the outskirts. Developers and real estate agents were quick to respond and to help people acquire and convert agricultural land to residential or commercial uses. They viewed land as a commodity to be bought and sold for a profit, rather than as a finite resource to be managed. As long as money could be made by converting agricultural land to other purposes, it was impossible to prevent such conversion.

North America remained essentially rural until industrial growth began in the last third of the 1800s, when the population began a trend toward greater urbanization. (See figure 12.3.) This large-scale migration to the cities had several causes. First, improvements in agriculture re-

quired less farm labor; therefore, people moved from the farm to the city. Second, job opportunities were available in the city because industry was developing. The average person was no longer a farmer but, rather, a factory worker, shopkeeper, or clerk living in a tenement or tiny apartment. This pattern of rural to urban migration occurred throughout North America, as it would in any developing nation. Urbanization was enhanced by immigrants. These new citizens tended to settle in towns and cities, where employment was more likely. Many immigrants congregated in certain parts of cities. As a result, today most large cities have ethnic sections, with populations of Irish, Italian, Chinese,

German, and Greek descent, that give North American cities a unique character. A third reason for the growth of cities was that they offered a greater variety of cultural, social, and artistic opportunities than did rural communities. Thus, cities were attractive for cultural as well as economic reasons.

During the last two decades, land development in North America has destroyed many natural areas that people had long enjoyed. The Sunday drive in the countryside has become a battle to escape the ever growing suburbs. The unique character of neighborhoods and communities has been changed by apartment complexes. Developments have sprung up along beaches, and recreational

PART FOUR   Human Influences on Ecosystems

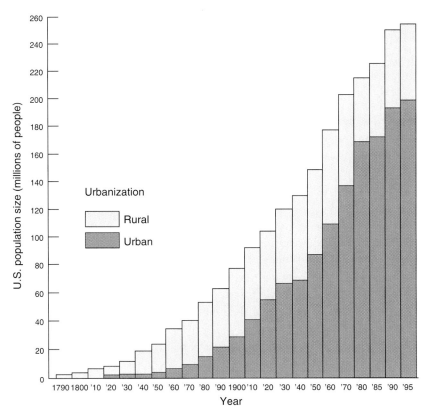

Figure 12.3

Rural to Urban Population Shift In 1800, the United States was essentially a rural country. Industrialization in the late 1800s began the shift to an increasing urban population. In 1995, 75 percent of the U.S. population was urban.

(Top) Sources: Data from the Statistical Abstract of the United States, *1993 and 1995 data from Population Reference Bureau, Inc., Washington, DC.*

GLOBAL PERSPECTIVE

State of the World's Lands

- Human activities are degrading much of the world's land surface, lowering the current and/or future capacity of the soil to support human life.
- In the past 45 years, an area about the size of China and India combined has suffered moderate to extreme soil degradation.
- Causes of soil degradation include overgrazing by livestock, inappropriate agricultural practices, deforestation, overexploitation of vegetation for fuelwood, and industrialization.
- Tropical deforestation has accelerated dramatically. Asia has the highest rate, followed by Latin America and Africa.
- Largely because of population growth, per capita cropland has declined in all regions of the world.

communities are common sights. Even the deserts and marshes have not been spared; they sprout the little red flags of subdividers. These changes have occurred without thought for other changes they might cause that may or may not be desirable.

The Rise of Suburbia

As cities continued to grow, certain sections within each city began to deteriorate. Industrial activity continued to be concentrated near water in the city's center. Industrial pollution and urban crowding turned the core of many cities into undesirable living areas. In the early 1900s, people who could afford to leave began to move to the outskirts. This trend continued after World War II, in the 1940s, 1950s, and 1960s, as more people were able to buy homes. Most of the homes were in the attractive suburbs, away from the population and congestion of the central city. These houses were built on large lots in decentralized patterns, which increased the cost of supplying services, increased energy needs, and made it very difficult to establish efficient

Figure 12.4

Transportation and Suburbia   This traffic congestion is a common experience for people who work in cities but live in the suburbs. The popularity of automobiles and the desire to live in the suburbs are closely tied.

public transportation networks. Rising automobile ownership and improved highway systems also encouraged suburban growth. The convenience of a personal automobile escalated decentralized housing patterns, which, in turn, required better highways, which led to further decentralization. (See figure 12.4.)

In some areas throughout North America, the growth of suburbs has been slowed due to the increased cost of housing and transportation. People have migrated back to some cities on a limited scale because of the lower cost of urban houses and the fact that public transportation is generally more efficient in the city than in the suburbs, thus freeing urban residents from the cost of daily commuting. This reverse migration, however, is still greatly offset by the continual growth of the suburban communities.

By 1960, unplanned surburban growth had become known as **urban sprawl.** Large housing tracts surrounded cities, which made it difficult for people to find open space. A city dweller could no longer take a bus to the city limits and enjoy the open space of the countryside.

Urban sprawl occurs in three ways. (See figure 12.5.) One type of growth is associated with the wealthier urbanites adjacent to the city. These subdivisions of homes are on large individual lots in the more pleasing geographic areas. A second form of urban sprawl is development along transportation routes. This is referred to as **ribbon sprawl** and usually consists of commercial and industrial buildings. Ribbon sprawl results in high costs for the extension of utilities and other public services. It also makes the extent of urbanization seem much larger than it actually is. A third development pattern is tract development. **Tract development** is the construction of similar residential units over large areas. Initially, these tracts are often separated from each other by farmland.

As suburbs continued to grow, cities (once separated by farmland) began to merge, and it became difficult to tell where one city ended and another began. This type of growth led to the development of regional cities. Although these cities maintain their individual names, they are really just part of one large urban area called a **megalopolis.** (See figure 12.6.) The eastern seaboard of the United States, from Boston, Massachusetts, to Washington, D.C., is an example of a continuous city. Other examples are London to Dover in England, the Toronto-Mississauga region of Canada, and the southern Florida coast from Miami northward.

a.

b.

c.

Figure 12.5

Urban Sprawl Note the three different types of growth depicted in these photos. (*a*) The wealthy suburbs with large lots are adjacent to the city. (*b*) Ribbon sprawl develops as a commercial strip along highways. (*c*) Tract development results in neighborhoods consisting of large numbers of similar houses on small lots.

1. Metropolitan belt
   A. Atlantic Seaboard (BoWash)
   B. Lower Great Lakes (ChiPitts)
2. California Region (SanSan)
3. Florida Peninsula (JaMi)
4. Gulf Coast
5. East central Texas—Red River
6. Southern Piedmont
7. North Georgia—South Tennessee
8. Puget Sound
9. Twin Cities Region
10. Colorado Piedmont
11. Saint Louis
12. Metropolitan Arizona
13. Willamette Valley
14. Central Oklahoma—Arkansas Valley
15. Missouri—Kaw Valley
16. North Alabama
17. Blue Grass
18. Southern Coastal Plain
19. Salt Lake Valley
20. Central Illinois
21. Nashville Region
22. East Tennessee
23. Oahu Island
24. Memphis
25. El Paso—Ciudad Juare
26. Toronto
27. Montreal

Figure 12.6

Regional Cities in the United States and Canada   By the year 2000, more than thirty major urban regional cities are expected to have developed. Each will have a population of at least one million people. Four super cities (BoWash, ChiPitts, SanSan, and JaMi) will each have a great deal more than one million people.

Source: J.P. Pickard, "U.S. Metropolitan Growth and Expansion, 1970–2000, with Population Projections" in Population Growth and the American Future, *Washington, DC. U.S. Government Printing Office, 1972.*

Some Problems Associated with Unplanned Urban Growth

Loss of Farmland

Most of the land that has recently been urbanized was previously used for high-value crops. Land that is flat, well drained, accessible to transportation, and close to cities is ideal farmland. However, it is also prime development land. Areas that once supported crops now support housing developments, shopping centers, and parking lots. (See figure 12.7.)

Urban development of farmland is proceeding at a rapid pace. One reason for this conversion is the way the land is taxed. Property is often taxed on what *can* be done with it, not necessarily what *is* being done with it. For example, if land can be used for both farming and residential development, it is taxed as if it were residential. If a farmer sells a portion of the farm to a developer who builds five houses, local taxing authorities would consider these houses to be the "highest and best use" of the land. All of the farmer's land would then probably be reassessed and the taxes increased substan-

tially. Farmers faced with this situation are often forced to sell their land because they are taxed on its commercial value rather than on its value as farmland. This policy encourages development and forces people out of farming. New policies that assess taxes based on current use of the land and not on the land's highest potential use must be explored.

Floodplain Problems

Cities may try to control their growth and maintain some open space, but the fact is that many cities are located in areas that are altogether unsuitable for large human populations. These areas are ripe for a

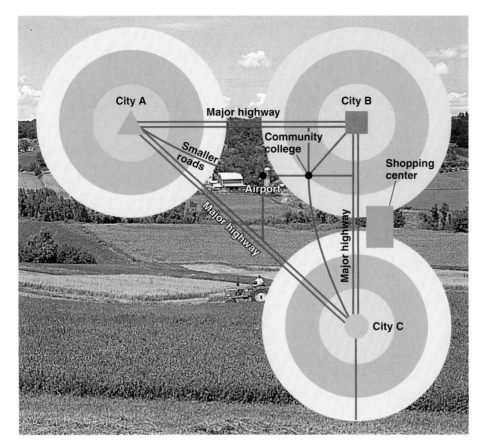

Figure 12.7

Loss of Farmland As cities grow outward, they eventually grow together to form a regional city. The land between them, once used for farming, becomes developed for residential and commercial purposes. Improved transportation routes and joint facilities (such as airports, shopping centers, and community colleges) hasten this loss of farmland.

natural disaster, such as an earthquake or a flood.

Many cities are located in areas called **floodplains,** the lowland area on either side of a river. Floodplains are generally flat, and so are inviting areas for residential development. But they are so named because they are periodically covered with water. Some may flood annually, while others flood less regularly. A better use of these areas is for open space or for recreation, yet developers continue to build houses and light industry there.

When a floodplain is developed for residential or commercial use, a retaining wall usually is built to prevent the periodic flooding natural to the area. This increases the cost of the development, increases the cost of insurance protection, and creates high-water problems downstream. Frequently, tax monies are used to repair the

damage that results from the unwise use of floodplains. Floodplain development is one of the insidious problems associated with a rapidly expanding population. As long as the population continues to increase, these less desirable areas are likely to be used for housing, whether they are subject to annual flooding or less frequent damage.

Floods are natural phenomena. Contrary to popular impressions, no evidence supports the premise that floods are worse today than they were one hundred or two hundred years ago, except perhaps on small, isolated watersheds. What has increased is the economic loss from the flooding. This loss reflects the fact that more cities, industries, railroads, and highways have been constructed on floodplains. Sometimes there are no alternatives to floodplain development, but too often,

risks are simply ignored. Because flooding causes loss of life and property, floodplains should no longer be developed for uses other than agriculture and recreation.

Many communities have enacted **floodplain zoning ordinances** to restrict future building in floodplains. Although such ordinances may prevent further economic losses, what happens to individuals who already live in floodplains? Floodplain building ordinances usually allow current residents to remain. However, these residents often find it extremely difficult to obtain property insurance. Relocation, usually at a financial loss, is the only alternative. Such situations are unfortunate; perhaps proper planning in the future will prevent these problems.

Wetlands Misuse

Since access to water was and continues to be important to industrial development, many cities are located in areas with extensive wetlands. **Wetlands** include swamps, tidal marshes, coastal areas, and estuaries. Because wetlands breed mosquitoes and are sometimes barriers to the free movement of people, they have often been considered useless or harmful. Most of them have been drained, filled, or used as dumps. Many modern cities have completely covered over extensive wetland areas and may even have small streams running under streets, completely enclosed in concrete.

Each kind of wetland has unique qualities and serves as a home to many kinds of plants and animals. Wetlands are frequently associated with the reproductive portion of the life cycle of the organisms that inhabit them. They provide nesting sites and spawning sites, and commercial and sport fisheries depend upon these habitats to produce and protect the young of the species they harvest. Because most wetlands receive constant inputs of nutrients from the water that drains from the surrounding land, they are highly productive and excellent places for aquatic species to grow rapidly. Human impact on wetlands has severely degraded or eliminated these spawning and nursery habitats.

Besides providing a necessary habitat for fish and other organisms, wetlands provide natural filters for sediments and

environmental
close-up

Wetlands Loss in Louisiana

The state of Louisiana has extensive coastal wetlands along the Gulf of Mexico. Approximately 10,000 hectares (100 square kilometers) are lost per year for a variety of reasons, most of them the result of human activity.

Much of the coastline consists of poorly compacted muds that are easily eroded if exposed to wind and wave action. Dams and reservoirs upstream on the Mississippi River have reduced the amount of sediment delivered to the region, making replacement of lost sediments impossible. Furthermore, as the muds settle and compact, the land subsides and erosion is more likely.

Another major contributing factor to the loss of these coastal wetlands is oil and gas exploration and the development of other resources from beneath the Gulf of Mexico. Typically, production companies cut canals through the marshes to give barges and other equipment easy access to

the drilling platforms and production facilities. When oil or gas is found, it must be transported to land through pipelines. Laying pipelines often requires cutting additional canals through the wetlands. These canals break the wall of vegetation that protects the soft muds, which accelerates the rate of erosion.

The nutria, a rodent introduced from South America, is also contributing to the problem. These animals reproduce rapidly and form such large populations that they eat all the vegetation in a local area. This exposes the muds to erosion and contributes to the loss of wetlands. Trapping these animals for their fur was at one time profitable for the local people and helped to control the animals. However, a shift in public opinion has reduced the acceptability of fur for clothing and depressed the fur market, making the trapping of nutria uneconomical.

---

runoff. This filtration process reduces sedimentation in water and allows time for it to be biologically cleaned before it enters larger bodies of water such as lakes and oceans. Wetlands also protect shorelines from erosion. When destroyed, wetlands must be replaced by costly artificial measures, such as breakwalls built to protect the shoreline.

Other Land-Use Problems

The geologic status of an area must also be considered in land-use decisions. Building cities on the sides of volcanos or on major earthquake-prone faults has led to much loss of life and property. Siting homes and villages on unstable hillsides or in areas subject to periodic fires is also unwise.

Another problem in some locations is lack of water. Southern California and metropolitan areas in Arizona must import water to sustain their communities. Wise planning would limit growth to whatever could be sustained by available resources. The risk of serious water shortages in Beijing, China, has forced the government to consider a massive 1,200-

kilometer diversion of water from the Yangtze River to the city. As these areas in California, Arizona, and China continue to grow, the strain on regional water resources will increase.

Multiple Land Use

Once a land-use decision is made, it may be irreversible. For example, when a highway is constructed, it is virtually impossible to use that land for any other purpose. If land is designated as a wilderness area, no human development is allowed. Other single exclusionary uses of land are agriculture, waste storage, housing, and industrial development.

Sometimes several different uses of the land can take place at the same time. They do not have to be exclusionary. Two or more uses of land occurring at the same time is called **multiple land use.** Perhaps the best example of multiple land use occurs in U.S. national forests. In 1960, the Multiple Use Sustained Yield Act divided use of national forests into four categories: wildlife habitat preservation, recreation, lumbering, and watershed protection.

This act encouraged both economic and recreational use of the forests.

Another example of multiple land use is the development of parks on floodplains. These parks provide open space, recreational land, and storage reservoirs for storm sewer runoff. Storage reservoirs can even be blended into the park landscape as ponds or lakes.

Once land is recognized as a nonrenewable resource, people may be more willing to engage in meaningful land-use planning. Such plans must consider the long-term needs of the region as well as the short-term goals of the population.

Recreational Land Use

Nearly three-fourths of the population of North America live in urban areas. These urban dwellers value open space because it breaks up the sights and sounds of the city and provides a place for recreation. Inadequate land-use planning in the past has rapidly converted urban open spaces to other uses. Until recently, creating a new park within a city was considered an uneconomical use of the land, but people

New York Central Park

Hyde Park London

Figure 12.8

Urban Open Space These photos show open space in two urban areas, New York's Central Park and Hyde Park in London. If the land had not been set aside, it would have been developed.

are now beginning to realize the need for parks and open spaces.

Some cities recognized the need for open space a long time ago and allocated land for parks. London, Toronto, and Perth, Australia, have centrally located and well-used parks. New York City set aside approximately 200 hectares for Central Park in the late 1800s. (See figure 12.8.) Boston has developed a park system that provides a variety of urban open spaces. Other cities have not dealt with this need for open space because they have lacked either the foresight or the funding.

Urban Recreation

Recreation is a basic human need. The most primitive tribes and cultures all engaged in games or recreational activities. New forms of recreation are continually being developed. In the congested urban center, cities often must construct special areas where recreation can take place.

A major problem with urban recreation is locating recreational facilities near residential areas. Facilities that are

not conveniently located may be infrequently used. For example, the hundreds of thousands of square kilometers of national parks in Alaska and the Yukon will be visited every year by a tiny proportion of the population of North America. Large urban centers are discovering that they must provide adequate, low-cost recreational opportunities within their jurisdiction. Some of these opportunities take the form of commercial establishments, such as bowling centers, amusement parks, and theaters. Others must be subsidized by the community. (See figure 12.9.) Playgrounds, organized recreational activities, and open space have usually been combined into an arm of the municipal government known as the parks and recreation department. Cities spend millions of dollars to develop and maintain recreation programs. Often, there is conflict over the allocation of financial and land resources. These are closely tied because open land is scarce in urban areas, and it is expensive. Riverfront property is ideal for park and recreational use, but it is also prime land for industry, commerce,

Figure 12.9

Urban Recreation In urban areas, recreation often takes the form of sports programs, playgrounds, and walking. Most cities recognize the need for such activities and develop extensive recreation programs for their citizens.

or high-rise residential buildings. Although conflict is inevitable, many metropolitan areas are beginning to see that recreational resources may be as important as economic growth for maintaining a healthy community.

An outgrowth of the trend toward urbanization is the development of **nature centers.** In many urban areas, there is so little natural area left that the people who live there need to be given opportunities to learn about nature. Nature centers are basically teaching institutions that provide a variety of methods for people to learn about and appreciate the natural world. Zoos, botanical gardens, and some urban parks, combined with interpretative centers, also provide recreational experiences. Nature centers are usually located near urban centers, in places where some appreciation of the natural processes and phenomena can be developed. They may be operated by municipal governments or by school systems or other nonprofit organizations.

Outdoor Recreation

Not all people enjoy the same recreation. Some people enjoy reading or watching television. Others prefer commercial activities such as golf, tennis, bowling, amusement areas, racetracks, and skiing. Another major area of recreation, usually classified as **outdoor recreation,** involves using the natural out-of-doors for hiking, camping, canoeing, and so forth. Millions of people want to use public lands for these activities.

Most recreational activities require the consumption of natural resources, such as minerals, fuels, and timber. Some activities require more resources than others, but even the backpacker, who has traditionally been considered an ecologically frugal individual, requires considerable equipment. Other activities, such as the use of off-road vehicles, require even more resources in the form of equipment and fuel. Table 12.1 lists various outdoor activities and the number of people who participate in them in the United States. As energy and other resources become less available, we may be forced to abandon some

Table 12.1

Number of People Who Participated in Selected Outdoor Recreational Activities in 1991

Activity	Number of participants
Bicycling	24.5 million
Fishing	24.3 million
Camping	20.4 million
Running	11.4 million
Hunting	9.5 million
Golf	9.3 million
Hiking	8.0 million
Tennis	7.8 million
Skiing	7.0 million

Source: Data from the Statistical Abstract of the United States, *1991.*

of our more extravagant recreational activities for those that are more ecologically conservative.

Conflicts over Recreational Land Use

Many people want to use the natural world for recreational purposes because nature can provide challenges that may be lacking in their day-to-day lives. Whether the challenge is hiking in the wilderness, underwater exploration, climbing mountains, or driving a vehicle through an area that has no roads, these activities offer a sense of adventure. Look at table 12.1 again. All of these activities use the out-of-doors, but not in the same way. Conflicts develop because some of these activities cannot occur in the same place at the same time. For example, wilderness camping and backpacking often conflict with off-road vehicles.

There is a basic conflict between those who prefer to use motorized vehicles and those who prefer to use muscle power in their recreational pursuits. (See figure 12.10.) This conflict is particularly strong because both groups would like to use the same publicly owned land. Both have paid taxes, both "own" the land, and both feel that it should be available for them to use as they wish. When everyone "owns" something, there is often little desire to regulate activities.

For example, much of the rangeland in the western United States is publicly owned. Based on "Animal Use Months" established by the Bureau of Land Management or the Forest Service, ranchers are allowed to graze cattle, but only on certain public lands. Technically, failure to comply can mean a loss of grazing rights. However, since the regulations are politically motivated and the regulatory agency is understaffed, there is no incentive for ranchers to limit the number of cattle or sheep on the land unless all ranchers agree to do so. The usual result is that a few individuals exploit the land in hopes of a short-term profit. This exploitation could cause permanent damage to the rangeland. Many people want to use the public lands for outdoor recreation and resent the control exercised by one category of user.

An obvious solution to this problem is to allocate land to specific uses and to regulate the use once allocations have been made. Several U.S. governmental agencies, such as the National Park Service, the Bureau of Land Management, the U.S. Forest Service, and the U.S. Fish and Wildlife Service, allocate and regulate the lands they control. However, these agencies have conflicting roles. The U.S. Forest Service has a mandate to manage forested public lands for timber production. This mandate often comes in conflict with recreational

Figure 12.10

Conflict over Recreational Use of Land   Land may be for both motorized and nonmotorized use. The people who participate in these two kinds of recreation are often antagonists over the allocation of land for recreational use.

uses. Similarly, the Bureau of Land Management has huge tracts of land that can be used for recreation, but it traditionally has been aligned with grazing interests.

A particularly sensitive issue is the designation of certain lands as wilderness areas. Obviously, if an area is to be wilderness, human activity must be severely restricted. This means that the vast majority of Americans will never see or make use of it. Many people argue that this is unfair because they are paying taxes to provide recreation for a select few. Others argue that if everyone were to use these areas, their charm and unique character would be destroyed and that, therefore, the cost of preserving wilderness is justifiable.

Areas designated as wilderness make up a very small proportion of the total public land available for recreation. (See figure 12.11.) The fact that there are relatively few wilderness areas has resulted in a further problem: The areas are being loved to death. People pressure on this resource has become so great that, in some cases, the wilderness quality is being tarnished. The designation of additional wilderness would relieve some of this pressure.

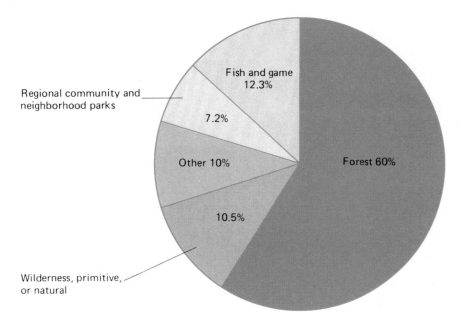

Figure 12.11

U.S. Federal Recreational Lands in 1995 Of the approximately 108 million hectares of federal recreational lands in the United States, approximately 10 percent is designated as wilderness, primitive, or natural.

Source: Data from the Statistical Abstract of the United States, *1995.*

Figure 12.12

Zoning Most communities have a zoning authority that designates areas for particular use. This sign indicates that decisions have been made about the "best" use for the land.

Land-Use Planning Principles

As we look back at the location and development of cities and metropolitan areas, we can see the mistakes that have been made. If we were in a position to start over, we might locate our population centers differently: we might regulate building and development for the least damaging but highest use of the land. But how does one go about determining the best use and least damaging use of an area? What goes into the process of land-use planning?

Land-use planning is the construction of an orderly list of priorities for the use of available land. Developing a plan involves gathering data on current use and geological, biological, and sociological information. From these data, projections are made about what human needs will be. All of the data collected are integrated with the projections, and each parcel of land is evaluated and assigned a best use under the circumstances. Finally, mechanisms for implementing the plan are divided into two categories: purchase of the land or regulation of its use.

Probably the simplest way to protect desirable lands is to purchase them. This method has often been used by government or by conservation-minded organizations or individuals. The major problem is the cost. Many communities are not in a financial position to purchase lands; therefore, they attempt to regulate land use by zoning laws.

Zoning is a common type of land-use regulation. When land is zoned, it is designated for specific potential uses. Common designations are agricultural, commercial, residential, recreational, and industrial. (See figure 12.12.)

Local governments often lack funds to hire professional planners. As a result, zoning regulations are frequently made by people who see only the short-term gain and not the possible long-term loss. The land is simply zoned for its current use. Even when well-thought-out land-use plans exist, they are usually modified to encourage local short-term growth rather than to provide for the long-range needs of the community. The community needs to be alert to variances from established land-use plans, because once the plan is compromised, it becomes easier to accept future deviations that may not be in the best interests of the community. Many times, individuals who make zoning decisions are real estate agents, developers, or local business people. These individuals wield significant local political power and are not always unbiased in their decisions. Concerned citizens must try to combat special interests by attending zoning commission meetings and by participating in the planning process.

Regional Planning

Regional planning is more effective than local land-use planning because political boundaries seldom reflect the geological and biological data base used in planning. Larger units can afford to hire professional planners, while local units cannot. The concept behind regional planning is coordination because problems do not respect political boundaries. For example, airport locations should be based on a regional plan that incorporates all local jurisdictions. Three cities only 30 kilometers apart should not build three separate airports when one regional airport could serve their needs better and at a lower cost to the taxpayers.

Although regional planning is increasing in North America, the majority of regional governmental bodies are presently voluntary and lack any power to implement programs. Their only role is to advise the member governments. Unfortunately, members of local governments still seem unwilling to give up power. They may view policy from a narrow perspective and put their own interests above the goals of the region. An elected, multipurpose, regional government, on the other hand, would be ideal for implementing land-use policy. Such governments exist in only a handful of places and show few signs of spreading.

One way to encourage regional planning is to develop policies at the state or provincial level. The first state to develop a comprehensive statewide land-use program was Hawaii. During the early 1960s, much of Hawaii's natural beauty was being destroyed to build houses and apartments for the increasing population. The same land that attracted tourists was being destroyed to provide hotels and supermarkets for them. Local governments had failed to establish and enforce land-use controls. Consequently, the Hawaii State Land-Use Commission was founded in 1961. This commission designated all land as urban, agricultural, or conservational. Each parcel of land could be used only for its designated purpose. Other uses were allowed only by special permit. To date, the record for Hawaii's action shows that it has been successful in controlling urban growth and preserving the islands' natural beauty, even though the population continues to grow. (See figure 12.13.)

Several states and provinces are attempting to follow Hawaii's lead in state land-use regulation. Some have passed legislation dealing with special types of land use. Examples include wetland preservation, floodplain protection, and scenic and historic site preservation. Although direct state involvement in land-use regulation is relatively new, it is expected to grow. Only large, well-financed levels of government can afford to pay for the growing cost of adequate land-use planning. State, provincial, and regional governments are also more likely to have the power to counter the political and economic influences of land developers, lobbyists, and other special-interest groups when conflicts over specific land-use policies arise.

Figure 12.13

Hawaii Land-Use Plan   Hawaii was the first state to develop a comprehensive land-use plan. This development in an agricultural area was stopped when the plan was implemented in the 1960s.

Urban Transportation Planning

A growing concern of city governments is to develop comprehensive urban transportation plans. While the specifics of such plans might vary from region to region, urban transportation planning usually involves four major goals:

1. Conserve energy and land resources.
2. Provide efficient and inexpensive transportation within the city for the elderly, young, poor, and handicapped.
3. Provide suburban people opportunities to commute efficiently.
4. Reduce urban pollution.

Any successful urban transportation plan should integrate all of these goals, but funding and intergovernmental cooperation are needed to achieve this. The problems associated with current urban transportation will certainly not disappear overnight, but comprehensive planning is the first step to solving them.

A primary method of transportation in most urban centers is the automobile. The automobile's advantages include convenience and freedom of movement.

However, many cities are beginning to recognize that the automobile's disadvantages may outweigh its advantages. The disadvantages include increased air pollution, substantial land use for highways and parking areas, and greater urban sprawl. Recognizing these problems, some cities, such as Toronto, London, San Francisco, and New York, have attempted to dissuade automobile use by developing mass transit systems and by allowing automobile parking costs to increase substantially.

The major urban mass transit systems are railroads, subways, trolleys, and buses. Such systems, however, are not without their problems. Mass transit is:

1. Economically feasible only along heavily populated routes
2. Less convenient than the automobile
3. Extremely expensive to build and operate
4. Often crowded and uncomfortable

Although mass transit meets a substantial portion of the urban transportation needs in some parts of the world, such as in Europe and Russia, its use in North America has declined since the 1940s.

(See figure 12.14.) A variety of forces has caused this decline. As people became more affluent, they could afford to own automobiles, which are a convenient, individualized method of transportation. Governments encouraged automobile use by financing highways and expressways, by maintaining a cheap energy policy, and by withdrawing support for most forms of mass transportation. They still encourage automobile transportation with hidden subsidies but maintain that rail and bus transportation should not be subsidized.

Rising gas prices and increasing competition for parking will prompt more North Americans to seek alternatives to private automobile use. Car and van pools and dial-a-ride systems are used by many individuals. Other types of transportation, such as the bicycle, should not be overlooked in urban transportation planning.

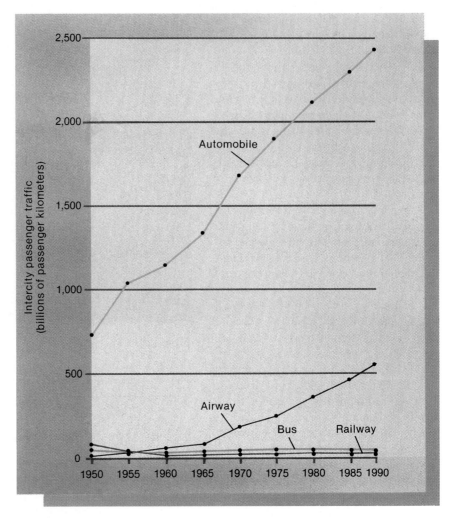

Figure 12.14

Decline of Mass Transportation   The automobile has consistently replaced rail and bus transit over the past 40 years.

Decision Making in Land-Use Planning: The Malling of America

The following situation has happened thousands of times during the past 20 years:

A developer has just announced plans to build a large shopping mall on the outskirts of your city in what is now prime farmland. Many jobs will be created by the construction and operation of the proposed mall, which will include stores and three new theaters. Presently, your city has some unemployment, only one theater in the downtown area, and little variety in its retail businesses. On the surface, the proposed mall seems to be only good news. Is this the case? Before answering, look at the entire situation.

- What will happen to the downtown area when the mall is opened?
- If you owned a downtown business, would you favor building the mall?
- What will happen to the taxes on the farms near the new mall?
- If you were a farmer, would you favor the project?
- What effect will paving prime farmland have?
- How will storm water runoff be affected?
- How will future housing development be influenced?
- What other problems can be associated with building the complex?
- If you were the mayor or city manager, would you favor construction of the mall? Why?
- Is the proposed project all good news after all?

SUMMARY

Land and landscape are considered natural resources. Of U.S. land, 45 percent is used for crops and livestock, about 50 percent is forests and natural areas, and the remainder is used for urban centers and transportation corridors. In Canada, 8 percent of the land is used for crops and livestock and 54 percent is forest and woodlands. A large percentage of northern Canada is wilderness. Historically, large urban centers began as small towns located near water. Water served the needs of the towns in many ways, especially as transportation. As towns became larger, the farmland surrounding them became suburbs surrounding industrial centers.

Many problems have resulted from unplanned urban growth. Current taxation policies encourage residential development of farmland, which results in a loss of valuable agricultural land. Floodplains and wetlands are often mismanaged. Loss of property and life results when people build on floodplains.

Wetlands protect our shorelines and provide a natural habitat for fish and wildlife.

Some examples of exclusionary land use are housing and highways. Multiple land use is possible where various uses can be integrated.

People in North America spend large amounts of money on leisure-time activities and services. Recreation is important socially and psychologically. People need recreation as a change from their normal working life. Conflicts sometimes arise over the use of resources for recreational purposes because some recreational activities cannot occur at the same time and place. These conflicts are usually resolved by the allocation and regulation of resources.

Land-use planning involves gathering data, projecting needs, and developing mechanisms for implementing the plan. Purchasing and zoning land are two ways to enforce land-use planning. The scale of local planning is often not large enough to be effective because problems may not be confined to political boundaries. Regional planning units can afford professional planners and are better able to withstand political and economic pressures. A growing concern of urban governments is to develop comprehensive urban transportation plans that seek to conserve energy and land resources, provide efficient and inexpensive transportation and commuting, and help to reduce urban pollution.

REVIEW QUESTIONS

1. Why did urban centers develop near waterways? Are they still located near water?
2. Describe the typical changes that have occurred in cities from the time they were first founded until now.
3. Why do people move to the suburbs?
4. Why do some farmers near urban areas sell their land for residential or commercial development? If you were in this position, would you sell?
5. What is a megalopolis?
6. What land uses are suitable on floodplains?
7. What is multiple land use? Can land be used for multiple purposes?
8. Why is it important to provide recreational space in urban planning?
9. How can recreational activities damage the environment? Do you engage in any of those activities?
10. What is the monetary impact of recreational activities?
11. What are some strictly urban-related recreational activities?
12. List some conflicts that arise when an area is designated strictly as recreational?
13. Describe the steps necessary to develop a land-use plan.
14. What are the advantages of regional or state planning?
15. List three benefits of land-use planning.

KEY TERMS

floodplain   231
floodplain zoning
   ordinances   231
land-use planning   236

megalopolis   229
multiple land use   232
nature centers   234

outdoor recreation   234
ribbon sprawl   229
tract development   229

urban sprawl   229
wetlands   231
zoning   236

RELATED INTERNET SITES

A site on toxic hot spots provides information about toxic chemicals in specific American communities.
   http://www.econet.apc.org/hotspots/
Department of the Interior, exhaustive pages on land management, fire, ecology, and the great basin.
   http://www.doi.gov/
Bureau of Land Management page.
   http://www.blm.gov/
Outdoor recreation planning and management.
   http://sfbox.vt.edu:10021/Y/yfleung/recres1.html#plain
Transportation Action Network: resources, information, and contacts on transportation, community, and the environment
   http://www.transact.org/

Wolf River Conservancy, dedicated to the protection of Wolf River and its floodplains.
   http://kesler.biology.rhodes.edu/wolf.html
National Wildlife Federation. Confronts worldwide threats to wildlife, wetlands, clean water, and more.
   http://www.nwf.org/international/
U.S. Fish and Wildlife national Wetlands Inventory.
   http://www.nwi.fws.gov/
The Swamp Project: an organization investigating cold-climate-constructed wetlands.
   http://www.computan.on.ca/~prodigal/ftgeo.htm

Soil and Its Uses

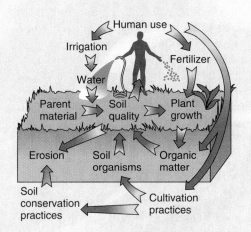

Objectives

After reading this chapter, you should be able to:

- List the physical, chemical, and biological factors responsible for soil formation.
- Explain the importance of humus to soil fertility.
- Differentiate between soil texture and soil structure.
- Explain how texture and structure influence soil atmosphere and soil water.
- Explain the role of living organisms in soil formation and fertility.
- Describe the various layers in a soil profile.
- Describe the processes of soil erosion by water and wind.
- Explain how contour farming, strip farming, terracing, waterways, windbreaks and conservation tillage reduce soil erosion.
- Understand that the misuse of soil reduces soil fertility, pollutes streams, and requires expensive remedial measures.
- Explain how land not suited for cultivation may still be productively used for other purposes.

Chapter Outline

Soil and Land
Soil Formation
Soil Properties
Soil Profile
Soil Erosion
Soil Conservation Practices
　Contour Farming
　Strip Farming
　Terracing
　Waterways
　Windbreaks
　Conservation Tillage
Environmental Close-Up: *Land Capability Classes*
Uses of Nonfarm Land
Issues and Analysis: *Soil Erosion in Virginia*

Soil and Land

Soil and land are often thought of as being the same, but they are not. **Land** is the part of the world not covered by the oceans, while **soil** is a mixture of minerals, organic material, living organisms, air, and water that together support the growth of plant life. Soil is a thin covering over the land. Farmers are particularly concerned with soil because the nature of the soil determines the kinds of crops that can be grown and which farming methods must be employed. Urban dwellers should also be concerned about soil because its health determines the quality and quantity of food they will eat. If the soil is so abused that it can no longer grow crops, or if it is allowed to erode, degrading air and water quality, both urban and rural residents suffer. To understand how soil can be protected, we must first understand its properties and how it is formed.

Soil Formation

A combination of physical and biological events forms soil. Soil building begins with the physical fragmentation of the **parent material,** which consists of ancient layers of rock or more recent geologic deposits from lava flows or glacial activity. The kind of parent material and the climate determine the kind of soil formed. Factors that can bring about fragmentation or chemical change of the parent material are known as **weathering.** Temperature changes and abrasion are two primary agents of **mechanical weathering.**

Heating a large rock can cause it to fracture because rock does not expand evenly. Pieces of the rock flake off. These pieces can be further reduced in size by other processes, such as the repeated freezing and thawing of water. Water that has seeped into rock cracks and crevices expands as it freezes, causing the cracks to widen. Subsequent thawing allows more water to fill the widened cracks, which are enlarged further by another period of freezing. Alternating freezing and thawing fragments large rock pieces into smaller

ones. (See figure 13.1). The roots of plants growing in cracks can also exert enough force to break rock.

The physical breakdown of rock is also caused by forces that move and rub rock particles against each other. For example, a glacier causes rock particles to grind against one another, resulting in smaller fragments and smoother surfaces. These particles are deposited by the glacier when the ice melts. In many parts of the world, the parent material from which soil is formed consists of glacial deposits. Wind and moving water also cause small particles to collide, resulting in further weathering. The smoothness of rocks and pebbles in a stream or on the shore is evidence that moving water has caused them to rub together, removing their sharp edges. Similarly, particles carried by wind collide with objects, fragmenting both the objects and the wind-driven particles.

Wind and moving water also remove small particles and deposit them at new locations, exposing new surfaces to the weathering process. For example, the landscape of the Painted Desert in the southwest United States was created by a combination of wind and moving water that removed easily transported particles, while rocks more resistant to weathering remained. (See figure 13.2.)

In addition to wind, moving water, glaciers, and changing temperature, certain chemical activities also alter the size and composition of parent material and participate in the soil-building process. This process is called **chemical weathering.** Small rock fragments exposed to the atmosphere may be oxidized; that is, they combine with oxygen from the air and chemically change to different compounds. Other kinds of rock may combine with water molecules in a process known as hydrolysis. Often, the oxidized or hydrolyzed molecules are more readily soluble in water and, therefore, may be removed by rain or moving water. Since rain is normally slightly acid, the acid content helps dissolve rocks.

The first organisms to gain a foothold in this modified parent material also contribute to fragmentation. Lichens often

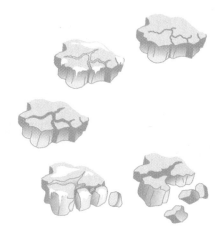

Figure 13.1

Physical Fragmentation by Freezing and Thawing The crack in the rock fills with water. As the water freezes and becomes ice, the pressure of the ice enlarges the crack. The ice melts, and water again fills the crack. The water freezes again and widens the crack. Alternate freezing and thawing splits the rock into smaller fragments.

form a pioneer community that traps small particles and chemically alters the underlying rock. As other kinds of organisms, like plants and small animals, become established, they contribute increasing amounts of organic matter, which are incorporated with the small rock fragments to form soil.

Decaying organic matter, called **humus,** becomes mixed with the top layers of rock particles and constitutes a very important ingredient of soil. Humus supplies some of the nutrients needed by plants and also increases the acidity of the soil so that inorganic nutrients, which are more soluble under acidic conditions, become available to plants. For example, wheat and corn grow best in soils with a pH between 5.5 and 7.0. A soil with a pH above 7.0 would be more productive if humus were added to increase the acidity. One problem with an acidic soil is that because nutrients are more soluble they can be carried away if there is heavy rainfall or if there are few plants to capture them. Humus can help to create a loose, crumbly soil texture that allows water to soak in and permits air to be incorporated into the soil. Compact soils have poor aeration, and the water runs off.

Figure 13.2

The Painted Desert This landscape was created by the action of wind and moving water. The particles removed by these forces were deposited elsewhere and may have become part of the soil in that new location.

Burrowing animals, soil bacteria, fungi, and the roots of plants are also part of the biological process of soil formation. One of the most important burrowing animals is the earthworm. One hectare of soil may support a population of 500,000 earthworms that can process as much as 9 metric tons of soil a year. These animals literally eat their way through the soil, resulting in further mixing of organic and inorganic material, which increases the amount of nutrients available for plant use. They often bring nutrients from the deeper layers of the soil up into the area where plant roots are more concentrated, thus improving the soil's fertility. Soil aeration and drainage are also improved by the burrowing of earthworms and other small soil animals. They also help to incorporate organic matter into the soil by collecting dead organic material from the surface and transporting it into burrows and tunnels.

Fungi and bacteria are decomposers and serve as important links in many mineral cycles. (See chapter 4.) They, along with animals, reduce organic mate-

rial to smaller particles and, therefore, improve the quality of the soil.

Over a period of time, this complex array of physical, chemical, and biological processes formed the soils we have today. Under ideal climatic conditions, soft parent material may develop into a centimeter of soil within 15 years. Under poor climatic conditions, a hard parent material may require hundreds of years to develop into that much soil. In any case, soil formation is a slow process.

Soil Properties

Soil properties include soil texture, structure, atmosphere, moisture, biotic content, and chemical composition. **Soil texture** is determined by the size of the rock particles within the soil. The largest soil particles are gravel, which consists of fragments larger than 2.0 millimeters in diameter. Particles between 0.05 and 2.0 millimeters are classified as sand. Silt particles range from 0.002 to 0.05 millimeters in diameter, and the smallest particles are clay parti-

cles, which are less than 0.002 millimeters in diameter.

Large particles, such as sand and gravel, have many tiny spaces between them, which allow both air and water to flow through the soil. Water drains from this kind of soil very rapidly, often carrying valuable nutrients to lower soil layers, where they are beyond the reach of plant roots. Clay particles tend to be flat and are easily packed together to form waterproof layers. Soils with a lot of clay do not drain well and are poorly aerated. Because water does not flow through clay very well, clay soils tend to stay moist for longer periods of time and do not easily lose minerals to infiltrating water.

Rarely does a soil consist of a single size of particle. Various particles are mixed in many different combinations, resulting in many different soil classifications. (See figure 13.3.) An ideal soil for agricultural use is a **loam,** which combines the good aeration and drainage properties of large particles with the nutrient-retention ability of clay particles.

Soil structure is different from its texture. **Soil structure** refers to the way various soil particles clump together. Sand particles do not clump, and therefore sandy soils lack structure, whereas clay soils tend to stick together in large clumps. A good soil forms small clumps that crumble easily. This ability to crumble is known as the soil's **friability.** The friability is determined by the soil structure and its moisture content. Sandy soils are very friable, while clay soils are not. If clay soil is worked when it is too wet, it can stick together in massive chunks that will not break up for years.

A good soil for agricultural use will crumble and has spaces for air and water. In fact, the air and water content depends upon the presence of these spaces. (See figure 13.4.) In good soil, about two-thirds of the spaces contain air after the excess water has drained. The air provides a source of oxygen for plant root cells. The relationship between the amount of air and water is not fixed. After a heavy rain, all of the spaces may be filled with water. If some of the excess water does not drain from the soil, the

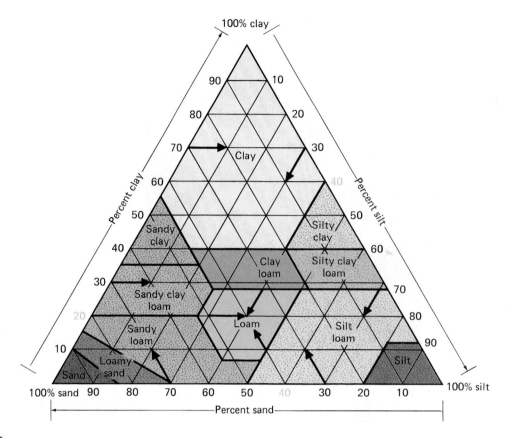

Figure 13.3

Soil Texture   Texture depends upon the percentage of clay, silt, and sand particles in the soil. Soils with the best texture for most crops are loams. As shown in the illustration, if a soil were 40 percent sand, 40 percent silt, and 20 percent clay, it would be a loam.

Source: Data from Soil Conservation Service.

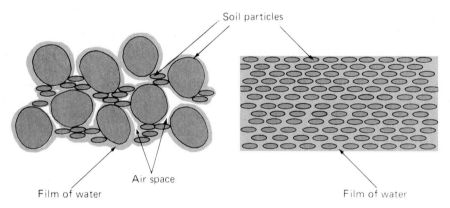

Figure 13.4

Pore Spaces and Particle Size   The soil on the left, which is composed of particles of various sizes, has spaces for both water and air. The particles have water bound to their surfaces (represented by the colored halo around each particle), but some of the spaces are so large that an air space is present. The soil on the right, which is composed of uniformly small particles, has less space for air. Since roots require both air and water, the soil on the left would be better able to support crops than that on the right.

plant roots may die from lack of oxygen. They are literally drowned. On the other hand, if there is not enough soil moisture, the plants wilt from lack of water. Soil moisture and air are also important in determining the numbers and kinds of soil organisms.

Protozoa, nematodes, earthworms, insects, bacteria, and fungi are typical inhabitants of soil. (See figure 13.5.) The role of protozoa in the soil is not firmly established, but they seem to act as parasites on other forms of soil organisms and, therefore, help to regulate the population size of those organisms. Nematodes, which are often called wireworms or roundworms, may aid in the recycling of dead organic matter. Some nematodes are parasitic on the roots of plants. Insects contribute to the soil by forming burrows and recycling organic materials, but they are also major crop

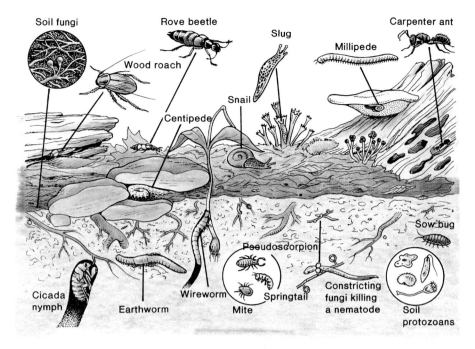

Figure 13.5

Soil Organisms All of these organisms occupy the soil and contribute to it by rearranging soil particles, participating in chemical transformation, and recycling dead organic matter.

pests that feed on plant roots. Bacteria and fungi are particularly important in the decay and recycling of materials. Their chemical activities change complex organic materials into nutrients that can be used by plants. For example, these microorganisms can convert the nitrogen contained in the protein component of organic matter into ammonia or nitrate, which are nitrogen compounds that can be utilized by plants. The amount of nitrogen produced varies with the type of organic matter, type of microorganisms, drainage, and temperature. All of these organisms are active within distinct layers of the soil, known as the soil profile.

Soil Profile

The **soil profile** is a series of horizontal layers of different chemical composition, particle size, and amount of organic matter. Each recognizable layer is known as a **horizon.** (See figure 13.6.) The uppermost layer of the soil contains more nutrients, and organic matter than do the deeper layers. This layer is known as the A horizon or topsoil. The thickness of the A horizon may vary from less than a centimeter on steep mountain slopes to over a meter in the rich grasslands of central North America. Most of the living organisms and nutrients are found near the top of the A horizon. The lower portions of the A horizon usually contain few nutrients because water flowing down through the soil dissolves and transports nutrients to the B horizon. This process is known as **leaching.** The B horizon, often called the subsoil, contains less organic material, fewer organisms, and accumulations of nutrients that were leached from higher levels. Because of this, the B horizon in many soils is a valuable source of nutrients for plants, and such subsoils support a well-developed root system. Because the amount of leaching depends on the available rainfall, areas of the world where rainfall is low may have a poorly developed B horizon.

The area below the subsoil is known as the C horizon, and it consists of weathered parent material. This parent material contains no organic materials, but it does contribute to some of the soil's properties. The chemical composition of the C horizon helps to determine the pH of the soil. The C horizon may also influence the soil's rate of water absorption and retention.

Soil profiles and the factors that contribute to soil development are extremely varied. Over 15,000 separate soil types have been classified in North America. However, most of the cultivated land in the world can be classified as either grassland soil or forest soil. (See figure 13.7.)

Grassland soils usually have a deep A horizon. The low amount of rainfall in grassland areas limits the amount of leaching from the topsoil. As a result, the A horizon is deep and supports most of the root growth. This lack of leaching also results in a thin layer of subsoil, which is low in mineral and organic content and supports little root growth.

Forest soils develop in areas of more abundant rainfall. The rainfall results in a layer of topsoil that is usually not as thick as that of grassland soil, but the material leached from the topsoil forms a subsoil that supports substantial root growth. One of the materials that accumulates in the B horizon is clay. In some soils, particularly forest soils, clay or other minerals may accumulate and form an impermeable "hardpan" layer that limits the growth of roots and may prevent water from reaching the soil's deeper layers.

Desert soils have very poorly developed horizons since there is little rainfall to leach materials from upper layers to lower layers. In addition, the low amount of plant growth results in little contribution of organic matter to the soil. In cold, wet climates, typical of the northern parts of Europe, Russia, and Canada, there may be considerable accumulations of organic matter since the rate of decomposition is reduced. The extreme acidity of these soils also reduces the rate of decomposition. Hot, humid climates also tend to have poorly developed soil horizons since the organic matter decays very rapidly, and soluble materials are carried away by the abundant rainfall.

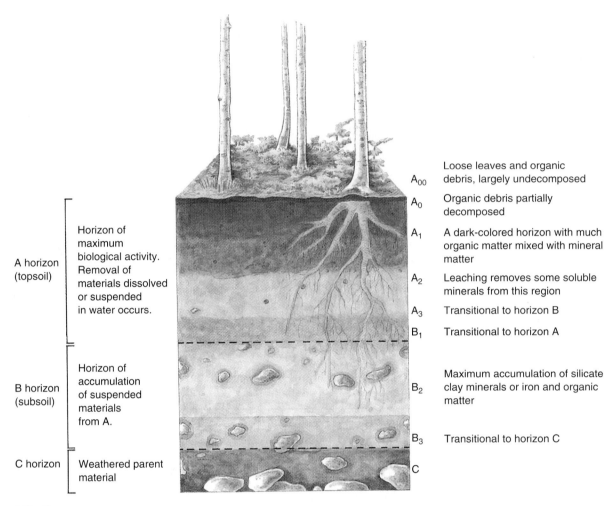

A horizon (topsoil) — Horizon of maximum biological activity. Removal of materials dissolved or suspended in water occurs.

B horizon (subsoil) — Horizon of accumulation of suspended materials from A.

C horizon — Weathered parent material

A_{00} — Loose leaves and organic debris, largely undecomposed

A_0 — Organic debris partially decomposed

A_1 — A dark-colored horizon with much organic matter mixed with mineral matter

A_2 — Leaching removes some soluble minerals from this region

A_3 — Transitional to horizon B

B_1 — Transitional to horizon A

B_2 — Maximum accumulation of silicate clay minerals or iron and organic matter

B_3 — Transitional to horizon C

C

Figure 13.6

Soil Profile A soil has layers that differ physically, chemically, and biologically. The top layer is known as the A horizon and contains most of the organic matter. The B horizon accumulates minerals and particles as water carries dissolved minerals downward from the A to the B horizon. The B horizon is often called the subsoil. Below the B horizon is a C horizon of weathered parent material.

In addition to the differences caused by the kind of vegetation and rainfall, topography influences the soil profile. (See figure 13.8.) On a relatively flat area, the topsoil formed by soil-building processes will collect in place and gradually increase in depth. The topsoil formed on rolling hills or steep slopes is often transported down the slope as fast as it is produced. On such slopes, the accumulation of topsoil may not be sufficient to support a cultivated crop. The topsoil removed from these slopes is eventually deposited in the flat floodplains. These regions serve as collection points for topsoil that was produced over extensive areas.

As a result, these river-bottom and delta regions have a very deep topsoil layer and are highly productive agricultural land.

Soil Erosion

Erosion is the wearing away and transportation of soil by water or wind. The Grand Canyon of the Colorado River, the floodplains of the Nile in Egypt, the little gullies on hillsides, and the deltas that develop at the mouths of rivers all attest to the ability of water to move soil. Anyone who has seen muddy water after a rainstorm has observed soil being moved by

water. (See figure 13.9.) The force of moving water allows it to carry large amounts of soil. Each year, the Mississippi River transports over 325 million metric tons of soil from the central region of North America to the Gulf of Mexico. This is equal to the removal of a layer of topsoil approximately 1 millimeter thick from the entire region. This movement of soil by water occurs in every stream and river in the world. Dry Creek, a small stream in California, has only 500 kilometers of mainstream and tributaries; however, each year, it removes 180,000 metric tons of soil from a 340-square-kilometer area. (See chapter 15.)

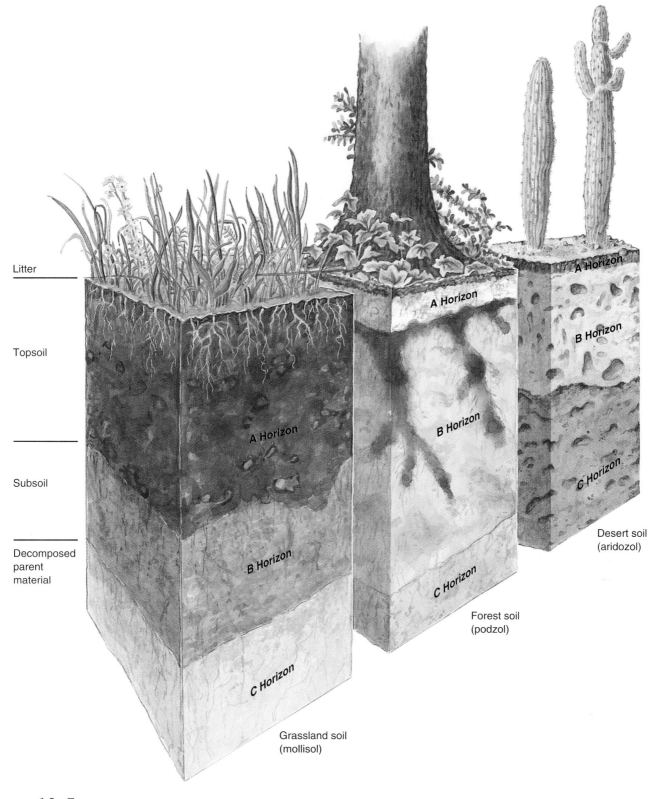

Litter

Topsoil

Subsoil

Decomposed
parent
material

A Horizon

B Horizon

C Horizon

Grassland soil
(mollisol)

A Horizon

B Horizon

C Horizon

Forest soil
(podzol)

A Horizon

B Horizon

C Horizon

Desert soil
(aridozol)

Figure 13.7

Major Soil Types   There are thousands of different soil types, but many of them can be classified into two broad categories. Soils formed in grasslands are known as chernozem soils and have a deep *A* horizon. The shallow *B* horizon does not have sufficient nutrients to support root growth. In forest soils, known as podzol soils, the *A* horizon is thinner, and leaching results in many nutrients in the *B* horizon. Thus roots are found in both the *A* and *B* horizons.

| A horizon | } 15 cm |
| B horizon | } 25 cm |

| A horizon | }150 cm |
| B horizon | } 15 cm |

Figure 13.8

The Effect of Slope on a Soil Profile   The topsoil formed on a large area of the hillside is continuously transported down the slope by the flow of water. It accumulates at the bottom of the slope and results in a thicker A horizon. The resulting "bottomland" is highly productive because it has a deep, fertile layer of topsoil, while the soil on the slope is less productive.

Soil erosion takes place everywhere in the world, but some areas are more exposed than others. Erosion occurs wherever grass, bushes, and trees are disappearing. Every year erosion carries away far more topsoil than is created. (See figure 13.10.)

Worldwide, erosion removes about 25.4 billion tons of soil each year. Deforestation and desertification both leave land open to erosion. In deforested areas, water washes down steep, exposed slopes, taking the soil with it. In desertified regions, exposed soils, cleared for farming, building, or mining, or overgrazed by livestock, simply blow away. Wind erosion is most extensive in Africa and Asia. Blowing soil not only leaves a degraded area behind but can bury and kill vegetation where it settles. It will also fill drainage and irrigation ditches. When high-tech farm practices are applied to poor lands, soil is washed away and chemical pesiticides and fertilizers pollute the runoff.

In Africa, soil erosion has reached critical levels, with farmers pushing farther onto deforested hillsides. In Ethiopia, for example, soil loss occurs at a rate of between 1.5 billion and 2 billion cubic meters a year, with some 4 million hectares of highlands considered irreversibly degraded. In Asia, in the eastern hills of Nepal, 38 percent of the land area

is fields that have been abandoned because the topsoil has washed away. In the Western Hemisphere, Ecuador is losing soil at 20 times the acceptable rate.

According to the International Fund for Agricultural Development (IFAD), traditional labor-intensive, small-scale soil conservation efforts that combine maintenance of shrubs and trees with crop growing and cattle grazing work best at controlling erosion. In parts of Pakistan, a program begun by IFA in 1980 to control rainfall runoff, erosion, and damage to rivers from siltation has increased crop yields and livestock productivity by 20 to 30 percent.

Badly eroded soil has lost all of the topsoil and some of the subsoil and is no longer productive farmland. Most current agricultural practices lose soil faster than it is replaced. Farming practices that reduce erosion, such as contour farming and terracing, are discussed later in the chapter.

Wind is also an important mover for soil. Under certain conditions, it can move large amounts. (See figure 13.11.) Wind erosion may not be as evident as water erosion, since it does not leave gullies. Nevertheless, it can be a serious problem. Wind erosion is most common in dry, treeless areas where the soil is exposed. In the Sahel region of Africa, much of the land has been denuded of

Figure 13.9

Water Erosion   The force of moving water is able to pick up soil particles and remove them. In cases of prolonged erosion, gullies (such as the one shown here) are likely to form.

vegetation because of drought, overgrazing, and improper farming practices. This has resulted in extensive wind erosion of the soil. (See figure 13.12.) In the Great Plains region of North America, there have been four serious periods of wind erosion since European settlement in the 1800s. If this area receives less than 30 centimeters of rain per year, there is not enough moisture to support crops. When this occurs for several years in a row, it is called a drought. Farmers plant crops, hoping for rain. When the rain does not come, they plow their fields again to prepare it for another crop. Thus, the loose, dry soil

PART FOUR   Human Influences on Ecosystems

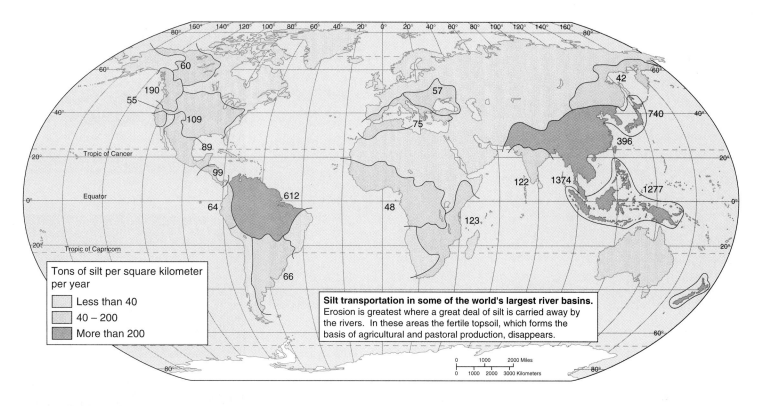

Tons of silt per square kilometer per year

Less than 40

40 – 200

More than 200

Silt transportation in some of the world's largest river basins. Erosion is greatest where a great deal of silt is carried away by the rivers. In these areas the fertile topsoil, which forms the basis of agricultural and pastoral production, disappears.

A tropical river

Figure 13.10

Soil Erosion and Silty River Soil erosion is widespread throughout the world. Silty rivers are evidence of poor soil conservation practices upstream.

Figure 13.11

Wind Erosion The dry, unprotected topsoil from this field is being blown away. The force of the wind is capable of removing all the topsoil and transporting it several thousand kilometers.

Figure 13.12

Wind Erosion in the Sahel The semiarid region just south of the Sahara Desert is in an especially vulnerable position. The rainfall is unpredictable, which often leads to crop failure. In addition, population pressure forces people in this region to try to raise crops in marginal areas. This often results in increased wind erosion.

a.

b.

Figure 13.13

Poor and Proper Soil Conservation Practices (*a*) This land is no longer productive farmland since erosion has removed the topsoil. (*b*) This rolling farmland shows strip contour farming to minimize soil erosion by running water. It should continue to be productive farmland indefinitely.

is left exposed, and wind erosion results. Because of the large amounts of dust in the air during those times, the region is known as the Dust Bowl. During the 1930s, wind destroyed 3.5 million hectares of farmland and seriously damaged an additional 30 million hectares in the Dust Bowl. Unfortunately, many agricultural practices adopted to combat erosion following the 1930s are being discontinued today for economic reasons, thus leaving the land vulnerable to erosion again.

Soil Conservation Practices

Whenever soil is lost by water or wind erosion, the topsoil, the most productive layer, is the first to be removed. When the topsoil is lost, the soil's fertility decreases, and expensive fertilizers must be used to restore the fertility that was lost. This raises the cost of the food we buy. In addition, the movement of excessive amounts of soil from farmland into streams has several undesirable effects.

(See chapter 14.) First, a dirty stream is less aesthetically pleasing than a clear stream. Second, a stream laden with sediment affects the fish population. Fishing may be poor because of unwise farming practices hundreds of kilometers upstream. Third, the soil carried by a river is eventually deposited somewhere. In many cases, this soil must be removed by dredging to clear shipping channels. We pay for dredging with our tax money, and it is a very expensive operation.

For all of these reasons, proper soil conservation measures should be employed to minimize the loss of topsoil. Figure 13.13 contrasts poor soil conservation practices with proper soil protection. When soil is not protected from the effects of running water, the topsoil is removed and gullies result. This can be prevented by slowing the flow of water over sloping land.

The kinds of agricultural activity that land can be used for are determined by soil structure, texture, friability, drainage, fertility, rockiness, slope of the land, amount and nature of rainfall, and other climatic conditions. A relatively large proportion—over 50 percent—of

U.S. land is suitable for crop cultivation. (See figure 13.14.) However, only 2 percent of that land does not require some form of soil conservation practice. This means that nearly all of the soil in the United States must be managed in some way to reduce the effects of soil erosion by wind or water.

Not all parts of the world are as well supplied as the United States with land that has agricultural potential. (See table 13.1.) For example, worldwide, approximately 11 percent of the land surface is suitable for crops, and an additional 24 percent is in permanent pasture. In the United States, about 21 percent is cropland, and 26 percent is in permanent pasture. Contrast this with the continent of Africa, in which only 6 percent is suitable for crops, and 26 percent can be used for pasture. Canada has only 5 percent suitable for crops and under 3 percent for

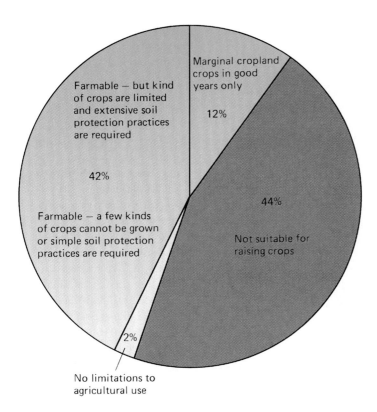

Figure 13.14

U.S. Land Suitable for Raising Crops
Only 2 percent of the land in the United States can be cultivated without some soil conservation practices. This 2 percent is primarily flatland, which is not subject to wind erosion. On 42 percent of the remaining land, some special considerations for protecting the soil are required, or the kinds of crops are limited. Twelve percent of the land is marginal cropland that can only provide crops in years when rainfall and other conditions are ideal. Forty-four percent is not suitable for cultivating crops but may be used for other purposes, such as grazing cattle or as forests.

Table 13.1

Percentage of Land Suitable for Agriculture

Country	Percent cropland	Percent pasture
World	11.0	26.0
Africa	6.3	28.8
Egypt	2.8	5.0
Ethiopia	12.7	40.7
Kenya	7.9	37.4
South Africa	10.8	66.6
North America	13.0	16.8
Canada	4.9	3.0
United States	19.6	25.0
South America	6.0	28.3
Argentina	9.9	51.9
Venezuela	4.4	20.2
Asia	15.2	25.9
China	10.3	30.6
Japan	12.0	1.7
Europe	29.9	17.1

Source: Data from World Resources 1996.

pasture. Europe has the highest percentage of cropland with 30 percent, but it has only 18 percent in permanent pasture.

Few reserves of arable land remain in the temperate regions of Europe and North America. Most undeveloped land that is suitable for cultivation and that receives enough rainfall for agriculture is found in the rainforests of South America and Africa.

Contour Farming

Contour farming, which is tilling at right angles to the slope of the land, is one of the simplest methods to prevent soil erosion. This practice is useful on gentle slopes and produces a series of small ridges at right angles to the slope. (See figure 13.15.) Each ridge acts as a dam to hold water from running down the incline. This allows more of the water to soak into the soil. Contour farming reduces soil erosion by as much as 50 percent and, in drier regions, increases crop yields by conserving water.

Strip Farming

When a slope is too steep or too long, contour farming alone may not prevent

Figure 13.15

Contour Farming   Tilling at right angles to the slope creates a series of ridges that slows the flow of the water and prevents soil erosion. This soil conservation practice is useful on gentle slopes.

Figure 13.16

Strip Farming   On rolling land, a combination of contour and strip farming prevents excessive soil erosion. The strips are planted at right angles to the slope, with bands of closely sown crops, such as wheat or hay, alternating with bands of row crops, such as corn or soybeans.

soil erosion. However, a combination of contour and strip farming may work. **Strip farming** is alternating strips of closely sown crops like hay, wheat, or other small grains with strips of row crops like corn, soybeans, cotton, or sugar beets. (See figure 13.16.) The closely sown crops retard the flow of water, which reduces soil erosion and allows more water to be absorbed into the ground. The type of soil, steepness, and length of slope dictate the width of the strips and determine whether strip or contour farming is practical.

Terracing

On very steep land, the only practical method of preventing soil erosion is to construct terraces. **Terraces** are level areas constructed at right angles to the slope to retain water and greatly reduce the amount of erosion. (See figure 13.17.) Terracing has been used for centuries in nations with a shortage of level farmland. The type of terracing seen in figure 13.17a requires the use of small machines and considerable hand labor and is not suitable for the mechanized farming typical in much of the world. Terracing is an expensive method of controlling erosion since it requires the moving of soil to construct the level areas, protecting the steep areas between terraces, and constant repair and maintenance. Many factors, such as length and steepness of slope, type of soil, and amount of precipitation, determine whether terracing is feasible.

Waterways

Even with such soil conservation practices as contour farming, strip farming, and terracing, farmers must often provide protected channels for the movement of water. **Waterways** are depressions on sloping land where water collects and flows off the land. When not properly maintained, these areas are highly susceptible to erosion. (See figure 13.18.) If a waterway is maintained with a permanent sod covering, the speed of the water is reduced, and soil erosion is decreased.

Windbreaks

Contour farming, strip farming, terracing, and maintaining waterways are all important ways of reducing water erosion, but wind is also a problem with certain soils, particularly in dry areas of the world. Wind erosion can be reduced if the soil is protected. (See chapter 17.) The best protection is a layer of vegetation over the surface. However, the process of preparing soil for planting and

a.

b.

Figure 13.17

Terraces   Since the construction of terraces requires the movement of soil and the protection of the steep slope between levels, terraces are expensive to build. (*a*) The terraces seen here are extremely important for people who live in countries that have little flatland available. They require much energy and hand labor to maintain but make effective agricultural use of the land without serious erosion. (*b*) This modification of the terracing concept allows the use of the large farm machines typical of farming practices in Canada, Europe, and the United States.

a.

b.

Figure 13.18

Protection of Waterways Prevents Erosion   (*a*) An unprotected waterway has been converted into a gully. (*b*) A well-maintained waterway is not cultivated; a strip of grass retards the flow of water and protects the underlying soil from erosion.

the method of planting often leave the soil exposed to wind. **Windbreaks** are plantings of trees or other plants that protect bare soil from the full force of the wind. Windbreaks reduce the velocity of the wind, thereby decreasing the amount of soil that it can carry away. (See figure 13.19.) In some cases, rows of trees are planted at right angles to the prevailing winds to reduce their force, while in others cases, a kind of strip farming is practiced in which strips of hay or grains are alternated with row crops that leave large amounts of the soil exposed. In some areas of the world, the only way to protect the soil is not to cultivate it at all, but to leave it in a permanent cover of grasses.

Conservation Tillage

Conventional tillage methods in much of the world require extensive use of farm machinery to prepare the soil for planting and to control weeds. Typically, a field is plowed and then disked or harrowed one to three times before the crop is planted. The plowing, which turns the soil over,

environmental
close-up

Land Capability Classes

Not all land is suitable for raising crops, or urban building. Such factors as the degree of slope, soil characteristics, rockiness, erodibility, and other characteristics determine the best use for a parcel of land. In an attempt to encourage people to use land wisely, the U.S. Soil Conservation Service has established a system to classify land-use possibilities. The table shows the eight classes of land and list the characteristics and capabilities of each.

Unfortunately, many of our homes and industries are located on type I and II land, which has the least restrictions on agricultural use. This does not make the best use of the land. Zoning laws and land-use management plans should consider the land-use capabilities and institute measures to assure that land will be used to its best potential.

Can you provide examples from your community where you feel there was not proper land use? In your opinion, should there be more or less land-use planning? What are some of the consequences of building or farming on land that is not suitable?

	Land class	Characteristics	Capability	Conservation measures
Land suitable for cultivation	I	Excellent, flat, well-drained land	Cropland	None
	II	Good land; has minor limitations, such as slope, sandy soil, or poor drainage	Cropland Pasture	Strip cropping Contour farming
	III	Moderately good land with important limitations of soil, slope, or drainage	Cropland Pasture Watershed	Contour farming Strip cropping Terraces Waterways
	IV	Fair land with severe limitations of soil, slope, or drainage	Pasture Orchards Urban Industry Limited cropland	Crops on a limited basis Contour farming Strip cropping Terraces Waterways
Land not suitable for cultivation	V	Use for grazing and forestry; slightly limited by rockiness, shallow soil, or wetness	Grazing Forestry Watershed Urban Industry	No special precautions if properly grazed or logged; must not be plowed
	VI	Moderate limitations for grazing and forestry because of moderately steep slopes	Grazing Forestry Watershed Urban Industry	Grazing or logging may be limited at times
	VII	Severe limitations for grazing and forestry because of very steep slopes vulnerable to erosion	Grazing Forestry Watershed Recreation Wildlife Urban Industry	Careful management is required when used for grazing or logging
	VIII	Unsuitable for grazing and forestry because of steep slope, shallow soil, lack of water, or too much water	Watershed Recreation Wildlife Urban Industry	Not to be used for grazing or logging; steep slope and lack of soil present problems

a.

b.

Figure 13.19

Windbreaks (*a*) In sections of the Great Plains, trees provide protection from wind erosion. The trees along the road are planted in a north-south direction to protect the land from the prevailing westerly winds. (*b*) In this field, temporary strips of vegetation serve as windbreaks.

has several desirable effects: Any weeds or weed seeds are buried, thus reducing the weed problem in the field. Crop residue from previous crops is incorporated into the soil where it will decay faster and contribute to soil structure. Nutrients that had been leached to deeper layers of the soil are brought near the surface. And the dark soil is exposed to the sun so that it warms up faster. This last effect is most critical in areas with short growing seasons. In many areas, fields are plowed in the fall, after the crop has been harvested, and the soil is left exposed all winter.

After plowing, the soil is worked by disks or harrows to break up any clods of earth, kill remaining weeds, and prepare the soil to receive the seeds. After the seeds are planted, there may still be weed problems. Farmers often must cultivate row crops to kill the weeds that begin to grow between the rows. Each trip over the field costs the farmer money, while at the same time increasing the amount of time the soil is exposed to wind or water erosion.

In recent years, several innovations in chemical herbicides and farm equipment have led to the development of tilling practices that protect the soil by reducing the amount of time it is exposed to erosion forces. Reduced tillage uses less cultivation to control weeds and to prepare the soil to receive seeds. Selective herbicides are used to kill unwanted vegetation prior to planting the new crop and to control weeds after-

ward. Several variations of conservation tillage are used:

1. Plowing followed by reduced secondary tillage. This usually involves plowing followed immediately by planting.
2. Strip tillage, a method that involves tilling only in the narrow strip that is to receive the seeds. The rest of the soil and the crop residue from the previous crop is left undisturbed.
3. No-till farming involves special planters that place the seeds in slits cut in the soil that still has on its surface the crop residue from the previous crop.

All of these types of reduced tillage use less fuel and less time but may require more herbicides. (See table 13.2). For many kinds of crops, yields are comparable to that produced by conventional tillage methods.

Other positive effects of reduced tillage, in addition to reducing erosion, are:

1. The amount of winter food, space between, and cover available for wildlife increases, which can lead to increased wildlife populations.
2. Since there is less runoff, siltation in streams and rivers is reduced. This results in clearer water for recreation and less dredging to keep waterways open for shipping.

3. Row crops can be planted on hilly land that cannot be converted to such crops under conventional tilling methods. This allows a farmer to convert low-value pasture land into cropland that gives a greater economic yield.
4. Since fewer trips are made over the field, petroleum is saved, even if the petrochemical feedstocks necessary to produce the herbicides are taken into account.
5. Two crops may be grown on a field in areas that had been restricted to growing one crop per field per year. In some areas, immediately after harvesting wheat, farmers have used reduced tillage methods to plant soybeans in the wheat stubble.
6. Because reduced tillage reduces the number of trips made over the field by farm machinery, the soil does not become compacted as quickly.

However, there are also some drawbacks to reduced-tillage methods:

1. The residue from previous vegetation may delay the warming of the soil, which may, in turn, delay planting some crops for several days.
2. The crop residue reduces evaporation from the soil and the upward movement of water and soil nutrients from deeper layers of the soil, which may retard the growth of plants.

Table 13.2

Comparison of Various Tilling Methods

Tillage method	Hours required per 100 hectares	Liters of fuel per 100 hectares	Cost of herbicide per 100 hectares (1993)
Conventional	200	2,915	$3,717
Reduced secondary tillage	125	1,390	$3,717
Strip tillage	95	1,020	$3,717
No tillage	62	375	$5,000

No-Till Farming Takes Root

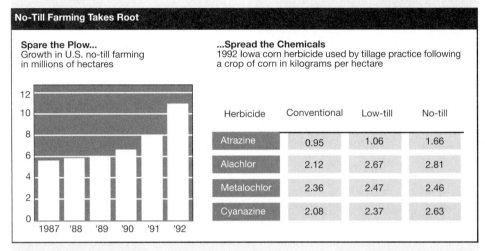

Spare the Plow...
Growth in U.S. no-till farming in millions of hectares

...Spread the Chemicals
1992 Iowa corn herbicide used by tillage practice following a crop of corn in kilograms per hectare

Herbicide	Conventional	Low-till	No-till
Atrazine	0.95	1.06	1.66
Alachlor	2.12	2.67	2.81
Metalochlor	2.36	2.47	2.46
Cyanazine	2.08	2.37	2.63

Source: (top) Data from J.E. Beuerlein and S.W. Bone, "Selecting a Tillage System" Extension Publication, Ohio State University; and D.H. Doster, "Economics of Alternative Tillage Systems" in Bulletin of the Entomological Society of America 22 (1976):297. (left) Data from Conservation Technology Information Center, West Lafayette, IN; (right) Data from U.S. Department of Agriculture.

Figure 13.20

Noncrop Use of Land to Raise Food As long as it is properly protected and managed, this land can produce food through grazing, but it should never be plowed to plant crops because the topsoil is too shallow and the rainfall is too low.

3. The accumulation of plant residue can harbor plant pests and diseases that will require more insecticides and fungicides.

Reduced tillage is not the complete answer to soil erosion problems but may be useful in reducing soil erosion on well-drained soils. It also requires that farmers pay close attention to the condition of the soil and the pests to be dealt with.

In 1972, 12 million hectares in the United States were under some form of conservation tillage. About 1.3 million hectares were being farmed using no-till methods. By 1992, this had risen to 60 million hectares of conservation-tilled land, of which 6.2 million hectares were being farmed using no-till methods. It is estimated that, by the year 2010, 95 percent of U.S. cropland will be under some form of reduced-tillage practice.

Uses of Nonfarm Land

Not all land is suitable for crops or continuous cultivation. Some must never be plowed for use as cropland, but people can still use it to provide for some of their needs. By using appropriate soil conservation practices, much of the land not usable for crops can be used for grazing, wood production, wildlife production, or scenic and recreational purposes. Figure 13.20 shows land that is not suitable for cultivation but that is capable of producing grass for cattle or sheep. Such land, if properly used, will produce food for many generations.

The land shown in figure 13.21 is not suitable for either crops or grazing. However, it is still a valuable and productive piece of land since it can be used to furnish lumber, wildlife habitats, and recreational opportunities.

All land is not equal. Each section has its own soil characteristics, climate, and degree of slope. When all these are taken into consideration, a proper use can be determined for each portion of the planet. Wise planning and careful husbandry of the soil is necessary if the land and its soil are to provide food and other necessities of life.

Figure 13.21

Forest and Recreational Use Although this land is not capable of producing crops or supporting cattle, it furnishes lumber, a habitat for wildlife, and recreational opportunities.

Soil Erosion in Virginia

In a court case in Virginia, a city attorney filed suit for damages against a land developer. The city contended that the developer was responsible for property damages as a result of erosion from his construction site. During the trial, it was shown that the developer had constructed his subdivision without considering the possible effects of soil erosion. As a result, during a heavy rain, mud from his project had covered the yards, sidewalks, and drives of homes located nearby. As much as 10 meters of soil had eroded from the exposed land of his development and been carried downhill, covering portions of the adjacent community.

Speaking to the developer, the city attorney stated: "After gambling and principally winning over a period of fifteen months, it is apparent that you are attempting to go with your winning streak; and flood, storms, and weather hold no terror for you." The city claimed that the developer had acted in a negligent manner and that he was, therefore, responsible for the damages caused by the deposited mud and should be ordered to pay for the cleanup. The developer claimed no responsibility. His position was that grading large tracts of land was a common construction procedure and that the weather was "an act of God" for which he could not be held accountable.

- If you were a member of the jury hearing the case, would you think the developer acted in a negligent manner by leaving the soil exposed for a long period of time?
- Do you believe that property owners can hold a person responsible for actions taken in an area some distance from their property?
- Do you believe that the erosion and subsequent deposit of mud was "an act of God"?
- Would you find the developer guilty or not guilty? Why?

SUMMARY

Soil is an organized mixture of minerals, organic material, living organisms, air, and water. Soil formation begins with the breakdown of the parent material by such physical processes as changes in temperature, freezing and thawing, and movement of particles by glaciers, flowing water, or wind. Oxidation and hydrolysis can chemically alter the parent material. Organisms also affect soil building by burrowing into and mixing the soil, by releasing nutrients, and by their decomposition.

Topsoil contains a mixture of humus and inorganic material, both of which supply soil nutrients. Soil fertility is determined by the inorganic matter, organic matter, water, and air spaces in the soil. The mineral portion of the soil consists of various mixtures of sand, silt, and clay particles.

A soil profile consists of the A horizon, which is rich in organic matter; the B horizon, which accumulates materials leached from the A horizon; and the C horizon, which consists of slightly altered parent material. Forest soils typically have a shallow A horizon and a deep, nutrient-rich B horizon with much root development. Grassland soils usually have a thick A horizon and very few nutrients in the thin B horizon. Therefore, most of the roots of the grasses are in the A horizon.

Soil erosion is the removal and transportation of soil by water or wind. Proper use of such conservation practices as contour farming, strip farming, terracing, waterways, windbreaks, and conservation tillage can reduce soil erosion. Misuse reduces the soil's fertility and causes air- and water-quality problems. Land unsuitable for crops may be used for grazing, lumber, wildlife habitats, or recreation.

REVIEW QUESTIONS

1. How are soil and land different?
2. Name the five major components of soil.
3. Describe the process of soil formation.
4. Name five physical and chemical processes that break parent material into smaller pieces.
5. In addition to fertility, what other characteristics determine the usefulness of soil?
6. How does soil particle size affect texture and drainage?
7. Describe a soil profile.
8. Define erosion.
9. Describe three soil conservation practices that help to reduce soil erosion.
10. Besides cropland, what are other possible uses of soil?

KEY TERMS

chemical weathering 242	humus 242	parent material 242	strip farming 252
contour farming 251	land 242	soil 242	terrace 252
erosion 246	leaching 245	soil profile 245	waterways 252
friability 243	loam 243	soil structure 243	weathering 242
horizon 245	mechanical weathering 242	soil texture 243	windbreak 253

RELATED INTERNET SITES

International Erosion Control, a global resource for the preservation and control of erosion.
   http://ieca.org/
Hydrodynamic and ocean engineering for researching erosion, shore protection, and coastal pollution.
   http://www.dl.stevens-tech.edu/
Information network focusing on soil and its conservation.
   http://www.vetiver.com/
USDA Natural Resources Conservation Service, working to protect natural resources, including soil, water, air, and animals.
   http://www.ncg.nrcs.usda.gov/

Searchable index for environmental science research materials. Lots of links.
   http://www.missouri.edu/~c465308/image/research.html
Soil Characteristics Laboratory, offers a description of processing.
   http://silva.snr.missouri.edu/~rficklin/labpage.html
National Society of Consulting Soil Scientists page.
   http://www.wolfe.net/~psmall/nscss.html
On-line journal for soil science.
   http://www.hintze-online.com/sos/

chapter 14

Agricultural Methods and Pest Management

Objectives

After reading this chapter, you should be able to:

- Explain how the invention of new farm machinery encouraged monoculture farming.
- List the advantages and disadvantages of monoculture farming.
- Explain why chemical fertilizers are used.
- Understand how fertilizers alter soil characteristics.
- Explain why modern agriculture makes extensive use of pesticides.
- Differentiate between hard pesticides and soft pesticides.
- Explain how chemicals can be used to delay or accelerate the harvesting of a crop.
- List four problems associated with pesticide use.
- Define biological amplification.
- Define organic farming.
- Explain why integrated pest management depends upon a complete knowledge of the pest's life history.

Chapter Outline

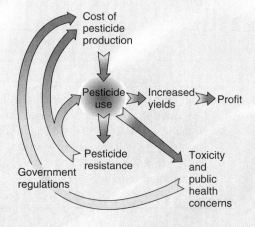

Differing Agricultural Methods

Around the world, people who grow food use a variety of agricultural methods. We will examine three of these.

In parts of the world with poor soil and low populations, a form of agriculture known as "slash and burn" is practiced. The forest in a small area is felled and the trees and other vegetation on the site are burned. (See figure 14.1.) This releases nutrients that are tied up in the biomass and allows a crop or two to be raised before the soil is exhausted. Once the soil is no longer suitable for raising crops, the site is abandoned. The surrounding forest quickly recolonizes the area, which returns over time to the original forest through the process of succession. This method is particularly useful on thin tropical soils and on steep slopes. The small size of the opening in the forest and its temporary existence prevent widespread damage to the soil, and erosion is minimized. The gardens that are planted are typically a mixture of plants, rather than rows of single species.

The traditional practices of the people who engage in slash-and-burn agriculture have been developed over hundreds of years and often are more effective than other methods of gardening. For example, mixing plants may have a beneficial effect, since shade-requiring species may be helped by taller plants, or nitrogen-fixing legumes may provide needed nutrients for species that have a nitrogen requirement. In addition, mixing species may reduce insect pest problems because some plants produce molecules that are natural insect repellants. The small, isolated, temporary nature of the gardens also reduces the likelihood of insect plagues. This form of agriculture is limited to a very small number of people in specialized habitats and is not the most efficient way to farm all kinds of habitats. Slash-and-burn farming also requires a long recovery time before the forest can be cleared for another cycle of agriculture.

In other parts of the world, extensive areas are farmed with a great deal of manual labor. Two situations favor this kind of farming: (1) when the growing site does

Figure 14.1

Slash-and-Burn Agriculture In many areas of the world where the soils are poor, crops can be raised if only small areas of the ecosystem are disturbed. The burning of vegetation releases nutrients that can be used by crops for one or two years before the soil is exhausted. The return of the natural vegetation prevents erosion and repairs the damage done by temporary agricultural use.

not allow for mechanization, and (2) when the kind of crop does not allow it. Soils or terrain that require that fields be small discourage mechanization, since large tractors and other machines cannot be used efficiently on small, oddly shaped fields. Many mountainous areas of the world fit into this category. In addition, some crops require such careful handling in planting, weeding, or harvesting that large amounts of hand labor are required. The planting of paddy rice and the harvesting of many fruits and vegetables are examples.

However, the primary reason for labor-intensive farming is economic. Many densely populated countries have numerous small farms that can be effectively managed with human labor, supplemented by that of draft animals and a few small gasoline-powered engines. (See figure 14.2.) In addition, in the less-developed regions of the world, the cost of labor is low, which encourages the use of hand labor rather than mechanization to do planting, weeding, and other activities. Mechanization requires large tracts of land that could only be accumulated by

the expenditure of large amounts of money or the development of larger cooperative farms from many small units. Even if social and political obstacles to such large land holdings could be overcome, there is still the problem of obtaining the necessary capital to purchase the machines. Large parts of the developing world fit into this category, including much of Africa, many areas in Central and South America, and many areas in Asia.

Mechanized agriculture is typical of North America, much of Europe, the republics of the former USSR, South America, and other parts of the world where money and land are available to support this form of agriculture. In large measure, machines and fossil-fuel energy replace the energy formerly supplied by human and animal muscles. Mechanization requires large expanses of fairly level land for the machines to operate efficiently. In addition, large expanses of land must be planted in the same crop for efficient planting, cultivating, and harvesting. Planting large tracts of land in the same crop is known as **monoculture.**

Figure 14.2

Labor-Intensive Agriculture   In many of the less-developed countries of the world the extensive use of hand labor allows for impressive rates of production with a minimal input of fossil fuels and fertilizers. This kind of agriculture is also necessary in areas that have only small patches of land suitable for farming.

(See figure 14.3.) Small sections of land with many kinds of crops require many changes of farm machinery, which takes time. Also, many crops interspersed with one another reduces the efficiency of farming operations because farmers must skip parts of the field, which increases travel time and uses expensive fuel.

Even though monoculture is an efficient method of producing food, it is not without serious drawbacks. When large farm machines are used to plow, plant, and harvest vast expanses of land at the same time, large tracts of land are left uncovered by vegetation, and soil erosion increases. (See chapter 13.) Mechanized farming also removes much of the organic matter each year. When the crop is harvested, a major portion of the plant typically is removed for food or some other purpose. This means that little organic matter is returned to the soil unless the farmer specifically plants a crop that is later plowed under to increase the soil's organic content.

To ensure that a crop can be planted, tended, and harvested efficiently, farmers rely on hybrid seed that provides uniform plants with characteristics suitable for mechanized farming. These hybrid plants have little genetic variety. When all the farmers in an area plant the same hyrids, pest control becomes a serious pest problem. If diseases or pests begin to spread, the magnitude of the problem becomes devastating because all the plants have the same characteristics and, thus, are susceptible to the same diseases. If genetically diverse crops are planted, or crops are rotated from year to year, this problem is not as great.

Because farm equipment is expensive, farmers tend to specialize in a few crops. This means that the same crop may be planted in the same field several years in a row. This lack of crop rotation may deplete certain essential soil nutrients, thereby requiring special attention to soil chemistry. In addition, planting the same crop repeatedly encourages the growth of insect and fungus pest populations because they have a huge food supply at their disposal. This requires the frequent use of insecticides and fungicides or some other methods of pest control.

Even though problems are associated with mechanized, monoculture agriculture, it has greatly increased the amount of food available to the world over the past one hundred years. Yields per hectare of land being farmed have increased over much of the world, particularly in the developed world, which includes the United States. (See figure 14.4.) This increase has come about because of improved varieties of crops, better farming methods, the use of agricultural chemicals, more efficent machines, and the use of energy-intensive as opposed to labor-intensive technology.

Agricultural scientists are also working in the area of genetic engineering and other forms of biotechnology. This work is often referred to as the green revolution. Over the next 20 to 30 years, scientists hope to breed high-yield plant strains that are more resistant to insects and disease, thrive on less fertilizer, make their own nitrogen fertilizer, do well in slightly salty soils, withstand drought, and use solar energy more efficiently during photosynthesis. To date, however, several factors have limited the success of these undertakings. First, without large amounts of fertilizer and water, most green revolution crop varieties produce yields no higher and often lower than those from traditional strains. Second, the cost of genetically engineered crop strains is too high for most of the world's subsistence farmers. There are also ethical issues regarding genetic engineering that farmers, consumers, and scientists are only beginning to debate.

Energy Versus Labor

Mechanized agriculture has substituted the energy stored in petroleum products for the labor of humans. For example, in

Figure 14.3

Monoculture This wheat field is an example of monoculture, a kind of agriculture that is highly mechanized and requires large fields for the efficient use of machinery. In this kind of agriculture, machines and fossil-fuel energy have replaced the energy of humans and draft animals.

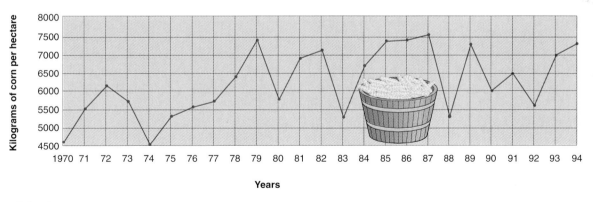

Figure 14.4

Increased Yields Resulting from Modern Technology The increased yields in the United States and many other parts of the world are the result of a combination of factors, including better genetic qualities in the seed, improved agricultural methods, the application of fertilizers and pesticides, and more efficient machinery.

Source: Data from Department of Agriculture, Agriculture Statistics, 1994 and previous years.

the United States in 1913, it required 135 hours of labor to produce 2,500 kilograms of corn. In 1980, it required about 15 hours of labor to produce 2,500 kilograms of corn. The energy supplied by petroleum products replaced the equivalent of 120 hours of labor. Energy is needed for tilling, planting, harvesting, and pumping irrigation water. Much of the energy is in the form of manufactured fertilizers, herbicides, fungicides, and insecticides that encourage crop growth while inhibiting weeds and reducing the impact of pests that feed on the crop or cause disease. For example, about 5 metric tons of fossil fuel are required to produce about 1 metric ton of fertilizer. Since the developed world is dependent on oil to provide energy to manufacture pesticides to support its agriculture, any change in the availability or cost of oil will have a major impact on the world's ability to feed itself.

The Impact of Fertilizer

Table 14.1 shows that, in some respects, fertilizer can be used to replace human labor. The United States, with only 3 percent of its work force in agriculture, uses nearly 30 percent more fertilizer than Brazil and over twice as much as India and has a cereal grain yield more than twice that of Brazil or India. The yields are similar for India and Brazil; however, Brazil uses less labor to accomplish the same yield. In this case, fertilizer appears to directly substitute for labor.

Approximately 25 percent of the world's crop yield is estimated to be directly attributed to the use of chemical fertilizers. Thus, if the world stopped using chemical fertilizers, food production would decline by 25 percent. At present, the world use of fertilizer is not increasing. Since fertilizer production relies on oil, the price and availability of chemical fertilizers is strongly influenced by the price of oil on the world market. If the price of oil increases, the price of fertilizer goes up, as does the cost of food. This is felt most acutely in parts of the world where money is in short supply, since the farmers are unable to buy fertilizer, and crop yields fall accordingly.

Table 14.1

Fertilizer and Food Production

Country	Percentage of the work force in agriculture (1991)	Fertilizer used (kilograms/hectare) (1993)	Yield in kilograms/hectare (cereal grains) (1994)
United States	3	108	5,092
Brazil	25*	85	2,256
India	70*	46	2,062

*Estimates based on various data.

Source: Data from World Resources Institute, 1994–95.

Chemical fertilizers are valuable because they replace the soil nutrients removed by plants. Some plant nutrients, such as carbon, hydrogen, and oxygen, are easily replaced by carbon dioxide from the air and water from the soil, but others are less easily replaced. The three primary soil nutrients often in short supply are potassium, phosphorus, and nitrogen compounds. They are often referred to as **macronutrients** and are the common ingredient of chemical fertilizers. Their replacement is important because, when the crop is harvested, the chemical elements that are a part of the crop are removed from the field. Since many of those elements originated from the soil, they need to be replaced if another crop is to be grown. Certain other elements are necessary in extremely small amounts and are known as **micronutrients.** Examples are boron, zinc, and manganese. As an example of the difference between macronutrients and micronutrients, harvesting a metric ton of potatoes removes 10 kilograms of nitrogen (a macronutrient) but only 13 grams of boron (a micronutrient). When the same crop is grown repeatedly in the same field, certain micronutrients may be depleted, resulting in reduced yields. These necessary elements can be returned to the soil in sufficient amounts by incorporating them into the fertilizer the farmer applies to the field.

Although chemical fertilizers replace inorganic nutrients, they do not replace soil organic matter. Organic material is important because its decay produces humus. Humus modifies the structure of the soil, preventing compaction and maintaining pore space, which allows water and air to move to the roots. Organic matter also maintains proper soil chemistry because it tends to lower the pH of the soil, which helps to release certain soil nutrients for use by plants. Soil bacteria and other organisms use organic matter as a source of energy. Since these organisms serve as important links in the carbon and nitrogen cycles, the presence of organic matter is important to their function. Thus, total dependency upon chemical fertilizers can change the physical, chemical, and biotic properties of the soil.

As water moves through the soil, it dissolves soil nutrients and carries them into streams and lakes, where they may encourage the growth of unwanted plants and algae. This is particularly true when fertilizers are applied at the wrong time of the year, just before a heavy rain, or in such large amounts that the plants cannot efficiently remove them from the soil before they are lost. These ideas are covered in greater detail in chapter 15 on water pollution.

Pesticides

In addition to chemical fertilizers, mechanized monoculture requires large amounts of other agricultural chemicals, such as pesticides, growth regulators, and preservatives. (See figure 14.5.) These chemicals have specific scientific names but are usually categorized into broad groups based on their effects. A **pesticide** is any chemical used to kill or control

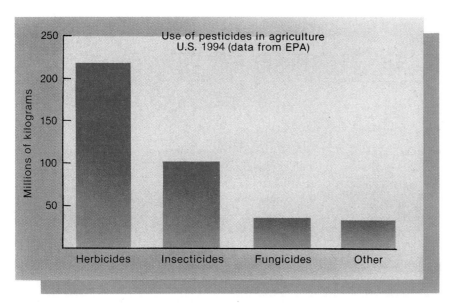

Figure 14.5

Pesticide Use in the United States   The most widely used pesticides are herbicides to control weeds and facilitate harvest of crops. Insecticides are also widely used.

Source: Data from Environmental Protection Agency.

populations of unwanted fungi, animals, or plants, often called **pests.** The term *pest* is not scientific but refers to any organism that is unwanted. Insects that feed on crops are pests, while others, like bees, are beneficial for pollinating plants. Unwanted plants are generally referred to as **weeds.**

Pesticides can be subdivided into several categories based on the kinds of organisms they are used to control. **Insecticides** are used to control insect populations by killing them. Unwanted fungal pests that can weaken plants or destroy fruits are controlled by **fungicides.** Mice and rats are killed by **rodenticides,** and plant pests are controlled by **herbicides.** A perfect pesticide is one that kills or inhibits the growth of the specific pest organism causing a problem. The pest is often referred to as the **target organism.** However, most pesticides are not very specific and kill many **nontarget organisms** as well. For example, most insecticides kill both beneficial and pest species, rodenticides kill other animals as well as rodents, and most herbicides kill a variety of plants, both pest and nonpests. Pesticides do not just kill pests but can kill a large variety of living things, including humans. Because of this, these chemicals might be

more appropriately called **biocides,** since they kill many kinds of living things.

Insecticides

For centuries, insects consumed a large proportion of the crops produced by farmers. In small garden plots, insects can be controlled by manually removing them and killing them. However, in large fields, this is not practical, so people have sought other ways to control pest insects. In addition, many insects harm humans because they spread diseases, such as sleeping sickness, bubonic plague, and malaria. Mosquitoes are known to carry over 30 diseases harmful to humans. In 1965, mosquitoes are estimated to be responsible for spreading diseases like malaria and sleeping sickness to 100 million people. The discovery of chemicals that could kill insects was a major advance in the control of disease and the protection of crops.

Nearly 3,000 years ago, the Greek poet Homer mentioned the use of sulfur to control insects. For centuries, it was known that natural plant products could repell or kill insect pests. Plants with insect-repelling abilities were interplanted with crops to help control the pests. Nicotine from tobacco, rotenone from

tropical legumes, and pyrethrum from chyrsanthemums were extracted and used to control insects. In fact, these compounds are still used today. However, because plant products are difficult to apply and have shortlived effects, other compounds were sought. In 1867, the first synthetic inorganic insecticide, Paris green, was formulated. It was a mixture of acetate and arsenide of copper and was used to control Colorado potato beetles.

The first synthetic organic insecticide to be used was DDT. It was originally thought to be the perfect insecticide. It was long-lasting, relatively harmless to humans, and very deadly to insects. During the first 10 years of its use (1942–1952), DDT is estimated to have saved 5 million lives, primarily because of its use to control disease-carrying mosquitoes. However, after a period of use, many mosquitoes and other insects became tolerant of DDT. The early success of DDT and its loss of effectiveness promoted a search for new synthetic organic compounds that could replace it. Since then, over 60,000 different compounds have been synthesized that have potential as insecticides. However, most of these have never been put into production because of cost, human health effects, or other drawbacks that make them unusable. Several categories of these compounds have been developed. Three that are currently used are chlorinated hydrocarbons, organophosphates, and carbamates.

Chlorinated Hydrocarbons

Chlorinated hydrocarbons are a group of pesticides of complex, stable structure that contain carbon, hydrogen, and chlorine, DDT (dichlorodiphenyltrichloroethane) was the first such pesticide manufactured, but several others have been developed. The chemical structure of DDT is shown in figure 14.6. Other chlorinated hydrocarbons are chlordane, aldrin, heptachlor, dieldrin, and endrin. It is not fully understood how these compounds work, but they are believed to affect the nervous systems of insects, resulting in their death.

One of the major characteristics of these pesticides is that they are very stable chemical compounds. This is both an advantage and a disadvantage. Since

Regulation of Pesticides

The Environmental Protection Agency (EPA) is charged by Congress to protect the nation's land, air, and water systems. Under a mandate of national environmental laws focused on air and water quality, solid wast management, and the control of toxic substances, pesticides, noise, and radiation, the agency strives to formulate and implement actions that lead to a compatible balance between human activities and the ability of natural systems to support and nurture life.

To fulfill this mandate, the EPA is charged with the enforcement of various federal laws and acts. One such act governs the registration of pesticides. Section 3 (c) (1) of the Federal Insecticide, Fungicide, and Rodenticide Act (FIFRA) states:

Procedure for Registration

---

1. *Statement required. Each applicant for registration of a pesticide shall file with the Administrator a statement which includes—*

   A. *the name and address of the applicant and of any other person whose name will appear on the labeling;*
   B. *the name of the pesticide;*
   C. *a complete copy of the labeling of the pesticide, a statement of all claims to be made for it, and any directions for its use;*
   D. *except as otherwise provided in subsection (c) (2) (D) of this section, if requested by the Administrator, a full description of the tests made and the results thereof upon which the claims are based, or alternately, a citation to data that appears in the public literature or that previously had been submitted to the Administrator and that the Administrator may consider in accordance with the following provisions;*
   E. *the complete formula of the pesticide; and*
   F. *a request that the pesticide be classified for general use, or restricted use, or for both.*

If a corporation wants to manufacture and market a pesticide within the United States, the pesticide must not adversely affect the environment. There is no specific format to test a product's effects on the environment, but some of the tests conducted include the following:

1. **Degradation**—A determination of the physical and chemical methods of decay and an identification of the decay products.
2. **Metabolism**—A determination of the organisms that act on the pesticide and the metabolic products released.

3. **Mobility**—A determination of where the molecules are likely to go and what route they will follow in the ecosystem.
4. **Accumulation**—Do the pesticide molecules accumulate in living tissue? If so, in what form and in what quantities?
5. **Hazard**—Evaluation of the possible hazard to plants, microorganisms, wildlife, and humans, whether target or nontarget organisms.

After receiving the information submitted in the procedure for registration, the Administrator of EPA may register the pesticide for use under the provisions of Section 3 (c) (5) of FIFRA:

Approval of Registration

The Administrator registers a pesticide if he or she determines that, when considered with any restrictions imposed under subsection (d),

1. *its composition is such as to warrant the proposed claims for it;*
2. *its labeling and other material required to be submitted comply with the requirements of this Act;*
3. *it will perform its intended function without unreasonable adverse effects on the environment; and*
4. *when used in accordance with widespread and commonly recognized practice, it will not generally cause unreasonable adverse effects on the environment.*

The Administrator also publishes all applications and supporting data so that various governmental agencies and the public have an opportunity to comment.

The FIFRA is designed to protect the environment from being damaged by the use of pesticides. The law places the burden upon the manufacturer to prove that a pesticide is safe. One source estimates that eight to twelve years and $50 million are needed to bring a major new pesticide from discovery to first registration. These figures do not include the capital cost of constructing a facility to manufacture the pesticide.

environmental close-up

A New Generation of Insecticides

The perfect insecticide would harm only target insect species, not be toxic to humans, not be persistent, and would break down into harmless materials. Most currently used insecticides were developed by screening a wide variety of compounds and modifying compounds that showed insect-killing properties. Most are highly toxic to many kinds of organisms and must be used with great care. However, a new generation of insecticides is being developed that starts with knowledge of insect biology and develops a compound that interferes with some essential aspect of the life-cycle. Some mimic normal hormones and prevent insects from maturing into adults. Therefore, the insects do not reproduce and the populations are greatly reduced.

Other insecticides attack specific chemical processes in cells and cause death. Nicotine is a normal product of some plants like tobacco, which is toxic to insects. It binds to specific receptors on nerve cells, causing them to fire uncontrollably. However, it is not stable in sunlight. By tinkering with the chemical structure of nicotine, a modified molecule, called imidacloprid, was developed. It is more stable, not toxic to mammals, because they do not have as many of the receptors as insects, and very effective against sucking insects.

Another new insecticide works by turning off the energy-producing processes of mitochondria in cells. All cells of higher organisms have mitochondria and would be injured by this insecticide, but a nontoxic form of the molecule was produced that is converted to a toxic form in the bodies of insects but is not converted in mammals. It is highly toxic to birds and some aquatic organisms, however, which would require strict control of its use.

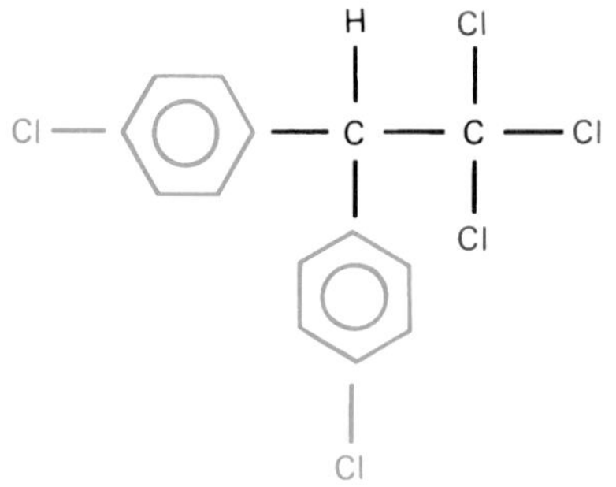

Figure 14.6

The Chemical Structure of DDT This diagram shows the arrangement of the atoms in a molecule of DDT. The two chlorophenyl portions of the molecule are shown in blue. The other chlorines are shown in red, and the remainder of the ethane molecule appears black.

they are long-lasting, they are called **persistent pesticides,** or **hard pesticides.** They can be applied once and be effective for a long time. However, since they do not break down easily, they tend to accumulate in the soil and in the bodies of animals in the food chain. Thus, they affect many nontarget organisms, not just the original target insects. In temperate regions of the world, DDT has a half-life

(the amount of time required for half of the chemical to decompose) of 10 to 15 years. This means that if 1 metric ton of DDT were sprayed over an area, 500 kilograms would still be present in the area 10 to 15 years later; 30 years from the date of application, 250 kilograms would still be present. The half-life of DDT varies depending on soil type, temperature, the kinds of soil organisms present, and other factors. In tropical parts of the world, the half-life may be as short as six months. An additional complication is that hard pesticides may break down into products that are still harmful.

Because of their negative effects, most of the chlorinated hydrocarbons are not used. DDT, aldrin, dieldrin, toxaphene, chlordane, and heptachlor have been banned in the United States and some other countries. However, many developing countries still use chlorinated hydrocarbons for insect control to protect crops and public health.

Organophosphates and Carbamates

Because of the problems associated with hard pesticides, scientists have developed a new generation of **soft pesticides** that

decompose to harmless products in a few hours or days. Like other insecticides, these are not species specific; they kill beneficial insects as well as harmful ones. Although the short half-life prevents the accummulation of toxic material in the environment, it is a disadvantage for farmers, since more frequent applications are required to control pests. This requires more labor and fuel and, therefore, is more expensive.

Both **organophosphates** and **carbamates** work by interfering with the ability of the nervous system to conduct impulses normally. Under normal conditions, a nerve impulse is conducted from one nerve cell to another by means of a chemical known as acetylcholine. When this chemical is produced at the end of one nerve cell, it causes an impulse to be passed to the next cell, thereby transferring the nerve message. As soon as this transfer is completed, an enzyme known as cholinesterase destroys acetylcholine, so the second nerve cell in the chain is stimulated for only a short time. Organophosphates and carbamates attach to the enzyme cholinesterase, preventing it from destroying acetylcholine. This results in nerve cells being continuously stimulated, causing uncontrolled spasms of

e n v i r o n m e n t a l
close-up

Politics and the Control of Ethylene Dibromide (EDB)

Since the 1940s, ethylene dibromide (EDB) has been used as a fumigant to protect stored grain from insect pests. More recently, it has been used to fumigate milling machinery and kill nematodes in the soil. It has also been used as a fumigant in fruit shipments to prevent the transportation of insect pests from one part of the world to another.

It was originally thought that EDB dissipated and was not a health hazard. However, residues were found in grain and fruits, and were also reported in drinking water from wells in areas where EDB was used as a soil fumigant. In the 1970s, it was shown that EDB causes cancer in rats and mice at levels of twenty parts per million.

In 1977, the U.S. Environmental Protection Agency first sought to control use of EDB, but manufacturers and users effectively fought its prohibition until September 1984, when the EPA succeeded in banning the use of EDB as a soil fumigant.

When the EPA banned the use of EDB, it was required by law to purchase the remaining stocks of the material from the manufacturers and to dispose of them at its own expense. (This feature of the law was lobbied for very strongly by pesticide manufacturers.) In March 1988, the EPA incinerated the remaining approximately 1.3 million liters of the EDB it had acquired when it banned the substance. The EPA had already paid $2.5 million for these stocks and disposing of them properly cost several million more. The requirement that the EPA acquire and dispose of banned pesticides has had a significant impact on the agency's ability to function.

---

nervous activity and uncoordination that result in death.

Although these soft pesticides are less persistent in the environment than are chlorinated hydrocarbons, they are generally much more toxic to humans and other vertebrates. Persons who use such pesticides must use special equipment and should receive special training because improper use can result in death. Since organophosphates bind more strongly to cholinesterase than do carbamates, they are considered more dangerous and for many applications have been replaced by carbamates.

Common organophosphates are malathion, parathion, and diazinon. Malathion is widely used for such projects as mosquito control, but parathion is a restricted organophosphate because of its high toxicity to humans. Diazinon is widely used in gardens. Sevin, aldicarb, and propoxur are examples of carbamates. Table 14.2 lists a variety of insecticides, what they are used for, and how they control the pest.

Herbicides

Herbicides are another major class of chemical control agents. In fact, about 60 percent of the pesticides used in U.S. agriculture are herbicides. (See figure 14.5.) They are widely used to control unwanted vegetation along power-line rights-of-way, railroad rights-of-way, and highways, as well as on lawns and cropland, where they are commonly referred to as weed killers. Weeds are plants growing in the wrong place. Bluegrass in a lawn is desirable, but bluegrass growing in a cornfield is a weed, so many herbicides are designed to kill beneficial as well as harmful plants.

Weed control is extremely important for agriculture since weeds take nutrients and water from the soil, making them unavailable to the crop species. In addition, weeds may shade the crop species and prevent it from getting the sunlight it needs for rapid growth. At harvest time, weeds reduce the efficiency of harvesting machines. Also, weeds generally must be sorted from the crop before it can be sold, which adds to the time and expense of harvesting.

Traditionally, farmers have expended much energy trying to control weeds. Initially, weeds were eliminated with manual labor and the hoe. Tilling the soil also helps to control weeds. Once the crop is planted, row crops like corn or soybeans may be cultivated to remove weeds from between the rows. All of these activities are expensive in time and fuel. Selective use of herbicides can have a tremendous impact on a farmer's profits.

Many of the recently developed herbicides can be very selective if used appropriately. Some are used to kill weed seeds in the soil before the crop is planted, while others are used after the weeds and the crop begin to grow. In some cases, a mixture of herbicides can be used to control weed species. Figure 14.7 shows the effects of using an herbicide that kills grasses but not other kinds of plants.

Several major types of herbicides are in current use. One type is synthetic plant-growth regulators that mimic natural-growth regulators known as **auxins.** Two common regulators are 2, 4-dichlorophenoxyacetic acid (2, 4-D) and 2, 4, 5-trichlorophenoxyacetic acid (2, 4, 5-T). When applied to broadleaf plants, these chemicals cause the metabolism of the plants to exceed their food-producing potential. The plants grow so rapidly that they run out of stored food and die. These kinds of herbicides can be used to control broadleaf weeds in grain crops, weed

Table 14.2

Various Kinds of Insecticides and Their Uses

Type of insecticide	Examples	Target organisms	Action on target organism
Chlorinated hydrocarbons	Aldrin Chlordane DDT Dieldrin Endosulfan Lindane	Various insect species	Interferes with nervous system by changing sodium and potassium ion balance

Essentially all chlorinated hydrocarbons have been banned or discontinued in the United States and many other countries because of their persistence and ability to accumulate in food chains. Lindane and Endosulfan have limited use in the United States, but others are used in many less-developed countries to control disease-spreading insects.

Organophosphates General use	Azinphosmethyl (Guthion®) Chlorpyrifos (Dursban®, Lorsban®) Diazinon (Spectracide®) Malathion Dimethoate (Cygon®) Parathion	Various insects and mite species	Interferes with cholinesterase enzyme of nervous system, causing uncontrollable muscle twitching
Flea collars and pest strips	Dichlorvos (Vapona®)	Fleas and ticks, household and barn insects	Vapors kill
Plant systemics	Demetron (Systox®) Dicrotophos (Bidrin®) Disulfoton (Di-Syston®) Mevinphos (Phosdrin®)	Insects that suck plant juices	Incorporates into plants; taken up when juices sucked from plant
Animal systemics	Crufomate (Ruelene©) Cythioate (Cyflee®)	Cattle grubs, fleas, ticks on dogs and cats	Taken up by pest from body tissues of host
Carbamates General use	Carbaryl (Sevin®) Propoxur (Baygon®) Methomyl (Lannate®, Nudrin®)	Many insects	Cholinesterase poisons
Plant systemics	Methomyl (Lannate®, Nudrin®) Aldicarb (Temik®) Carbonfuran (Furadan®)	Plant-sucking insects	Cholinesterase poisons
Formamidines	Formetanate (Carzol®) Amitraz (Taktic®, Mitac®)	Insect eggs, mites, ticks, and young caterpillars	Nervous system poisons
Organosulfurs	Aramite® Genite®	Mites	Nervous system poisons
Plant products	Nicotine Rotenone Pyrethins	Various insects	Nerve poison inhibits ATP paralysis
Synthetic pyrethrins	Fenvalarate (Pydrin®, Tribute®) Permethrin (Ambush®, Dragnet®, Pramex®, Pounch®, Torpedo®) Bifenthrin (Capture®, Talstar®)	Many insect pests	Interferes with nervous system by changing sodium and potassium balance

Source: Data from George W. Ware, The Pesticide Book, *3rd edition, 1989, Thomson Publications.*

Figure 14.7

The Effect of Herbicides   The grasses in this photograph have been treated with herbicides. The soybeans are unaffected and grow better without competition from grasses.

Figure 14.8

Herbicide Use to Maintain Rights-of-Way   Power-line rights-of-way are commonly maintained by herbicides that kill the woody vegetation, which might grow so tall that it interferes with the power lines.

species in coniferous forests, and brush along rights-of-way. (See figure 14.8.)

A second kind of herbicide disrupts photosynthetic activity of plants, causing their death. Others inhibit enzymes, precipitate proteins, stop cell division, or destroy cells directly. Depending on the concentration of the herbicide used, some

are toxic to all plants, while others are very selective as to which species of plants they affect. One such herbicide is diuron. In proper concentrations and when applied at the appropriate time, it can be used to control annual grasses and broadleaf weeds in over 20 different crops. However, at higher concentrations, it kills all vegetation in an area. Fenuron is a herbicide that kills woody plants. In low concentrations, it is used to control woody weed plants in cropland. In high concentrations, it is used on noncroplands, such as power-line rights-of-way. Table 14.3 lists several kinds of herbicides, their target species, and how they work.

Other Agricultural Chemicals

In addition to herbicides, other agrochemicals are used for special applications. For example, a synthetic auxin sprayed on cotton plants prior to harvest causes the leaves to drop off, which facilitates the harvesting process by reducing clogging of the mechanical cotton picker.

NAA (naphthaleneacetic acid) is used by fruit growers to prevent immature apples from dropping from the trees and being damaged. This chemical can keep the apples on the trees for up to 10 extra days, which allows for a longer harvest period and fewer lost apples.

Under other conditions, it may be valuable to get fruit to fall more easily. Cherry growers use ethephon to promote loosening of the fruit so that the cherries will fall more easily from the tree when shaken by the mechanical harvester. This method lowers the cost of harvesting the fruit. (See figure 14.9.)

Fungicides and Rodenticides

Fungus pests can be divided into two categories. Some are natural decomposers of organic material, but when the organic material being destroyed happens to be a crop or other product useful to humans, the fungus is considered a pest. Other fungi are parasites on crop plants and weaken or kill the plants, thereby reducing the yield. Fungicides are used as fumigants (gases) to protect agricultural products from spoilage, as sprays and dusts to

Table 14.3

Various Kinds of Herbicides and Their Uses

Type of herbicide	Examples	Target organisms	Action on target organisms
Acetanilides	Metolachlor (Dual®) Alachlor (Lasso®)	Preemergent for many kinds of weeds	Meristematic growth inhibitor
Petroleum oils	Motor oil, gasoline, etc.	All plants	Disrupts cell membranes
Arsenic compounds	Arsonic acid, MSMA, arsinic acid, DSMA, cacodylic acid	All plants	Inhibits enzymes
Phenoxy herbicides	2,4-D; 2,4,5-T; 2,4-DB; MCPA, silvex	Broadleaf weeds	Stimulates uncontrolled growth
Substituted amides	Diphenamid (Dymid®, Enide® Propanil (Stampede®)	Preemergent for all plants Postemergent for many weeds	Inhibits seedlings Inhibits photosynthesis Inhibits growth
Nitroanilines	Trifluralin (Treflan®) Benefin (Balfin®, Quilan®)	Preemergent for all plants	Inhibits root and shoot growth
Substituted ureas	Diuron (Karmex®, Krovar®) Linuron (Lorox®)	Preemergent	Inhibits photosynthesis
Carbamates	Propham (Chem Hoe®) Chlorpropham (Furloe®)	All plants	Stops cell division
Pyridinoxy and picolinic acids	Picloram (Tordon®) Triclopyr (Garlon®)	Broadleaf weeds	Stimulates growth
Aliphatic acid	TCA, Dalapon	Quack grass, Bermuda grass	Precipitates proteins
Arylaliphatic acids	Dicamba (Banvel®) Chloramben (Amiben®)	Preemergent against germinating seeds and seedlings	Inhibits growth
Phenol derivatives	DNOC, dinoseb Petachlorophenol	All plants Defoliant	Prevents ATP formation Destroys cells
Bipyriodyliums	Diquat, paraquat	All plants	Destroys cell membranes
Benzonitriles	Dichlobenil (Casoron®) Bormoxynil (Brominal®, Buctril®)	Contact against broadleaf weeds	Poisons respiration
Benzothiadiazoles	Bentazon (Basagran®)	Postemergent against broadleaf weeds	Inhibits photosynthesis
Diphenyl ethers	Nitrofen Acifluorfen (Blazer®)	Contact against many weeds	Inhibits respiration
Imidazoles	Imazapyr (Arsenal®) Imazaquin (Scepter®, Image®)	Broadleaf weeds	Inhibits growth
Oxyphenoxy acid esters	Fluazifop-butyl (Fusilade®) Fenoxaprop-ethyl (Whip®)	Grasses	Inhibits growth
Phosphono amino acids	Glyphosate (Roundup®) Glufosinate (Ignite®)	Grasses	Inhibits growth
Phthalic acids	Chlorothal (Dacthal®) Endothall (Aquathol®, Endothal®)	Many weeds	Inhibits growth
Pyridazinones and pyridinones	Amitrole Pyrazon (Pyramin®)	All plants	Prevents chlorophyll formation
Sulfonylureas	Chlorsulfuron (Glean®) Chlorimuron-ethyl (Classic®)	Broadleaf weeds	Inhibits growth
Thiocarbamates	EPTC (Eptam®) Pebulate (Tillam®)	Preemergent against grasses and some broadleaf weeds	Inhibits seedling growth
Triazines	Atrazine (Aatrex®) Simazine (Princep®)	All plants	Inhibits photosynthesis
Uracils	Lenacil (Venzar®) Bromacil (Hyvar®)	All plants	Inhibits photosynthesis

Source: Data from George W. Ware, The Pesticide Book, 3rd edition, 1989, Thomson Publications.

Figure 14.9

Chemical Loosening of Cherries to Allow Mechanical Harvesting By using chemicals and machinery, this farmer can rapidly harvest the cherry crop. This practice reduces the amount of labor required to pick the cherries but requires the application of chemicals to loosen the fruit.

prevent the spread of diseases among plants, and as seed treatments to protect seeds from rotting in the soil before they have a chance to germinate. Methylmercury is often used on seeds to protect them from spoilage prior to germination. However, since methylmercury is extremely toxic to humans, these seeds should never be used for food. To reduce the chance of a mixup, treated seeds are usually dyed a bright color.

Like fungi, rodents are harmful because they destroy food supplies. In addition, they can carry disease and damage corps in the field. In many parts of the world, such as India, the government pays a bounty to people who kill rats because this is an inexpensive way to protect the food supply. Several kinds of rodenticides have been developed to control rodents. One of the most widely used is warfarin, a chemical that causes internal bleeding in animals that consume it. It is usually incorporated into a food substance so that rodents eat warfarin along with the bait. Because it is effective in all mammals, including humans, it must be used with care to prevent nontarget animals from having access to the chemical. As with many kinds of pesticides, some populations of

rodents have become tolerant of warfarin, while others avoid baited areas. In many cases, rodent problems can be minimized by building storage buildings that are rodent proof, rather than relying on rodenticides to control them.

Problems with Pesticide Use

A perfect pesticide would have the following characteristics:

1. It would be inexpensive.
2. It would affect only the target organism.
3. It would have a short half-life.
4. It would break down into harmless materials.

However, the perfect pesticide has not been invented. Many of the more recently developed pesticides have fewer drawbacks than the early hard pesticides, but none are without problems.

Although there has been a trend away from using hard pesticides in the United States, some are still allowed for special purposes, and they are still in

common use in other parts of the world. Because of their stability, these chemicals have become a long-term problem. Hard (persistent) insecticides become attached to small soil particles, which are easily moved by wind or water to any part of the world. Hard pesticides and other pollutants have been discovered in the ice of the poles and are present in detectable amounts in the body tissues of animals, including humans, throughout the world. Thus, chemicals originally sprayed on a cornfield in Iowa or a sugarcane field in Brazil may become a problem in Arctic or Antarctic ecosystems.

Another problem associated with persistence is that pesticides may accumulate in the bodies of animals without killing the animals. If an animal receives small quantities of hard pesticides or other persistent pollutants in its food and is unable to eliminate them, the concentration within the animal increases. This process of accumulating higher and higher amounts of material within the body of an animal is called **bioaccumulation.** Many of the hard pesticides and their breakdown products are fat soluble and build up in the fat of animals. When an affected animal is eaten by a carnivore, these toxins are concentrated, causing disease or death of the carnivores at higher trophic levels, even though lower-trophic-level organisms are not injured. This phenomenon of increasing levels of a substance in the bodies of higher-trophic-level organisms is known as **biological amplification.**

The well-documented case of DDT is an example of how biological amplification occurs. Since DDT is not very soluble in water, it is usually dissolved in an oil or fatty compound, which is then mixed with water before application. This mixture is then sprayed over an area and falls on plants that the insect population uses for food, or it may fall directly on the insect. The insect takes the DDT into its body, where the chemical interferes with normal metabolic activities. If the insect obtains a high enough concentration, it dies. If it takes in small amounts, it digests and breaks down the DDT as it would any other organic chemical compound. Since DDT is soluble in fat or oil, the insect stores DDT or its

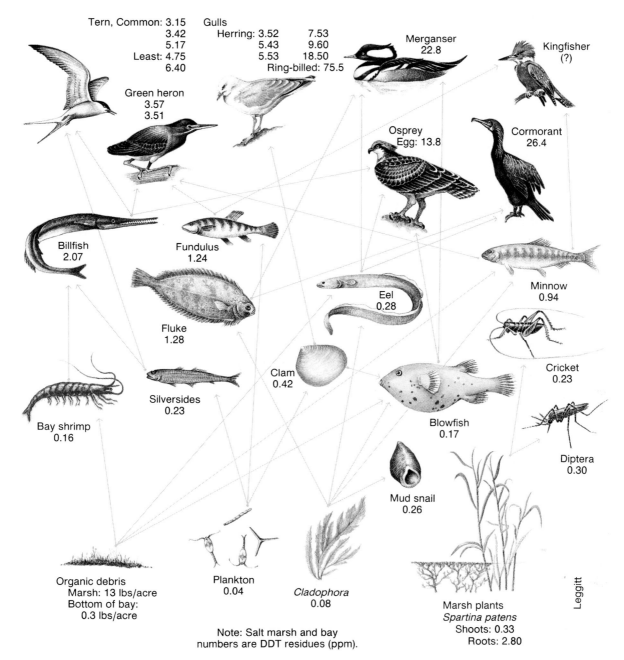

Tern, Common: 3.15
3.42
5.17
Least: 4.75
6.40

Gulls
Herring: 3.52   7.53
5.43   9.60
5.53   18.50
Ring-billed: 75.5

Merganser
22.8

Kingfisher
(?)

Green heron
3.57
3.51

Osprey
Egg: 13.8

Cormorant
26.4

Billfish
2.07

Fundulus
1.24

Minnow
0.94

Eel
0.28

Fluke
1.28

Cricket
0.23

Clam
0.42

Blowfish
0.17

Silversides
0.23

Diptera
0.30

Bay shrimp
0.16

Mud snail
0.26

Organic debris
Marsh: 13 lbs/acre
Bottom of bay:
0.3 lbs/acre

Plankton
0.04

Cladophora
0.08

Marsh plants
Spartina patens
Shoots: 0.33
Roots: 2.80

Leggitt

Note: Salt marsh and bay
numbers are DDT residues (ppm).

Figure 14.10

Biological Amplification   Note how the concentration of DDT increases as it passes through the food chain. At each successive trophic level, the amount of DDT increases because the animals are accumulating the DDT from the bodies of the animals they are eating as food.

breakdown products in the fatty tissues of its body.

If an area has been lightly sprayed with DDT, some insects die, but others will be able to tolerate the low DDT levels. They may contain as much as one part per billion of DDT in their tissues. This is not very much, but it can have a tremendous effect on the animals that feed on the insects.

If an aquatic habitat is sprayed with a small concentration of DDT, or received DDT from runoff, small aquatic organisms may accumulate a concentration that is up to 250 times greater than the concentration of DDT in the surrounding water. These organisms are eaten by shrimp, clams, and small fish, which are, in turn, eaten by larger fish. DDT concentrations of large fish can be as much as

2,000 times the original concentration sprayed on the area. (See figure 14.10.)

What was a very small initial concentration has now become so high that it could be fatal to animals at higher trophic levels. This has been of particular concern for birds, since DDT interferes with the production of eggshells, making them much more fragile. This problem is more common in carnivorous birds because

China's Ravenous Appetite

History has shown that when people rise above subsistence level one of the first things they want is to eat a little better. They tend to leave the basic grains to the animals and eat the animals themselves or their products in the form of meat, fish, eggs, and butter. They eat more vegetables and use more edible oils. In short, they move a bit higher up the food chain.

A sizable percentage of China's 1.2 billion citizens are moving up the food chain. In the early 1980s, the typical urban Chinese diet consisted of rice, porridge, and cabbage. By the middle 1990s the diet had dramatically changed to include meat, eggs, or fish at least once a day.

Diet is not the only thing changing in China. Even with population control, China's population grows by 14 million people a year, and young people are leaving the countryside and moving to the cities. China's urban population, currently over 300 million, should double by 2010. China has 21 percent of the world's population but only 7 percent of the arable land; in other words, it has less arable land than the United States but five times the population. China's farm sector is simply unable to keep up with the surging demand for what people regard as better food.

Total meat consumption in China is growing 10 percent a year; feed consumption is growing 15 percent. Demand for poultry, which requires 2 to 3 pounds of feed per pound of bird, has doubled in five years. China's new diet will make it more dependent on the United States, Canada, and Australia for feed grains. By 1997 China will have gone from being a net exporter of grain to importing 16 million tons. The switch in corn is even more dramatic. As recently as 1995, China was the second-largest corn exporter in the world. But with the chickens and pigs eating so much corn, in one year China moved from exporting 12 million tons of corn to importing 4 million tons.

China has also become the world's largest importer of fertilizer. In addition, while the government is reclaiming some wasteland in the north for agriculture, this is more than offset by the loss of fertile, multiple-cropped farmland in the southern coastal provinces. Overall, China's farmland is shrinking by at least 0.5 percent per year.

China also faces a severe water shortage in the arid north. In the Yellow River Valley's large grain belt, irrigation projects tripled crop yields during the 1950s and 60s, resulting in severe overpumping of ground water. Today the water table is falling, aquifers are vanishing, and farmers now have to compete with industry and households for water. The era of the flush toilet has arrived.

China is only a part—but certainly the biggest part—of a larger story. What's true for China is also happening throughout much of Asia. Large, populous countries such as Indonesia, Thailand, and the Philippines are rapidly urbanizing. They are gaining purchasing power and losing farmland, and the people are adding more animal proteins and processed foods to their diets. Given the magnitude of the changes, the world's ability to feed itself in the future is difficult to predict.

- What do you see as the future of food consumption in China in the next decade?
- Is China capable of maintaining a diet similar to that of North Americans? Is such a diet desirable?

**KFC and McDonald's in Beijing
Demand for poultry and beef is soaring**

GLOBAL PERSPECTIVE

Contaminated Soils in the Former Soviet Union

The newly independent republics of the former Soviet Union have inherited many kinds of environmental problems generated by the failed policies of the Soviet government. One of those policies was a push for increased agricultural production. New land was brought under cultivation, irrigation projects were developed, and extensive use of fertilizer and pesticides stimulated production.

This single-minded approach has resulted in serious environmental damage. In many areas, the soil has been contaminated by excessive fertilizer and pesticide use, poisoning groundwater and tainting food products. It is estimated that 30 percent of the food produced throughout the former Soviet Union is contaminated and should not be eaten. Forty-two percent of all baby food is contaminated by nitrates and pesticides. Latvian ecologists estimate that excessive pesticide use currently results in 14,000 deaths per year and that 700,000 people become ill from pesticides throughout the former Soviet Union annually.

One problem pesticide is DDT. It was officially banned in the Soviet Union in the 1970s, but the drive to improve agricultural production allowed its continued use. Waivers to the ban were allowed, or DDT was simply renamed by government bureaucrats and used under its new label. It is estimated that 10 million hectares of agricultural land in the former Soviet Union are contaminated with DDT.

The widespread use of agricultural chemicals has also affected the groundwater. Persistent pesticides, nitrates, and heavy metals have entered the groundwater, which many rely on for drinking water. Some areas have cadmium levels three to ten times higher than the amount permitted.

It appears that the fall of Communism in the former Soviet Union did not solve all the problems of the people. These environmental problems are surfacing after years of secrecy and denial under the shortsighted, production-oriented agricultural policy of the failed political system.

---

they are at the top of the food chain. Although all birds of prey have probably been affected to some degree, those that rely on fish for food seem to have been affected most severely. Eagles, osprey, cormorants, and pelicans are particularly susceptible species.

DDT is a well-known case of biological amplification in ecosystems. Other persistent molecules are known to behave in similar fashion. Mercury, aldrin, chlordane, and other chlorinated hydrocarbons, such as polychlorinated biphenols (PCB's) used as insulators in electric transformers, are all known to accumulate in ecosystems.

Because of their persistence, their effects on organisms at higher trophic levels, and concerns about long-term human health problems, most chlorinated hydrocarbon pesticides have been banned from use in the United States and some other countries. The use of DDT was prohibited in the United States in the early 1970s. Aldrin, dieldrin, heptachlor and chlordane have also been prohibited from use on crops, although heptachlor and chlor-

dane were still used for termite control until recently. In 1987, Velsicol Chemical Corporation agreed to stop selling chlordane in the United States.

Another problem associated with insecticides is the ability of insect populations to become resistant to them. All organisms within a given species are not identical. Each individual has a slightly different genetic composition and slightly different characteristics. When an insecticide is used for the first time on an insect, it kills all the individuals that are susceptible. Individuals with characteristics that allow them to tolerate the insecticides may live to reproduce.

If, in a population of insects, only 5 percent of the individuals possess genes that make them resistant to an insecticide, the first application of the insecticide will kill 95 percent of the population and so will be of great benefit in controlling the size of the insect population. However, the surviving individuals that are tolerant of the insecticide will constitute the majority of the breeding populations. Since these individuals possess ge-

netic characteristics for tolerating the insecticide, so will most of their offspring. Therefore, in the next generation, the number of individuals able to tolerate the insecticide will increase, and the second use of the insecticide will not be as effective as the first. Since some species of insect pests can produce a new generation each month, this process of selecting individuals capable of tolerating the insecticide can result in resistant populations in which 99 percent of the individuals are able to tolerate the insecticide within five years. As a result, that particular insecticide is no longer as effective in controlling insect pests, and increased dosages or more frequent spraying may be necessary. Figure 14.11 indicates that over 500 species of insects have populations resistant to insecticides.

Cotton growers have become extremely dependent on the use of insecticides. About 40 percent of the insecticides used in the United States are for controlling pests in cotton. Because of this intensive use, many populations of pest insects have become resistant to the

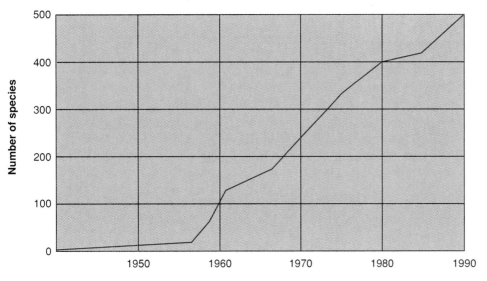

Pest species resistant to insecticides

Figure 14.11

Resistance to Insecticides   The continued use of insecticides has constantly selected for genes that give resistance to a particular insecticide. As a result, many species of insects and other arthropods are now resistant to many kinds of insecticides, and the number continues to increase.

Source: Data from George P. Georghiou, University of California at Riverside.

Table 14.4

Average Dose Necessary to Kill Two Cotton Pests

Compound	Average dose necessary to kill (milligrams per gram of larva)			
	Bollworm		Tobacco budworm	
	1960	1965	1961	1965
DDT	0.03	1,000+	0.13	16.51
Endrin	0.01	0.13	0.06	12.94
Carbaryl	0.12	0.54	0.30	54.57
Strobane and DDT	0.05	1.04	0.73	11.12
Toxaphene and DDT	0.04	0.46	0.47	3.52

Reprinted with permission from P.L. Adkisson, "Controlling Cotton's Insect Pests: A New System" in Science, 216:19–22, April 1982. Copyright © 1982 American Association for the Advancement of Science.

commonly used insecticides. Table 14.4 shows the effect of continual use of insecticides on two pests, the bollworm and the tobacco budworm. The size of the dose necessary to kill the pest increased greatly in a five-year period. In some cases, the dose increased tenfold and in others a hundredfold or more. By 1965, these insecticides were no longer able to control bollworm or tobacco budworm.

A third problem associated with pesticide use is that most of them kill beneficial species as well as pest species. In fact, they may kill predator and parasitic insects that normally control the pest insects. This allows pest species to increase rapidly because there are no natural checks to their population growth. Additional applications of insecticides are necessary to prevent the pest population from

rebounding to levels even higher than the initial one. Once the decision is made to use pesticides, it often becomes an irreversible tactic, because stopping their use would result in extensive crop damage.

A related concern is that the use of insecticides may cause an insect that is not a problem to become one. For example, when synthetic organic insecticides came into common use with cotton in the 1940s, the insect parasite and predators were eliminated, and the bollworm and tobacco budworm became major pests. The use of insecticides caused a different pest problem to develop.

A fourth problem associated with pesticides is their short-term and long-term health effects. If properly applied, most pesticides are safe. However, in many cases, people applying the pesticide are unaware of how it works and what precautions should be used in its application. In many parts of the world, subsistence farmers may not be able to read the caution labels on the packages. Consequently, deaths from the misuse of pesticides are common.

For most people, however, the most critical health problem related to pesticide use is inadvertent exposure to very small quantities of pesticides. Many pesticides have been proven to cause mutations, produce cancers, or cause abnormal births in experimental animals. There are questions about the effects of chronic, minute exposures to pesticide residues in food or through contamination of the environment.

Although the risks are very small, many people find pesticides unacceptable and seek to prohibit their sale and use. For example, in 1986, the U.S. Environmental Protection Agency banned the use of dinoseb, a herbicide, because tests by the German chemical company Hoechst AG indicated that dinoseb causes birth defects in rabbits. Other studies indicate that dinoseb causes sterility in rats. Similarly, in 1987, Velsicol Chemical Corporation signed an agreement with the U.S. EPA to stop producing and distributing chlordane in the United States. Chlordane had been banned previously for all applications except for termite control. After it was shown that harmful levels of chlordane could exist in treated homes,

environmental close-up

Food Additives

Food additives are chemicals added to food before its sale. They have several purposes:

1. To prolong the storage life of the food.
2. To make the food more attractive by adding color or flavor
3. To modify nutritive value

Many kinds of molecules are added to foods to prolong shelf life. Calcium propionate is added to baked goods because they can become contaminated with airborne spores of molds and bacteria, which grow on the food and spoil it. BHT, TBHQ, and other commonly seen alphabetic mixtures have a similar function.

Sometimes additives are used just to make the food appear more attractive. For years, Red Dye II was used to color a variety of foods. Its use was discontinued when it was found to be carcinogenic. Other food colorings are still widely used to increase appeal to the consumer. Many kinds of artificial flavors are added to products as well. Commonly used flavor enhancers are monosodium glutamate, table salt, and citric acid.

Other food additives are used to modify the nutritive value of the food product. Iodine in table salt is a good example. Iodine is a trace element required for proper thyroid functioning. Individuals suffering from a lack of iodine often develop an enlargement of the thyroid gland known as goiter. The addition of iodine to table salt has eliminated goiter in the United States. Most cereals and baked goods and many other products have various vitamins and minerals added to improve their nutritive value. Some additives, like Nutrasweet™, reduce the calories while giving the consumer the sensation of tasting something sweet.

Some additives are an unavoidable residue of some step of the food production industry and could more properly be called contaminants. Pesticide residues are an example. Pesticides are used to grow foods, but they are also used to eliminate pests in the storing, processing, and transportation steps of the food industry.

Diethylstilbestrol (DES) was at one time used in the poultry industry to produce fatter birds. Because there were indications that DES is carcinogenic, the U.S. Food and Drug Administration banned the use of DES in chickens and declared that it was potentially hazardous to humans. Further studies were conducted to determine if DES was safe to use in raising beef. It was used with cattle because an animal gains weight more rapidly with DES in its feed. Because DES was eventually linked to breast cancer in women, it has been banned from all animal feed use in the United States and Europe.

---

its use was finally curtailed. Also in 1987, the Dow Chemical Company announced that it would stop producing the controversial herbicide 2, 4, 5-T. In both of these cases, new pesticides were developed to replace the older types, so that discontinuing them did not result in an economic hardship for farmers or other consumers.

Why Are Pesticides So Widely Used?

If pesticides have so many drawbacks, why are they used so extensively? There are three primary reasons. First, the use of pesticides has increased, at least in the short term, the amount of food that can be grown in many parts of the world. In the United States, pests are estimated to consume 33 percent of the crops grown. On a worldwide basis, pests consume approximately 35 percent of crops. This represents an annual loss of $18.2 billion in the United States alone. Farmers, grain-storage operators, and the food industry continually seek to reduce this loss. A retreat from dependence on pesticides would certainly reduce the amount of food produced. Agricultural planners in most countries are not likely to suggest changes in pesticide use that would result in malnutrition and starvation for many of their inhabitants.

A second reason pesticides are used so extensively is economic. The cost of pesticides is more than offset by increased yields and profits for the farmer. In addition, the production and distribution of pesticides is big business. Companies that have spent millions of dollars developing a pesticide are going to argue very strongly for its continued use. Since farmers and agrochemical interests have a powerful voice in government, they have successfully prevented certain pesticides from being prohibited.

A third reason for extensive pesticide use is that many health problems are currently impossible to control without insecticides. This is particularly true in areas of the world where insect-borne diseases would cause widespread public health consequences if insecticides were not used.

Organic Farming

Before the invention of synthetic fertilizers, insecticides, herbicides, fungicides, and other agrochemicals, all farming was organic. Animal manure and crop rotation provided soil nutrients; a mixture of crops prevented regular pest problems; and manual labor killed insects and weeds. With the development of mechanization, larger areas could be farmed, draft animals were no longer needed, and many farmers changed from mixed agriculture, in which animals were an important ingredient, to monoculture. Chemical fertilizers replaced manure as a source of soil nutrients and crop rotation was no longer important since hay and grain were no longer grown for draft animals and cattle. The larger fields of crops like corn, wheat, and cotton presented opportunities for pest problems to develop. Chemical pesticides "solved" this problem.

As a result of the problems and costs of chemicals, some farmers are returning to an earlier form of agriculture in which they use traditional methods to produce crops without using pesticides and chemical fertilizers. They are even willing to accept lower yields because they do not have to pay for fertilizers and pesticides. Thus, they can still make a profit. In addition, farmers often receive premium prices for products that are organically grown.

A study by Lockeretz, Shearer, and Kohl lists some comparisons between conventional and organic farming. (See table 14.5.) Conventional farms had a higher yield than organic farms and also netted $384 per hectare, as compared to $333 per hectare for organic farms. Organic farms used only 40 percent of the energy used on conventional farms, thus reducing their costs and raising their profit margins.

This transition to organic farming does present some problems, however. The use of legumes, such as beans, soybeans, clover, or alfalfa, in crop rotation reduces the amount of land available for cash crops and requires that cattle be a part of the farmer's operation. Crop rotation also requires a greater investment in farm machinery, since certain crops require specialized equipment. Also, the raising of cattle requires additional expenditures for feed supplements and veterinary care. Critics say that organic farming cannot produce the amount of food required for today's population and that it

Table 14.5

Yield Difference Between Organic and Conventional Farming

Crop	Metric tons per hectare	
	Organic farming	Conventional farming
Corn	6.45	7.00
Soybeans	2.44	2.57
Wheat	1.88	3.28

Source: Data from William Lockeretz, et al., "Organic Farming in the Corn Belt," in Science, 211:540–47, February 6, 1981.

can only be successful in specific cases. Proponents disagree and stress that organic farming reduces chemicals in the environment and soil erosion. As fuel and chemical costs increase, organic farming or some modification of it may become an alternative approach to food production.

Integrated Pest Management

Another way of reducing dependence on pesticides is to use integrated pest management. **Integrated pest management** uses a variety of methods to control pests, rather than relying on pesticides alone. It depends on a complete understanding of all ecological aspects of the crop and the particular pests to which it is susceptible. It requires information about the metabolism of the crop plant, the biological interactions between pests and their predators or parasites, the climatic conditions that favor certain pests, and techniques for encouraging beneficial insects. It may involve the selective use of pesticides. Much of the information necessary to make integrated pest management work goes beyond the knowledge of the typical farmer. The metabolic and ecological studies necessary to pinpoint weak points in the life cycles of pests can usually only be carried out at universities or government research institutions. These studies are expensive and must be completed for each kind of pest, since each pest has a unique biology. Once a viable technique has been developed, an educational program is necessary to interest farmers in us-

ing integrated pest management rather than the "spray and save" techniques that they understand and that pesticide salespersons continually encourage.

Several methods are employed in integrated pest management. These include sex attractants, male sterilization, the release of natural predators or parasites, the development of resistant crops, the use of natural pesticides, the modification of farming practices, and the selective use of pesticides. In some species of insects, a chemical called a **pheromone** is released by females to attract males. Males of some species of moths can detect the presence of a female from a distance of up to 3 kilometers. Since many moths are pests, synthetic odors can be used to control them. Spraying an area with the pheromone confuses the males and prevents them from finding females which results in a reduced moth population the following year. In a similar way, a synthetic sex attractant molecule known as Gyplure is used to lure male gypsy moths into traps, where they become stuck. Since the females cannot fly and the males are trapped, the reproductive rate drops, and the insect population may be controlled.

Another technique that reduces reproduction is male sterilization. In the southern United States and Central America, the screwworm fly weakens or kills large grazing animals, such as cattle, goats, and deer. The female screwworm fly lays eggs in open wounds on these animals, where the larvae feed. However, it was discovered that the female mates only once in her lifetime. Therefore, the fly population can be controlled by raising and releasing large numbers of sterilized male screwworm flies. Any female that mates with a sterile male fails to produce fertilized eggs and cannot reproduce. In Curaçao, an island 65 kilometers north of Venezuela, a program of introducing sterile male screwworm flies eliminated this disease from the 25,000 goats on the island. In parts of the southwestern United States, the sterile male technique has also been very effective. The screwworm fly has been eliminated from the United States and northern Mexico, and much of Central America may become free of them as well. In 1990, sterile males were released in Libya to begin eliminating screwworm flies that had been introduced with a South American cattle shipment.

During an epidemic of Mediterranean fruitflies in southern California and northern Mexico, in the early 1980s, a similar technique was employed. Unfortunately, the X-ray technique used to sterilize the males was ineffective, and most of the flies released were not sterile, which made the problem worse rather than better. Pesticides were eventually used to control the fruitflies. Recent concern about the Mediteranean fruitfly in California has resulted in the controversial aerial spraying of malathion.

The manipulation of predator-prey relationships can also be used to control pest populations. For instance, the ladybird beetle, commonly called a ladybug, is a natural predator of aphids and scale insects. (See figure 14.12.) Artificially increasing the population of ladybird beetles reduces aphid and scale populations. In California during the late 1800s, scale insects on orange trees damaged the trees and reduced crop yields. The introduction of a species of ladybird beetle from Australia quickly brought the pests under control. Years later, when chemical pesticides were first used in the area, so many ladybird beetles were accidentally killed that scale insects once again became a serious problem. When pesticide use was discontinued, ladybird beetle populations rebounded, and the scale insects were once again brought under control. (See figure 14.13.)

In 1961, grape growers in California's San Joaquin Valley were troubled by a grape leafhopper. To combat the pest, growers applied DDT. However, the leafhopper quickly developed resistance

Figure 14.12

Beneficial Insect   The ladybird beetle is a predator of many kinds of pest insects, including aphids.

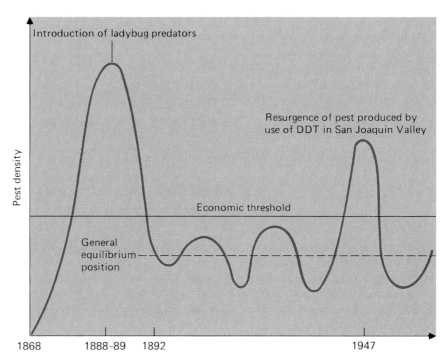

Figure 14.13

Insect Control with Natural Predators   In 1889, the introduction of ladybird beetles (ladybugs) brought the cottony cushion scale under control in the orange groves of the San Joaquin Valley. In the 1940s, DDT reduced the ladybird beetle population and the cottony cushion scale population increased. Stopping the use of DDT allowed the ladybird population to increase, reducing the pest population and allowing the orange growers to make a profit.

From Man and the Environment, *2/e by Arthur S. Boughey, © 1975. Reprinted by permission of Prentice-Hall, Inc., Upper Saddle River, NJ.*

to DDT, and other insecticides had to be employed. Grape growers spent over $8 million to control leafhoppers but created a resistant pest instead. Research into the biology of leafhoppers revealed that a particular species of parasitic wasp is a natural control of the leafhopper. The female wasp deposits an egg on the egg of a leafhopper. When the wasp egg hatches, the larva uses the leafhopper egg as food. In one growing season, the wasp produces nine or ten generations, compared to three for the leafhopper. This is sufficient to keep the leafhopper under control.

However, it was also discovered that the leafhopper spent its entire life in the vineyards, while wasps lived in the vineyard only during the summer. In the winter, the wasp required an alternate host, a noneconomically important leafhopper that normally lived on blackberry bushes. (See figure 14.14.) In a natural ecosystem where wild grapes and blackberries were interspersed, the leafhoppers never reached problem levels. With the establishment of large vineyards, most of the blackberry bushes were destroyed, and the remaining bushes were so far from the grapes that the wasps were unable to migrate from their winter habitat to their summer habitat. Thus, the wasps were no longer able to keep the leafhopper population under control.

The use of specific strains of the bacterium Bacillus thuringiensis to control mosquitoes and moths is another example of the use of one organism to control another. A crystalline material produced by the bacterium destroys the lining of the gut of the feeding insect, resulting in its death. One strain of *B. thuringiensis* is used to control mosquitoes while another is primarily effective against the caterpillars of leaf-eating moths, including the gypsy moth.

A method receiving increasing attention is that of producing resistant crop species through selective breeding or genetic engineering. These techniques seek to develop characteristics in major crop plants that allow them to resist fungi, insects, and other pests. For a long time, wheat rust, a fungus parasite that weakens wheat and reduces yields, was a major problem. Modern strains of wheat, and other crops, have now been

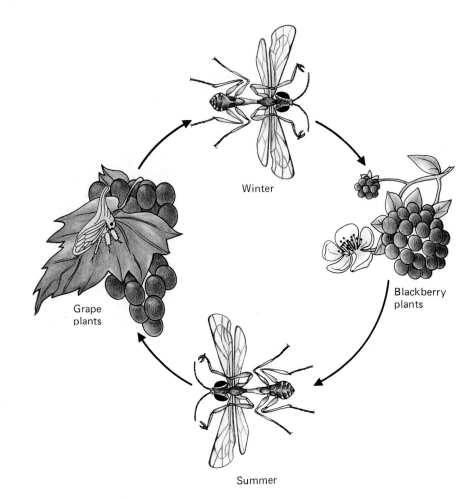

Figure 14.14

Life Cycle of a Parasitic Wasp During the summer, the population of leafhoppers that feed on grape leaves is controlled by a parasitic wasp. These wasps migrate to blackberry bushes to overwinter but return to the grapes during the summer. If the distance is too great, the wasps are unable to migrate between grapes and the blackberry bushes. As a result, the grape leafhoppers are unchecked, and they damage the grape plants. Therefore, the removal of blackberry patches to clear land for the growing of grapes increases the amount of damage that the leafhoppers inflict on the grapes.

developed that are able to resist such common plant pests. Another development is strains of crop plants with shorter growing seasons that complete their entire life cycle before pest populations become large enough to cause problems. Another recent development could allow herbicide-resistant genes to be introduced into crop plants by genetic engineering techniques. This would allow herbicides to be used on the crop without harming it, while the competing weeds would be destroyed.

Naturally occurring pesticides found in plants also can be used to control pests. For example, marigolds are planted to reduce the number of soil nematodes, and garlic plants are used to check the spread of Japanese beetles.

Often, modification of farming practices can reduce the impact of pests. In some cases, all crop residues are destroyed to prevent insect pests from finding overwintering sites. For example, shredding and plowing under the stalks of cotton in the fall reduces overwintering sites for boll weevils and reduces their numbers significantly, thereby reducing the need for expensive insecticide applications. Many farmers are also returning to crop rotation, which tends to prevent the buildup of specific pests that typically

PART FOUR   Human Influences on Ecosystems

Herring Gulls as Indicators of Contamination in the Great Lakes

Herring gulls nest on islands and other protected sites throughout the Great Lakes region. Since they feed primarily on fish, they are near the top of aquatic food chains and tend to accumulate toxic materials from the food they eat. Eggs taken from nests can be analyzed for a variety of contaminants.

Since the early 1970s, the Canadian Wildlife Service has operated a monitoring program to assess trends in the levels of various contaminants in the eggs of herring gulls. In general, the contaminant levels have declined as both the Canadian and U.S. governments have taken action to stop new contaminants from entering the Great Lakes. The figure shows the trends for PCBs. PCBs are a group of organic compounds; some molecules are much more toxic than others. They were used as fire retardants, lubricants, insulation fluids in electrical transformers, and in some printing inks. Both Canada and the United States have eliminated most uses of PCBs, reducing the levels found in herring gull eggs.

- Should the health of a bird species be used to develop policy?
- If PCBs are no longer being used, why are herring gulls still being contaminated?

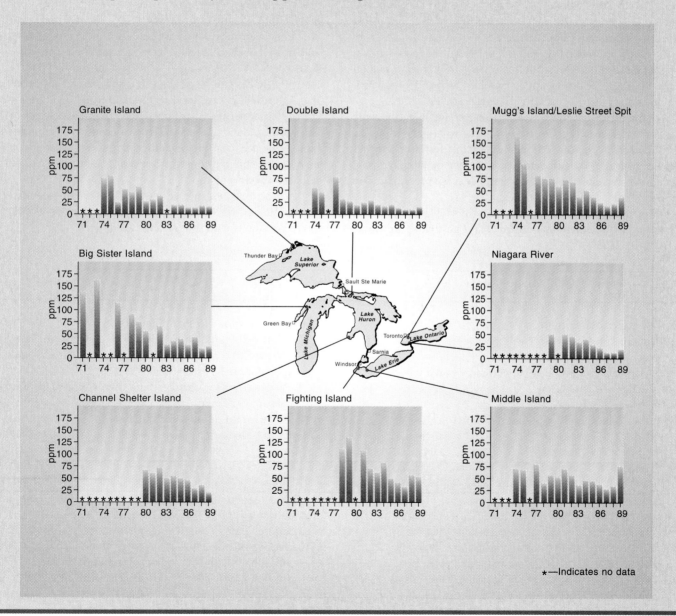

★—Indicates no data

occurs when the same crop is raised in a field year after year.

Integrated pest management also makes selected use of pesticides. Identifying the precise time when the pesticide will have the greatest effect at the lowest possible dose has several advantages: It reduces the amount of pesticide used, and may still allow the parasites and predators of pests to survive. Such precise applications often require the assistance of a trained professional who can correctly identify the pests, measure the size of the population, and time pesticide applications for maximum effect. In several instances, pheromone-baited traps capture insect pests from fields, and an assessment of the number of insects caught can be a guide to when insecticides should be applied.

Integrated pest management will become increasingly popular as the cost of pesticides increases and knowledge about the biology of specific pests becomes available. However, as long as humans raise crops, there will be pests that will outwit the defenses we develop. Integrated pest management is just another approach to a problem that began with the dawn of agriculture.

SUMMARY

Although small slash-and-burn garden plots are common in some parts of the world, most of the food in the world is raised on more permanent farms. In countries where population is high and money is in short supply, much of the farming is labor intensive, making use of human labor for many of the operations necessary to raise crops. However, the majority of the world's food is grown on large, mechanized farms that use energy rather than human muscle for tilling, planting, and harvesting crops and for the production and application of fertilizers and pesticides.

Monoculture involves planting large areas of the same crop year after year. This causes problems with plant diseases, pests, and soil depletion. Although chemical fertilizers can replace soil nutrients that are removed when the crop is harvested, they do not replace the organic matter necessary to maintain soil texture, pH, and biotic richness.

Mechanized monoculture is heavily dependent on the control of pests by chemical means. Hard pesticides are stable and persist in the environment, where they may be amplified in ecosystems. Consequently, many of the older hard pesticides have been quickly replaced by soft pesticides that decompose much more quickly and present less of an environmental hazard. However, most soft pesticides are more toxic to humans and must be handled with greater care than the older hard pesticides.

Pesticides can be divided into several categories based on the organism they are used to control. Insecticides are used to control insects, herbicides are used for plants, fungicides for fungi, and rodenticides for rodents. Because of the problems of biological amplification, resistance of pests to pesticides, and human health concerns, many people are seeking pesticide-free alternatives to raising food.

Integrated pest management makes use of a complete understanding of an organism's ecology to develop pest-control strategies that use no or few pesticides.

REVIEW QUESTIONS

1. What is monoculture?
2. List three reasons why fossil fuels are essential for mechanized agriculture.
3. Describe why pesticides are commonly used in mechanized agriculture.
4. Why are fertilizers used?
5. How do hard and soft pesticides differ?
6. What is biological amplification?
7. How do organic farms differ from conventional farms?
8. Name three nonchemical mehods of controlling pest populations?
9. What are the advantages and disadvantages of integrated pest management?
10. List three uses of food additives.

KEY TERMS

auxin   268
bioaccumulation   272
biocide   265
biological amplification   272
carbamate   267
chlorinated hydrocarbon   265
fungicide   265

hard pesticide   267
herbicide   265
insecticide   265
integrated pest
   management   278
macronutrient   264

micronutrient   264
monoculture   261
nontarget organism   265
organophosphate   267
persistent pesticide   267
pest   265

pesticide   264
pheromone   278
rodenticide   265
soft pesticide   267
target organism   265
weed   265

Education and research on the development of organic foods and farming for human and environmental health.

http://www.earthfoods.co.uk/

VegEdge, provides a vegetable pest fact sheet and pest management data.

http://www.mes.umn.edu/~vegipm/

Environmental working group examines pesticide contamination threats to humans and the environment.

http://www.ewg.org/

Pesticide Action Network, advances ecological alternatives to pesticides and promotes sustainable agriculture.

http://www.panna.org/panna/

Information access to rural development agencies.

http://www.rurdev.usda.gov/

Agricultural Network Information Center, provides information and links related to agriculture.

http://www.agnic.org/

Searchable index, from agriculture to wildlife.

http://www.agview.com/

USDA site dedicated to biological control of pests and pest management.

***http://rsru2.tamu.edu/

chapter 15

Water Management

Objectives

After reading this chapter, you should be able to:

- Explain how water is cycled through the hydrologic cycle.
- Explain the significance of groundwater, aquifers, and runoff.
- Explain how land use affects infiltration and surface runoff.
- List the various kinds of water use and the problems associated with each.
- List the problems associated with water impoundment.
- List the major sources of water pollution.
- Define biochemical oxygen demand (BOD).
- Explain how nutrients cause water pollution.
- Differentiate between point and nonpoint sources of pollution.
- Explain how heat can be a form of pollution.
- Differentiate between primary, secondary, and tertiary sewage treatments.
- Describe some of the problems associated with storm-water runoff.
- List sources of groundwater pollution.
- Explain how various federal laws control water use and prevent misuse.
- List the problems associated with water-use planning.
- Explain the rationale behind the federal laws that attempt to preserve certain water areas and habitats.
- List the problems associated with groundwater mining.
- Explain the problem of salinization associated with large-scale irrigation in arid areas.
- List the water-related services provided by local governments.

Chapter Outline

The Water Issue

The Hydrologic Cycle

Kinds of Water Use

   Domestic Use of Water

   Agricultural Use of Water

Environmental Close-Up: *Is It Safe to Drink the Water?*

   In-Stream Use of Water

   Industrial Use of Water

Kinds and Sources of Water Pollution

Environmental Close-Up: *New Optimism for the New River*

   Municipal Water Pollution

   Industrial Water Pollution

Global Perspective: *Comparing Water Use and Pollution in Industrialized and Developing Countries*

   Thermal Pollution

   Marine Oil Pollution

Wastewater Treatment

Runoff

Groundwater Pollution

Water-Use Planning Issues

Global Perspective: *Can Lake Victoria Be Rescued?*

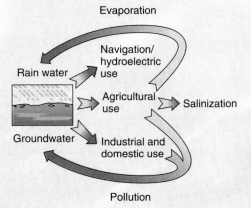

   Preserving Scenic Water Areas and Wildlife Habitats

   Groundwater Mining

Global Perspective: ECOPARQUE

Environmental Close-Up: *Groundwater Mining in Garden City, Kansas*

   Salinization

   Water Diversion

   Managing Urban Water Use

Environmental Close-Up: *Is It Too Late for the Everglades?*

Global Perspective: *Death of a Sea*

Environmental Close-Up: *The Blue Crabs of Chesapeake Bay*

Issues and Analysis: *The California Water Plan*

Global Perspective: *The Cleanup of the Holy Ganges*

The Water Issue

All living organisms are composed of cells that contain at least 60 percent water. Organisms can exist only where there is access to adequate supplies of water. Water is also a unique and necessary resource because it has remarkable physical properties. Its ability to act as a solvent and its capacity to store heat are perhaps the most useful.

As a solvent, water can dissolve and carry substances ranging from nutrients to industrial and domestic wastes. A glance at any urban sewer will quickly point out the value of water in dissolving and transporting wastes. Water can also store or contain heat. Because water heats and cools more slowly than most other substances, it is used in large quantities for cooling in electric power generation plants and in other industrial processes. Water's ability to retain heat modifies local climatic conditions in areas near large bodies of water. These areas do not have the wide temperature changes characteristic of other areas.

For most human uses, as well as some commercial ones, the quality of the water is as important as its quantity. Water must be substantially free of salinity, plant and animal waste, and bacterial contamination to be suitable for human consumption. Unpolluted freshwater supplies that are suitable for drinking are known as **potable waters.** Early human migration routes and settlement sites were influenced by the availability of drinking water. Today, despite advances in drilling, irrigation, and purification, the location, quality, quantity, ownership, and control of potable waters remains an important human concern.

Over 70 percent of the earth's surface is covered by water. The vast majority of it, however, is ocean salt water, which has limited use. (See figure 15.1.) Salt water cannot be consumed by humans or used for many industrial processes. Clean freshwater supplies have always been considered inexhaustible. Only now do we understand that we will probably exhaust our usable water supplies in some areas because of both human and natural factors. One human factor that has affected usable water supplies is a steadily increasing

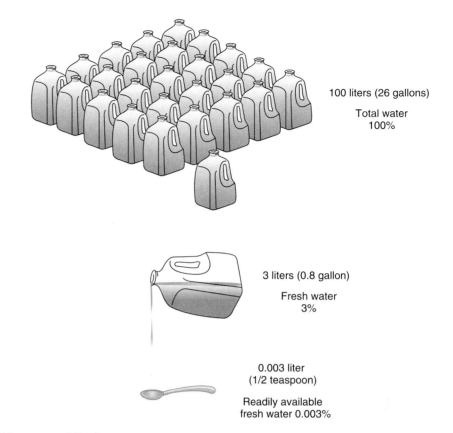

100 liters (26 gallons)
Total water
100%

3 liters (0.8 gallon)
Fresh water
3%

0.003 liter
(1/2 teaspoon)
Readily available
fresh water 0.003%

Figure 15.1

Only a tiny fraction of the world's water supply is available as fresh water for human use.

demand for freshwater for industrial, agricultural, and personal needs.

Shortages of potable water throughout the world can also be directly attributed to human abuse in the form of pollution. Water pollution has negatively affected water supplies in almost all of the world's densely populated industrialized nations, including Japan, Europe, the former Soviet Union, Canada, and the United States.

Unfortunately, the outlook for the world's freshwater supply is not very promising. According to studies by the United Nations and the International Joint Commission, many sections of the world will experience shortages of potable waters by the year 2000. Although the world's supply of water is continually being replenished by rainfall, this rainfall varies significantly. (See figure 15.2.) Parts of the world, particularly Mexico, India, Europe, and sections of Africa, continue to suffer massive droughts because of long periods of ex-

tremely limited rainfall. In other parts of the world, floods result from too much rain.

The Hydrologic Cycle

All water is locked into a constant recycling process called the **hydrologic cycle.** (See figure 15.3.) Solar energy evaporates water from the ocean surface. Water also evaporates from the soil and from the surfaces of plants. The water loss from plants is called **transpiration.** The air containing water vapor moves across the surface of the earth. As this warm, moist air cools, water droplets form and fall to the land as precipitation. Although some precipitation may simply stay on the surface until it evaporates, most will either sink into the soil or flow downhill until it eventually returns to the ocean. The water that infiltrates the soil may be stored for long periods in underground reservoirs. This water is called **groundwater,**

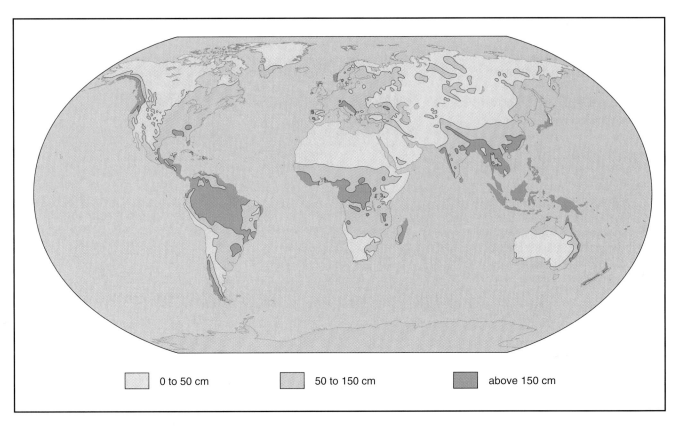

| | 0 to 50 cm | | 50 to 150 cm | | above 150 cm |

Figure 15.2

Global Average Annual Precipitation   The world's precipitation is not uniformly distributed. Although some areas of the world, such as Southeast Asia and Central America, have extremely high levels of precipitation, other areas, such as northern Africa and the Middle East, have very little annual precipitation.

as opposed to the surface water that enters a river system as **runoff.**

The way in which land is used has significant impact on evaporation, runoff, and infiltration. When water is used for cooling in power plants or to irrigate crops, the rate of evaporation is increased. Water impounded in reservoirs also evaporates rapidly. This rapid evaporation can affect local atmospheric conditions. Water infiltration is also not constant and is greatly influenced by human activity. Urban complexes with paved surfaces and storm sewers increase runoff and reduce infiltration. Demand for underground water in urban areas is usually high, but urban developments actually increase the gap between supply and demand.

Water that enters the soil and is not picked up by plant roots moves slowly downward until it reaches an impervious layer of rock. This water accumulates in porous strata called an **aquifer.** There are

two basic kinds of aquifers: **unconfined** and **confined.** An unconfined aquifer usually occurs near the land's surface and may be called a water-table aquifer because the upper surface of its water is the **water table.** The lower boundary is an impermeable layer of clay or rock. Unconfined aquifers are replenished (recharged) primarily by rain that falls on the ground directly above the aquifer and percolates down to the water table. The water in such aquifers is at atmospheric pressure and flows in the direction of the water table's slope.

A confined aquifer, also known as an **artesian aquifer,** is bounded on the top and bottom by confining layers and is saturated with water under greater-than-atmospheric pressure. The artesian aquifer is primarily replenished by rain and surface water from a recharge zone that may be many kilometers from where the aquifer is tapped for use. If water can-

not pass through the confining layer of an artesian aquifer, the layer is called an **aquiclude.** If water can pass in and out of the confining layer, the layer is called an **aquitard.** The vadose zone (also known as the unsaturated zone or zone of aeration) is an unsaturated area below the ground surface. (See figure 15.4.)

Aquifers are extremely important in supplying water for industrial, agricultural, and municipal use. Most of the larger urban areas in the western part of the United States depend upon underground water for their water supply. This groundwater supply exists as long as it is not used faster than it can be replaced. Determining how much water can be used and what the uses should be is often a problem.

Water uses can be measured by either the amount withdrawn or the amount consumed. Water withdrawn for use is diverted from its natural course. It may be

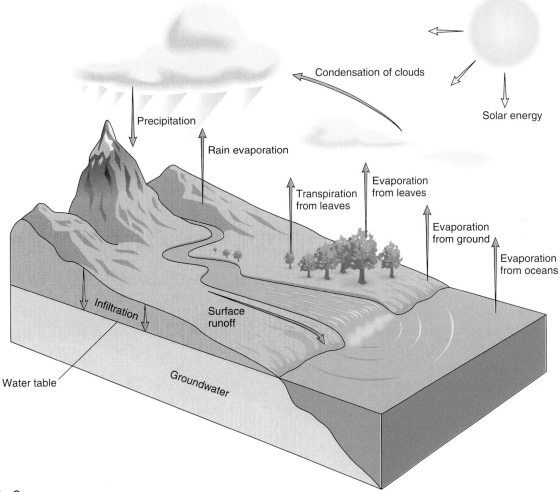

Figure 15.3

The Hydrologic Cycle   The cycling of water through the environment follows a simple pattern. Moisture in the atmosphere condenses into droplets that fall to the earth as rain or snow, supplying all living things with its life-sustaining properties. Water, flowing over the earth as surface water or through the soil as groundwater, returns to the oceans, where it evaporates back into the atmosphere to begin the cycle again.

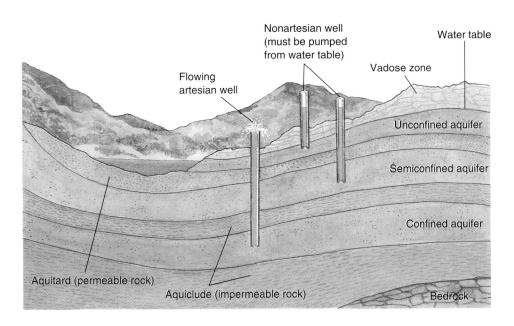

Figure 15.4

Types of Aquifers   The figure depicts the aquitard and aquiclude layers, the water table, and the unconfined, confined, and semiconfined aquifers.

withdrawn and then later returned to its source so it can be used again in the future. For example, when a factory removes water from a river for cooling purposes, it returns the water to the river so it can be used later. Water that is incorporated into a product or lost to the atmosphere through evaporation and transpiration cannot be reused in the same geographic area and is so said to be consumed. Much of the water used for irrigation evaporates. Some of this water is removed with the crop when it is harvested. Therefore, irrigation both withdraws and consumes water.

Kinds of Water Use

Water use can be classified into four categories: (1) domestic use, (2) agricultural use, (3) in-stream use, and (4) industrial use.

Domestic Use of Water

Many rural residents obtain safe water from untreated private wells. Urban residents usually get water from complex and costly water purification facilities. Urban growth has created problems in the development, transportation, and maintenance of quality water supplies. A relatively small amount of freshwater—roughly 8 percent of the global total—is withdrawn for domestic and municipal requirements. In regions experiencing rapid population growth, such as Asia, domestic use is expected to increase sharply by the year 2000. About 60 percent of water used for domestic purposes is returned to rivers as wastewater.

Domestic activities in highly developed nations require a great deal of water. This domestic use includes drinking, air conditioning, bathing, washing clothes, washing dishes, flushing toilets, and watering lawns and gardens. On average, each person in a North American home uses 300 to 400 liters of water each day. Most of this **domestic water** is used as a solvent to carry away wastes, with only a small amount used for drinking. (See figure 15.5.) Yet all water that enters the house has been purified and treated to make it safe for consumption. Until re-

Water use by a typical North American family of four

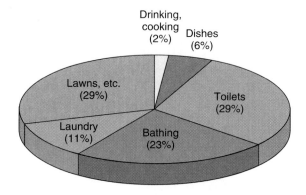

Figure 15.5

Urban Domestic Water Uses   Over 150 billion liters of water are used each day for urban domestic purposes in North America.

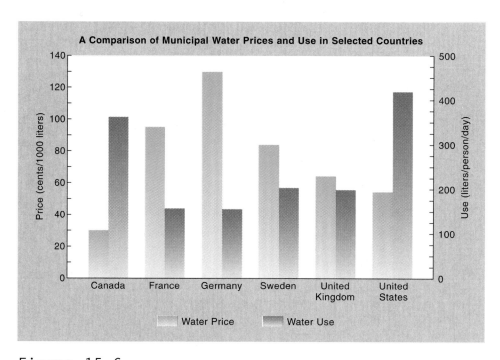

Figure 15.6

Water Use Decreases as Water Price Increases   There is a direct correlation between the amount of water that is used and the increase in price.

cently, the cost of water in almost every community has been so low that there was very little incentive to conserve, but increasing purification costs have raised the price of domestic water and it is becoming evident that increased costs do tend to reduce use. (See figure 15.6.)

Natural processes cannot cope with the highly concentrated wastes typical of a large urban area. The unsightly and smelly waste presents a potential health problem for the municipality. Cities and towns must provide domestic water supplies and treat the wastewater following its use. Both processes are expensive and require trained personnel.

The major problem associated with domestic use of water is maintaining an

Figure 15.7

Urban Expansion   Maintaining a suitable supply of water for growing metropolitan areas can pose major problems, especially in arid areas.

adequate, suitable supply for growing metropolitan areas. (See figure 15.7.) Demand for water in urban areas sometimes exceeds the immediate supply, particularly when the supply is local surface water. This is especially true during the summer, when water demand is high, and precipitation is often low. Many communities have begun public education campaigns designed to help reduce water usage. (See figure 15.8.)

More domestic water is wasted than consumed. The loss of domestic water represents nearly 20 percent of the water withdrawn from public supplies. This loss occurs mainly through leaking water pipes and mains. Another major cause of water loss has been public attitudes. As long as water is considered a limitless, inexpensive resource, little effort will be made to conserve it. As the cost of water rises and attitudes toward

water change, so will usage and efforts to conserve.

Agricultural Use of Water

The major consumptive use of water in most parts of the world is for agriculture, and principally for **irrigation.** In the 1980s, for example, irrigation acounted for nearly 80 percent of all the water consumed in North America. The amount of water used for irrigation and livestock continues to increase throughout the world. Future agricultural demand for water will depend on the cost of water for irrigation; the demand for agricultural products, food, and fiber; governmental policies; and the development of new technology.

In some areas, irrigation is a problem because there is not a supply of water nearby. This is particularly true in the

western United States, where about 14 million hectares of land are irrigated. (See figure 15.9.) In some places, water must be piped hundreds of kilometers for irrigation.

Because most of the world's consumptive use of water is for irrigation, it is becoming increasingly important to modify irrigation practices to use less water. Water loss from irrigation can be reduced in many ways. Increasing the cost of water will stimulate farmers to conserve, just as it does homeowners. Another method is to reduce the amount of water-demanding crops grown in dry areas, or change from high-water-demanding to lower-water-demanding crops. For example, wheat or soybeans require less water than do potatoes or sugar beets.

Switching to trickle irrigation also reduces water loss. With trickle irrigation, a series of pipes with strategically placed

Is It Safe to Drink the Water?

Roughly 1,000 contaminants have been detected in the public water supply in the United States, and virtually every major water source is vulnerable to pollution. About half the U.S. population relies on surface water from rivers, lakes, and reservoirs that may contain industrial wastes and pesticides washed off fields by rain. The other half uses groundwater that may be tainted by chemicals slowly seeping in from toxic-waste dumps. In some areas where groundwater supplies are being gradually depleted, the chemical pollutants are becoming more concentrated.

Most pollutants are probably not concentrated enough to pose significant health hazards; however, there are exceptions. The most widespread danger in water is lead, which can cause high blood pressure and an array of other health problems. Lead is especially hazardous to children, since it impairs the development of brain cells. The U.S. EPA estimates that at least 42 million Americans are exposed to unacceptably high levels of lead, and the U.S. Public Health Service estimates that perhaps 9 million children are at least slightly affected by it.

The contamination comes from old lead poisoning and solder that have been used in plumbing for years. These materials are gradually being replaced in homes and water systems. Individuals may want to have their water tested for lead by an official lab. If the level is too high, they can investigate ways to deal with the problem or switch to bottled water for drinking and cooking. Even then, caution is called for: Some bottled waters contain many of the same contaminants that tap water does.

The other four types of contamination in the U.S. water supply, along with their source and risk, are shown in the table. Regardless of the problems, however, the water supply in the United States is among the cleanest in the world.

Sources of lead
- Lead pipes between water mains and homes
- Lead pipes in building
- Leaded solder in copper pipes
- Brass faucets

Evidence of lead
Lead is invisible and odorless, but here are some signs to watch for:
- Rust-colored water
- Stained dishes or laundry
- Other signs of corrosion

People affected
Those who live in:
- Houses built before 1930
- Newer homes with lead solder
- High-rise buildings

Ways to reduce lead
- Before using water, let it run cold for two minutes. This is especially important if it has been standing in pipes for six hours or more.
- Don't drink, cook, or prepare baby formula with hot water; it dissolves lead more quickly.

Toxins from the tap

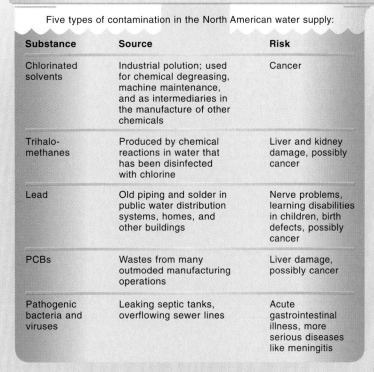

Five types of contamination in the North American water supply:

Substance	Source	Risk
Chlorinated solvents	Industrial polution; used for chemical degreasing, machine maintenance, and as intermediaries in the manufacture of other chemicals	Cancer
Trihalo-methanes	Produced by chemical reactions in water that has been disinfected with chlorine	Liver and kidney damage, possibly cancer
Lead	Old piping and solder in public water distribution systems, homes, and other buildings	Nerve problems, learning disabilities in children, birth defects, possibly cancer
PCBs	Wastes from many outmoded manufacturing operations	Liver damage, possibly cancer
Pathogenic bacteria and viruses	Leaking septic tanks, overflowing sewer lines	Acute gastrointestinal illness, more serious diseases like meningitis

Effects of lead poisoning
The absorption rate of lead in children is about 50 percent, compared with 8 percent for adults.

Lead concentration
Micrograms of lead per deciliter of blood

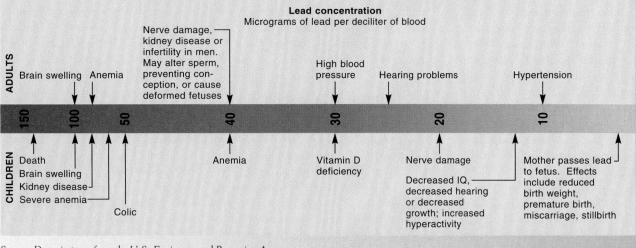

ADULTS

Brain swelling   Anemia   Nerve damage, kidney disease or infertility in men. May alter sperm, preventing conception, or cause deformed fetuses   High blood pressure   Hearing problems   Hypertension

150   100   50   40   30   20   10

CHILDREN

Death
Brain swelling
Kidney disease
Severe anemia
Colic   Anemia   Vitamin D deficiency   Nerve damage   Decreased IQ, decreased hearing or decreased growth; increased hyperactivity   Mother passes lead to fetus. Effects include reduced birth weight, premature birth, miscarriage, stillbirth

Source: Data, in part, from the U.S. Environmental Protection Agency.

Water savings guide

Conservative use will save water		Normal use will waste water
Wet down, soap-up, rinse off 4 gallons	**Shower**	Regular shower 25 gallons
May we suggest a shower?	**Tub bath**	Full tub 36 gallons
Minimize flushing Each use consumes 5-7 gallons	**Toilet**	Frequent flushing is very wasteful
Fill basin 1 gallon	**Washing hands**	Tap running 2 gallons
Fill basin 1 gallon	**Shaving**	Tap running 20 gallons
Wet brush Rinse briefly 1/2 gallon	**Brushing teeth**	Tap running 10 gallons
Take only as much as you require	**Ice**	Unused ice goes down drain
Please report immediately	**Leaks**	A small drip wastes 25 gallons a day
Turn off light, TV, heaters, and air conditioning when not in room	**Energy**	Wasting energy also wastes water

Thank you
for using
this column....................................and not this one

Figure 15.8

Ways to conserve water. Note: 1 gallon equals approximately 3.785 liters.

openings lies on the ground so that water is delivered directly to the roots of the plants, rather than flooding entire fields. Although used extensively in greenhouses, trickle irrigation is generally too costly for large agricultural operations. Methods that do not use as much water as flooding irrigation and that do not cost as much in labor and equipment as the trickle method include furrow irrigation, corrugation irrigation, overhead irrigation, and subirrigation. Each of these methods has its drawbacks and advantages as well as conditions under which it works well.

Irrigation requires a great deal of energy. Researchers estimate that 40 percent of the energy devoted to agriculture in Nebraska is used for irrigation. Increasing energy costs may force

some farmers to reduce or discontinue irrigation. In addition, much of western Nebraska relies on groundwater for irrigation, and the water table is dropping rapidly. If a water shortage develops, land values will decline. Land use and water use are interrelated and cannot be viewed independently.

In-Stream Use of Water

When a flow of water in streams is interrupted or altered, the value of the stream is changed. Major **in-stream uses** of water are for hydroelectric power, recreation, and navigation. Electricity from hydroelectric power plants is an important energy resource. Presently, hydroelectric power plants produce about 13 percent of the total electricity, generated

in the United States. They do not consume water and do not add waste products to it. However, the dams needed for the plants have definite disadvantages, including the high cost of construction and the resulting destruction of the natural habitat in streams and surrounding lands. While dams reduce flooding, they do not eliminate it. (See figure 15.10.) In fact, the building of a dam often encourages people to develop the floodplain. As a result, when flooding occurs, the loss of property and lives may be greater.

The sudden discharge of impounded water from a dam can seriously alter the downstream environment. If the discharge is from the top of the reservoir, the stream temperature rapidly increases. Discharging the colder water at the bottom of the reservoir causes a sudden decrease in the stream's water temperature. Either of these changes is harmful to aquatic life.

The impoundment of water also reduces the natural scouring action of a flowing stream. If water is allowed to flow freely, the silt accumulated in the river is carried downstream during times of high water. This maintains the river channel and carries nutrient materials to the river's mouth. But if a dam is constructed, the silt deposits behind the dam, eventually filling the reservoir with silt.

In addition, impounded water has a greater surface area, which increases evaporation. In areas where water is scarce, the amount of water lost can be serious. This is particularly evident in hot climates. Furthermore, flow is often intermittent below the dam, which alters the water's oxygen content and interrupts fish migration. The populations of algae and other small organisms are also altered. Because of all these impacts, dam construction requires careful planning.

Water tends to be a focal point for recreational activities. (See figure 15.11.) Sailing, waterskiing, swimming, fishing, and camping all require water of reasonably good quality. Water is used for recreation in its natural setting and often is not physically affected. Even so, it is necessary to plan for recreational use, because overuse or inconsiderate use can degrade water quality. For example, waves gener-

Figure 15.9

Irrigation Many areas require irrigation to be farmed economically. The fields in the top photograph have a long pipe, which releases water while slowly rotating about a central pivot. In this way, the field is automatically irrigated. In contrast to the rotating pipe system, trickle irrigation delivers the water directly to the roots of the plants, thus reducing water loss.

ated by powerboats can accelerate shore-line erosion and cause siltation.

Dam construction creates new recreational opportunities because reservoirs provide sites for boating, camping, and related recreation. However, these opportunities come at the expense of a previously free-flowing river. Some recreational pursuits, such as river fishing, are lost.

Most major rivers are used for navigation. North America currently has more than 40,000 kilometers of commer-

cially navigable waterways. These waterways must have sufficient water to ensure passage of transport vessels. Canals, locks, and dams guarantee this volume. Often, dredging is necessary to maintain the proper channel depth. Dredging can resuspend in the water contaminated sediments that had been covered over. In addition, the flow within the hydrologic system is changed, which, in turn, affects the water's value for other uses.

Most large urban areas rely on water to transport resources. During recent years, the inland waterway system has carried about 10 percent of goods such as grain, coal, ore, and oil. In North America, expenditures for the improvement of the inland waterway system have totalled billions of dollars.

In the past, almost any navigation project was quickly approved and funded, regardless of the impact on other uses. Today, however, such decisions are not made until the impacts on other uses are carefully analyzed.

Industrial Use of Water

Water for **industrial use** accounts for more than half of total water withdrawals. Ninety percent of the water used by industry is for cooling. Most industrial processes involve heat exchanges. Water is a very effective liquid for carrying heat away from these processes. For example, electric power generating plants use water to cool steam so that it changes back into water. If the water heated in an industrial process is dumped directly into a watercourse, it significantly changes the water temperature. This affects the aquatic ecosystem by increasing the metabolism of the organisms and reducing the water's ability to hold dissolved oxygen.

Industry also uses water to dissipate and transport waste materials. In fact, many streams are now overused for this purpose, especially in urban centers. The use of watercourses for waste dispersal degrades the quality of the water and may reduce its usefulness for other purposes. This is especially true if the industrial wastes are toxic.

During the past 30 years, many nations have passed laws that severely restrict industrial discharges of wastes

Figure 15.10

Dams Interrupt the Flow of Water The flow of water in most large rivers is controlled by dams. Most of these dams provide electricity. In addition, they prevent flooding and provide recreational areas. However, dams destroy the natural river system.

Figure 15.11

Recreational Use of Water Marinas provide recreation; however, wetlands destruction and large dredging operations may be necessary to build them.

into watercourses. In the United States, the federal role in maintaining water quality began in 1948 with the passage of the Federal Water Pollution Control Act. This act provided federal funds and technical assistance to strengthen local, state, and interstate water-quality programs. Through amendments to the act in 1956, 1965, 1972, and 1987, the federal role in water-pollution control was increased to include establishing area-wide waste-treatment management plans, and establishing the framework for a national program of water-quality regulation.

Kinds and Sources of Water Pollution

Water pollution occurs when the use of water by one segment of society interferes with the health and well-being of other members. In an industrialized society, maintaining unpolluted water in all drains, streams, rivers, and lakes is probably impossible. Water quality is related to the use intended for the water. Adding material to water may make it unfit for some uses but not others. If silt is added to a lake, the water may still be drinkable, but the lake may no longer be an acceptable place to swim. If salts are added to a lake, the water may be less acceptable for drinking but not for recreation. It may not be necessary to maintain absolutely pure water. (See table 15.1 and figure 15.12)

There are also economic considerations. The cost of removing the last few percentages of some materials from the water may not be justified. This is certainly true of organic matter, which is biodegradable. However, radioactive wastes and toxins that may accumulate in living tissue are a different matter. Attempting to remove these materials is often justified because of their potential harm to humans and other living organisms.

Municipal Water Pollution

Municipalities are faced with the double-edged problem of providing suitable drinking water for the population and disposing of wastes. These wastes consist of

environmental
close-up

New Optimism for the New River

The New River that flows from Mexico into the United States is pea-green in color and texture, laden with fecal matter and carcinogens, topped with detergent foam, and carrying the virus that causes polio and the bacteria that cause typhoid and cholera. For half a century, the river has been a binational disaster, carrying the human and industrial wastes of Mexicali, Mexico, into the Imperial Valley of California and earning an ignominious distinction as the dirtiest river in the United States.

In Mexico, the river runs for 35 kilometers past schools and neighborhoods before reaching the U.S. border at Calexico. Then it winds an additional 10 kilometers north through some of the richest farmland in the world before emptying its load—now including agricultural pesticides—into the Salton Sea in California. Five U.S. presidential administrations and their counterparts in Mexico City have promised to do something about the New River, but until recently little has been done. (See insert.)

Although it is too early to say that the river is getting cleaned up, there are encouraging signs that help is finally on the way, for in this era of the North American Free Trade Agreement border environmental problems are being taken more seriously in both the United States and Mexico. The U.S. Congress has appropriated funding for repair jobs on the Mexicali sewer system aimed at significantly reducing the flow of raw sewage into the river. In addition, the International Boundary and Water Commission has made the New River a priority in its anti-pollution efforts along the border.

Part of the credit for the burst of official interest in the river lies with NAFTA and the binational agencies created by it. But some of the credit also belongs to the Imperial County Board of Supervisors and its strategy to get the attention of the Environmental Protection Agency (EPA) in Washington, D.C.

At first, the supervisors thought of using the federal government for letting the New River fester for so long.

A Tide of Broken Promises

Here are some highlights of the half-century history of accords that have done little to clean up the New River:

1944: Washington and Mexico City agree to work cooperatively to clean up the river.

1950: Congress authorizes the secretary of state to work with the Mexican government on cleanup.

1972: After meeting with President Richard Nixon on the matter, Mexico's President Luis Echeverria Alvarez orders an end to the dumping of raw sewage in the New River, promising that most of it will be curbed within the year.

1979: U.S. President Jimmy Carter and Mexican President Jose Lopez Portillo, meeting in Mexico City, discuss the continuing problems of the river. The same year, a Mexican federal official is quoted in the Los Angeles Times: "We have the solution already planned. We just need the time to finish it."

1986: U.S. President Ronald Reagan and Mexican President Roberto de la Madrid, meeting in Mexicali, discuss river problems.

1987: A Mexican federal official quoted in the Los Angeles Times says "give us one year" to make major improvements.

1992: The International Boundary and Water Commission adopts a plan to clean up the river. A U.S. commissioner promises to swim in it in 1995.

1994: The Imperial County Board of Supervisors sends a plea for help to the Environmental Protection Agency: "The New River has been an environmental disgrace for over 50 years."

1995: Imperial County Supervisor Wayne Van De Graaff says, "When all is said and done about the New River, there's been a lot more said than done."

From Los Angeles Times, November 9, 1995. Copyright, 1995, Los Angeles Times.

Instead, they hired a Los Angeles attorney who had once been the general counsel to the EPA. The attorney recommended a bureaucratic procedure called a "petition for rulemaking" in which Imperial County laid out its grievances against the polluted river and asked that something be done.

The EPA responded by promising to deal more aggressively with the New River. Officials and citizens in Imperial County are optimistic that this time Washington and Mexico City really mean what they say about cleaning up the river.

Table 15.1

Sources and Impacts of Selected Pollutants

Pollutant	Source	Effect on humans	Effect on ecosystem
Acids	Atmospheric deposition; mine drainage; decomposing organic matter.	Reduced availability of fish and shellfish.	Death of sensitive aquatic organisms; increased release of trace metals from soils, rock, and metal surfaces such as water pipes.
Chlorides	Runoff from roads treated for removal of ice or snow; irrigation runoff; brine produced in oil extraction; mining.	Reduced availability of drinking water supplies; reduced availability of shellfish.	At high levels, toxic to freshwater organisms.
Disease-causing organisms	Dumping of raw and partially treated sewage; runoff of animal wastes from feed lots.	Increased costs of water treatment; death and disease; reduced availability and contamination of fish, shellfish, and associated species.	Reduced survival and reproduction of aquatic organisms due to disease.
Elevated temperatures	Heat trapped by cities that is transferred to water; unshaded streams; solar heating of reservoirs; warm water discharges from power plants and industrial facilities.	Reduced availability of fish.	Elimination of cold-water species of fish and shellfish; less oxygen; heat-stressed animals susceptible to disease.
Heavy metals	Atmospheric deposition; road runoff; discharges from sewage treatment plants and industrial sources; creation of reservoirs; acidic mine effluents	Increased costs of water treatment; disease and death; reduced availability and healthfulness of fish and shellfish.	Lower fish populations due to failed reproduction; death of invertebrates leading to reduced prey for fish.

Source: Data, in part, from World Resources 1994–95.

storm-water runoff, wastes from industry, and wastes from homes and commercial establishments.

Wastes from homes consist primarily of organic matter from garbage, food preparation, cleaning of clothes and dishes, and human wastes. Human wastes are mostly undigested food material and a concentrated population of bacteria, such as *Escherichia coli* and *Streptococcus faecalis*, collectively called coliform bacteria. These particular bacteria normally grow in the large intestine of humans, where they are responsible for some food digestion and for the production of vitamin K. Low numbers of these bacteria in water are not harmful to healthy people. Because they can be easily identified, their presence in the water is used to indicate the amount of pollution from human waste. The numbers of these types of bacteria present in water are directly related to the amount of human waste entering the water. When human wastes are disposed of in water systems, other potentially harmful bacteria from the human large intestine may be present in amounts too small to detect by sampling. The greater the amount of wastes deposited in the water,

Pollutant	Source	Effect on humans	Effect on ecosystem
Nutrient enrichment	Runoff from agricultural fields; pastures, and livestock feedlots; landscaped urban areas; dumping of raw and treated sewage and industrial discharges.	Increased water treatment costs; reduced availability of fish, shellfish and associated species; impairment of recreational uses.	Algal blossoms resulting in low oxygen levels and reduced diversity and growth of large plants; reduced diversity of animals; fish kills.
Organic molecules	Runoff from agricultural fields and pastures; landscaped urban areas; logged areas; discharges from chemical manufacturing and other industrial processes; combined sewers.	Increased costs of water treatment; reduced availability of fish, shellfish, and associated species; odors.	Reduced oxygen; fish kills; reduced numbers and diversity of aquatic life.
Sediment	Runoff from agricultural pastures and livestock feed lots; logged hillsides; degraded stream banks; road construction.	Increased water treatment costs; reduced availability of fish, shellfish, and associated species; filling in of lakes, streams, and artificial reservoirs and harbors, requiring dredging.	Covering of spawning sites for fish; reduced numbers of insect species; reduced plant growth and diversity; reduced prey for predators; clogging of gills and filters.
Toxic chemicals	Urban and agricultural runoff; municipal and industrial discharges; leachate from landfills; atmospheric deposits.	Increased costs of water treatment; increased risk of certain cancers; reduced availability and healthfulness of fish and shellfish.	Reduced growth and survivability of fish eggs and young; fish diseases; death of carnivores due to bioamplification in the food chain.

the more likely it is that there will be small populations of disease-causing bacteria. Therefore, the harmless bacteria are indicators that other organisms may be present.

The nonliving organic matter in sewage presents a different kind of pollution problem because it decays in the water. As microorganisms metabolize the organic matter, they use up available oxygen.

The amount of oxygen required to decay a certain amount of organic matter is called the **biochemical oxygen demand (BOD).** (See figure 15.13.) Measuring the BOD of a body of water is one way to determine how polluted it is. If too much organic matter is added to the water, all of the available oxygen will be used up. Then anaerobic (not requiring oxygen) bacteria begin to break down wastes. Anaerobic

respiration produces chemicals that have a foul odor and an unpleasant taste and that generally interfere with the well-being of humans.

Although the wastewater from cleaning dishes and clothing may contain some organic material, the more important group of contaminants found in this water are soaps and detergents. Soaps and detergents are useful because one end of the

Acidification

Decline of Water Level

Extinction of Indigenous Ecosystem and Biota

Eutrophication

Contamination with Toxic Chemicals

Accelerated Siltation

Figure 15.12

Six Major Environmental Problems in World Lakes and Reservoirs. The report of UNEP/ILEC joint project "Survey of the State of World Lakes" points out six major problems of world lakes/reservoirs.

PART FOUR   Human Influences on Ecosystems

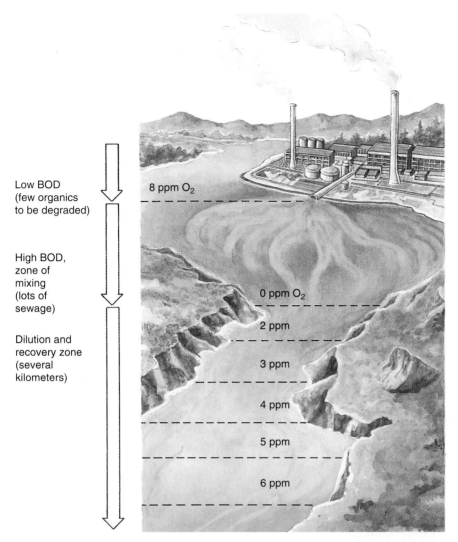

Low BOD
(few organics
to be degraded)

High BOD,
zone of
mixing
(lots of
sewage)

Dilution and
recovery zone
(several
kilometers)

8 ppm O_2

0 ppm O_2

2 ppm

3 ppm

4 ppm

5 ppm

6 ppm

Figure 15.13

Effect of Organic Wastes on Dissolved Oxygen Sewage contains a high concentration of organic materials. When these are degraded by organisms, oxygen is removed from the water. Therefore, there is an inverse relationship between sewage and oxygen in the water. The greater the BOD, the less desirable the water is for human use. The more the pollution, the greater the BOD.

molecule dissolves in dirt or grease and the other end dissolves in water. When the soap or detergent molecules are rinsed away by the water, the dirt or grease goes with them.

Until recently, many detergents contained phosphates as a part of their chemical structure. Phosphates are plant nutrients and therefore are a limiting factor of plant growth. (A **limiting factor** is a necessary material that is in short supply, and because of the lack of it, an organism cannot reach its full potential. See chapter 4.) Thus, when

phosphates from detergents are added to the surface water, they act as a fertilizer and promote the growth of undesirable algae populations. Algae and larger plants may interfere with the use of the water by fouling boat propellers, clogging water intake pipes, changing the taste and odor of water, and causing the buildup of organic matter on the bottom. As this organic matter decays, oxygen levels decrease, and fish and other aquatic species die.

Because these problems are all associated with the addition of plant nutri-

ents such as phosphates to the water, many states have banned the sale of detergents with high phosphate content. Detergent manufacturers point out that nutrients from **agricultural runoff** are probably more significant than detergents in adding phosphate to lakes and streams. While such runoff is important in adding to the total phosphate load of surface water, this defense does not alter the fact that detergents also add to this surface water load. Most domestic wastewater goes through a sewage treatment plant, so it is easier to measure and control the phosphate content. In areas where agriculture is uncommon, detergents can be the major source of phosphate pollution.

Industrial Water Pollution

Factories and industrial complexes frequently dispose of some or all of their wastes into municipal sewage systems. Depending on the type of industry involved, these wastes may be a combination of organic materials, petroleum products, metals, acids, and so forth. The organics and oil add to the BOD of the water. The metals, acids, and other ions need special treatment, depending on their nature and concentration. As a result, municipal sewage treatment plans must be designed with their industrial customers in mind. In most cases, cities prefer that industries take care of their own wastes. This allows industries to segregate and control toxic wastes and design wastewater facilities that meet their specific needs.

Most companies, when they remodel their facilities, include wastewater treatment as a necessary part of an industrial complex. However, many older facilities continue to pollute. These companies discharge acids, particulates, heated water, and noxious gases into the water. Generally, these plants are easily identified because the pollution comes from a single effluent pipe or series of pipes. This pollution is said to come from a **point source.** Diffuse pollutants, such as agricultural runoff, road salt, and acid rain, are said to come from **nonpoint sources.** Economic pressure and adverse publicity can affect companies that continue to pollute from

Comparing Water Use and Pollution in Industrialized and Developing Countries

Characteristic	Industrialized countries	Developing countries
Water use		
Water use per capita	Heavy per capita usage. Highest usage per capita is in U.S., Canada, Switzerland. Usage is stabilizing.	Small per capita use. Water usage increases as living standards go up.
Where water is used	Mostly for industry and agriculture, followed by domestic use.	Mostly for irrigated agriculture, especially in Asia, followed by industry, then domestic use.
Access to water and sanitation	Water and sanitation generally available. Only small population increases are expected.	Large segments of the population do not have safe drinking water and sanitation services. A rapidly growing urban population will create an increasing demand for water and sanitation.
Pollution Control		
Domestic wastewater	Most countries treat domestic wastes. In Central European countries, lack of sewage treatment systems is a serious problem.	Almost all urban sewage is discharged without treatment.
Industrial wastes	Strictly regulated in most industrialized countries, but pollution exists from discharges made up to 100 years ago. Some accidental spills. In Central Europe, industrial wastes laced with heavy metals and toxic chemicals have caused river quality to deteriorate dramatically and the Black and Baltic seas to become heavily polluted. Acid lakes and rivers are a problem in North America and Europe.	Largely untreated. A significant and growing problem. problems with acidification developing in Southern China and tropical Africa.
Land use runoff	Fertilizers and pesticides are a continuing problem. Soil washed from agricultural and urban areas and oil from city streets pollute rivers.	Uncontrolled use of fertilizers and pesticides is creating a growing health problem. In areas of heavy deforestation, soil erosion leads to sediment-clogged rivers.

point sources. However, pollutants that come from nonpoint sources are very difficult to control. In addition, pollution legislation relating to them is very difficult to enforce.

Amendments to the U.S. Federal Water Pollution Control Act of 1972 (PL 92-500) have mandated changes in how industry treats water. Industries are no longer allowed to use water and return it to its source in poor condition. One of the standards regulates the temperature of the water that is returned to its source. Because many industries use water for cooling, thermal pollution has become a problem.

Thermal Pollution

Thermal pollution occurs when an industry removes water from a source, uses the water for cooling purposes, and then returns the heated water to its source.

Power plants heat water to convert it into steam, which drives the turbines that generate electricity. For steam turbines to function efficiently, the steam must be condensed into water after it leaves the turbine. This condensation is usually accomplished by taking water from a lake or stream to absorb the heat. This heated water is then discharged.

Cooling water used by industry does not have to be released into aquatic ecosystems. There are three other methods of discharging the heat. One method is to construct a large, shallow pond. Hot water is pumped into one end of the pond, and cooler water is removed from the other end. The heat is dissipated from the pond into the atmosphere and substrate.

A second method is to use a cooling tower. In a cooling tower, the heated water is sprayed into the air and cooled by evaporation. The disadvantage of cooling towers and shallow ponds is that large amounts of water are lost by evaporation. The release of this water into the air can also produce localized fogs.

The third method of cooling, the dry tower, does not release water into the atmosphere. In this method, the heated water is pumped through tubes, and the heat is released into the air. This is the same principle used in an automobile radiator. The dry tower is the most expensive to construct and operate.

The least expensive and easiest method of discharging heated water is to return the water to the aquatic environment, but this can create problems for the inhabitants of the area. Although an increase in temperature of only a few degrees may not seem significant, some aquatic ecosystems are very sensitive to minor temperature changes. Many fish are triggered to spawn by temperature changes. For example, lake trout will not spawn in water above 10° C. If a lake has a temperature of 8° C, the lake trout will reproduce, but an increase of 3° C would prevent spawning and result in this species' eventual elimination from that lake.

Ocean estuaries are very fragile. The discharge of heated water into an estuary may alter the type of plant food available. As a result, animals with specific food habits may be eliminated because the warm water supports different food organisms. The entire food web in the estuary may be altered by only slight temperature increases.

Marine Oil Pollution

Marine oil pollution has many sources. One source is accidents, such as oil-drilling blowouts or oil-tanker accidents. Examples include the Torrey Canyon supertanker that broke up off the coast of England in 1967 and the Amoco Cadiz supertanker that broke up off the coast of France in 1978, resulting in a spill of more than 254 million liters of oil.

In 1989, the Exxon Valdez ran aground in Prince William Sound, Alaska, causing an oil spill of over 42 million liters of oil. The Exxon Valdez spill affected nearly 1,500 kilometers of Alaskan coastline and caused extensive damage to native wildlife. Long-term effects on such species as the salmon are still uncertain. Salmon are of particular concern in Alaska because of the over $700 million-a-year salmon-fishing industry.

By the end of 1990, Exxon Corporation had spent close to $2.5 billion in an effort to clean up the worst-ever oil spill in the United States. In addition, in March 1991, Exxon agreed to a settlement of $1 billion—$900 million to complete the cleanup and $100 million to pay criminal fines. How successful the cleanup was is open to debate. Some argue that a greater effort is needed, while others contend that the problems resulting from the spill have, for the most part, been resolved. Whatever the outcome of the debate, one fact is certain: Cleaning up an oil spill the size of the Alaskan spill will never be a simple task; court challenges continue.

By 1994, there was little visible sign of the oil that had covered the coastline in 1989, A U.S. National Oceanic and Atmospheric Administration (NOAA) study estimates that 50 percent of the oil biodegraded on beaches or in the water; 20 percent evaporated; 14 percent was recovered; 12 percent is at the bottom of the sea, mostly in the Gulf of Alaska; 3 percent lies on shorelines; and less than 1 percent still drifts in the water column.

Seabirds were the biggest casualties of the spill, dying en masse up to 800 kilometers from the spill site. More than 36,000 bird corpses were collected, but many scientists now believe between 375,000 and 435,000 died, the most ever documented in the history of oil pollution at sea. Commercially important fish species, including salmon and herring stocks, fared comparatively well. However, in the summer of 1993, fishing boats blockaded the entrance to the oil-loading areas in Prince William Sound to protest the damage they felt was done to the fishery. Many commercial fishers maintain that low catches are the result of the pollution caused by the spill.

A question still unresolved is how some cleanup techniques affected the coast. Exxon hired a fleet of local fishing boats and assembled giant, barge-mounted spray guns to scrub miles of contaminated beaches with hot water during the summer of 1989. NOAA researchers now believe the unusual high-pressure, hot-water spraying may have done more harm than good. The water, at temperatures up to 60° C, basically cooked small organisms and greatly slowed the recovery in areas where it was used, compared to a few oiled shorelines left unscrubbed at the request of the researchers.

Opinions differ widely as to the effect of the spill and the ultimate value of the cleanup. Some researchers have argued that the spill simply was not as bad as previously thought. Prince William Sound was so clean and healthy to begin with that it was able to recover from the massive spill far more quickly than could a more polluted marine system, such as the Chesapeake Bay.

In 1993, the tanker Brae, loaded with 75 million liters of oil, ran into rocks off Scotland's remote Shetland Islands. While the initial outlook for the island's environment was bleak, the effects of the oil spill were not as serious as first thought. About 40 percent of the oil was scattered by the very heavy seas and another 40 percent evaporated, thus limiting environmental damage.

The greatest sources of all marine oil pollution are not accidental. Nearly two-thirds of all human-caused marine oil pollution comes from three sources: (1) runoff from streets, (2) improper disposal of lubricating oil from machines or automobile crankcases, and (3) intentional oil discharges that occur during the loading and unloading of tankers. With regard to the latter, pollution occurs when the tanks are cleaned or oil-contaminated ballast water is released. Oil tankers use seawater as ballast to stabilize the craft after they have discharged their oil. This oil-contaminated water is then discharged back into the ocean when the tanker is refilled.

As the number of offshore oil wells and the number and size of oil tankers have grown, the potential for increased oil pollution has also grown. Many methods for controlling marine oil pollution have been tried. Some of the more promising methods are recycling and reprocessing used oil and grease from automobile service stations and industries, and enforcing stricter regulations on the offshore drilling, refining, and shipping of oil.

Wastewater Treatment

Because water must be cleaned before it is released, most companies and municipalities maintain wastewater treatment facilities. The percentage of sewage that is treated, however, varies greatly throughout the world. (See table 15.2.) Treatment of sewage is usually classified as primary, secondary, or tertiary. **Primary sewage treatment** removes larger particles by filtering water through large screens and then settling it in ponds or lagoons. Water is removed from the top of the settling lagoon and released. Water thus treated does not have any sand or grit, but it still carries a heavy load of organic matter, dissolved salts, bacteria, and other microorganisms. The organisms use the organic material for food, and as long as there is sufficient oxygen, they will continue to grow and reproduce. If the receiving body of water is large enough and

Table 15.2

Percent of Sewage Treated in Selected Areas	
Area	Percent
Europe	72
Mediterranena Sea	30
Caribbean Basin	less than 10
Southeast Pacific	almost zero
South Asia	almost zero
South Pacific	almost zero
West and Central Africa	almost zero

Source: World Resources Institute 1994–95.

the organisms have enough time, the organic matter will be degraded. In crowded areas, where several municipalities take water and return it to a lake or stream, within a few kilometers of each other, primary water treatment is not adequate.

Secondary sewage treatment usually follows primary treatment and involves holding the wastewater until the organic material has been degraded by the bacteria and other microorganisms. To encourage this action, the wastewater is mixed with large quantities of highly oxygenated water, or the water is aerated directly, as in a trickling filter system. In this system, the wastewater is sprayed over the surface of rock to increase the amount of dissolved oxygen. The rock also provides a place for the bacteria and other microbes to attach so they are exposed simultaneously to the organic material and to oxygen. Most secondary treatment facilities try to promote the growth of microorganisms. These microorganisms feed on the dissolved organic matter and small suspended particles, which then become incorporated into their bodies as part of their cell structure. The bodies of the microorganisms are larger than the dissolved and suspended organic matter, so this process concentrates the organic wastes into particles that are large enough to settle out. The sludge that settles consists of living and dead microorganisms and their waste products. In **activated sludge sewage treatment** plants, some of the sludge is returned to aeration tanks, where it is mixed with incoming wastewater. This kind of process uses less land than a trickling filter.

(See figure 15.14.) Both processes produce a sludge that settles out of the water.

The sludge that remains is concentrated and often dried before disposal. Sludge disposal is a major problem in large population centers. In the San Francisco Bay area, 2,500 metric tons of sludge are produced each day. Most of this is carried to landfills and lagoons, and some is composted and returned to the land as fertilizer.

Primary and secondary facilities are the most common types of sewage treatment in North American cities. The water discharged from these sewage treatment plants must be disinfected. The least costly method of disinfection is chlorination of the wastewater after it has been filtered and the organic materials have been allowed to settle. Using ultrasonic energy to mechanically break down waste may be less harmful and more effective, but it is also more expensive.

A growing number of larger sewage treatment plants use additional processes called tertiary sewage treatment. **Tertiary sewage treatment** involves a variety of techniques to remove dissolved pollutants left after primary and secondary treatments. The tertiary treatment of wastewater removes phosphorous and nitrogen that could increase aquatic plant growth. Tertiary treatment is very costly because it requires specific chemical treatment of the water to eliminate specific problem materials. (See table 15.3.) Certain industries are beginning to maintain their own secondary or tertiary wastewater facilities because of the specific nature of their waste products.

a.

Figure 15.14

Primary and Secondary Wastewater Treatment Primary treatment is physical; it includes filtrating and settling of wastes. Photograph (*a*) is a settling tank in which particles settle to the bottom. Secondary treatment is mostly biological and includes the concentration of dissolved organics by microorganisms. Two major types of secondary treatment are trickling filter and activated sludge methods. Photograph (*b*) shows an activated sludge system.

b.

Table 15.3

Tertiary Treatment		
Kind of tertiary treatment	**Problem chemicals**	**Methods**
Biological	Phosphorous and nitrogen compounds	1. Large ponds are used to allow aquatic plants to assimilate the nitrogen and phosphorous compounds from the water before the water is released. 2. Columns containing denitrifying bacteria are used to convert nitrogen compounds into atmospheric nitrogen.
Chemical	Phosphates and industrial pollutants	1. Water can be filtered through calcium carbonate. The phosphate substitutes for the carbonate ion, and the calcium phosphate can be removed. 2. Specific industrial pollutants, which are nonbiodegradable, may be removed by a variety of specific chemical processes.
Physical	Primarily industrial pollutants	1. Distillation. 2. Water can be passed between electrically charged plates to remove ions. 3. High-pressure filtration through small-pored filters. 4. Ion-exchange columns.

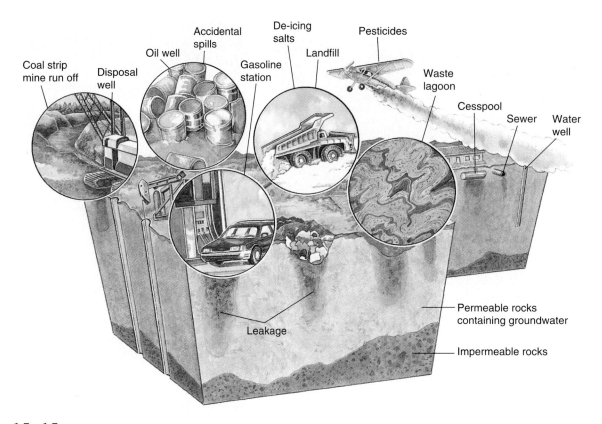

Figure 15.15

Sources of Groundwater Contamination   A wide variety of activities have been identified as sources of groundwater contamination.

Runoff

Storm-water runoff from streets and buildings is often added directly to the sewer system and sent to the municipal wastewater treatment facility. During heavy precipitation or spring thaws, the sewage treatment plant may be unable to handle the volume of wastewater, so some of it might be discharged directly to surface water without treatment. Modern wastewater treatment facilities have holding tanks to contain storm runoff and domestic wastes during heavy rains until they can be treated. However, many older plants lack adequate storage and must release untreated wastewater directly into receiving waters.

Many cities along the Mississippi River have recently completed sewer separation projects. Rainwater runoff now empties directly into the river without passing through treatment plants. This system is easier on the treatment facilities, but city residents must be more careful about their use of lawn chemicals and disposal of wastes to avoid polluting the river.

Agricultural runoff and mine drainage are nonpoint sources of pollution and thus are difficult to detect and control. One of the largest water pollution problems is agricultural runoff from large expanses of open fields. Precipitation dissolves materials in these areas and carries the materials away. Either the water runs over the surface and carries away exposed topsoil and nutrients (which are deposited in drains, streams, and rivers), or water seeps into the soil and carries dissolved nitrogen and phosphorous compounds into the groundwater. (See chapter 13.)

Farmers can reduce runoff in several ways. One is to leave a zone of undisturbed land near drains or stream banks. This retards surface runoff because soil covered with vegetation tends to slow the movement of water and allows the silt to be deposited on the surface of the land rather than in the streams. This can be costly because farmers may need to remove valuable cropland from cultivation. To retard leaching, farmers can keep the soil covered with a crop as long as possible and carefully con-

trol the amount and the timing of fertilizer application. This makes good economic sense because any fertilizer that runs off or leaches out of the soil has to be replaced.

Another nonpoint source of water pollution is mining. When coal is mined, water that drains from the mines is often very acidic. In addition, fine coal dust particles are suspended in the water, which makes the water chemically and physically less valuable as a habitat. Dissolved ions of iron, sulfur, zinc, and copper also are present in mine drainage. Control involves containing mine drainage so that it does not mix with surface water.

Groundwater Pollution

A wide variety of activities, some once thought harmless, have been identified as potential sources of groundwater contamination. In fact, possible sources of human-induced groundwater contamination span every facet of social, agricultural, and industrial activities. (See figure 15.15.) Major sources include:

1. **Agricultural Products**

   Pesticides contribute to unsafe levels of organic contaminants in groundwater. Seventy-three different pesticides have been detected in the groundwater in Canada and the United States. Accidental spills or leaks of pesticides pollute groundwater sources with 10 to 20 additional pesticides. Other agricultural practices contributing to groundwater pollution include animal feeding operations, fertilizer applications, and irrigation practices.

2. **Underground storage tanks**

   Up to 350,000 of the 1.4 million underground storage tanks containing gasoline and other hazardous substances in North America may already be leaking. Four liters of gasoline can contaminate the water supply of a community of 50,000 people.

3. **Landfills**

   Approximately 90 percent of the active landfills in North America have no liners to stop leaks to underlying groundwater, and 96 percent have no system to collect the leachate that seeps from the landfill. Sixty percent of landfills place no restrictions on the waste accepted, and many landfills are not inspected even once a year.

4. **Septic tanks**

   Poorly designed and inadequately maintained septic systems have contaminated groundwater with nitrates, bacteria, and toxic cleaning agents. Over 20 million septic tanks are in use, and up to a third have been found to be operating inproperly.

5. **Surface impoundments**

   Over 225,000 pits, ponds, and lagoons are used in North America to store or treat wastes. Seventy-one percent are unlined, and only 1 percent use a plastic or other synthetic, nonsoil liner. Ninety-nine percent of these impoundments have no leak-detection systems. Seventy-three percent have no restriction on the waste placed in the impoundment. Sixty percent are not even inspected annually. Many of these ponds are located near groundwater supplies.

Other sources of groundwater contamination include mining wastes, salting for snow control, land application of treated wastewater, open dumps, cemeteries, radioactive disposal sites, urban runoff, construction excavation, and animal feedlots.

Water-Use Planning Issues

In the past, wastes were discharged into waterways with little regard to the costs imposed on other users by the resulting decrease in water quality. With today's increasing demands for high-quality water, unrestrained waste disposal could lead to serious conflicts about water uses and cause social, economic, and environmental losses at both local and international levels. (See table 15.4.)

Water-use planning will need to deal with a number of issues, such as the following:

- Increased demand for water will force increased reuse of existing water supplies.
- In many areas where water is used for irrigation, both the water and the soil become salty because of evaporation. When this water

Table 15.4

International Water Disputes

River	Countries in dispute	Issues
Nile	Egypt, Ethiopia, Sudan	Siltation, flooding, water flow/diversion
Euphrates, Tigris	Iraq, Syria, Turkey	Reduced water flow, salinization
Jordan, Yarmuk Litani, West Bank aquifer	Israel, Jordan, Syria, Lebanon	Water flow/diversion
Indus, Sutlei	India, Pakistan	Irrigation
Ganges-Brahmaputra	Bangladesh, India	Siltation, flooding, water flow
Salween	Myanmar, China	Siltation, flooding
Mekong	Cambodia, Laos, Thailand, Viet Nam	Water flow, flooding
Paraná	Argentina, Brazil	Dam, land inundation
Lauca	Bolivia, Chile	Dam, salinization
Rio Grande, Colorado	Mexico, United States	Salinization, water flow, agrochemical pollution
Rhine	France, Netherlands, Switzerland, Germany	Industrial pollution
Maas, Schelde	Belgium, Netherlands	Salinization, industrial pollution
Elbe	Czechoslovakia, Germany	Industrial pollution
Szamos	Hungary, Romania	Industrial pollution

From Michael Renner, "National Security: The Economic and Environmental Dimensions" in Worldwatch Paper 89, 1989. Reprinted by permission of Worldwatch Institute.

Can Lake Victoria Be Rescued?

Lake Victoria—shared by Uganda, Kenya, and Tanzania in Africa, and the world's second largest freshwater lake after Lake Superior—is in trouble. The lake is undergoing profound environmental changes, the result of rapid population growth along its shores, rising algae levels within the lake, the disappearance of important native fish species, and a growing fish-export industry. Scientists studying the lake fear that without aggressive efforts to stop overfishing, protect wetlands and forests along the shore, reduce runoff from farms and factories, and rescue endangered fish species, this region of 30 million people is in danger of losing a major source of food and livelihood.

Twenty years ago, Lake Victoria contained more than 400 fish species. The majority have now disappeared. Lake Victoria's plight holds dangerous implications for Africa's other great lakes, which many scientists say are comparable to rainforests in their diversity of species.

Saving the 50,000-square-kilometer Lake Victoria will require a coordinated effort by Uganda, Kenya, and Tanzania. It will require significant foreign aid for conservation, research, and educational campaigns to help change the behavior of fishers, farmers, and other local people whose activities affect the lake. Scientists from Africa and elsewhere have developed the Lake Victoria Environmental Management Plan, a $20 million conservation project that is under consideration by the governments of the three lakeshore countries and by the World Bank. Conservation plans also call for stricter protection of forests and wetlands bordering the lake, which prevent erosion and help filter out runoff and pollution. The demand for fuel and farmland in this thickly populated region with a population growth of almost 4 percent per year makes conservation measures difficult to enforce.

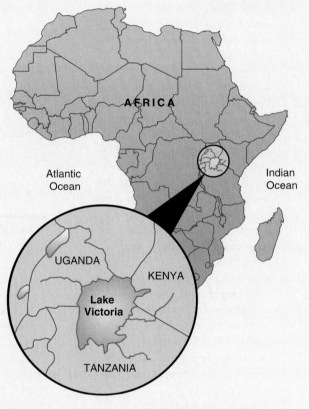

returns to a stream, the quality of the water is lowered.
- In some areas, wells provide water for all categories of use. If the groundwater is pumped out faster than it is replaced, the water table is lowered.

- In coastal areas, seawater may intrude into the aquifers and ruin the water supply.
- The demand for water-based recreation is increasing dramatically and requires high-quality water, especially for water recreation involving total body contact, such as swimming.

Preserving Scenic Water Areas and Wildlife Habitats

As mentioned earlier, the use of land influences water quality. This is particularly true along shorelines and riverbanks where the land and water meet. Some bodies of water have unique scenic value.

Fishing has been a livelihood of people along the lake for centuries, but in the last decade it also has become a major export industry for the three lakeshore countries. Nile perch, a large predatory fish introduced into Lake Victoria around 1960, sells for high prices from processors who ship it to Europe and Israel. The presence of the non-native fish has damaged the lake's ecosystem by contributing to the decline of many native species of fish that had been a major source of protein for area residents.

Fisheries officials consider the growing export industry an economic triumph, but researchers worry that it cannot be maintained at current levels. Lake fish, once the cheapest source of protein in the region, is becoming too expensive for many local people to afford, and fishers are catching too many young Nile perch, a sign of overfishing that could cause an eventual collapse of the fish population.

International research teams are collecting sediment cores from the lake bottom as a way of reading the chronicle of environmental changes in Lake Victoria. The samples show that carbon, nitrogen, and other nutrients from soil and wood smoke have been deposited in the lake at increasingly rapid rates since the 1920s, probably because of population growth and intense agricultural activity along the shore.

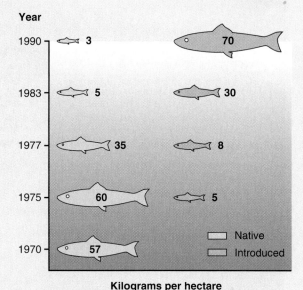

Kilograms per hectare

Source: Data from Kenyan Marine and Fisheries Research Institute.

While environmental awareness and concern is growing in the three countries bordering Lake Victoria, there is a great deal of work to do if the problems facing the lake are to be resolved. Having read this short description of the issues, discuss how they are all interrelated. How would you begin to design a plan that would address the issues? What problem would you address first?

To protect these resources, the way in which the land adjacent to the water is used must be consistent with preserving these scenic areas.

The U.S. Federal Wild and Scenic Rivers Act of 1968 established a system to protect wild and scenic rivers from development. All federal agencies must con-sider the wild, scenic, or recreational values of certain rivers in planning for the use and development of the rivers and adjacent land. The process of designating a river or part of a river as wild or scenic is complicated. It often encounters local opposition from businesses dependent on growth. Following reviews by state and federal agencies, rivers may be designated as wild and scenic by action of either congress or the Secretary of the Interior. Sections of over 65 streams in the United States have been designated as wild or scenic.

Many unique and scenic shorelands have also been protected from future

The value of wetlands

A wetland is an area covered by water enough of the year to support aquatic plant and animal life. There are many different kinds of wetlands, both freshwater and saltwater. Some are along rivers and streams, others are adjacent to lakes and oceans. Wetlands:

•Filter toxic wastes, excess nutrients, sediment and other pollutants
•Help prevent erosion
•Reduce flooding by storing storm water
•Reduce storm damage by absorbing waves

•Are a feeding, resting spot for migratory waterfowl
•Provide nesting grounds for a number of species, including oysters, clams, crabs, and shrimp
•Provide food, habitat for other aquatic species

Uplands | Wetlands | Body of water | Wetlands | Uplands

Figure 15.16

The Value of Wetlands We once thought of wetlands as only a breeding site for mosquitoes. Today we are beginning to appreciate their true value.

development. Until recently, estuaries, and shorelands have been subjected to significant physical modifications, such as dredging and filling, which may improve conditions for navigation and construction but destroy fish and wildlife habitats. Recent actions throughout North America have attempted to restrict the development of shorelands. Development has been restricted in some particularly scenic areas, such as Cape Cod National Seashore in Massachusetts and the Bay of Fundy in the Atlantic provinces of Canada.

Historically, poorly drained areas were considered worthless. Subsequently, many of these wetlands were filled or drained and used for development. The natural and economic importance of wetlands has been recognized only recently. In addition to providing spawning and breeding habitats for many species of wildlife, wetlands act as natural filtration systems by trapping nutrients and pollutants and preventing them from entering adjoining lakes or streams. Wetlands also slow down floodwaters and permit nutrient-rich particles to settle out. In addition, wetlands can act as reservoirs and release water slowly into lakes, streams, or aquifers, thereby preventing floods. (See figure 15.16) Coastal estuarine zones and adjoining sand dunes also provide significant natural flood control. Sand dunes act as barriers and absorb damaging waves caused by severe storms.

Groundwater Mining

Groundwater mining means that water is removed from an aquifer faster than it is replaced. When this practice continues for a long time, the water table eventually declines. Groundwater mining is common in areas of the western United States due to growing cities and increasing irrigation. In aquifers with little or no recharge, virtually any withdrawal constitutes mining, and sustained withdrawals will eventually exhaust the supply. This problem is particularly serious in commu-

ECOPARQUE

The city of Tijuana is home to approximately 1 million people. Given its rapid growth rate over the past three decades, the population will continue to increase. This growth has occurred on arid land where water is becoming a scarce resource. Both the city and the adjacent U.S.-Mexico border region also have severe water-pollution problems. One source of pollution is the direct discharge of residential wastewater into the Tijuana River and the nearby coastal waters. Because of the water scarcity and the pollution, scientists in Tijuana began thinking of alternative technologies for treating wastewater stating that "water is too valuable to be called waste." One answer is an alternative treatment facility called ECOPARQUE.

ECOPARQUE began in 1986 as a study of decentralized wastewater treatment by El Colegio de la Frontera Norte (COLEF), a college in Tijuana. By 1996, ECOPARQUE had developed into a successful model of sustainable and integrated urban wastewater treatment. The central feature of ECOPARQUE is the basic treatment unit for recapturing wastewater. The unit includes a stainless steel fine screen, a plastic biofilter, and a passive sediment deposit clarifier. These three elements operate without electrical energy or any mechanical parts. For this reason, the process falls under the category of appropriate technology for construction and use in developing countries.

ECOPARQUE and its treated wastewater are being used for a variety of sustainable purposes, including:

- Urban reforestation on hillsides and canyons in Tijuana.

- A nursery that produces a variety of plants, including vegetables and fruits.

- Food production in an urban center that includes an expanding fruit harvest, soon to be joined by beekeeping and an agricultural production unit. There are also plans for raising some species of fish as a food source.

- A meterological station that will provide local authorities with useful urban planning information.

- An experimental solar field for use at ECOPARQUE and the surrounding neighborhoods.

- A site for environmental and technological education for schoolchildren and scientists.

Given the range of activities at ECOPARQUE, it is easy to see why the scientists who planned the facility believed that water is too valuable to be called waste. While it may not be feasible to build facilities similar to ECOPARQUE in every city in the developing world, the design plans for ECOPARQUE are being made available to other countries. There is hope that many other cities will turn what is a water pollution and health problem into a multifaceted and sustainable environmental and community success story.

---

nities that depend heavily upon groundwater for their domestic needs.

Groundwater mining can also lead to problems of settling or subsidence of the ground surface. Removal of the water allows the ground to compact, and large depressions may result. For example, in the San Joaquin Valley of California, groundwater has been withdrawn for irrigation and cultivation since the 1850s. In the last 40 years, groundwater levels have fallen over 100 meters. More than 1,000 hectares of ground have subsided, some as much as 6 meters. Currently, the ground surface in that area is sinking 30 centimeters per year. In 1981, in Winter Park, Florida, a large area of subsidence—a "sinkhole"—occurred because of excessive water withdrawal. (See figure 15.17.)

Groundwater mining is a serious problem in western Texas. This area depends on irrigation for agriculture, and the population has increased dramatically over the past 25 years. The ever-increasing demand for groundwater has led to its rapid depletion. Precipitous declines in agricultural production are forecast within the next 10 years. As the groundwater is depleted, the land values will decline. Land use then

environmental
close-up

Groundwater Mining in Garden City, Kansas

For 20 million years, much of the precipitation that fell on the Great Plains infiltrated the sand and gravel aquifer surrounding Garden City, Kansas. For nearly one hundred years, wheat (a low water-demanding crop) was the predominant crop. But in 1960, farmers began to tap the groundwater in the Ogallala Aquifer, and because of this availability of water, corn (which requires more water than wheat) quickly replaced wheat as the main crop. For $4.50, a farmer could distribute 1 meter of water over 1 hectare of land.

An economic boom resulted when farmers began to irrigate and grow corn. Today, nearly 25,000 wells are irrigating 1.4 million hectares of land, and the corn is being fed to feedlot cattle. Five large packing plants in the area process enough cattle in a single day to feed a million people.

The amount of irrigation in the area is the subject of much controversy. Although irrigation has been in operation for over 30 years, the cost of pumping enough water to cover 1 hectare of land with 1 meter of water could increase to $250. The increase is a result of higher fuel costs, inflation, and the decreased amount of water in the aquifer. In over 30 years of pumping, the water level in the aquifer has dropped 4 meters. At the present rate of pumping, the aquifer is estimated to be dry early in the next century.

Soil conservationists predict that, when the water is no longer available from the aquifer, farmers will not be able to produce a crop, and that conditions will be similar to those of the dust bowls of the 1930s. Farmers, cattle growers, and slaughterhouse owners are resisting any attempts to limit the amount of water allowed for irrigation. They are certain that new sources of water will be found in unexploited aquifers or by diverting water from water-rich regions like the Great Lakes.

Figure 15.17

Development of a "Sinkhole" If the water table is lowered as a result of groundwater mining or a severe drought, the land surface can subside. The space formerly occupied by water can collapse and create sinkholes. This photo shows only one of many sinkholes that occurred in Florida in 1981 as the result of the combined effects of excessive groundwater use and a long drought.

directly affects water use, and water use and availability directly influence land use. London, Mexico City, Venice, Houston, and Las Vegas are some other cities that have experienced subsidence as a result of groundwater withdrawal. (See table 15.5.)

Groundwater mining poses a special problem in coastal areas. As the fresh groundwater is pumped from wells along the coast, the saline groundwater moves inland, replacing fresh groundwater with unusable salt water. (See figure 15.18.) Saltwater intrusion is a serious problem in heavily populated coastal areas throughout the world.

Salinization

Another water-use problem results from the salinity caused by increasing salt concentrations in soil. When plants extract the water they need, the salts present in all natural waters become concentrated. Irrigation of arid farmland makes this problem more acute because so much water is lost due to evaporation. Irrigation is most common in hot,

PART FOUR Human Influences on Ecosystems

Table 15.5

Groundwater Depletion in Major Regions of the World, Circa 1995

Region/aquifer	Estimates of depletion
California	Groundwater overdraft exceeds 1.7 billion cubic meters per year. The majority of the depletion occurs in the Central Valley, which is referred to as the vegetable basket of the United States.
Southwestern United States	In parts of Arizona, water tables have dropped more than 120 meters. Projections for parts of New Mexico indicate that water tables will drop an additional 22 meters by 2020.
High Plains Aquifer System, United States	This aquifer underlies nearly 20 percent of all the irrigated land in the United States. To date the net depletion of the aquifer is in excess of 350 billion cubic meters, or roughly 15 times the average annual flow of the Colorado River. Most of the depletion has been in the Texas high plains, which has witnessed a 26 percent decline in irrigated land from 1979 to 1989. Current depletion is estimated to be in excess of 13 billion cubic meters a year.
Mexico City and Valley of Mexico	Use exceeds natural recharge by 60 to 85 percent, causing land subsidence and falling water tables.
African Sahara	North Africa has vast nonrecharging aquifers where current depletion exceeds 12 billion cubic meters a year.
India	Water tables are declining throughout much of the most productive agricultural land in India. In parts of the country, groundwater levels have declined 90 percent during the past two decades.
North China	The water table underneath portions of Beijing has dropped 40 meters during the past 40 years. A large portion of northern China has significant groundwater overdraft.
Arabian Peninsula	Groundwater use is nearly three times greater than recharge. At the depletion rates projected into the beginning of the next century, exploitable groundwater reserves could be exhausted within the next 50 years. Saudi Arabi depends on nonrenewable groundwater for roughly 75 percent of its water. This includes irrigation of 2 million to 4 million tons of wheat per year.

dry areas, which normally have high rates of evaporation. This results in a concentration of salts in the soil and in the water that runs off the land. (See figure 15.19.) Every river increases in salinity as it flows to the ocean. The salinity of the Colorado River water increases 20 times as it passes through irrigated cropland between Grand Lake in north central Colorado and the Imperial Dam in southwest Arizona. The problem of salinity will continue to increase as irrigation increases.

Water Diversion

Water diversion is the physical process of transferring water from one area to an-

other. An early example of water diversion is the aqueducts of ancient Rome. Thousands of diversion projects have been constructed since then. New York City, for example, diverts water from Pennsylvania (250 kilometers away), and Los Angeles obtains part of its water supply from the Colorado River.

While diversion is necessary in many parts of the world, it can be misused or detrimentally affect an area. An example of this is the proposed Garrison Diversion Unit, which is designed to divert water from the Missouri River system for irrigation. (See figure 15.20.) This action would have enormous environmental consequences. Wildlife refuges, native grasslands, forests, and waterfowl breeding

marshes would be damaged or destroyed. In addition, hundreds of kilometers of streams could be seriously damaged.

There is considerable opposition to this plan, both by private groups and governmental agencies. In fact, the Canadian government has fought against the project. In contrast, the people who need the water for irrigation see the Garrison Diversion Unit plans as beneficial. If this plan was implemented, they could raise corn (a more profitable crop), rather than the dryland crop of wheat. The corn could then be used for cattle feed.

The initial plan to irrigate this portion of the Great Plains was thought of during the Dust Bowl era of the 1930s. The Federal Flood Control Act of 1944

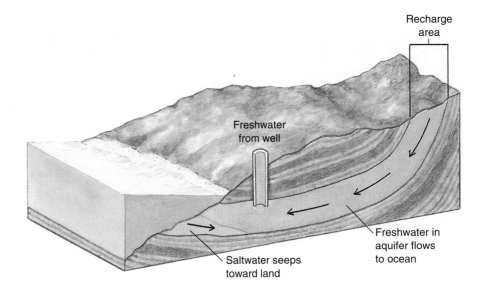

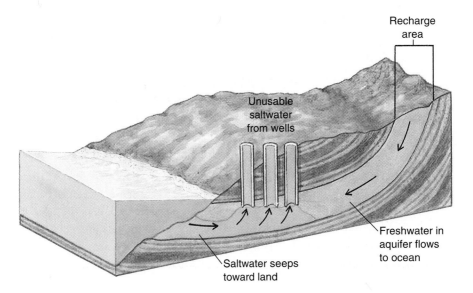

Figure 15.18

Saltwater Intrusion   When saltwater intrudes on fresh groundwater, the groundwater becomes unusable for human consumption and for many industrial processes.

authorized the construction of the Garrison Diversion Unit, but controversy concerning it continues 50 years after its proposal.

Managing Urban Water Use

Providing water services for metropolitan areas is another serious water-planning issue. Metropolitan areas must provide three basic water services:

1. Water supply for human and industrial needs

2. Wastewater collection and treatment

3. Storm-water collection and management

Water for human and industrial use must be properly treated and purified. It is then pumped through a series of pipes to consumers. After the water is used, it flows through a network of sewers to a wastewater treatment plant before it is released. Metropolitan areas must also deal with great volumes of excess water during storms. Because urban areas are paved and little

rainwater can be absorbed into the ground, managing storm water is a significant problem. Cities often have severe local flooding because the water is channeling along streets to storm sewers. If these sewers are overloaded or blocked with debris, the water cannot escape and flooding occurs.

Many cities have a single system to handle both sewage and storm-water runoff. During heavy runoff, the flow is often so large that the wastewater treatment plant cannot handle the volume. The wastewater is then diverted directly into the receiving body of water without

Figure 15.19

Salinization   As water evaporates from the surface of the soil, the salts it was carrying are left behind. In some areas of the world, this has permanently damaged cropland. Other areas must flush the soil to rid it of salt to maintain its fertility.

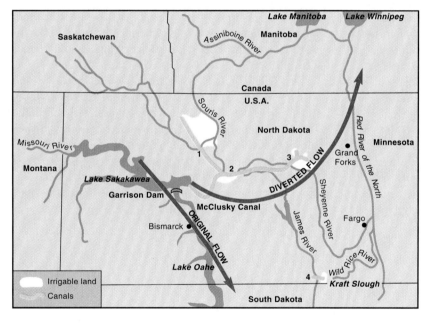

Figure 15.20

The Garrison Diversion Unit   This unit would divert water that flows into the Missouri River to the McClusky Canal and the Sheyenne River and eventually into Hudson Bay.

being treated. Some cities have areas in which to store this excess water until it can be treated. This is expensive and, therefore, is done only if federal or state funding is available.

Water services are expensive. They must be provided with an understanding that water supplies are limited. Also, water's ability to dilute and degrade pollutants is limited. Proper land-use planning is essential if metropolitan areas are to provide services and limit pollution.

In pursuing these objectives, city planners encounter many obstacles. Large metropolitan areas often have hundreds of local jurisdictions (governmental and bureaucratic areas) that divide responsibility for management of basic water services. The Chicago metropolitan area is a good example. This area is composed of six counties and approximately two thousand local units of government. It has 349 separate water-supply systems and 135 separate wastewater disposal systems. Efforts to implement a water-management plan when so many layers of government are involved are complicated and frustrating.

Is It Too Late for the Everglades?

Close to a million people visit the Everglades Park each year to view a landscape that is part African veldt and part tropical swamp. While the Everglades may look healthy, parts of its finely tuned ecosystem are breaking down. According to some scientists, the park is dying.

Years of mismanagement of the water supply have dessicated much of the park's wetlands. Agriculture and urban development have increasingly diverted the flow that sustains the park. Water policies favoring farmers and urban dwellers have forced wading birds to relocate their colonies and cut their population in the southern Everglades from 300,000 to 15,000. Alligators, whose population in the drying park is also drastically shrinking, are cannibalizing their young. Exotic plants, thriving on nutrients from agricultural runoff, are forcing out natural vegetation. Bass and catfish in the park should not be eaten because they are laced with natural mercury leached into ponds from soil dried to dust. Panthers, snail kites, and wood storks, three of the park's 13 endangered wildlife species, are on the verge of extinction.

After Everglades Park was established in 1947, about 800,000 hectares of wetlands to the north were drained for farms and urban development. South Florida boomed. Some 4.5 million people now live in the horseshoe crescent around the Everglades region, and 600 new residents arrive each day. The once-clear water of Lake Okeechobee, which is north of the park, are now being polluted with dairy-farm runoff and choked with algae. Cattails that proliferate in the phosphorus-laden runoff from the sugar plantations are choking the waterways. Some 175,000 hectares of drained Everglades are in sugar production. Since the water is free, South Florida's farmers take as much as they need and have little incentive to conserve.

There are plans to help the Everglades. The Kissimmee River, which flows into Lake Okeechobee, was channelized into an arrow-straight river in the late 1960s. This project destroyed the marshes and allowed nutrients to pollute the lake and thus the park. Beginning in 1990, the Kissimmee was being returned to its natural state, with twisting oxbow curves and extensive wetlands. In 1989, the U.S. Congress approved the purchase of 43,000 hectares to be added to the east section of the park. The state of Florida has obtained an additional 60,000 hectares as a buffer zone for the park. In addition, canals, spillways, and pumping stations are being modified to assure a steady supply of cleaner water for the park. These ambitious attempts to put a little wild back in the wilderness of the Everglades project have a good chance to succeed.

In 1993, the U.S. Department of the Interior proposed an agreement with the state of Florida and the state's vegetable farmers and sugar industry to restore the Everglades. The agreement would begin an ambitious plan to reduce phosphorous pollution and return natural water flow. The cost of the restoration program would be paid for out of a $465 million settlement with the sugar industry. A scaled down plan was approved in 1996 over the objections of many in the agricultural community.

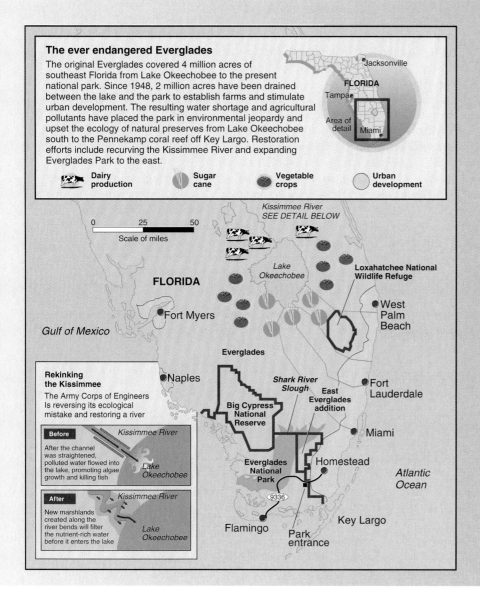

The ever endangered Everglades

The original Everglades covered 4 million acres of southeast Florida from Lake Okeechobee to the present national park. Since 1948, 2 million acres have been drained between the lake and the park to establish farms and stimulate urban development. The resulting water shortage and agricultural pollutants have placed the park in environmental jeopardy and upset the ecology of natural preserves from Lake Okeechobee south to the Pennekamp coral reef off Key Largo. Restoration efforts include recurving the Kissimmee River and expanding Everglades Park to the east.

FLORIDA

Jacksonville

Tampa

Area of detail

Miami

Dairy production   **Sugar cane**   **Vegetable crops**   **Urban development**

Kissimmee River
SEE DETAIL BELOW

0   25   50
Scale of miles

FLORIDA

Lake Okeechobee

Loxahatchee National Wildlife Refuge

Gulf of Mexico

Fort Myers

West Palm Beach

Everglades

Rekinking the Kissimmee

The Army Corps of Engineers Is reversing its ecological mistake and restoring a river

Naples

Shark River Slough

Fort Lauderdale

Before

Kissimmee River

After the channel was straightened, polluted water flowed into the lake, promoting algae growth and killing fish

Lake Okeechobee

Big Cypress National Reserve

East Everglades addition

Miami

After

Kissimmee River

New marshlands created along the river bends will filter the nutrient-rich water before it enters the lake

Lake Okeechobee

Everglades National Park

9336

Homestead

Atlantic Ocean

Flamingo

Park entrance

Key Largo

Construction threatens to destroy the fragile ecosystem of the Everglades.

Everglades wading birds once numbered more than a million, but drought and pollution have decimated their number.

Death of a Sea

The Aral Sea lies on the border between Uzbekistan and Kazakhstan in the former Soviet Union. It was once larger than any of the Great Lakes except Superior, and now it is disappearing. Since the 1920s, Soviet agricultural planners have used up the Aral Sea, diverting its waters for irrigation. The two rivers feeding the Aral were drawn off to irrigate millions of hectares of cotton. The irrigation canal, the world's longest, stretches over 1,300 kilometers, paralleling the boundaries of Afghanistan and Iran. The cotton production plan worked, and by 1937, the Soviet Union was a net exporter of cotton. The success of the cotton program, however, spelled the end for the Aral Sea.

For a long time, the ecological impact on the sea and surrounding area was largely hidden from public view. Since the 1960s, however, the Aral has lost about 40 percent of its surface area, or almost 20,000 square kilometers of what are now largely dry, salt-encrusted wastelands. The once-thriving fishing industry that depended on the water is all but gone. Some 20 of the 24 fish species there have disappeared. The fish catch, which totalled 44,000 tons a year in the 1950s and supported some 60,000 jobs, has dropped to zero. Abandoned fishing villages dot the sea's former coastline.

Another apparent consequence of the dried up sea is a host of human illnesses. A high rate of throat cancer is attributed to dust from the drying sea. Each year, winds pick up 40 million to 150 million tons of a toxic dust–salt mixture from the dry sea bed and deposit them on the surrounding farmland, harming or killing crops. The low river flows have concentrated salts and toxic chemicals, making water supplies hazardous to drink and contributing to disease. In the northwest part of the Republic of Uzbekistan, the infant-mortality rate is the highest in the former Soviet Union.

The former fishing center of the sea was a town named Muynak. The town is now landlocked more than 30 kilometers from the water. Less than 25 years ago, Munyak was a seaport. The population of Muynak is down from 40,000 citizens in 1970 to 12,000 today. In 1990, the mayor and last harbormaster of Muynak commented:

The water continued to go away while the salinity increased. The weather changed for the worse, with the summers getting hotter and the winters colder. The people feel salt on their lips and in their eyes all the time. It's getting hard to open your eyes here.

What has happened in the Aral Sea basin shows vividly how damage to economy, community, and human health can follow close on the heels of ecological destruction.

The Blue Crabs of Chesapeake Bay

Scientists are conducting the first accurate count of blue crabs in the Chesapeake Bay, an inlet of the Atlantic Ocean in Virginia, Maryland, and Delaware. The count is an attempt to produce a forecasting tool that could be used to set limits on the dwindling numbers of crabs being harvested. While no one believes that the blue crab is about to become extinct, there are concerns over the decline in the number of crabs that have been harvested over the past decade.

Anxiety about the crab is high because the bay's supply of other species also once thought to be boundless— such as shad, sturgeon, oyster, and striped bass—has been severely depleted by disease, pollution, and overfishing. The crab is a vital link in the ecological chain of the Chesapeake Bay. It lives in a wide variety of habitats, is the major consumer of bottom-dwelling fish, and is itself consumed by predators.

Until recently, little was known about how many crabs were in the bay and whether their numbers had fallen to dangerous levels. For years, the crab seemed invulnerable because each female produces millions of eggs whose survival was thought to depend solely on physical forces such as currents. Recent studies, however, have shown that human impact is perhaps more directly related to the crab population. Pollution, construction of shoreline homes, wetland destruction, and a growing appetite for crabs are considered bigger problems for the crab than once believed.

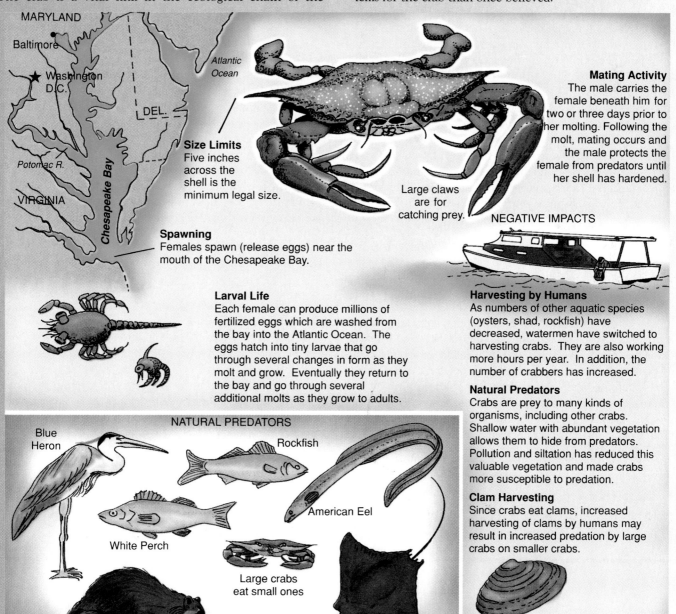

MARYLAND
Baltimore
Washington D.C.
Atlantic Ocean
DEL.
Potomac R.
Chesapeake Bay
VIRGINIA

Size Limits
Five inches across the shell is the minimum legal size.

Spawning
Females spawn (release eggs) near the mouth of the Chesapeake Bay.

Large claws are for catching prey.

Mating Activity
The male carries the female beneath him for two or three days prior to her molting. Following the molt, mating occurs and the male protects the female from predators until her shell has hardened.

NEGATIVE IMPACTS

Larval Life
Each female can produce millions of fertilized eggs which are washed from the bay into the Atlantic Ocean. The eggs hatch into tiny larvae that go through several changes in form as they molt and grow. Eventually they return to the bay and go through several additional molts as they grow to adults.

Harvesting by Humans
As numbers of other aquatic species (oysters, shad, rockfish) have decreased, watermen have switched to harvesting crabs. They are also working more hours per year. In addition, the number of crabbers has increased.

Natural Predators
Crabs are prey to many kinds of organisms, including other crabs. Shallow water with abundant vegetation allows them to hide from predators. Pollution and siltation has reduced this valuable vegetation and made crabs more susceptible to predation.

Clam Harvesting
Since crabs eat clams, increased harvesting of clams by humans may result in increased predation by large crabs on smaller crabs.

NATURAL PREDATORS
Blue Heron
Rockfish
White Perch
American Eel
Large crabs eat small ones
Raccoon
Cownose Ray

The California Water Plan

The management of freshwater is often a controversial subject, involving social, ecological, and economic aspects. A good example is the California Water Plan.

In the early 1900s, it became clear that the growth of Los Angeles, which was then a small coastal town, would be encouraged by irrigating the surrounding land. Los Angeles looked to the Owens Valley, 40 kilometers north, for a source of water. The Los Angeles Aqueduct connecting these two areas was completed in 1913.

After the Owens Valley project, California developed a statewide water program known as the California Water Plan. This plan, being implemented over many years, is necessary because most of the state's population and irrigated land are found in the central and southern regions, but most of the water is in the north. The plan details the construction of aqueducts, canals, dams, reservoirs, and power stations to transport water from the north to the south In addition to supplying water for southern regions, the aqueducts provide irrigation for the San Joaquin Valley. Eventually, the new land made available for agriculture will amount to about 400,000 hectares.

The California Water Plan has been one of the most controversial programs ever undertaken in California. Its adoption instigated a sectional feud between the moist "north" and the dry "south." Southern California was accused of trying to steal northern water, but because the large population in the southern part of the state carried the vote, the plan was adopted. Environmentalists still claim that the project has irreparably scarred the countryside and upset natural balances of streams, estuaries, vegetation, and wildlife. They argue that providing water to southern California promotes population growth, which leads to further urbanization and land development.

Many questions and controversies center on whether the water is really needed. Ninety percent of the water used in southern California is for irrigation; there is evidently abundant water for domestic and industrial use. A few of the crops raised in the San Joaquin Valley account for a large percentage of the irrigation water. Most notable is rice. Rice is not a native crop and demands intensive irrigation. The cost is borne by all the rate payers, most of whom are urban. California is now one of the nation's most productive agricultural areas.

Allocation of water resources is a matter of economics as well as technology. The California water project has been criticized for using public funds to increase the value of privately held farmland. Furthermore, technological advances in desalination plants may give a new dimension (unforeseen when the water plan was devised) to the problem of water resources. Although more aqueducts, canals, and pumping plants are planned, whether they will be completed is uncertain.

Southern California has been in a drought emergency for several years. By 1991, Los Angeles was in its eighth straight year without adequate rainfall. A number of unusual proposals for emergency supplies of water were reviewed, including:

- Bringing in Colorado water by train
- Importing Canadian water by barge
- Building a desalination plant for Pacific Ocean water
- Using an oil pipeline to carry water from northern California

In addition to finding new sources of water, conservation practices are enforced. In Santa Barbara, water use decreased 40 percent in three months once the city raised water rates and banned landscape irrigation. In Los Angeles, 1 million homeowners had to cut water use by 10 percent. Residents of Marin County, north of San Francisco, were limited to 200 liters of water per day. Santa Monica mandated a 25 percent cutback for residents and businesses.

Water for irrigation was also targeted for big cutbacks, with some farmers receiving little, if any, water for crops. In Marin County, irrigation accounts were reduced 85 percent. By the summer of 1991, nearly a dozen California counties had proclaimed drought disasters and were seeking aid from the state and federal governments. Most California communities had water rationing, conservation programs, and depleted reservoirs.

- What are the major advantages of having a water plan?
- What problems develop when water is transported to arid regions?
- Should water from northern California be sent to southern California?
- Should crops like rice be raised in California?
- Aside from lack of rainfall, why do you think southern California is in a water crisis?

Oregon

Shasta Lake

Lake Oroville

Sacramento
River

Feather
River

Lake Tahoe

Sacramento

North Bay Aqueduct

South Bay Aqueduct

San Fransisco

California Aqueduct

San Luis Reservoir

Los Angeles Aqueduct

San
Joaquin
Valley

Nevada

Coastal Branch

Techachapi
Mts.

Silverwood Lake

Lake Perris

Colorado River
Aqueduct

Castaic Lake

Los Angeles

Colorado River

San Diego

Salton Sea

Arizona

Northern California
< 1/3 of the state's
population
> 2/3 of the state's
water resources

Southern California
> 2/3 of the state's
population
< 1/3 of the state's
water reservoir

CALIFORNIA
AQUEDUCT

STATE WATER PROJECT
DEPARTMENT OF WATER RESOURCES

The Cleanup of the Holy Ganges

Every day, thousands of Hindus flock to the banks of the Holy Ganges River in India. There they drink and bathe in what they believe to be holy water, as partially cremated corpses float past them and nearby drains emit millions of liters of raw sewage. Clean water is one of India's most scarce resources, but as in many developing nations, the money and technology needed to properly treat sewage are not available in most cities and villages. With the country's 900 million population expected to double in 34 years, officials are concerned that Indians will have no choice but to continue dumping raw waste into local waterways, contributing to epidemics of diarrhea and other diseases that kill thousands of people annually.

Keeping the Ganges clean is made especially difficult because faith in the river's incorruptible purity has generated complacency and ambivalence about its pollution among many of the 300 million people who live in the Ganges Basin. More than 1,600 million liters of untreated municipal sewage, industrial waste, agricultural runoff, and other pollutants are discharged into the river every day. At the same time, officials estimate that more than 1 million people a day bathe or take a "holy dip" in the Ganges, and thousands drink straight from its banks.

The Hindu belief in cremation has led to several environmental problems. A wood cremation takes more than 50 pounds of wood, costing two week's wages for the typical Indian. There are complaints that wood cremations are helping to devour India's forests. Given the high cost of wood, many bodies are not completely cremated, and the partially burned corpses are disposed of in the Ganges.

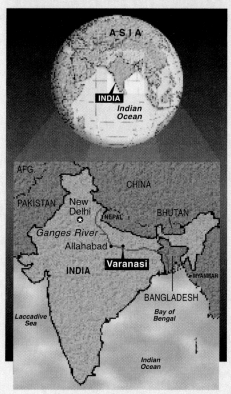

One solution to this problem has been the introduction of 25,000 specially raised snapping turtles that are attracted to the rotten smell of corpses. While this solution may seem extreme, there is another that is more acceptable in the long run. In 1992, as part of the Ganges Action Plan, an electric crematorium was built at the city of Varanasi that charges less than $2 per body and does a thorough job of turning the bodies to ash. There is some concern however, that many Hindus will not want to abandon the traditional ritual of a wood cremation for the more efficient and less costly electric cremation.

Although this entire scenario may seem somewhat unusual to North Americans, it is important to keep in mind how culture and religion affect our environment. To many Indians, the cultural and religious practices of North Americans are equally puzzling.

SUMMARY

Water is a renewable resource that circulates continually between the atmosphere and the earth's surface. The energy for the hydrologic cycle is provided by the sun. Water loss from plants is called transpiration. Water that infiltrates the soil and is stored in underground reservoirs is called groundwater, as opposed to surface water that enters a river system as runoff. The two basic kinds of aquifers are unconfined and confined. The way in which land is used has a significant impact on rates of evaporation, runoff, and infiltration.

The four human uses of water are domestic, agricultural, in-stream, and industrial. Water use is measured by either the amount withdrawn or the amount consumed. Domestic water is in short supply in many metropolitan areas. Most domestic water is used for waste disposal and washing, with only a small amount used for drinking. The largest consumptive use of water is for agricultural irrigation. Major in-stream uses of water are for hydro-electric power, recreation, and navigation. Most industrial uses of water are for cooling and for dissipating and transporting waste materials.

Water can be heavily used in many ways without permanent damage. However, adding material to water may make it unfit for many uses though not for all.

Major sources of water pollution are municipal sewage, industrial wastes, and agricultural runoff. Organic matter in water requires oxygen for its decomposition and therefore has a large biochemical oxygen demand (BOD). Oxygen depletion can result in fish death and changes in the normal algae community, which leads to visual and odor problems. Nutrients, such as nitrates and phosphates from detergents and agricultural runoff, enrich water and stimulate algae and aquatic plant growth.

Point sources of pollution are easy to identify and resolve. Nonpoint sources of pollution, such as agricultural runoff and mine drainage, are more difficult to detect and control than those from municipalities or industries.

Thermal pollution occurs when an industry returns heated water to its source. Temperature changes in water can alter the kinds and numbers of plants and animals that live in it. The methods of controlling thermal pollution include cooling ponds, cooling towers, and dry cooling towers.

Wastewater treatment consists of primary treatment, a physical settling process; secondary treatment, biological degradation of the wastes; and tertiary treatment, chemical treatment to remove specific components. Two major types of secondary wastewater treatments are the trickling filter and the activated sludge sewage methods.

Groundwater pollution comes from a variety of sources, including agriculture, landfills, and septic tanks. Marine oil pollution results from oil drilling and oil-tanker accidents, runoff from streets, improper disposal of lubricating oil from machines and car crankcases, and intentional discharges from oil tankers during loading or unloading.

Reduced water quality can seriously threaten land use and in-place water use. In the United States and other nations, legislation helps to preserve certain scenic water areas and wildlife habitats. Shorelands and wetlands provide valuable services as buffers, filters, reservoirs, and wildlife areas. Water management concerns of growing importance are groundwater mining, increasing salinity, water diversion, and managing urban water use. Urban areas face several problems, such as providing water suitable for human use, collecting and treating wastewater, and handling storm-water runoff in an environmentally sound manner. Water planning involves many governmental layers, which makes effective planning difficult.

REVIEW QUESTIONS

1. Describe the hydrologic cycle.
2. List several uses of water that are nonconsuming and nonwithdrawing.
3. What is the major industrial use of water?
4. What are the similarities between domestic and industrial water use? How are they different from in-stream use?
5. How is land use related to water quality and quantity? Can you provide local examples?
6. Is the definition of water pollution related to the water's intended use? Give some examples to support your answer.
7. What is biochemical oxygen demand? How is it related to water quality?
8. How is water pollution related to agricultural activities? Why is this a growing concern?
9. Differentiate between point and nonpoint sources of water pollution?
10. How are most industrial wastes disposed of? How has this changed over the past 25 years?
11. What is thermal pollution? How can it be controlled?
12. Describe primary, secondary, and tertiary sewage treatment.
13. What are the types of wastes associated with agriculture?
14. Why is storm-water management more of a problem in an urban area than in a rural area?
15. What is the Federal Wild and Scenic Rivers Act? Why is it important?
16. Define groundwater mining.
17. How does irrigation increase salinity?
18. What are the three major water services provided by metropolitan areas?

KEY TERMS

activated sludge sewage treatment   302
agricultural products   305
agricultural runoff   299
aquiclude   286
aquifer   286
aquitard   286
artesian aquifer   286
biochemical oxygen demand (BOD)   297

confined aquifer   286
domestic water   288
groundwater   285
groundwater mining   308
hydrologic cycle   285
industrial uses   293
in-stream uses   292
irrigation   289
landfills   305
limiting factor   299

nonpoint source   299
point source   299
potable waters   285
primary sewage treatment   302
runoff   286
secondary sewage treatment   302
septic tanks   305
storm-water runoff   304

surface impoundments   305
tertiary sewage treatment   302
thermal pollution   301
transpiration   285
unconfined aquifer   286
underground storage tanks   305
water diversion   311
water table   286

RELATED INTERNET SITES

Frequently asked questions on water treatment and tastes and smells of drinking water.
   http://www.siouxlan.com/water/faq.html
Water treatment information source.
   http://www.culligan.ca/
Information on the water quality of rivers and lakes in North America.
   http://www.trader.com/users/5012/0614/water.htm
Everything you want to know about solid waste treatment and management.
   http://www.the-space.com/sws/
Award-winning site focusing on humanity's attempt to adapt to desert environments.
   http://www.dri.edu/

Environmental geochemistry and mineralogy. Topics concerning surface water, soils, and the mining industry.
   http://www.mineral.tu-freiberg.de/index_en.html
Pacific Islanders and their ongoing struggle to rehabilitate their islands after phosphate mining.
   http://www.ion.com.au/~banaban/
Workers to Protect and Conserve Idaho's Forestry, Grazing, Mining, Water, and Endangered Species.
   http://www.desktop.org/iru

PART five

Most people have strong feelings about the quality of the environment. However, these feelings vary greatly because people do not agree on what is good or bad for environmental quality. Economics, perception of risk, and politics have become interwoven into the environmental quality equation. There are great differences of opinion about what environmental quality should be or what the current quality of the environment is. The amount of pollution can often be quantified, and pollution may be more or less serious depending on local conditions. Although pollution, by definition, is harmful, some amounts of pollution can be tolerated, depending on the situation.

Chapter 16 provides background information on risk assessment and economic principles useful in evaluating the costs of pollution and the value gained from eliminating pollution.

Chapters 17, 18, and 19 discuss air pollution, solid waste disposal, and hazardous and toxic wastes, respectively. These chapters also describe approaches used to deal with these problems. Throughout the selection, economic and political realities are discussed as a part of the pollution equation.

Chapter 20 discusses how environmental decisions are made. It is a complex process, integrating public input, economic demands, political posturing, and legal requirements. As complex as the process is, the decisions we have made have significantly improved the quality of our environment over the past several decades.

Risk and Cost: Elements of Decision Making

Objectives

After reading this chapter, you should be able to:

- Describe why the analysis of risk has become an important tool in environmental decision making.
- Understand the difference between risk assessment and risk management.
- Describe the issues involved in risk management.
- Understand the difference between true and perceived risks.
- Define what an economic good or service is.
- Understand the relationship between the available supply of a commodity or service and its price.
- Understand how and why cost-benefit analysis is used.
- Understand the concept of sustainable development.
- Understand environmental external costs and the economics of pollution prevention.
- Understand the market approach to curbing pollution.

Chapter Outline

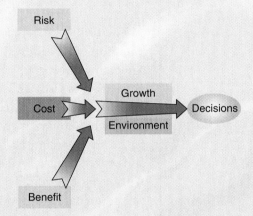

Measuring Risk

Two factors are primary in many decisions in life: risk and cost. We commonly ask such questions as "How likely is it that someone will be hurt?" and "What is the cost of this course of action?" Environmental decision making is no different. If a new air-pollution regulation is contemplated, industry will be sure to point out that it will cost a considerable amount of money to put these controls in place and will reduce profitability. Citizens will point out that their tax money will have to support another governmental bureaucracy. On the other side, advocates will point out the reduced risk of illnesses and the reduced cost of health care for people who live in areas of heavy air pollution.

Risk analysis has become an important decision-making tool at all levels of society. In the area of environmental concerns, assessing and managing risks help us determine what environmental policies are appropriate. The analysis of risk generally involves a probability statement. **Probability** is a mathematical statement about how likely it is that something will happen. Probability is often stated in terms like "The probability of developing a particular illness is 1 in 10,000," or "The likelihood of winning the lottery is 1 in 5,000,000." It is important to make a distinction between *probability* and *possibility*. When we say something is *possible*, we are just saying that it could occur. It is a very inexact term. *Probability* defines how likely *possible* events are.

Another important consideration is the consequences of an event. If a disease is likely to make 50 percent of the population ill (the probability of becoming ill is 50 percent) but no one dies, that is very different from analyzing the safety of a dam, which if it failed would cause the deaths of thousands of people downstream. We would certainly not accept a 50 percent probability that the dam would fail. Even a 1 percent probability in that case would be unacceptable. The assessment and management of risk involve an understanding of probability and the consequences of decisions. (See figure 16.1.)

Risk assessment involves analyzing a risk to determine the probability of an ad-

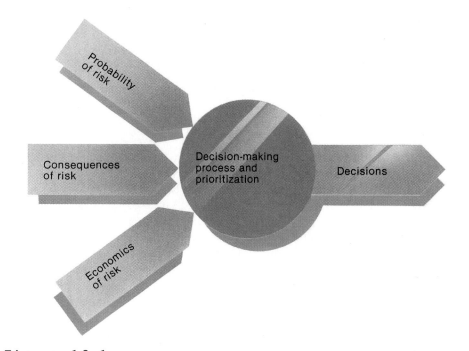

Figure 16.1

Decision-Making Process   The assessment, cost, and consequences of risks are all important to the decision-making process.

verse effect. Risk management is a much broader task that includes assessment and the consequences of risks in the decision-making process. We will look at both these tasks in the next two sections.

Risk Assessment

Environmental **risk assessment** is the use of facts and assumptions to estimate the probability of harm to human health or the environment that may result from exposures to pollutants, toxic agents, or management decisions. What risk assessment provides for environmental decision makers is an orderly, clearly stated, and consistent way to deal with scientific issues when evaluating whether a hazard exists and what the magnitude of the hazard may be.

Calculating the hazardous risk to humans of a particular activity, chemical, or technology is difficult. If a technology is well known, scientists use probabilities based on past experience to estimate risks. For example, the risk of developing black lung disease from coal dust in mines is well established. To predict the risks associated with new technology, much less

accurate statistical probabilities, based on models rather than real-life experiences, must be used.

Risks associated with new chemicals are difficult to quantify. While animal tests are widely accepted in predicting whether or not a chemical will cause cancer in humans, their use in predicting how many cancers will be caused in a group of exposed people is still very controversial. Most risk assessments are *estimates* of the probability that a person who is exposed to certain chemicals will develop cancer or other negative effects. (See figure 16.2.)

Such estimates typically are based on broad assumptions to ensure that a lack of complete knowledge does not result in an underestimation of the risk. For example, people may be more or less sensitive to the effects of certain chemicals than the laboratory animals studied. Also, people vary in their sensitivity to cancer-causing compounds. Thus, what may present no risk to one person may be a high risk to others. Persons with breathing difficulties are more likely to be adversely affected by high levels of air pollutants than are healthy individuals. In addition,

environmental
close-up

What's In a Number?

Risk values are often stated, shorthand-fashion, as a number. When the risk concern is cancer, the risk number represents a probability of occurrence of additional cancer cases. For example, such an estimate for Pollutant X might be expressed as 1×10^{-6}, or simply 10^{-6}. This number can also be written as 0.000001, or one in a million—meaning one additional case of cancer projected in a population of one million people exposed to a certain level of Pollutant X over their lifetimes. Similarly, 5×10^{-7}, or 0.0000005, or five in *100* million, indicates a potential risk of five additional cancer cases in a population of 100 million people exposed to a certain level of the pollutant. These numbers signify incremental cases above the background cancer incidence in the general population. American Cancer Society statistics indicate that the background cancer incidence in the general population is one in three over a lifetime.

If the effect associated with Pollutant X is not cancer but another health effect, perhaps neurotoxicity (nerve damage) or birth defects, then numbers are not typically given as probability of occurrence, but rather as levels of exposure estimated to be without harm. This often takes the form a reference dose (RfD). An RfD is typically expressed in terms of milligrams (of pollutant) per kilogram of body weight per day, e.g., 0.004 mg/kg-day. Simply described, an RfD is a rough estimate of daily exposure to the human population (including sensitive subgroups) that is likely to be without appreciable risk of deleterious effects during a lifetime. The uncertainty in an RfD may be one or several orders of magnitude (i.e., multiples of 10).

What's in a number? The important point to remember is that the numbers by themselves don't tell the whole story. For instance, even though the numbers are identical, a cancer risk value of 10^{-6} for the "average exposed person" (perhaps someone exposed through the food supply) is not the same thing as a cancer risk of 10^{-6} for a "most exposed individual" (perhaps someone exposed from living or working in a highly contaminated area). It's important to know the difference. Omitting the qualifier "average" or "most exposed" incompletely describes the risk and would mean a failure in risk communication.

A numerical estimate is only as good as the data it is based on. Just as important as the *quantitative* aspect of risk characterization (the risk numbers), then, are the *qualitative* aspects. How extensive is the data base supporting the risk assessment? Does it include human epidemiological data as well as experimental data? Does the laboratory data base include less data on more than one species? If multiple species were tested, did they all respond similarly to the test substance? What are the "data gaps," the missing pieces of the puzzle? What are the scientific uncertainties? What science policy decisions were made to address these uncertainties? What working assumptions underlie the risk assessment? What is the overall confidence level in the risk assessment? All of these qualitative considerations are essential to deciding what reliance to place on a number and to characterizing a potential risk.

Source: Data from EPA Journal.

the estimate of human risk is based on extrapolation from animal tests in which high, chronic doses are used. Human exposure is likely to be lower or infrequent. Because of all these uncertainties, government regulators have decided to err on the side of safety to protect the public health. That approach has been criticized by those who say it carries protection to the extreme, usually at the expense of industry.

Over the past decade, risk assessment has had its largest impact in regulatory practices involving cancer-causing chemicals called **carcinogens.** In the United States, for example, the decisions to continue registration of pesticides, to list substances as hazardous air pollutants under the Clean Air Act, and to regulate water contaminants under the Safe Drinking Water Act depend to a large degree on the risk assessments for the substances in question.

Risk assessment analysis is also being used to help set regulatory priorities and support regulatory action. Those chemicals or technologies that have the highest potential to cause damage to health or the environment receive attention first, while those perceived as having minor impacts receive less immediate attention. Medical waste is perceived as

having high risk, and laws have been enacted to minimize the risk, while the risk associated with the use of fertilizer on lawns is considered minimal and is not regulated.

The science supporting environmental regulatory decisions is complex and rapidly evolving. Many of the most important threats to human health and the environment are highly uncertain. Risk assessment quantifies risk and states the uncertainty that surrounds many environmental issues. This can help institutions research and plan in a way that is consistent with scientific and public concern for environmental protection.

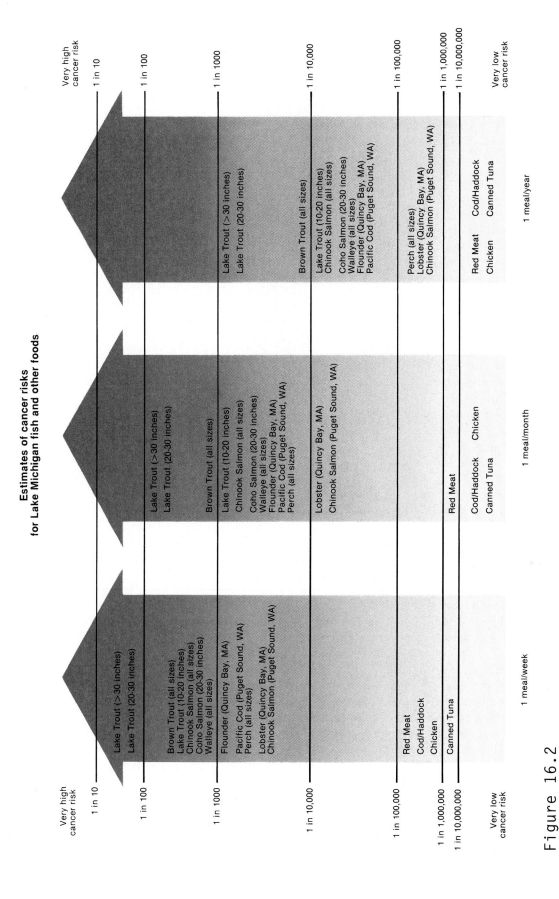

Figure 16.2

Cancer Risk and Fish Consumption A study by the National Wildlife Federation provides estimates for cancer risks associated with the consumption of sport fish.

Risk Management

Risk management is a decision-making process that involves risk assessment, technological feasibility, economic impacts, public concerns, and legal requirements. Risk management includes:

1. Deciding which risks should be given the highest priority
2. Deciding how much money will be needed to reduce each risk to an acceptable level
3. Deciding where the greatest benefit would be realized by spending limited funds
4. Deciding how much risk is acceptable
5. Deciding how the plan will be enforced and monitored

Risk management raises several issues. With environmental concerns such as acid rain, ozone depletion, and hazardous waste, the scientific basis for regulatory decisions is often controversial. As was previously mentioned, hazardous substances can be tested, but only on animals. Are animal tests appropriate for determining impacts on humans? There is not an easy answer to this question. Dealing with global warming, ozone depletion, and acid rain require projecting into the future and estimating the magnitude of future effects. Will sea level rise? How many lakes will become acidified? How many additional skin cancers will be caused by depletion of the ozone layer? Estimates from equally reputable sources vary widely. Which ones do we believe?

The politics of risk management frequently focus on the adequacy of the scientific evidence. The scientific basis can be thought of as a kind of problem definition. Science determines that some threat or hazard exists, but because scientific facts are open to interpretation, there is controversy. For example, it is a fact that dioxin is a highly toxic material known to cause cancer in laboratory animals. It is also very difficult to prove that human exposure to dioxin has led to the development of cancer, although high exposures have resulted in acne in exposed workers. Acne is a common result of exposure to molecules like dioxin.

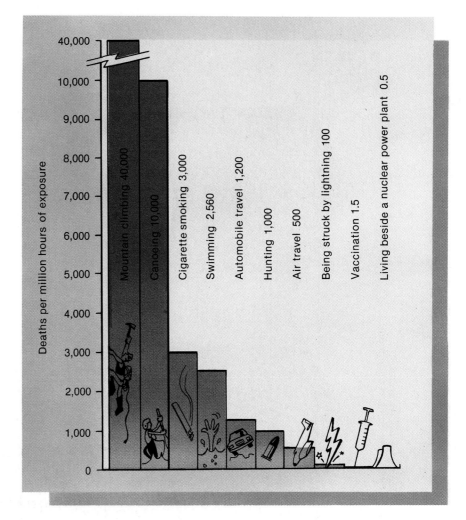

Figure 16.3

Risks of Death In many cases, historical data can be used to predict the risk of exposure to specific activities, chemicals, or technologies.

This is why problem definition is so important. Defining the problem helps to determine the rest of the policy process (making rules, passing laws, or issuing statements). If a substance poses little or no risk, then policy action is unnecessary. For example, some observers believe chemicals pose many threats that need to be addressed. Others believe chemicals pose little threat; instead, they see scare tactics and government regulations as unnecessary attacks on businesses. Logging forests poses risks of soil erosion and the loss of some animal species. The timber industry sees these risks as minimal and as a threat to its economic well-being, while many environmentalists consider the risks unacceptable. These and similar disagreements are often serious public-relations problems for both government and business because most of the public has a poor understanding of the risks they accept daily.

True and Perceived Risks

People often overestimate the frequency and seriousness of dramatic, sensational, well-publicized causes of death and underestimate the risks from more familiar causes that claim lives one by one. Risk estimates by "experts" and by the "public" on many environmental problems differ significantly. This discrepancy and the reasons for it are extremely important because the public generally does not trust experts to make important risk decisions alone.

While public health and environmental risks can be minimized, eliminating all risks is impossible. Almost every daily activity—driving, walking, working—involves some element of risk. (See figure 16.3.)

From a risk management standpoint, whether one is dealing with a site-specific situation or a national standard, the deciding question ultimately is: What degree of risk is acceptable? In general, we are not talking about a "zero risk" standard, but rather a concept of **negligible risk:** At what point is there really no significant health or environmental risk? At what point is there an adequate safety margin to protect public health and the environment?

Risk management involves comparing the estimated true risk of harm from a particular technology or product with the risk of harm perceived by the general public. The public generally perceives involuntary risks, such as nuclear power plants or nuclear weapons, as greater than voluntary risks, such as drinking alcohol or smoking. In addition, the public perceives newer technologies, such as genetic engineering or toxic-waste incinerators, as greater risks than more familiar technologies, such as automobiles and dams. Many people are afraid of flying for fear of crashing; however, automobile accidents account for a far greater number of deaths—about 45,000 in the United States each year, compared to less than a thousand annually from plane crashes. (See figure 16.4.)

A fundamental problem facing governments today is how to satisfy people who are concerned about a problem that experts state presents less hazard than another less visible problem, especially when economic resources to deal with the problems are limited. Debate continues over the health concerns raised by asbestos, dioxin, radon, and Alar. (See Environmental Close-Up: The Alar Controversy.)

Some researchers argue that the public is frequently misled by the politics of public health and environmental safety. This is understandable since many prominent people become involved in such issues and use their public image to encourage people to look at issues from a particular point of view.

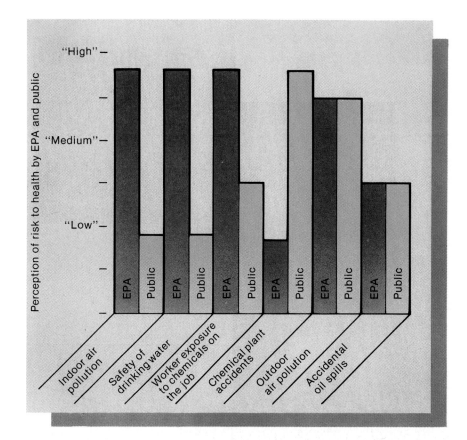

Figure 16.4

Perception of Risk   Professional regulators and the public do not always agree on what risks are.

Whatever the issue, it is hard to ignore the will of the people, particularly when sentiments are firmly held and not easily changed. A fundamental issue surfaces concerning the proper role of a democratic government and other organizations in a democracy when it comes to matters of risk. Should the government focus available resources and technology where they can have the greatest tangible impact on human and ecological well-being, or should it focus them on problems about which the public is most upset? What is the proper balance? For example, adequate prenatal health care for all pregnant women would have a greater effect on the health of children than would removing asbestos from all school buildings.

Obviously, there are no clear answers to these questions. However, experts and the public are both beginning to realize that they each have something to offer concerning how we view risk. Many risk experts who have been accustomed to looking at numbers and probabilities are now conceding that there is rationale for looking at risk in broader terms. At the same time, the public is being supplied with more data to enable them to make more informed judgments.

Throughout this discussion of risk assessment and management we have made numerous references to costs and economics. It is not economically possible to eliminate all risk. As risk is eliminated, the cost of the product or service increases. Many environmental issues are difficult to evaluate from a purely economic point of view, but economics is one of the tools useful to analyzing any environmental problems.

Economics and the Environment

Environmental problems are primarily economic problems. While this may be

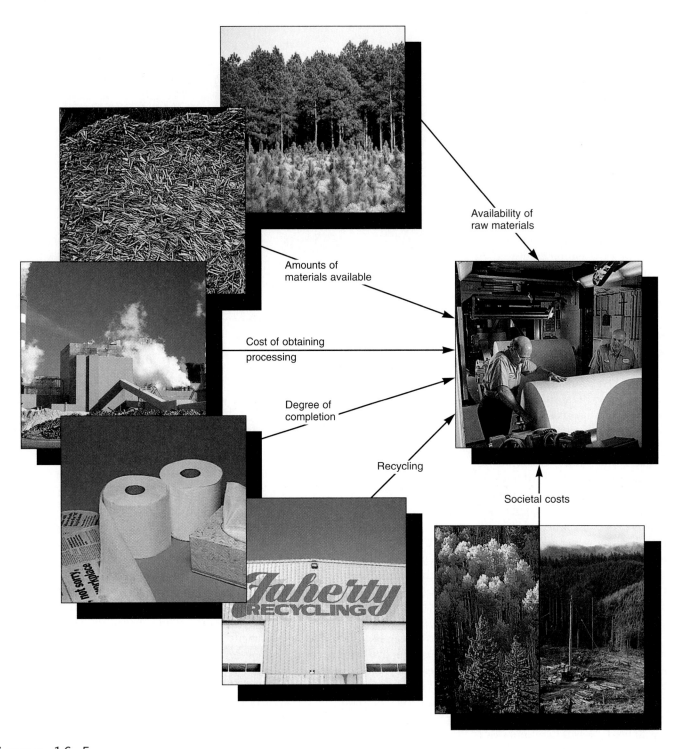

Figure 16.5

Factors That Determine Supply   The supply of a good or service is dependent on the several factors shown here.

an overstatement, it often is difficult to separate economics from environmental issues or concerns. Basically, economics deals with resource allocation. It is a description of how we value goods and services. We are willing to pay for things or services we value highly and are unwilling to pay for things we think there is plenty of. For example, we will readily pay for a warm, safe place to live but would be offended if someone suggested that we pay for the air we breathe.

Our goal as a society is to seek long-term economic growth that creates jobs while improving and sustaining the environment. Achieving this goal requires an environmental strategy that repairs past environmental damage; helps us shift from waste management to pollution prevention; and uses valuable resources more efficiently.

In the nearly three decades since the first Earth Day, many nations have made considerable progress in responding to threats to public health and the environment. Yet major challenges remain. We can put in place a set of policies and programs that will establish a new course for the development and use of environmental technologies into the next century. We must, however, broaden our environmental tool kit, replacing those instruments that are no longer effective with a new set of tools designed to meet today's challenges and tomorrow's needs.

Economic Concepts

An economic good or service can be defined as anything that is scarce. Scarcity exists whenever the demand for anything exceeds its supply. We live in a world of general scarcity. **Resources** are anything that contributes to making desired goods and services available for consumption. Resources are limited, relative to the desires of humans to consume. The **supply** is the amount of a good or service available to be purchased. **Demand** is the amount of a product that consumers are willing and able to buy at various prices. In economic terms, supply depends on:

1. The raw materials available to produce a good or service using present technology
2. The amounts of those materials available
3. The costs of extracting, shipping, and processing the raw materials
4. The degree of competition for those materials among users
5. The feasibility and cost of recycling already used material
6. The social and institutional arrangements that might have an impact (See figure 16.5.)

The relationship between available supply of a commodity or service and its price is known as a **supply/demand curve.** (See figure 16.6.) The price of a product or service reflects the strength of

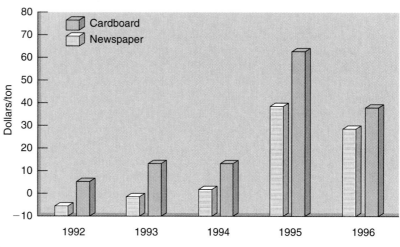

a.

b.

Figure 16.6

Supply and Demand (*a*) As a result of increased interest in recycling, the supply of recycled newspapers has increased substantially between 1970 and 1993. The demand changed very little; therefore, the price in 1993 was extremely low. (*b*) Beginning in 1994, the demand for recycled newspaper and cardboard increased, thus the price rose significantly.

Source: (*b*) *Data from* Recycling Times.

the demand for and the availability of the commodity. When demand exceeds supply, the price rises. Cost increases cause people to seek alternatives or to decide not to use a product or service, which results in lower demand.

For example, food production depends heavily on petroleum for the energy to plant, harvest, and transport food crops. In addition, petrochemicals are used to make fertilizer and chemical pest-control agents. As petroleum prices rise, farmers reduce their petroleum use. Perhaps they farm less land or use less fertilizer or pesticide. Regardless, the price of food must rise as the price of petroleum rises. As the prices of certain foods rise, consumers seek less costly forms of food.

When the supply of a commodity exceeds the demand, producers must lower their prices to get rid of the product, and eventually, some of the producers go out of business. Ironically, this happens to farmers when they have a series of good years. Production is high, prices fall, and some farmers go out of business.

Market-Based Instruments

With the growing interest in environmental protection during the past decade, policy makers are examining new methods to reduce harm to the environment. One area of growing interest is Market-Based Instruments (MBIs). MBIs provide an alternative to the common command-and-control legislation because they use economic forces and the ingenuity of entrepreneurs to achieve a high degree of environmental protection at a low cost. Instead of dictating how industry should conduct its activities, MBIs provide incentives by imposing costs on pollution-causing activities. This approach allows companies to decide for themselves how best to achieve the required level of environmental protection.

MBIs can be grouped into five basic categories:

1. **Information Programs**
   These programs rely on informed consumer's market choices to reduce environmental problems. Information about the environmental or risk consequences of choices make clear to consumers that it is in their personal interest to change their decisions or behavior. Examples include radon or lead testing or labeling pesticide products. Another type of program, such as the Toxic Release Inventory in the United States, discloses information on environmental releases by polluters. This provides corporations with incentives to improve their environmental performance to enhance their public image.

2. **Tradable emissions permits**
   These permits give companies the right to emit specified quantities of pollutants. Companies that emit less than the specified amounts can sell their permits to other firms or "bank" them for future use. Businesses responsible for pollution have an incentive to internalize the external cost they were previously imposing on society: If they clean up their pollution sources, they can realize a profit by selling their permit to pollute. Once a business recognizes the possibility of selling its permit, it sees that pollution is not costless. The creation of new markets (the permits) to reduce pollution reduces the external aspects to waste disposal and makes them internal costs, just like costs of labor and capital.

3. **Emission fees, taxes, and charges**
   These fees provide incentives for environmental improvement by making environmentally damaging activity or products more expensive. Businesses and individuals control the level of pollution to where it is cheaper to abate the pollution than to pay the charges. Emissions fees can be useful when pollution is coming from many small sources, such as vehicular emissions or agricultural runoff, where direct regulation or trading schemes are impractical. Taxes contribute to government revenue; charges are used to fund environmental cleanup programs.

4. **Deposit-refund programs**
   These programs place a surcharge on the price of a product, which is refunded when the used product is returned for reuse or recycling.

5. **Subsidies**
   Subsidies may include consumer rebates for purchases of environmentally friendly goods, soft loans for businesses planning to implement environmental products, and other monetary incentives designed to reduce the costs of improving environmental performance.

A **subsidy** is a gift from government to a private enterprise that is considered important to the public interest. This is particularly true when the private enterprise is having ecnomic difficulty. Agriculture, transportation, space technology, and communication are frequently subsidized by governments. These gifts, whether loans, favorable tax situations, or direct grants, are all paid for by taxes on the public.

Subsidies are costly in two ways: First, the bureaurcracy necessary to administer a subsidy costs money, and the subsidy is an indirect way of keeping the market price of a product low. The actual cost is higher because these subsidy costs must be added to the market price to arrive at the product's true cost. Second, in many cases, subsidies encourage activities that in the long term may be detrimental to the environment. For example, the transportation subsidies for highway construction encourage use of inefficient individual automobiles. Higher taxes on automobile use to cover the cost of building and repairing highways would encourage the use of more energy-efficient public transport.

While research is still needed to determine how MBIs should be structured and used, there has been sufficient experience to justify their further deployment as a means of attaining environmental objectives at least possible cost. Successful programs so far include a sulfur dioxide trading market, lead phase-down banking and trading, and hundreds of pay-by-the-bag trash collection programs in many countries.

Although the various economic instruments discussed have their own niches, they can be used effectively in combination. For example, trading or pricing approaches work better if supported by information programs: Communities that adopted pay-by-the-bag systems of trash

environmental close-up

The Alar Controversy

In 1989, U.S. consumers experienced a public-health scare concerning the pesticide Alar (daminozide), a plant growth regulator that was used on apples to postpone fruit drop, enhance the fruit's color and shape, and extend storage life. The scare was raised by several scientists and covered extensively by the media.

The public was alarmed to the point of panic. Worried mothers dumped apple juice down the drain, despite the EPA's repeated assurances that such measures were unnecessary. Apples were banished from school cafeterias. With each new event reported by the media, consumers were further unsettled and confused as points of disagreement between the EPA and outside groups were raised on the risks of Alar.

The initial panic concerning Alar has subsided somewhat, but an uneasy confusion remains, and consumers have been left with a lingering doubt about the safety of their food. In general, the public has limited patience with extended deliberations by scientists and regulators over a chemical's potential harmful effects. While scientists and regulatory officials are concerned with questions of scientific uncertainty and statistical risk assessment, consumers, who generally do not speak the language of risk assessment, tend to ask very direct questions: Is it safe to eat apples? Is it safe for my child to consume apple products? Does Alar cause cancer?

Two questions are implicit in the question: Does Alar cause cancer? First, is Alar a known human carcinogen? The answer is no; scientists do not have direct evidence in humans that traces actual cancer cases to Alar exposure. In fact, comparatively few chemicals in the world have been demonstrated beyond doubt, on the basis of epidemiological data, to cause cancer in humans.

Second, does Alar cause cancer in laboratory animals? The answer is yes; Alar and its breakdown product, called unsymmetrical dimethylhydrazine (UDMH), have increased the incidence of malignant tumors in mice.

It is difficult to understand cancer risks, or any kind of risk, without a meaningful frame of reference. For perspective, one of the key phrases consumers should keep in mind in the Alar case, and generally in cases of chemicals said to pose cancer risks, is "long term." In evaluating the risks of pesticides to consumers, the EPA uses the working assumption that dietary exposure to the pesticide occurs over a lifetime (70 years). This is just one of many assumptions that the EPA factors into its chemical risk analysis.

The truth, however, is that hard evidence on the effects of pesticides is generally limited to cases where short-term, highly concentrated exposure has caused acute toxic poisoning in humans or killed important nontarget organisms in significant numbers. Such acute toxic effects are immediately apparent.

Most risk scenarios are not so easy to assess, and this is especially true of chronic or delayed health effects, such as cancer, reproductive dysfunctions, or effects on the newborn. Such chronic or delayed effects do not become apparent for a long time, and when they do occur, it is almost always impossible to trace them with certainty to exposures to specific chemicals. Instead, the evidence at hand consists of the raw materials of risk assessment: animal data tabulations, cancer potency estimates based on animal study results, food consumption statistics, and exposure estimates. As the Alar case demonstrated, such data may be used selectively and inappropriately to make calculations that misrepresent pesticide risks.

Consumer education and risk communication will become increasingly necessary if pesticide decision making is to take place in an atmosphere that is relatively free of fear, confusion, and unnecessary economic disruption.

---

disposal had fewer problems if households were given adequate information well in advance. Environmental tax systems can incorporate trading features (e.g., taxes can be levied on net emissions after trades). It will be increasingly important to use the various MBI methods together. The challenge is to design the most appropriate instruments to deal with environmental problems, bearing in mind the relevant policy objectives: steady progress in reducing risks, cost-effectiveness, encouragement of technological innovation, fairness, and administrative simplicity.

Cost-Benefit Analysis

People use **cost-benefit analysis** to determine whether a policy generates more social costs than social benefits, and if benefits outweigh costs, how much spending would obtain optimal results. Steps in cost-benefit analysis include:

1. Identification of the project to be evaluated
2. Determination of all impacts, favorable and unfavorable, present and future, on all of society
3. Determination of the value of those impacts, either directly through market values or indirectly through price estimates
4. Calculation of the net benefit, which is the total value of positive

impacts less the total value of negative impacts

For example, the cost of reducing the amount of lead in drinking water in the United States to acceptable limits is estimated to be about $125 million a year. The benefits to the nation's health from such a program are estimated at nearly $1 billion per year. Thus under a cost-benefit analysis, the program is economically sound. Recycling of solid waste, which was once cost-prohibitive, is now cost-effective in many communities because the costs of landfills have risen dramatically. Table 16.1 gives examples of the kinds of costs and benefits involved in improving air quality. Although not a complete list, the table indicates the kinds of considerations that go into a cost-benefit analysis. Some of these are easy to measure in monetary terms; others are not.

Concerns about the Use of Cost-Benefit Analysis

Does everything have an economic value? Critics of cost-benefit analysis raise this point, among others. Some people argue that, if economic thinking pervades society, many simple noneconomic values like beauty or cleanliness can survive only if they prove to be "economic." (See figure 16.7.)

It has long been the case in many developed countries that major projects, especially those undertaken by the government, require some form of cost-benefit analysis with respect to environmental impacts and regulations. In the United States, for example, such requirements were established by the National Environmental Policy Act of 1969, which requires environmental impact statements for major government-supported projects. Increasingly, similar analyses are required for projects supported by national and international lending institutions such as the World Bank.

There are clearly benefits to requiring such analysis. Although environmental issues must be considered at some point during project evaluation, efforts to do so are hampered by the difficulty of assigning specific value to environmental resources.

Table 16.1

Costs and Benefits of Improving Air Quality

Costs	Benefits
Installation and maintenance of new technology 1. Scrubbers on smokestacks 2. Automobile emissions control	Reduced deaths and disease
Redesign of industries and machines	Reduced plant and animal damage
Additional energy costs to industry and public	Lower cleaning costs for industry and public
Unemployment as some industries go out of business	More clear, sunny days; better visibility
Retraining of employees to use new technology	Less eye irritation and fewer respiratory problems
Costs associated with monitoring and enforcement	Fewer odor problems

A flowchart for a cost/benefit analysis.

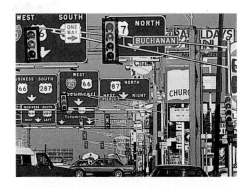

Figure 16.7

Does Everything Have an Economic Value? The use of water and land is often based on the economic benefits obtained. Are there other benefits that cannot be measured economically?

In cases of Third World development projects, these already difficult environmental issues are made more difficult by cultural and socioeconomic differences. A less-developed country, for example, may be less inclined to insist on or be able to afford expensive emissions-treatment technology on a project that will provide jobs and economic development.

One particularly compelling critique of cost-benefit analysis is that for analysis to be applied to a specific policy, the analyst must decide which preferences count—that is, which preferences have "standing" in cost-benefit analysis. In theory, cost-benefit analysis should count all benefits and costs associated with the policy under review, regardless of who benefits or bears the costs. In practice, however, this is not always done. For example, if a cost is spread thinly over a great many people, it may not be recognized as a cost at all. The cost of air pollution in many parts of the world could fall into such a category. Debates over how to count benefits and costs for future generations, inanimate objects such

as rivers, and nonhumans, such as endangered species, are also common.

A noted critic of cost-benefit analysis is E. F. Schumacher, who states in his book *Small Is Beautiful* (Harper and Row, 1973):

> Cost-benefit analysis is a procedure by which the higher is reduced to the level of the lower and the priceless is given a price. All it can do is lead to self-deception or the deception of others; for to undertake to measure the immeasurable is absurd and constitutes but an elaborate method of moving from preconceived notions to foregone conclusions; all one has to do to obtain the desired results is to impute suitable values to the immeasurable costs and benefits. The logical absurdity, however, is not the greatest fault of the undertaking; what is worse, and destructive of civilization, is the pretense that everything has a price or, in other words, that money is the highest of all values.

While Schumacher's objectives to cost-benefit analysis may be somewhat idealistic, he does raise some issues that should be discussed. Is it possible to separate economic issues from environmental issues? (See table 16.2.)

Economics and Sustainable Development

The past president of the Japan Economic Research Center, Saburo Okita, once stated that a slowdown of economic growth is needed to prevent further deterioration of the environment. Whether or not a slowdown is necessary provokes sharp differences of opinion.

One school of thought argues that economic growth is essential to finance the investments necessary to prevent pollution and to improve the environment by a better allocation of resources. A good school of thought, which is also progrowth, stresses the great potential of science and technology to solve problems and advocates relying on technological

Table 16.2

Environmental Industry Revenue

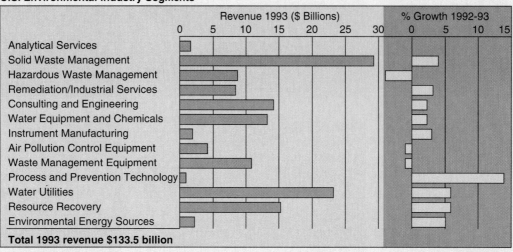

U.S. Environmental Industry Segments

Revenue 1993 ($ Billions): Analytical Services, Solid Waste Management, Hazardous Waste Management, Remediation/Industrial Services, Consulting and Engineering, Water Equipment and Chemicals, Instrument Manufacturing, Air Pollution Control Equipment, Waste Management Equipment, Process and Prevention Technology, Water Utilities, Resource Recovery, Environmental Energy Sources

% Growth 1992-93

Total 1993 revenue $133.5 billion

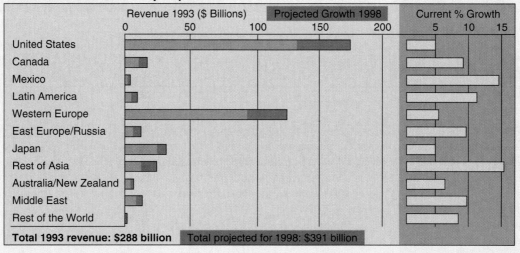

Global Environmental Industry Projections

Revenue 1993 ($ Billions) — Projected Growth 1998 / Current % Growth: United States, Canada, Mexico, Latin America, Western Europe, East Europe/Russia, Japan, Rest of Asia, Australia/New Zealand, Middle East, Rest of the World

Total 1993 revenue: $288 billion   Total projected for 1998: $391 billion

Pollution cleanup and control has developed into a major business.

From Environmental Business Journal, *Environmental Business International, Environmental Business Publishing, San Diego, CA. Reprinted by permission.*

---

advances to solve environmental problems. Neither of these schools of thought sees any need for fundamental changes in the nature and foundation of economic policy. Environmental issues are viewed mainly as a matter of setting priorities in the allocation of resources.

A newer school of economic thought believes that economic and environmental well-being are mutually reinforcing goals that must be pursued simultaneously if either one is to be reached. Economic growth will create its own ruin if it con-

tinues to undermine the healthy functioning of Earth's natural systems or to exhaust natural resources. It is also true that healthy economies are most likely to provide the necessary financial investments in environmental protection. For this reason, one of the principal objectives of environmental policy must be to ensure a decent standard of living for all. The solution, at least in the broad scope, would be for a society to manage its economic growth in such a way as to do no irreparable damage to its environment.

This concept is referred to as **sustainable development.** The World Commission on Environment and Development defined sustainable development as "development that meets the needs of the present without compromising the ability of future generations to meet their own needs," The term has been criticized as ambiguous and open to a wide range of interpretations, many of which are contradictory. The confusion arises because "sustainable growth" and "sustainable use" have been used interchangeably, as

environmental
close-up

"Green" Advertising Claims— Points to Consider

Like many consumers, you may be interested in buying products that are less harmful to the environment. You have probably seen products with such "green" claims as "environmentally safe," "recyclable," "degradable," or "ozone friendly." But what do these claims really mean? How can you tell which products really are less harmful to the environment? Here are some pointers to help you decide.

1. Look for environmental claims that are specific. Read product labels to determine whether they have specific information about the product or its packaging. For example, if the label says "recycled," check how much of the product or packaging is recycled. Labels with "recyclable" claims mean that these products can be collected and made into useful products. This is relevant to you, however, only if this material is collected for recycling in your community.

2. Be wary of overly broad or vague environmental claims. These claims provide little information to help you make purchasing decisions. Labels with unqualified claims that a product is "environmentally friendly," "eco-safe," or "environmentally safe" have little meaning, for two reasons. First, all products have some environmental impact, though some may have less impact than others. Second, these phrases alone do not provide the specific information needed to compare products and packaging on their environmental merits.

3. Some products claim to be "degradable." Degradable materials will not help save landfill

space. Biodegradable materials, like food and leaves, break down and decompose into elements found in nature when exposed to air, moisture, and bacteria or other organisms. Photodegradable materials, usually plastics, disintegrate into smaller pieces when exposed to enough sunlight. Either way, however, degradation of any material occurs very slowly in landfills, where most solid waste is sent. That is because modern landfills are designed to minimize the entry of sunlight, air, and moisture. Even organic materials like paper and food may take decades to decompose in a landfill.

GENUINE
RECYCLED
PAPER

30% POST
CONSUMER
WASTE

a.   b.   c.

Three types of recycling symbols are commonly used in the United States. (a) This symbol simply means that the object is potentially recyclable, not that it has been or will be recycled. (b) This symbol indicates that a product contains recycled material, but it does not indicate how much recycled material is in the product (it could be only a very small amount). (c) This symbol states explicitly the percentage of recycled content found in the product.

if their meanings were the same. They are not. "Sustainable growth" is a contradiction in terms: Nothing physical can grow indefinitely. "Sustainable use" is applicable only to renewable resources: it means using them at rates within their capacity for renewal.

By balancing economic requirements with ecological concerns, the needs of the people are satisfied without jeopardizing the prospects of future generations. While this concept may seem to be common sense, the history of the world shows that

it has not been a common practice. A major obstacle to sustainable development in many countries is a social structure that gives most of the nation's wealth to a tiny minority of its people. It has been said that a person who is worrying about his next meal is not going to listen to lectures on protecting the environment. What to residents in the Northern Hemisphere seem like some of the worst environmental outrages—cutting rain forests to make charcoal for sale as cooking fuel, for example—are often commit-

ted by people who have no other form of income.

The disparities that mark individual countries are also reflected in the planet as a whole. Most of the wealth is concentrated in the Northern Hemisphere. From the Southern Hemisphere's point of view, it is the rich world's growing consumption patterns—big cars, refrigerators, and climate-controlled shopping malls—that are the problem. The problem for the long term is that people in developing countries now want those consumer items

Pollution Prevention Pays!

Extended product responsibility (EPR) is an emerging principle of resource conservation and pollution prevention. EPR advocates using a lifecycle perspective to identify pollution prevention and resource conservation opportunities that maximize eco-efficiency. Under this principle, there is assumed responsibility for the environmental impacts of a product throughout its life cycle, including impacts on the selection of materials for the product, impacts from the manufacturer's production process, and downstream impacts from the use, recycling, or disposal of the product.

While this concept is relatively new, successful examples of it are in operation. For example, several years ago the Minnesota Mining and Manufacturing (3M) Company's European chemical plant in Belgium switched from a polluting solvent to a safer but more expensive water-based substance to make the adhesive for its Scotch™ Brand Magic™ Tape. The switch was not made to satisfy any environmental law in Belgium or the 12-nation European Community. 3M managers were complying with company policy to adopt the strictest pollution-control regulations that any of its subsidiaries is subject to—even in countries that have no pollution laws at all.

Part of the policy is founded on corporate public relations, a response to growing customer demand for "green" products and environmentally responsible companies. But as many North American multinationals with similar global environmental policies are discovering, cleaning up waste, whether voluntarily or as required by law, can cut costs dramatically.

Since 1975, 3M's "Pollution Prevention Pays" program—or 3P—has cut the company's air, water, and waste pollution around the world in half and, at the same time, has saved nearly $600 million in the last 18 years on changes in its manufacturing process, including $100 million overseas. Less waste has meant less spending to comply with pollution-control laws. But, in many cases, 3M actually has made money selling wastes it formerly hauled away. And, because of recycling prompted by the 3P program, it has saved money by not having to buy as many raw materials.

AT&T followed a similar path. In 1990, it set voluntary goals for the company's 40 manufacturing and 2,500 nonmanufacturing sites worldwide. According to its latest estimates, AT&T has (1) reduced toxic air emissions, many caused by solvents used in the manufacture of computer circuit boards, by 73 percent; (2) reduced emissions of chlorofluorocarbons—gases blamed for destroying the ozone layer in the Earth's atmosphere—by 76 percent; and (3) reduced manufacturing waste 39 percent.

Xerox Corporation had focused on recycling materials in its global environment efforts. It provides buyers of its copiers with free United Parcel Service pickup of used copier cartridges, which contain metal-alloy parts that otherwise would wind up in landfills. The cartridges and other parts are now cleaned and used to make new ones. Other recycled Xerox copier parts include power supplies, motors, paper transport systems, printed wiring boards, and metal rollers. In all, 1 million parts per year are remanufactured. The initial design and equipment investment was $10 million. Annual savings total $200 million.

The philosophy of pollution prevention is that pollution should be prevented or reduced at the source whenever feasible. It is increasingly being shown that preventing pollution can cut business costs and thus increase profits. Pollution prevention, then, does make cents!

CONTROL AND PREVENTION TECHNOLOGIES—SOME EXAMPLES

Control/Treatment/Disposal

- Sewage treatment
- Industrial wastewater treatment
- Refuse collection
- Incineration
- Off-site recovery and recycling of wastes
- Landfilling
- Catalytic conversion and oxidation
- Particulate controls
- Flue-gas desulfurization
- Nitrogen oxides control technology
- Volatile organic compound control and destruction
- Contaminated site remediation

Prevention

- Improved process control to use energy and materials more efficiently
- Improved catalysis or reactor design to reduce by-products, increase yield, and save energy in chemical processes
- Alternative processes (e.g., low or no-chlorine pulping)
- In-process material recovery (e.g., vapor recovery, water reuse, and heavy metals recovery)
- Alternatives to chlorofluorocarbons and other organic solvents
- High-efficiency paint and coating application
- Substitutes for heavy metals and other toxic substances
- Cleaner or alternative fuels and renewable energy
- Energy-efficient motors, lighting, heat exchangers, etc.
- Water conservation
- Improved "housekeeping" and maintenance in industry

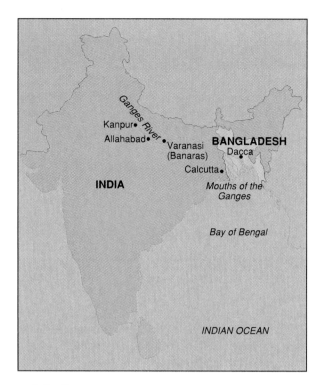

Figure 16.8

Indian Deforestation Causes Floods in Bangladesh Because the Ganges River drains much of India and the country of Bangladesh is at the mouth of the river, deforestation and poor land use in India can result in devastating floods in Bangladesh.

that make life in the industrial world so comfortable as well as environmentally costly. If the standard of living in China and India were to rise to that of Germany or the United States, the environmental impact on the planet would be significant.

If sustainable development is to become feasible, it will be necessary to transform our approach to economic policy. Steps in that direction would include changing the definition of Gross National Product (GNP) to include environmental costs and benefits and redefining productivity to include environmental improvement or decline. The concept of sustainable development may seem simple, but implementing it will be a very complex process.

Historically, rapid exploitation of resources has provided only short-term economic growth, and the environmental consequences in some cases have been incurable. For example, 40 years ago, forests covered 30 percent of Ethiopia. Today, forest covers only 1 percent, and deserts are expanding. One-half of India once was covered by trees; today, only 14 per-

cent of the land is in forests. As the Indian trees and topsoil disappear, the citizens of Bangladesh drown in India's runoff. (See figure 16-8.)

Sustainable development requires choices based on values. Both depend upon information and education, especially regarding the economics of decisions that affect the environment. A. W. Clausen, in his final address as president of the World Bank, noted the

> increasing awareness that environmental precautions are essential for continued economic development over the long run. Conservation, in its broadest sense, is not a luxury for people rich enough to vacation in scenic parks. It is not just a motherhood issue. Rather, the goal of economic growth itself dictates a serious and abiding concern for resource management.

High-income developed nations, such as the United States, Japan, and much of Europe, are in a position to promote sustainable development. They

have the resources to invest in research and the technologies to implement research findings. Some believe that the world should not impose environmental protection standards upon poorer nations without also helping them move into the economic mainstream.

Gaylord Nelson, the founder of the first Earth Day, lists five characteristics that define sustainability:

1. *Renewability:* A community must use renewable resources, like water, topsoil, and energy sources, no faster than they can replace themselves. The rate of consumption of renewable resources cannot exceed the rate of regeneration.
2. *Substitution:* Whenever possible, a community should use renewable resources instead of nonrenewable resources. This can be difficult, because there are barriers to substitution. To be sustainable, a community has to make the transition before the nonrenewable resources become prohibitively scarce.

Table 16.3

Economic Solutions to Pollution and Resource Waste

Solution	Internalizes external costs	Innovation	International competitiveness	Administrative costs	Increases government revenue
Regulation	Partially	Can encourage	Decreased*	High	No
Subsidies	No	Can encourage	Increased	Low	No
Withdrawing harmful subsidies	Yes	Can encourage	Decreased*	Low	Yes
Tradable rights	Yes	Encourages	Decreased*	Low	Yes
Green taxes	Yes	Encourages	Decreased*	Low	Yes
User fees	Yes	Can encourage	Decreased*	Low	Yes
Pollution-prevention bonds	Yes	Encourages	Decreased*	Low	No

Unless more cost-effective and productive technologies are developed.

3. *Interdependence:* A sustainable community recognizes that it is a part of a larger system and that it cannot be sustainable unless the larger system is also sustainable. A sustainable community does not import resources in a way that impoverishes other communities, nor does it export its wastes in a way that pollutes other communities.

4. *Adaptability:* A sustainable community can absorb shocks and can adapt to take advantage of new opportunities. This requires a diversified economy, educated citizens, and a spirit of solidarity. A sustainable community invests in, and uses, research and development.

5. *Institutional commitment:* A sustainable community adopts laws and political processes that mandate sustainability. Its economic systems supports sustainable production and consumption. Its educational systems teach people to value and practice sustainable behavior.

External Costs

Many of the important environmental problems facing the world today arise because modern production techniques and consumption patterns transfer waste disposal, pollution, and health costs to society. Such expenses, whether they are measured in monetary terms or in diminished environmental quality, are borne by someone other than the individuals who use a resource. They are referred to as **external costs.** For example, consider an individual who attempts to spend a day of leisure fishing in a lake. Transportation, food, fishing equipment, and bait for the day's activities can be purchased in readily accessible markets. But suppose that upon arrival at the lake, the individual finds a lifeless, polluted body of water. Let us further suppose that a chemical plant situated on the lakeshore is responsible for degrading the water quality on the lake to the point that all or most of the fish are destroyed. In this example, fishers collectively bear the external costs of chemical production in the form of lost recreational opportunities.

Pollution-control costs include pollution-prevention costs and pollution costs. **Pollution-prevention** costs are those incurred either in the private sector or by government to prevent, either entirely or partially, the pollution that would otherwise result from some production or consumption activity. The cost incurred by local government to treat its sewage before dumping it into a river is a pollution-prevention cost; so is the cost incurred by a utility to prevent air-pollution by installing new equipment.

Pollution costs can be broken down into two categories:

1. The private or public expenditures to avoid pollution damage once pollution has already occurred

2. The increased health costs and loss of the use of public resources because of pollution

The large cost of cleaning up spills, such as that from the Exxon Valdez and from the war in the Persian Gulf, is an example of a pollution cost, as is the increased health risk to humans from eating seafood contaminated from the oil. By the middle 1990s there were several new economic solutions being utilized to both internalize external costs and to prevent pollution. (See table 16.3.)

Common Property Resource Problems

Economists have stated that, when everybody shares ownership of a resource, there is a strong tendency to overexploit and misuse that resource. Thus, common public ownership could be better described as effectively having no owner.

For example, common ownership of the air makes it virtually costless for any industry or individual to dispose of wastes by burning them. The air-pollution cost is not reflected in the economics of the polluter but becomes an external cost to society. Common ownership of the ocean makes it inexpensive for cities to use the ocean as a dump for their wastes. (See figure 16.9.)

Similarly, nobody owns the right to harvest whales. If any one country delays in getting its share of the available supply of whales, other countries may beat that

617 million

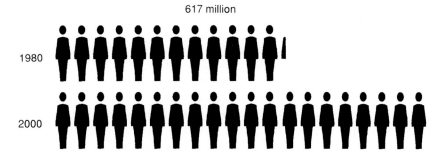

1980

2000

997 million

a.

b.

c.

Figure 16.9

Abuse of the Ocean   Since the oceans of the world are a shared resource that nobody owns, there is a tendency to use the resource unwisely. (a) Growing population in coastal areas leads to more marine pollution and destruction of coastal habitats. (b and c) Many countries use the ocean as a dump for unwanted trash. This barge will dump its waste in the ocean, and much of this waste will ultimately end up on the shore as currents and waves move it from where it was dumped. Some 6.5 million tons of litter finds its way into the sea each year.

Source: (a) *Data from UN Environmental Programme; World Resources Institute.*

---

country to it. Thus, there is a strong incentive to overharvest the whale population and create a consequent threat to the survival of the species. What is true on an international scale for whales is true also for other species within countries. Note that endangered species are wild and undomesticated; the survival of privately owned livestock is not a concern.

Finally, common ownership of land resources, such as parks and streets, is the source of other environmental problems. People who litter in public parks do not generally dump trash on their own property. The lack of enforceable property rights to commonly owned resources explains much of what economist John Kenneth Galbraith has termed "public squalor amid private affluence."

Economic Decision-Making and the Biophysical World

For most natural scientists, current crises like biodiversity loss, climate change and many other environmental problems are symptoms of an imbalance between the socioeconomic system and the natural world. While it is true that humans have always changed the natural world, it is also clear that this imprint is currently much greater than anything experienced in the past. One reason for the profound effect of human activity on the natural world is the fact that there are so many of us.

One of the most serious consequences of the growing human impact on the natural world is the loss of biodiversity. Biological diversity is thought to

affect the stability of ecosystems and the ability to cope with crises. In the 570-million-year history of complex life on earth, there have been several major extinction events. The loss of biodiversity after major extinction episodes ranged from 20 percent to more than 90 percent. The loss in biodiversity caused by human activity since the Industrial Revolution alone is somewhere between 10 and 20 percent. If current trends continue, losses are likely to reach 50 percent by the end of the next century. After each extinction event, it took between 20 million and 100 million years for biodiversity to recover to previous levels, a length of time between 100 and 500 times longer than the 200,000-year history of *Homo sapiens*.

The greatest single cause of the loss of biodiversity is habitat destruction, that is, the destruction of the web of organisms and functions that support individual species. The recognition that individual species are supported by others within the ecosystem is foreign to the way markets view the world.

The example of biodiversity illustrates the conflicting frameworks of economics and ecology. Market decisions fail to account for the context of a species or the interconnections between resource quality and ecosystem functions. For example, the value of land used for beef production is measured according to its contribution to output. Yet long before output and the use value of land decreases, the diversity of grass varieties, microorganisms in the soil, or groundwater quality may be affected by intensive beef production. As long as yields are maintained, these changes go unnoticed by markets and are unimportant to land-use decisions.

Another obvious difference between economics and ecology is the relevant time frame in markets and ecosystems. The biophysical world operates in tens of thousands and even millions of years. The time frame for market decisions is short. Particularly where economic policy is concerned, two- to four-year election cycles are the frame of reference; for investors and dividend earners, performance time frames of three months to one year are the rule.

Space or place is another issue. For ecosystems, place is critical. Take groundwater as an example. Soil quality, hydrogeological conditions, regional precipitation rates, plants that live in the region, and losses from evaporation, transpiration, and groundwater flow all contribute to the size and location of groundwater reservoirs. These capacities are not simply transferable from one location to another. For economic activities, place is increasingly irrelevant. Topography, location and function within a bioregion, or local ecological features do not enter into economic calculations except as simple functions of transportation costs or comparative advantage. Production is transferable, and the preferred location is anywhere production costs are the lowest.

Another difference between economics and ecology is that they are measured in different units. The unifying measure of market economics is money. Progress is measured in monetary units that everyone uses and understands to some degree. Biophysical systems are measured in physical units such as calories of energy, CO_2 absorption, centimeters of rainfall, or parts per million of nitrate contamination. Focusing only on the economic value of resources while ignoring biophysical health may mask serious changes in environmental quality or function.

Economics, Environment, and Developing Nations

As previously mentioned, the earth's "natural capital," on which humankind depends for food, security, medicines, and for many industrialized products, is its biological diversity. The majority of this diversity is in the developing world and much of it is threatened by exploitation and development. In order to pay for development projects, many economically poorer nations are forced to borrow money from banks in the developed world.

So great is the burden of external debt that many developing nations see little option but to overexploit their natural resource base. By 1995, the debt in the developing nations had risen to over US$1,435 billion, a figure almost half their collective gross national product. The debt burdens have led what investment there is in many developing countries to projects with safe, short-term returns and programs absolutely necessary for immediate survival. Environmental impacts are often neglected, the view being that severely indebted countries cannot afford to pay attention to environmental costs until other problems are resolved. This strategy suggests that environmental problems can be "corrected" once a country has reached a higher income level, but it ignores the growing realization that environmental impacts frequently cause international problems. Many countries under pressure from their debt crisis feel forced to overexploit their natural resources, rather than manage them sustainably.

One new method of helping manage a nation's debt crises is referred to as debt-for-nature exchange. Debt-for-nature exchanges are an innovative mechanism for addressing the debt issue while encouraging investment in conservation and sustainable development. The exchanges, or swaps, allow debt to be bought at discount but redeemed at a premium, in local currency, for use in conservation and sustainable development projects. Debt-for-nature originated in 1987, when a nonprofit organization, Conservation International, bought $650,000 of Bolivia's foreign debt in exchange for Bolivia's promise to establish a national park. By 1996, at least 16 debtor countries—in the Caribbean, Africa, Eastern Europe, and Latin America—had made similar deals with official and nongovernmental organizations. By 1992, nearly US$100 million of debt around the world had been purchased at a cost of some US$16 million but redeemed for the equivalent of US$60 million. This money was used to establish biosphere reserves and national parks, develop watershed protection programs, build inventories of endangered species, and develop environmental education.

In debt-for-nature exchanges, debtor countries benefit from the reduction of their foreign-currency debt obligations and add to their expenditure invested at home. The conservation investor receives a premium on the investment. This can be used for conservation and to establish sustainable development projects. Creditor banks gain by converting their non-paying debts. Although they receive only part of their initial loan, some return is better than a total loss.

The primary goal of debt-for-nature exchanges has not been debt reduction but the funding of natural-resource-management investment. The contribution made by exchanges could increase, as in the case of the Dominican Republic, where 10 percent of the country's outstanding foreign commercial debt is to be redeemed by exchanges. Although eliminating the debt crisis alone is no guarantee of investment in environmentally sound projects, instruments like debt-for-nature exchanges can, on a small scale, reduce the mismanagement

The Tragedy of the Commons

The problems inherent in common ownership of resources were outlined by biologist Garrett Hardin in a now classic essay entitled "The Tragedy of the Commons" (1968). The original "commons" were areas of pastureland in England that were provided free by the king to anyone who wished to graze cattle.

There are no problems on the commons as long as the number of animals is small in relation to the size of the pasture. From the point of view of each herder, however, the optimal strategy is to enlarge his or her herd as much as possible: If my animals do not eat the grass, someone else's will. Thus, the size of each herd grows, and the density of stock increases until the commons becomes overgrazed. The result is that everyone eventually loses as the animals die of starvation. The tragedy is that, even though the eventual result should be perfectly clear, no one acts to avert disaster. In a democratic society, there are few remedies to keep the size of herds in line.

The ecosphere is one big commons stocked with air, water, and irreplaceable mineral resources—a "people's pasture," but a pasture with very real limits. Each nation attempts to extract as much from the commons as possible while enough remains to sustain the herd. Thus, the United States and other industrial nations consume far more than their share of the total world resource harvest each year, much of it imported from less-developed nations. The nations of the world compete frantically for all the fish that can be taken from the sea before the fisheries are destroyed. Each nation freely uses the commons to dispose of its wastes, ignoring the dangers inherent in overtaxing the waste-absorbing capacity of rivers, oceans, and the atmosphere.

The tragedy of the commons also operates on an individual level. Most people are aware of air pollution, but they continue to drive their automobiles. Many families claim to need a second or third car. It is not that these people are antisocial; most would be willing to drive smaller or fewer cars if everyone else did, and they could get along with only one small car if public transport were adequate. But people frequently get "locked into" harmful situations, waiting for others to take the first step, and many unwittingly contribute to tragedies of the commons. After all, what harm can be done by the birth of one more child, the careless disposal of one more beer can, or the installation of one more air conditioner?

---

of natural resources and encourage sustainable development.

Attitudes of banks in the industrialized nations also seem to be changing. For example, the World Bank, which lends money for Third World development projects, has long been criticized by environmental groups for backing large, ecologically unsound programs, such as a cattle-raising project in Botswana that led to overgrazing. During the past few years, however, the World Bank has been factoring environmental concerns into its programs. One product of this new approach is an environmental action plan for Madagascar. The 20-year plan, which has been drawn up jointly with the World Wildlife fund, is aimed at heightening public awareness of environmental issues, setting up and managing protected areas, and encouraging sustainable development.

Lightening the Load

Ship captains pay careful attention to a marking on their vessels called the Plim-soll line. If the water level rises above the Plimsoll line, the boat is too heavy and is in danger of sinking. When the line is submerged, rearranging items on the ship will not help much. The problem is the total weight, which has surpassed the carrying capacity of the ship.

This analogy points out that human activity can reach a scale that the earth's natural systems can no longer support. In 1992, more than 1,600 scientists, including 102 Nobel laureates, underscored this point by collectively signing a "Warning to Humanity." Their warning stated in part that "a new ethic is required, a new attitude towards discharging our responsibility for caring for ourselves and for the earth. . . . This ethic must motivate a great movement, convincing reluctant leaders and reluctant governments and reluctant peoples themselves to effect the needed changes."

Such a new successful global effort to lighten humanity's load on the earth would need to directly address three major driving forces of environmental decline:

the inequitable distribution of income, resource consumptive economic growth, and rapid population growth. It would redirect technology and trade to buy time for this great change to occur. Although there is much to say about each of these challenges, some key points bear noting.

Wealth inequality may be the most difficult problem, since it has existed for centuries. The difference today, however, is that the future of rich and poor alike depends on reducing poverty and thereby eliminating this driving force of global environmental decline. In this way, self-interest joins ethics as a motive for redistributing wealth, and raises the chances that it might be done.

Important actions to narrow the income gap must include reducing Third World debt. This was talked about a great deal in the 1980s, but little was accomplished. In addition, the developed nations must focus more foreign aid, trade, and international lending policies directly on improving the living standards of the world's poor.

Shrimp and Turtles

Each year, shrimp boats traveling the South Atlantic and Gulf of Mexico accidentally catch an estimated 45,000 sea turtles in their nets. More than 12,000 of them drown, including hundreds of the endangered Kemp's ridley sea turtle, as well as threatened loggerhead and endangered green sea turtles.

The National Marine Fisheries Service, looking for a technological innovation to stop the accidental netting and killing of the turtles, developed a device to keep turtles out of shrimp nets. The turtle excluder device, or TED, attaches to standard shrimp nets. Data compiled by the National Marine Fisheries Service show that TEDs can reduce turtle captures in shrimp nets by 97 percent.

Beginning in 1982, conservationists and shrimp industry representatives joined forces in a program to encourage shrimpers to use TEDs voluntarily. TEDs, however, were not widely accepted by the shrimpers. After the failure of the voluntary plan, the U.S. Department of Commerce mandated the use of TEDs. Congressional delegations from states bordering the Gulf of Mexico are constantly pressuring for a review or change of the regulations. Many shrimpers have broken the law by refusing to use TEDs.

Shrimpers who oppose the TEDs claim that the devices are expensive and dangerous, and cut down on their catches. Redesigning the TEDs has led to a model that is less bulky and does not reduce catch as much; however, even this version worries the shrimpers. Early in the debate, they feared that Mexico would dominate the industry because it was not required to use TEDs, but Mexico's policy changed in 1993.

Biologists argue that any delay in bringing TEDs into widespread use will mean the continued destruction of tens of thousands of turtles and bring the Kemp's ridley even closer to extinction.

- Are the benefits of saving the turtles greater than the costs to the shrimpers?
- Can you arrive at a solution to this issue? Are there opportunities for compromise?

Trawling for shrimp also kills several endangered and threatened species of turtles in the South Atlantic and Gulf of Mexico.

A key description for reducing the kinds of economic growth that harm the environment is the same as that for making technology and trade more sustainable: internalizing environmental costs. If this is done through the adoption of environmental taxes, such as taxing based on pollution emitted, governments could avoid imposing heavier taxes overall by lowering income taxes accordingly. In addition, establishing better measures of economic accounting is critical. Since the calculations used to produce the gross national product do not account for the destruction or depletion of natural resources, this popular economic measure is extremely misleading. It tells us we are making progress even as our ecological foundations are being diminished. A better guide toward a sustainable path is essential. The United Nations and several governments have been working to develop better accounting methods, and while the progress has been slow, there is growing hope in the heightened awareness that a change is necessary.

As our discussion has shown, the economics of environmental problems is complex and difficult. The single most difficult problem to overcome is the assignment of an appropriate economic value to resources that have not previously been examined from an economic perspective. When air, water, scenery, and wildlife are assigned an economic value, they are looked at from an entirely different point of view.

SUMMARY

Risk assessment is the use of facts and assumptions to estimate the probability of harm to human health or the environment that may result from exposures to pollutants, toxic agents, or management decisions. While it is difficult to calculate risks, risk assessment is used in risk management, which analyzes factors in decision making. The politics of risk management focus on the adequacy of scientific evidence, which is often open to divergent interpretations. In assessing risk, people frequently overestimate new and unfamiliar risks, while underestimating familiar ones.

To a large degree, environmental problems can be viewed as economic problems. Economic policies and concepts, such as supply/demand and subsidies, play important roles in environmental decision making. Another important economic tool is cost-benefit analysis. Cost-benefit analysis is concerned with whether a policy generates more social benefits than social costs. Criticism of cost-benefit analysis is based on the question of whether everything has an economic value. It has been argued that, if economic thinking dominates society, then even noneconomic values, like beauty, can survive only if a monetary value is assigned to them.

A newer school of economic thought is referred to as sustainable development. Sustainable development has been defined as actions that address the needs of the present without compromising the ability of future generations to meet their own needs. Sustainable development requires choices based on values.

Pollution is extremely costly. When the costs are imposed on society, they are referred to as external costs. Costs of pollution control include pollution-prevention costs and pollution costs. Prevention costs are less costly, especially from a societal perspective.

Economists have stated that, when everyone shares ownership of a resource, there is a strong tendency to overexploit and misuse the resource. This concept was developed by Garrett Hardin in "The Tragedy of the Commons."

Recently, a market approach to curbing pollution has been proposed that would assign a value to not polluting, thereby introducing a profit motive to pollution reduction. Critics of this approach argue that pollution "permits" are nothing more than an acceptance of pollution of public resources.

Economic concepts are also being applied to the debt-laden developing countries. One such approach is the debt-for-nature swap. This program, which involves transferring loan payments for land that is later turned into parks and wildlife preserves, is gaining popularity.

REVIEW QUESTIONS

1. How is risk assessment used in environmental decision making?
2. What is incorporated in a cost-benefit analysis? Develop a cost-benefit analysis for a local issue.
3. What are some of the concerns about the use of cost-benefit analysis in environmental decision making?
4. What concerns are associated with sustainable development?
5. What are some examples of environmental external costs?
6. Define what is meant by pollution-prevention costs.
7. Define the problem in common property resource ownership. Provide some examples.
8. Describe the concept of debt for nature.

KEY TERMS

carcinogens 328
cost-benefit analysis 335
demand 333
external costs 342

negligible risk 331
pollution costs 342
pollution prevention 342
probability 327

resources 333
risk assessment 327
risk management 330
subsidy 334

supply 333
supply/demand curve 333
sustainable development 338

RELATED INTERNET SITES

Searchable index focusing on stratospheric ozone and human health. Includes information on ultraviolet rays and ozone depletion.
   ***http://sedac.ciesin.org/ozone
Institute for Agricultural and Trade Policies, focusing on domestic and international policy-making.
   ***http://www.iatp.org/iatp
Award-winning site on international development and the environment. Negotiations on habitat, climate change, and biodiversity.
   ***http://www.iisd.ca/linkages/
Discovery Channel. Updates television programs regarding science and the environment. Great graphics.
   http://www.school.discovery.com/
Clean Air Action Committee, focusing on Atlanta's willingness to experiment with alternative fuels.
   http://www.4cleanair.com/

Recycler's World, an award-winning, practical site for all forms of information related to the recycling industry.
   http://www.recycle.net/recycle/index.html
The Earth Times, aimed at opinion and policy makers and community and business leaders.
   http://www.earthtimes.org/
People for the Ethical Treatment of Animals. News releases, updates, and in-depth information on wildlife and animal life.
   http://www.envirolink.org/arrs/peta/
University of Oregon Environmental Studies Department. Combines theory and practice about environmental science and public policy.
   http://zebu.uoregon.edu/~ambiente/enviro/

c h a p t e r 17

Air Pollution

Objectives

After reading this chapter, you should be able to:

- Explain why air can accept and disperse significant amounts of pollutants.
- List the major sources and effects of the five primary pollutants.
- Describe how photochemical smog is formed and how it affects humans.
- Explain how PCV valves, APC valves, catalytic converters, scrubbers, precipitators, filters, and changes in fuel types reduce air pollution.
- Explain how acid rain is formed.
- Understand that humans can alter the atmosphere in such a way that the climate may change.
- Understand the link between chlorofluorocarbon use and ozone depletion.
- Recognize that enclosed areas can trap air pollutants that are normally diluted in the atmosphere.

Chapter Outline

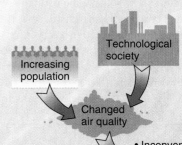

- Inconvenience
- Health problems
- International conflict
- Reduced freedoms
- Increasing costs
- New technologies

The Atmosphere

The atmosphere, or air, is normally composed of 79 percent nitrogen, 20 percent oxygen, and a 1 percent mixture of carbon dioxide, water vapor, and small quantities of several other gases. Most of the atmosphere is held close to the earth by the pull of gravity, so it gets thinner with increasing distance from the earth. (See figure 17.1.)

Even though gravity keeps the air near the earth, the air is not static. As it absorbs heat from the earth, it expands and rises. When its heat content is radiated into space, the air cools, becomes more dense, and flows toward the earth. As the air circulates due to heating and cooling, it also moves horizontally over the surface of the earth because the earth rotates on its axis. The combination of all air movements creates the wind patterns characteristic of different regions of the world. (See figure 17.2.)

As we discussed in chapter 11, pollution is something produced by humans that interferes with our well-being. Because we cause pollution, we may be able to do something to prevent it. There are several natural sources that degrade the quality of the air, such as gases and particles from volcanos, dust from natural sources, or odors from decomposition of dead plants and animals. However, since these activities are not controlled by humans, they do not fit our definition of pollution. Automobile emissions, chemical odors, and factory smoke are considered air pollution, however, and we will focus on these and similar examples.

The problem of air pollution is directly related to the number of people living in an area and the kinds of activities they engage in. When a population is small and its energy use is low, the impact of people is minimal. Their pollution is diluted and the overall negative effect is slight. However, our urbanized, industrialized civilization has a growing population and a history of increasing use of fossil fuels and technological aids. We release large quantities of polluting byproducts into our environment as we manufacture products demanded by the population.

Gases or small particles released into the atmosphere are likely to be mixed,

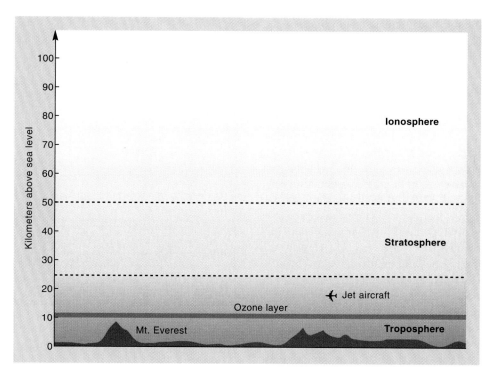

Figure 17.1

The Atmosphere   The atmosphere is divided into the troposphere, the relatively dense layer of gases close to the surface of the earth; the stratosphere, more distant with similar gases but less dense; and the ionosphere, composed of ionized gases.

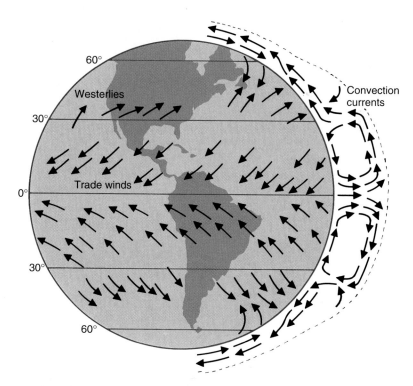

Figure 17.2

Global Wind Patterns   Wind is the movement of air caused by temperature differences and the rotation of the earth. Both of these contribute to the patterns of world air movement. In North America, most of the winds are westerlies (from the west to the east).

diluted, and circulated, but they are likely to stay near the earth due to gravity. When we put a material into the air, we do not get rid of it; we just dilute it and move it out of the immediate area. When people lived in small groups, the smoke from their fires was diluted. It was produced in such low concentrations that it did not interfere with neighboring groups downwind. In industrialized urban areas, pollutants cannot always be diluted before the air reaches another city. The polluted air from Chicago is further polluted by Gary, Indiana, supplemented by the wastes of Detroit and Cleveland, and finally moves over southeastern Canada and New England to the ocean. (See figure 17.3.) While not every population center adds the same kind or amount of waste, each adds to the total load carried.

Air pollution is not just an aesthetic problem, It also causes health problems. Hundreds of deaths have been directly related to poor-quality air in cities. A well-documented case of pollution that was harmful to human health occurred in Donora, Pennsylvania. (See figure 17.4.) The city of Donora is located in a valley. In October 1948, the pollutants from a zinc plant and steel mills became trapped in the valley, and a dense smog formed. Within five days, seventeen people died and 5,910 persons became ill. The polluted atmosphere affected nearly 50 percent of the city's 12,300 inhabitants. The technology that provided jobs for the people was also killing them.

Many of the megacities of the developing world have extremely poor air quality. Beijing, Seoul, Mexico City, and Cairo exceed World Health Organization guidelines for air quality for at least two pollutants. The causes of this air pollution are open fires, large numbers of poorly maintained motor vehicles, and poorly regulated industrial plants. The World Health Organization estimates that particulates in Mexico City contribute to 6,400 deaths each year. Not only does poor air quality in such cities increase the death rate, but the general health of the populace is lowered. Chronic coughing and susceptibility to infections are common in these cities. Deaths from air pollution occur

primarily among the elderly, the infirm, and the very young. Bronchial inflammations, allergic reactions, and irritation of the mucous membranes of the eyes and nose all indicate that air pollution must be reduced.

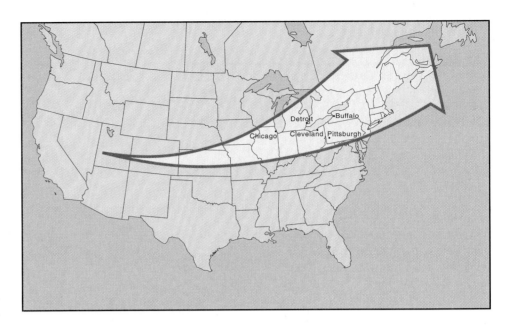

Figure 17.3

Accumulation of Pollutants   As an air mass moves across the continent from west to east, each population center adds its pollutants to the total load in the atmosphere.

Figure 17.4

Air Pollution   Donora, Pennsylvania, was the scene of a serious air-pollution incident. The pollutants from industry were trapped by an inversion, and almost half of the population became ill.

Primary Air Pollutants

Around the world, five major types of materials are released directly into the atmosphere in their unmodified forms and in sufficient quantities to pose a health

Air Pollution in Mexico City

Mexico City has been labeled the city with the worst air pollution ever recorded. The air over Mexico City exceeded ozone limits set by the World Health Organization on more than 300 days in one year. With about 20 million people, Mexico City is one of the largest cities in the world and grows larger each day. Open sewers and garbage dumps contribute dust and bacteria to the atmosphere. The city has about 35,000 factories and 3.6 million vehicles. Most of these are older models that are poorly tuned and, therefore, pollute the air with a mixture of hydrocarbons, carbon monoxide, and nitrogen oxides. A large proportion of the air pollution is the result of automobiles. A major source of hydrocarbons is from leakage of LP gas. Fixing the leaks would significantly reduce the hydrocarbon level in the atmosphere. The high altitude (over 2,000 meters) results in even greater air pollution from automobiles because automobile engines do not burn fuel efficiently at such high altitudes. Mexico City's location in a valley also allows for conditions suitable for thermal inversions during the winter.

Many foreign companies and governments give special "hazard pay" for working in Mexico City because of the air pollution. Pediatricians estimate that 85 percent of childhood illnesses are related to air pollution and say that the only way to improve the health of many of the children is to get them out of the city. The government has responded with a comprehensive program to clean up the air in the city. Twenty-five million trees have been planted to help clean the air, and 4,400 hectares of land have been purchased to provide green space for the city.

Public information campaigns encourage people to keep their automobiles tuned to reduce air pollution, lead-free gasoline has been made available and is being used, and catalytic converters are now required on all automobiles manufactured after 1991. Taxis manufactured before 1985 have been banned, and by the year 2000, all pre-1991 taxis must be replaced.

Personal automobiles and taxis are prohibited from being driven on the streets one day a week. This encourages people to carpool or use public transportation. The government is planning to improve public transportation to make it more attractive for people to switch from private automobiles to public transport. Both actions should reduce the number of automobiles releasing pollutants into the air. A polluting, government-owned oil refinery was shut down, and power plants and many industries have switched from oil to natural gas, which pollutes less.

These actions have had significant effects on the quality of the air. Although ozone in the air continues to be a major problem, as it is in metropolitan areas around the world, lead levels, carbon monoxide, and sulfur dioxide have been brought under control. In order to further improve air quality, increasingly strong restrictions on polluting industries and further restrictions on the use of private automobiles will be necessary. The Mexican government is considering such actions.

risk. They are carbon monoxide, hydrocarbons, particulates, sulfur dioxide, and nitrogen compounds. This group of pollutants is known as **primary air pollutants.** (See table 17.1.) These materials may interact with one another in the presence of an energy source to form new **secondary air pollutants** such as ozone and other very reactive materials.

Carbon Monoxide (CO)

Carbon monoxide (CO) is produced when organic materials, such as gasoline, coal, wood, and trash, are incompletely burned. The single largest source of carbon monoxide is the automobile. (See figure 17.5.) Although increased fuel efficiency and the use of catalytic converters have reduced carbon monoxide emissions

Table 17.1

Sources of Primary Air Pollutants	
Pollutant	Sources
Carbon monoxide	Incomplete burning of fossil fuels
	Tobacco smoke
Hydrocarbons	Incomplete burning of fossil fuels
	Tobacco burning
	Chemicals
Particulates	Burning fossil fuels
	Farming operations
	Construction operations
	Industrial wastes
	Building demolition
Sulfur dioxide	Burning fossil fuels
	Smelting ore
Nitrogen compounds	Burning fossil fuels

Figure 17.5

Carbon Monoxide The major source of carbon monoxide, hydrocarbons, and nitrogen oxides is the internal combustion engine, which is used to provide most of our transportation. The more concentrated the number of automobiles, the more concentrated the pollutants. Carbon monoxide concentrations of a hundred parts per million are not unusual in rush-hour traffic in large metropolitan areas. These concentrations are high enough to cause fatigue, dizziness, and headaches.

per kilometer driven, carbon monoxide remains a problem because the number of automobiles and the number of kilometers driven have increased. In many parts of the world, automobiles are poorly maintained and may have inoperable pollution control equipment, resulting in even greater amounts of carbon monoxide.

The next largest source of carbon monoxide is smoking tobacco. Currently in the United States and some other countries, there is a great deal of pressure to restrict areas where smoking is permitted to minimize exposure to secondhand cigarette smoke. Restaurants designate nonsmoking sections (some even advertise themselves as smoke-free), public buildings have designated nonsmoking areas, and corporations and colleges are designating their buildings as smoke-free. Smoking is decreasing in the industrialized world today, but in the developing nations, smoking retains its image of glamour and sophistication as a result of extensive marketing campaigns by cigarette companies.

Several hours of exposure to air containing 0.001 percent of carbon monoxide can cause death. Because carbon monoxide remains attached to hemoglobin for a long time, even small amounts tend to accumulate and reduce the blood's oxygen-carrying capacity. The amount of

carbon monoxide produced in heavy traffic can cause headaches, drowsiness, and blurred vision. A heavy smoker in congested traffic is doubly exposed and may experience severely impaired reaction time compared to nonsmoking drivers.

Fortunately, carbon monoxide is not a persistent pollutant. Natural processes convert carbon monoxide to other compounds that are not harmful. Therefore, the air can be cleared of its carbon monoxide if no new carbon monoxide is introduced into the atmosphere.

Hydrocarbons (HC)

In addition to carbon monoxide, automobiles emit a variety of **hydrocarbons (HC).** Hydrocarbons are a group of organic compounds consisting of carbon and hydrogen atoms. They are either evaporated from fuel supplies or are remnants of fuel that did not burn completely. The internal combustion engine is the major source of hydrocarbons, although refineries and other industries add hydrocarbons to the total atmospheric burden. Hydrocarbons in the atmosphere are no great problem. Most of them are washed out of the air when it rains and run off into surface water. They cause an oily film on surfaces, but hydrocarbons do

not generally cause more than nuisance problems, except when they react to form secondary pollutants.

Many modifications to automobile engines have reduced the loss of hydrocarbons to the atmosphere. Recycling some gases through the engine, using higher oxygen concentrations in the fuel-air mixture, and using valves to prevent the escape of gases are three of these modifications. In addition, catalytic converters burn exhaust gases more completely so that fewer hydrocarbons leave the tail pipe.

Particulates

Particulates, small pieces of solid materials dispersed into the atmosphere, constitute the third largest category of air pollutants. Smoke particles from fires, bits of asbestos from brake linings and insulation, dust particles, and ash from industrial plants contribute to the particulate load. Particulates cause problems ranging from the annoyance of soot settling on a backyard picnic table to the **carcinogenic** (cancer-causing) effects of asbestos. Particulates frequently get attention because they are so readily detected by the public. Heavy black smoke from a factory can be seen without expensive monitoring equipment and generally causes an outcry, whereas the production of colorless gases like carbon monoxide and sulfur dioxide goes unnoticed.

Particulates can accumulate in the lungs and interfere with the ability of the lungs to exchange gases. However, this lung damage usually happens to people who are repeatedly exposed to large amounts of particulate matter on the job. Miners and others who work in dusty conditions are most likely to be affected. For most of the population, particulates affect health by acting as centers for the deposition of moisture and gases from the atmosphere. As we breathe air containing particulates, we come in contact with concentrations of other potentially more harmful materials that have accumulated on the particulates. Sulfuric, nitric, and carbonic acids, which irritate the lining of our respiratory system, frequently form on particulates.

Sulfur Dioxide (SO₂)

Sulfur Dioxide (SO₂) is a compound of sulfur and oxygen that is produced when sulfur-containing fossil fuels are burned. Coal and oil were produced from organisms that had sulfur in their living structure. When the coal or oil was formed, some of the sulfur was incorporated into the fossil fuel. The sulfur is released as sulfur dioxide when the fuel is burned. Sulfur dioxide has a sharp odor and irritates respiratory tissue. It also reacts with water, oxygen, and other materials in the air to form sulfur-containing acids. The acids can become attached to particles, which, when inhaled, are very corrosive to lung tissue. In 1306, Edward I of England banned the burning of "sea coles," coal found on the sea shore, in the city of London. This coal was high in sulfur content and was, in part, responsible for the city's noxious odors. Edward's ban might very well be the earliest environmental legislation concerning air quality.

London, England, was also the site of one of the earliest killer fogs. In 1952, the city was covered with a dense fog for several days. During this time, air over the city failed to mix with layers of air in the upper atmosphere due to temperature conditions. The factories continued to release smoke and dust into this stagnant layer of air, and it became so full of the fog, smoke, and dust that people got lost in familiar surroundings. This combination of smoke and fog has become known as smog. Many London residents developed respiratory discomfort, headaches, and nausea. Four thousand people died in a few weeks. Their deaths have been associated with the high levels of sulfur compounds in the smog. Thousands of others suffered from severe bronchial irritation, sore throats, and chest pains. The 1948 Donora, Pennsylvania, incident already mentioned also involved symptoms related to the particles and sulfur dioxide in the air.

Oxides of Nitrogen (NO and NO₂)

Oxides of nitrogen (NO and NO₂) are the fifth category of primary air pollutants. Several compounds contain nitrogen and oxygen in different combinations; nitrogen oxide (NO) and nitrogen dioxide (NO_2) are the most common. When combustion takes place in air, nitrogen and oxygen molecules from the air may react with each other, and oxides of nitrogen result:

Figure 17.6

Photochemical Smog   The interaction among hydrocarbons, oxides of nitrogen, and sunlight produces new compounds that are irritants to humans. The visual impact of this smog is shown in these photographs.

$$N_2 + O_2 \rightarrow 2NO \text{ (nitrogen oxide)}$$
$$2NO + O_2 \rightarrow 2NO_2 \text{ (nitrogen dioxide)}$$

A mixture of nitrogen oxide and nitrogen dioxide is called NO_X. The nitrogen dioxide in the mixture reacts with other compounds to produce photochemical smog, discussed in the next section.

The primary source of nitrogen oxides is the automobile engine. Catalytic converters reduce the amount of nitrogen oxides released from the internal combustion engine, but increased automobile traffic has resulted in significant levels of NO_X in many metropolitan areas. Nitro-gen oxides are noteworthy because they are involved in the production of secondary air pollutants.

Photochemical Smog

Secondary air pollutants are compounds that result from the interaction of various primary air pollutants. **Photochemical smog** is a mixture of pollutants resulting from the interaction of nitrogen oxide and nitrogen dioxide with ultraviolet light. (See figure 17.6.) The two most destructive components of photochemical smog are ozone (O_3) and peroxyacetylnitrates. Both of these materials are excellent oxidizing agents, which means that they react readily with many other compounds, including those found in living things, causing destructive changes. Ozone is particularly harmful because it destroys chlorophyll in plants and injures lung tissue in humans and other animals. Peroxyacetylnitrates, in addition to being oxidizing agents, are eye irritants. Both ozone and peroxyacetylnitrates are secondary air pollutants that result from chemical reactions assisted by light.

A typical photochemical smog incident involves an interesting series of events.

1. Morning rush-hour traffic produces large amounts of nitrogen oxide (NO):

$$N_2 + O_2 \rightarrow 2NO$$

2. The nitrogen oxide reacts with molecular oxygen (O_2) from the atmosphere to form nitrogen dioxide (NO_2). It is the nitrogen dioxide in the atmosphere that gives photochemical smog its reddish-brown haze.

$$2NO + O_2 \rightarrow 2NO_2$$

3. Later in the morning, nitrogen dioxide reacts with ultraviolet light to form atomic oxygen (O):

$$NO_2 \xrightarrow[\text{light}]{\text{ultraviolet}} NO + O$$

4. Abundant molecular oxygen in the atmosphere reacts with atomic oxygen to form ozone:

$$O_2 + O \rightarrow O_3$$

5. Hydrocarbons in the atmosphere react with ozone to produce peroxyacetylnitrates. Various pollution-control devices now reduce the amount of hydrocarbons escaping to the atmosphere, thereby decreasing the amount of peroxyacetylnitrates formed.

6. As the ozone and peroxyacetylnitrates react with living things, they cause damage and are converted to less reactive molecules, and the smog eventually clears.

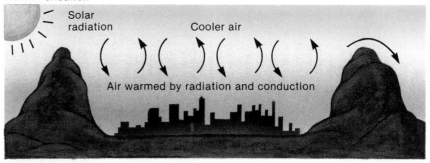

Figure 17.7

Thermal Inversion   Under normal conditions, the air at the earth's surface is heated by the sun and rises to mix with the cooler air above it. When a thermal inversion occurs, a layer of warm air is formed above the cooler air at the surface. The cooler air is then unable to mix with the warm air above and cannot escape because of surrounding mountains. The cool air is trapped, sometimes for several days, and accumulates pollutants. If the thermal inversion continues, the levels of pollution can become dangerously high.

Due to their climate and the geographic features of their locations, such large metropolitan areas as Los Angeles, Salt Lake City, Phoenix, and Denver have more trouble with photochemical smog than do East Coast metropolitan areas. Each of these cities is ringed by mountains. The prevailing winds are from the west. As cool air flows into these valleys, it pushes the warm air upward. This warm air becomes sandwiched between two layers of cold air and acts like a lid on the valley, a condition known as a **thermal inversion.** The air is trapped in the valley. (See figure 17.7.) The lid of warm air cannot rise further because it is covered by a layer of cooler air pushing down on it. It cannot move out of the area because of the ring of mountains. Without normal air circulation, smog accumulates. Harmful chemicals continue to increase in concentration until a major weather change causes the lid of warm air to move up and over the mountains. Then the underlying cool air can begin to circulate, and the polluted air is diluted.

Smog problems could be substantially decreased by reducing the use of internal combustion engines (perhaps eliminating them completely) or by moving population centers away from the valleys that produce thermal inversions. Both of these solutions would require expenditures of billions of dollars and major changes in lifestyle; therefore, people will probably continue to live with the problem.

Other Significant Air Pollutants

In recent years, two other air pollutants have been recognized as significantly affecting the health and welfare of people: lead and toxic chemicals. The primary sources of lead are gasoline and paint. For many years lead was added to gasoline to help engines run more effectively. Recognition that lead emissions were hazardous resulted in the lead additives being removed from gasoline in North America and Europe. This has resulted in a decline in the amount of lead in the atmosphere (See figure 17.8) However, most other countries in the world still use leaded gasoline.

Another major source of lead is paints. Many older homes have paints that contain lead, since various lead compounds are colorful pigments. Dust from flaking paint or remodeling or demolition is released into the atmosphere. Although the amount of lead may be small, its presence in the home can result in significant exposure to inhabitants, particularly young children who chew on painted surfaces and often eat paint chips.

Air toxics are harmful chemicals that are released into the atmosphere on purpose or are released accidentally as a result of leaks or poorly designed manufacturing processes. Materials such as pesticides are purposely released to kill insects or other pests. However, the majority of air toxics are released as a result of manufacturing processes. Although air toxics are important to the entire public, they are most critical for people who are exposed on the job since they are likely to be exposed often and to higher concentrations. There are literally hundreds of different air toxics.

Control of Air Pollution

All of the air pollutants we have examined thus far are produced by humans. That means their release into the atmosphere can be controlled. Methods of controlling air pollution depend upon the type of pollutant and the willingness or ability of industries, governments, and individuals to make changes.

Eliminating photochemical smog completely would require large-scale changes in driving habits and other aspects of our lifestyle and culture. That is not likely to happen, but the problem can be lessened by reducing the particular primary air pollutants that contribute to photochemical smog: NO_x and hydrocarbons. Government regulations have pressured the automobile industry to re-

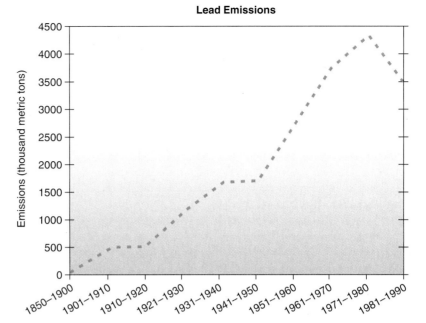

Lead Emissions

Figure 17.8

Lead Emissions The amount of lead released into the atmosphere has declined significantly since the early 1980s when the lead additive was removed from gasolines in North America and much of Europe.

Source: Data from Science, *Vol. 272, April 12, 1996.*

duce emissions. The positive crankcase ventilation valve (PCV) and gas caps with air-pollution control valves (APC) reduce hydrocarbon loss. Catalytic converters reduce carbon monoxide, oxides of nitrogen, and hydrocarbons in emissions and necessitate the use of lead-free fuel. This lead-free fuel requirement, in turn, significantly reduces the amount of lead (and other metal additives) in the atmosphere.

Particulates are produced primarily in automobiles or by industries burning fuels. In industry, particulate release can be controlled with scrubbers, precipitators, and filters. (See figure 17.9.) These devices are effective, but expensive. They can be retrofitted to the smokestack, or they can be designed into the combustion system. They filter the products of combustion and remove some, while allowing others to be released into the atmosphere. Electric charge differentials may also be used to remove the unwanted materials from the emission.

Another major source of particulates are dusts produced by a variety of industrial activities. Transfer of materials such

Figure 17.9

Control of Particulates The installation of proper pollution-control devices can significantly reduce the amount of particulate material released into the air. These photographs illustrate the difference before and after the installation of control devices.

environmental
close-up

The 1990 Clean Air Act

In November 1990, U.S. President George Bush signed the 1990 Clean Air Act, the first major revision to the Clean Air Act of 1977. This bill is surprisingly strong, considering that several powerful interest groups lobbied Congress intensively to weaken provisions of the bill that would apply to them. Although electrical utilities, the automobile industry, and certain urban areas with severe air-pollution problems won special concessions, the bill signed by President Bush man-

dates major changes in the amounts of air pollutants that can be released. The major consequences are economic. Ultimately, consumers will be asked to pay more for products and services as the producers pass on their air-pollution control costs in the prices of their products. The table lists some of the major provisions and expected consequences of the 1990 Clean Air Act.

Major Provisions	Consequences
1. Reduce urban smog by 15 percent by 1996 and 3 percent per year until federal air-quality standards are met. Some cities with severe smog problems, like Los Angeles, have up to twenty years to comply.	Many changes needed, which will require improved technology and additional costs.
2. Utilities must reduce release of sulfur dioxide by one-half by the year 2000.	Utilities will need to spend $3 billion a year for low-sulfur coal and pollution-control devices.
3. Utilities must reduce release of nitrogen oxides by one-third by the year 2000, starting in 1992.	
4. Utilities can buy and sell "pollution credits." Dirty plants can purchase permits to pollute from utilities that are not using their full allotment of pollution credits.	Economic incentives will exist for utilities to clean up their emissions. In special cases, they may be able to prolong the use of a dirty plant by purchasing permits to pollute.
5. Passenger cars must emit 60 percent less nitrogen oxide and 40 percent less hydrocarbons by the year 2003.	Increased cost of fuel.
6. Auto-emission controls must last for 100,000 miles.	Increased cost for new cars.
7. On-board canisters required to capture vapors during refueling.	
8. Cleaner-burning fuels will be required in the most polluted cities (Baltimore, Chicago, Hartford, Houston, Los Angeles, Milwaukee, New York, Philadelphia, and San Diego).	Oil companies will need to develop new fuels.
9. One million special low-emission vehicles (primarily in fleets) must be present by the year 2000. Three hundred thousand private low-emission vehicles must be in California by 1999.	New automotive technology will need to be developed. Costs will be passed on to the consumer.
10. Toxic emissions must be reduced by 90 percent by the year 2000. Factories must install maximum-achievable control technology. This includes small businesses.	Many small businesses may need help to meet this standard. Additional costs will be passed on to the consumer as higher prices.
11. Production of chlorofluorocarbons (CFCs), halons, and carbon tetrachloride will be banned by the year 2000, methyl chloroform by the year 2002. Recycling of chemicals will be required.	Alternative chemicals will need to be developed. Additional costs will be passed on to the consumer for recycling equipment and higher-priced substitutes.

environmental close-up

Secondhand Smoke

Smoking of tobacco has long been associated with a variety of respiratory diseases, including lung cancer, emphysema, and heart disease. Because of these strong links, tobacco products carry warning labels. Many nonsmoking people, however, are exposed to environmental tobacco smoke (secondhand smoke) because they live and work in spaces where people smoke. The U.S. Environmental Protection Agency estimates that approximately 3,000 nonsmokers die each year of lung cancer as a result of breathing air that contains secondhand smoke. Young children who are exposed to secondhand smoke are much more likely to have respiratory infections.

In July 1993, the U.S. Environmental Protection Agency recommended several actions to prevent people from being exposed to secondhand indoor smoke. They include recommendations that:

people not smoke in their homes or permit others to do so;

all organizations that deal with children have policies that protect children from secondhand smoke;

every company have a policy that protects employees from secondhand smoke;

smoking areas in restaurants and bars be placed so that the smoke will have little chance of coming into contact with nonsmokers.

---

as grain or coal from one container to another generates dust. Mining and other earth-moving activities, including farming operations, cause dust.

Industry and automobiles are not the only producers of particulate pollutants. Many people in the world use wood as their primary source of fuel for cooking and heat. In developed countries like the United States and Canada, some people use fireplaces and wood-burning stoves as a primary source of heat, but most use them for supplemental heat or for aesthetic purposes. However, the use of large numbers of wood-burning stoves and fireplaces can generate a significant air-pollution problem, called a brown cloud. Many municipalities, such as Boise, Idaho; Salt Lake City, Utah; and Denver, Colorado (and some entire states), use fines to enforce a ban on wood burning during severe air-pollution episodes. Many other communities issue pollution alerts and request that people not use wood-burners. Many of these communities have regulations

about the number and efficiency of wood-burning stoves and fireplaces. Some communities, such as Castle Rock, Colorado, prohibit the construction of houses with fireplaces or wood-burning stoves. Many people readily switch to gas fireplaces or high-efficiency wood-burners when they understand that their enjoyment of a rustic stove may be leading to a degraded environment, but most need to be forced to comply by official regulations and the threat of fines. Obviously, accumulations of many small sources of air pollution may cause as big a problem as one large emission source and are frequently more difficult to control.

To control sulfur dioxide, which is produced primarily by electric power generating plants, several possibilities are available. One alternative is to change from high-sulfur fuel to low-sulfur fuel. Switching from a high-sulfur coal to a low-sulfur coal reduces the amount of sulfur released into the atmosphere by 66 percent. Switching to oil, natural gas, or

nuclear fuels would reduce sulfur dioxide emissions even more. However, these are not long-term solutions because low-sulfur fuels are in short supply, and nuclear power plants pose a different set of pollution problems.

A second alternative is to remove the sulfur from the fuel before the fuel is used. Chemical or physical treatment of coal before it is burned can remove nearly 40 percent of the sulfur. This is technically possible, but it increases the cost of electricity to the rate payer.

Scrubbing the gases emitted from a smokestack is a third alternative. The technology is available, but, of course, these control devices are costly to install, maintain, and operate. As with auto emissions, governments have required the installation of these devices, but when industries install them, the cost of construction and operation is passed on to the consumer. The cost of installing scrubbers on a typical power plant is about $200 million. In the United States, an estimated 20 million or more metric tons of sulfur oxides are released into the atmosphere each year. Installing scrubbers on just 50 of the largest coal-burning plants would reduce this amount by over one-third. Some utility companies question the effect of this pollutant and even deny that they are the main source of sulfur emissions.

In the past, a common solution was to build taller smokestacks. Tall stacks release their gases above the inversion layers and, therefore, add the sulfur dioxide to the upper atmosphere, where it is diluted before it comes in contact with the population downwind. This works as long as the stack is tall enough and as long as not so much pollution is added to the air that it cannot be diluted to an acceptable level. However, sulfur dioxide reacts with oxygen and dissolves in water in the atmosphere to form sulfuric acid. This acid is washed from the air when it rains or snows. It damages plants and animals and increases corrosion of building materials and metal surfaces.

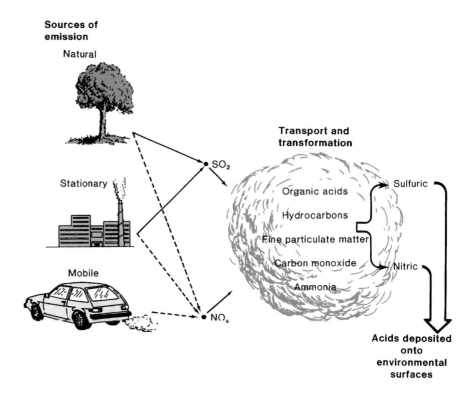

Figure 17.10

Acid Deposition Molecules from natural sources, power plants, and internal combustion engines react to produce the chemicals that are the source of acid deposition.

Acid Deposition

Acid deposition is the accumulation of potential acid-forming particles on a surface. Acids result from natural causes, such as vegetation, volcanoes, and lightning; and from human activities, such as coal burning and use of the internal combustion engine. (See figure 17.10.) These combustion processes produce sulfur dioxide (SO_2) and oxides of nitrogen (NO_X). Oxidizing agents, such as ozone, hydroxyl ions, or hydrogen peroxide, along with water, are necessary to convert the sulfur dioxide or nitrogen oxides to sulfuric or nitric acid. Various reactive hydrocarbons (HC) encourage the production of oxidizing agents.

The acid-forming reactants are classified as wet or dry. Wet reactions occur in the atmosphere and come to earth as some form of precipitation: acid rain, acid snow, or acid dew. Dry deposition occurs with the settling of the precursors of the acid on a surface. An acid does not actually form until these materials mix with water. Even though the acids are formed and deposited in several different ways, all of these processes usually are referred to as **acid rain.**

Acid rain is a worldwide problem. Reports of high acid-rain damage have come from Canada, England, Germany, France, Scandinavia, and the United States. Rain is normally slightly acidic, with a pH between 5.6 and 5.7 due to atmospheric carbon dioxide that dissolves to produce carbonic acid. But acid rains sometimes have a concentration of acid a thousand times higher than normal. In 1969, New Hampshire had a rain with a pH of 2.1. In 1974, Scotland had a rain with a pH of 2.4. The average rain in much of the northeastern part of the United States and adjoining parts of Ontario has a pH between 4.0 and 4.5.

Acid rain can cause damage in several ways. Buildings and monuments are often made from materials that contain limestone (calcium carbonate, $CaCO_3$), because limestone is relatively soft and easy to work. Sulfuric acid (H_2SO_4), a major component of acid rain, converts limestone to gypsum ($CaSO_4$), which is more soluble and is eroded over many years of contact with acid rain. (See figure 17.11.) Metal surfaces can also be attacked by acid rain.

Figure 17.11

Damage Due to Acid Deposition Sulfuric acid (H_2SO_4), which is a major component of acid deposition, reacts with limestone ($CaCO_3$) to form gypsum ($CaSO_4$). Since gypsum is water-soluble, it washes away with rain. The damage to this monument is the result of such acid reacting with the stone.

Acid Rain: Canada Versus the United States

"U.S. foot-dragging and interference in the development of scientific information has reached frustrating proportions." These words, spoken by the Honorable John Roberts, Canadian Minister of the Environment, sum up the confrontation between Canada and the United States regarding the question of acid deposition. Canadians have long contended that much of the acid deposition in their country originates in the United States. They are very concerned because 2.5 million square kilometers of Canada are highly susceptible to acid deposition.

With the signing of the Memorandum of Intent on Transboundary Air Pollution in 1980, the United States and Canada took the first step to cooperatively reduce the amount of acid deposition. The memorandum created scientific groups to study the problems of air pollution. After two years of study, there was still no accord. The Canadians accused the Reagan administration of delaying the studies, and the United States accused the Canadians of acting too rapidly.

In 1982, Canada suggested a mutual 50 percent reduction of sulfur dioxide by 1990. Citing a lack of research, the United States did not agree with the plan. This prompted the Environment Minister to state, "Always the constant refrain rings out from the administration that nothing is proven, and that an indefinite amount of further study is needed, not prompt action. Well, we can't wait. Our lakes and forests are literally dying."

In fact, the dispute reached such proportions that in 1983 the U.S. Department of Justice ruled that a Canadian-produced film on acid rain had to be labeled as political propaganda before it could be shown in the United States. The U.S. Department of Justice also required that the names of U.S. groups viewing the film be reported to the Justice Department. This attitude prompted the Canadian Minister of the Environment to observe, "It sounds like something you would expect from the Soviet Union, not the United States."

In his 1984 State of the Union Address, President Reagan affirmed that the United States would take no direct action regarding the question of acid deposition other than to continue to research the problem. Later that year, the Canadians announced a goal of reducing acid deposition by 50 percent and trusted that the United States would join them. However, the possibility seemed remote, for in May 1984, the House subcommittee voted against a bill to reduce the emissions of sulfur dioxide by 10 million metric tons by 1993. This killed any U.S. action regarding reduced acid deposition for 1984.

During the Bush administration, there was a softening of attitudes toward transboundary air pollution. The 1990 Clean Air Act set in place a set of rules that is significantly lowering the amount of acid precipitation in the United States and is resulting in less acid rain crossing the border into Canada. As evidence of this change in attitude, President Bush and Prime Minister Mulroney signed an agreement in 1991 to cooperate in reducing transboundary air pollution.

The effects of acid rain on ecosystems are often more difficult to quantify. Intense sulfur dioxide pollution around smelters is known to cause the death of many kinds of trees and other vegetation. But this is an extreme case and may not be directly comparable to less intense acid rain. However, in many parts of the world, acid rain is suspected of causing the death of many forests and reducing the vigor and rate of growth of others. (See figure 17.12.) In Central Europe, many forests have declined significantly, resulting in the death of about 6 million hectares of trees. Northeastern North America has also seen significant tree death and reduction in vigor, particularly at higher elevations. Some areas have had 50 percent mortality of red spruce trees.

A clear link between the decline of the forests and acid rain is difficult to

Figure 17.12

Forest Decline   Many forests at high elevations in northeastern North America have shown significant decline, and dead trees are common.

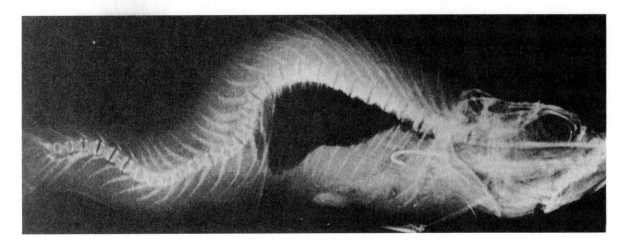

Figure 17.13

Effects of Acid Deposition on Organisms   The low pH of the water in which this fish lived caused the abnormal bone development that ultimately resulted in the death of the fish.

establish, but several hypotheses have been formulated. Molecules like sulfur dioxide and ozone are known sources of air pollution and cause direct damage to plants. The sulfur dioxide also contributes to acid formation. As soil becomes acidic, aluminum is released from binding sites and may interfere with the plant roots' ability to absorb nutrients. A recent long-term study in New Hampshire strongly suggests that the many years of acid precipitation have reduced the amount of calcium in the soil, which is needed by plants for growth. Because there are no easy ways to replace the calcium even if acid rain were to stop, it would still take many years for the forests to return to health. Reduction in the pH of the soil may also change the kind of bacteria in the soil and reduce the availability of nutrients for plants. While none of these factors alone would necessarily result in plant death, each could add to the stresses on the plant and may allow other factors, such as insect infestations, extreme weather conditions (particularly at high elevations), or drought, to further weaken trees and ultimately cause their death.

The effects of acid rain on aquatic ecosystems are much more clear-cut. In several experiments, lakes were purposely converted to acid lakes and the changes in the ecosystems recorded. The experiments showed that, as lakes become more acidic, there is a progressive loss of many kinds of organisms. The food web be-

comes less complicated, many organisms fail to reproduce, and many others die. Most healthy lakes have a pH above 6. At a pH of 5.5, many desirable species of fish have been eliminated; at a pH of 5, only a few starving fish may be found, and none are reproducing. Lakes with a pH of 4.5 are nearly sterile.

There are several reasons for these changes. Many of the early reproductive stages of insects and fish are more sensitive to acid conditions than are the adults. In addition, the young often live in shallow water, which is most affected by a flood of acid into lakes and rivers during the spring snowmelt. The snow and its acids have accumulated over the winter, and the snowmelt releases large amounts of acid all at once. Crayfish need calcium to form their external skeleton. As the pH of the water decreases, the crayfish are unable to form new skeletons and so they die. Reduced calcium availability also results in some fish with malformed skeletons. (See figure 17.13.) As mentioned earlier, increased acidity also results in the release of aluminum, which impairs the function of a fish's gills.

The extent to which lakes have been acidified is great. About 14,000 lakes in Canada and 11,000 in the United States have been seriously altered by becoming acidic. Many lakes in Scandinavia are similarly affected. The extent to which acid deposition affects an ecosystem depends on the nature of the bedrock in the

area and the ecosystem's proximity to acid-forming pollution sources. (See figure 17.14.) Parent material derived from igneous rock is not capable of buffering the effects of acid deposition, while soils derived from sedimentary rocks such as limestone release bases that neutralize the effects of acids. Because of this, eastern Canada and the U.S. Northeast are particularly susceptible to acid rain. These areas have high amounts of granite rock and are downwind from the major air-pollution sources of North America. Scandinavian countries have a similar geology and receive pollution from industrial areas in the United Kingdom and Europe. Thousands of kilometers of streams and up to 200,000 lakes in eastern Canada and the northeastern United States are estimated to be in danger of becoming acidified because of their location and geology.

Global Warming and Climate Change

During the 1980s, scientists, governments, and the public became concerned about the possibility that the world may be getting warmer. The United Nations Environment Programme established an Intergovernmental Panel on Climate Change (IPCC) to study the issue and make recommendations. Its First Assess-

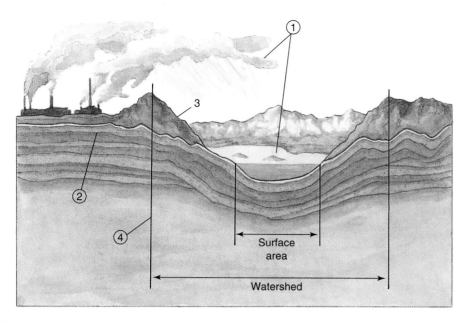

Figure 17.14

Factors That Contribute to Acid Rain Damage   In an aquatic ecosystem, the following factors increase the risk of damage from acid deposition: (1) a lake is located downwind from a major source of air pollution; (2) the area around the lake is hard, insoluble bedrock covered with a layer of thin infertile soil; (3) the soil has a low buffering capacity; and (4) there is also a low watershed to lake surface area ratio.

Source: From U.S. Environmental Protection Agency's "Acid Rain."

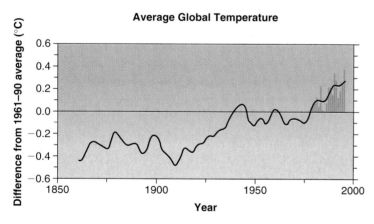

Figure 17.15

Changes in Average Global Temperature   Despite considerable variation, there has been a general trend toward increasing temperatures. Nineteen ninety-five was the warmest year on record.

Data from U.K. Meteorological Office/University of East Anglia published in Science 12 January 1996.

ment was published in 1990. In 1996 the IPCC published its Second Assessment and concluded that climate change is occurring and that it is highly probable that human activity is an important cause of the change. The IPCC has reached several important conclusions.

1. The average temperature of the earth has increased 0.3–0.6° C

(1995 was the warmest year on record) and sea level has risen 10–25 cm in the last 100 years. (See figure 17.15.)

2. A continued increase in temperature and sea level will occur. Various models suggest an increase in temperature of 1–3.5° C and a sea level rise of 15–95 cm by the year 2100.

3. There is a strong correlation between the increase in temperature and the amount of greenhouse gases present in the atmosphere.

4. Human activity greatly increases the amounts of these greenhouse gases.

It is important to recognize that, although a small increase in the average temperature of the earth may seem trivial, this increase could set in motion changes that could significantly alter the climate of major regions of the world. (See figure 17.16) Sea level would rise for two reasons. Because of the warmer temperatures, water stored in polar glaciers would melt, adding to the amount of water in the oceans. In addition, an increase in temperature would cause the waters currently in the oceans to take up more space, resulting in a rise in sea level. This would result in flooding of coastal areas. The land area of some island nations and countries such as Bangladesh would change dramatically as flooding occurred.

An average increase in the temperature of the earth could also have profound effects on land. An obvious conclusion is that there will be more "hot" days and fewer "cold" days. This has quite different consequences depending on where one lives. More hot days in areas that already are hot could add significantly to the stress on plants, humans, and other animals. Temperate climates of the world could shift northward, and many areas that are currently temperate could be converted to hot, dry regions. Another outcome would be changes in the weather a region would experience. Rain patterns may change; droughts may become more severe in some regions while excessive rainfall may cause problems in other regions. Changes in climate would influence the kinds of plants and animals that could live in an area. Thus climate

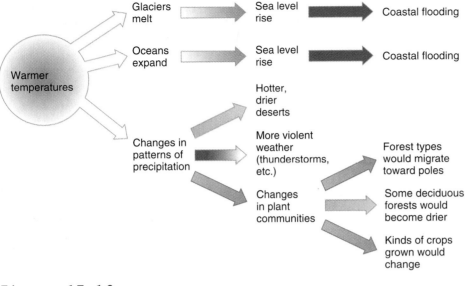

Figure 17.16

Effects of Global Warming   Global warming would have several effects on the climate of the world. The climate changes would have important impacts on human and other living things.

Table 17.2

Major Greenhouse Gases	
Gas	Contribution to global warming (percentage)
Carbon dioxide	57
Chlorofluorocarbons	25
Methane	12
Nitrous oxide	6

change would cause changes in natural communities of organisms and affect agriculture as well.

These predictions are based on computer models of climate. Some scientists have criticized the predictions as being inaccurate and constructed from sketchy data. However, as more accurate information is gathered and inserted into the models, the general conclusions remain the same. It is difficult for the general public to comprehend these changes or see evidence of them since each of us experiences only our own local weather and climate. Also, there will always be short-term variations in weather patterns. The models, however, are attempting to predict the long-term trends. The consequences of a global warming would be so

great that many are suggesting we alter our lifestyles regardless of whether the phenomenon is true, just to be on the safe side.

What actually causes global warming? An explanation is relatively straightforward. Several gases in the atmosphere are transparent to light but absorb infrared radiation. These gases allow sunlight to penetrate the atmosphere and be absorbed by the earth's surface. This sunlight energy is reradiated as infrared radiation (heat), which is absorbed by the gases. Because the effect is similar to what happens in a greenhouse (the glass allows light to enter but retards the loss of heat), these gases are called greenhouse gases and the warming thought to occur from their increase is called the **greenhouse**

effect. The most important greenhouse gases are carbon dioxide (CO_2), chlorofluorocarbons (primarily CCl_3F and CCl_2F_2), methane (CH_4), and nitrous oxide (N_2O). Table 17.2 lists the relative contribution of each of these gases to the potential for global warming.

Carbon dioxide (CO_2) is the most abundant of the greenhouse gases. It occurs as a natural consequence of respiration. However, much larger quantities are put into the atmosphere as a waste product of energy production. Coal, oil, natural gas, and biomass are all burned to provide heat and electricity for industrial processes, home heating, and cooking. These sources are increasing the amount of carbon dioxide in the atmosphere. Measurement of carbon dioxide levels at the Mauna Loa Observatory in Hawaii show that the carbon dioxide level has increased from about 315 ppm (parts per million) in 1958 to about 360 ppm in 1995. (See figure 17.17.)

A major step toward slowing global warming would be to increase the efficiency of energy utilization. This would also be of value in conserving the shrinking supplies of energy resources. It makes sense to increase energy efficiency, thus reducing carbon dioxide production, even if global warming is not a concern. One way to stimulate a move toward greater efficiency would be the imposition of a carbon tax. A carbon tax would increase the cost of fuels by taxing the amount of carbon put into the atmosphere by their use. This would increase the demand for fuel efficiency because the cost of fuel would rise. It would also stimulate the development of alternative fuels with a lower carbon content and generate funds for research in many aspects of fuel efficiency and alternative fuel technologies.

Another approach to the problem is to increase the amount of carbon dioxide removed from the atmosphere. If enough biomass is present, the excess carbon dioxide can be utilized by vegetation during photosynthesis, thereby reducing the impact of carbon dioxide released by fossil-fuel burning. Australia, the United States, and several other countries have announced plans to plant billions of trees to help remove carbon dioxide from the atmosphere. Many critics argue that this

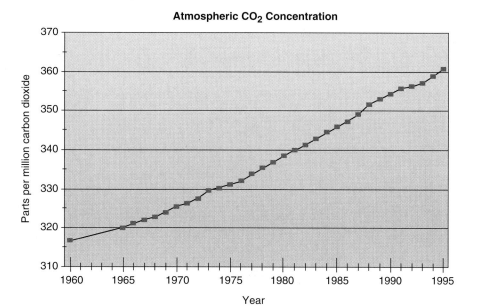

Atmospheric CO₂ Concentration

Figure 17.17

Change in Atmospheric Carbon Dioxide   Since the establishment of a carbon dioxide monitoring station at Mauna Loa Observatory in Hawaii, a steady increase in carbon dioxide levels has been observed. Since 1960 the concentration of carbon dioxide in the atmosphere has increased by nearly 14 percent.

approach will only provide a short-term benefit since, eventually, the trees will mature and die, and their decay will release carbon dioxide into the atmosphere at some later time.

An associated concern is the destruction of vast areas of rainforest in tropical regions of the world. These ecosystems are extremely efficient at removing carbon dioxide and storing the carbon atoms in the structure of the plant. The burning of tropical rainforests to provide farm or grazing land not only adds carbon dioxide to the atmosphere but also reduces the ability to remove carbon dioxide from the atmosphere, since the grasslands or farms created do not remove carbon dioxide as efficiently as do the original rainforests. Furthermore, the grazing lands and farms in such regions of the world are often abandoned after a few years and do not return to their original forest condition.

Chlorofluorocarbons are entirely the result of human activity. They were widely used as refrigerant gases in refrigerators and air conditioners, as cleaning solvents, as propellants in aerosol containers, and as expanders in foam products.

Although they are present in the atmosphere in minute quantities, they are extremely efficient as greenhouse gases (about 15,000 times more efficient at retarding heat loss than is carbon dioxide). Since all chlorofluorocarbons are made by people for specific purposes, their levels can be easily controlled.

The use of chlorofluorocarbons as propellants in aerosol spray cans or as expanders in foam products is not necessary. Other more benign materials, such as hydrocarbons, could be used. Since the 1970s, when chlorofluorocarbons were linked to the depletion of the ozone layer in the upper atmosphere, their use in aerosol cans has been banned in the United States, Canada, Norway, and Sweden, and the European Economic Community agreed to reduce use of chlorofluorocarbons in aerosol cans. However, worldwide, chlorofluorocarbons are still widely used as aerosol propellants. In foam and solvents, care can be taken to recover the chlorofluorocarbons for reuse rather than allow them to escape into the atmosphere. Alternative refrigerants are available, and care could be taken to recycle refrigerant gases rather than just

venting them into the air. In January 1991, DuPont announced the development of new refrigerants that would not harm the ozone layer. These will be installed in refrigerators and air conditioners in the future. We will need to exploit all these options if we want to reduce chlorofluorocarbon production. Until substitutes become available, the mandatory recycling of chlorofluorocarbons will reduce the rate at which new gases are added to the atmosphere.

In 1987, several industrialized countries, including Canada, the United States, the United Kingdom, Sweden, Norway, Netherlands, the Soviet Union, and West Germany, agreed to freeze production of chlorofluorocarbons at present levels and reduce production by 50 percent by the year 2000. This document, known as the Montreal Protocol, was ratified by the U.S. Senate in 1988. In May 1989, 82 nations signed the Helsinki Declaration, pledging to phase out the use of most chlorofluorocarbons by the year 2000. Several companies in the United States and Japan (e.g., General Motors, AT&T, Nissan) have announced plans to either phase out chlorofluorocarbon use or reduce use significantly. In 1990 in London, international agreements were reached to further reduce the use of chlorofluorocarbons. A major barrier to these negotiations was the reluctance of the developed countries of the world to establish a fund to help less-developed countries implement technologies that would allow them to obtain refrigeration and air conditioning without the use of chlorofluorocarbons. As a result of these international efforts and rapid changes in technology, the use of chlorofluorocarbons has dropped rapidly, and concentrations of chlorofluorocarbons in the atmosphere have stabilized and are expected to decline in the future.

Methane enters the atmosphere primarily from biological sources. Several kinds of bacteria that are particularly abundant in wetlands and rice fields release methane into the atmosphere. Methane-releasing bacteria are also found in large numbers in the guts of termites and various kinds of ruminant animals such as cattle. Some methane enters the atmosphere from fossil-fuel sources. Control of

environmental
close-up

Aesthetic Pollution

Most pollution can be measured by the amount of a particular chemical in water, air, or soil. Scientists measure these amounts and their effects on human, animal, and plant health. But some forms of pollution affect our aesthetic senses and so are more difficult to define.

Visual pollution is a sight that offends us. This type of pollution is highly subjective and is, therefore, difficult to define or control. To most people, an open garbage dump is a form of visual pollution. A dilapidated home or building may also be offensive, especially if located in an area of higher-priced homes. A heavily littered highway or street is aesthetically offensive to most people, and litter along a wilderness trail is even more unacceptable. Some sources of visual pollution are not so clear-cut, however. To many people, roadside billboards are offensive, but they can be helpful to advertisers and to travelers looking for information. This difference of opinion is typical when it comes to judging aesthetic pollution. People do not always agree on what is offensive. That makes regulation of aesthetic pollution difficult.

Taste is another subjective area. Some of the chemicals discharged into waterways have a definite taste. If chemicals do no biological injury but have an unpleasant taste, they are taste pollutants. Minute quantities of certain chemicals can affect the taste of our drinking water and our food. In fish, a concentration of 100 ppm of phenol can be tasted. Although this small amount of chemical is not harmful biologically, it does make the fish very unappetizing.

Odor pollution may originate from many sources. Various airborne chemicals have a specific odor that may be offensive. People who live near stockyards, paper mills, chemical plants,

steel mills, and other industries may be offended by the odors originating from these sources. Many of these industries discharge into the water waste materials that decompose or evaporate and cause odor pollution. People who are constantly exposed to an odor are usually not as offended by it as are people who are newly exposed to it. When an odor is constantly received, the brain ceases to respond to the stimulus. In other words, the person is not aware of the odor.

As we have seen, aesthetic pollutants—sights, tastes, and odors—can be extremely difficult to define. Each of us has our own idea of what is offensive, which makes it very difficult to establish aesthetic pollution standards. This type of pollution can be controlled by educating the public so that generally unacceptable levels of aesthetic pollution are eliminated.

---

methane sources is unlikely since the primary sources involve agricultural practices that would be very difficult to change. For example, nations would have to convert rice paddies to other forms of agriculture and drastically reduce the number of animals used for meat production. Neither is likely to occur since food production in most parts of the world needs to be increased, not decreased.

Nitrous oxide, a minor component of the greenhouse gas picture, primarily enters the atmosphere from fossil fuels and fertilizers. It could be reduced by more careful use of nitrogen-containing fertilizers.

Ozone Depletion

In the 1970s, various sectors of the scientific community became concerned about the possible reduction in the ozone layer in the upper atmosphere surrounding the earth. **Ozone** is a molecule of three atoms of oxygen (O_3). In 1985, it was discovered that a significant thinning of the ozone layer over the Antarctic occurred during

the Southern Hemisphere spring. Some regions of the ozone layer showed 95 percent depletion. Ozone depletion also has been found to be occurring farther north. Measurements in arctic regions suggest a thinning of the ozone layer there also. These findings have caused several countries to become involved in efforts to protect the ozone layer.

The presence of ozone in the outer layers of the atmosphere, approximately 12–25 kilometers from the earth's surface, shields the earth from the harmful effects of ultraviolet light radiation. Ozone (O_3)

absorbs ultraviolet light and is split into an oxygen molecule and an oxygen atom:

$$O_3 \xrightarrow{\text{Ultraviolet light}} O_2 + O$$

Oxygen molecules are also split by ultraviolet light to form oxygen atoms:

$$O_2 \xrightarrow{\text{Ultraviolet light}} 2O$$

Recombination of oxygen atoms and oxygen molecules allows ozone to be formed again and to be available to absorb more ultraviolet light.

$$O_2 + O \rightarrow O_3$$

This series of reactions results in the absorption of 99 percent of the ultraviolet light energy coming from the sun and prevents it from reaching the earth's surface. Less ozone in the upper atmosphere would result in more ultraviolet light reaching the earth's surface, causing increased skin cancers and cataracts in humans and increased mutations in all living things.

Chlorofluorocarbons are strongly implicated in the ozone reduction in the upper atmosphere. Chlorine reacts with ozone in the following way to reduce the quantity of ozone present:

$$Cl + O_3 \rightarrow ClO + O_2$$
$$ClO + O \rightarrow Cl + O_2$$

These reactions both destroy ozone and reduce the likelihood that it will be formed because atomic oxygen (O) is removed as well.

The use of chlorofluorocarbons in air conditioners, in refrigerators, and as propellants has resulted in the release of large amounts into the atmosphere. Because chlorofluorocarbons are implicated in both global warming and ozone depletion, considerable international attention is focused on controlling their manufacture and release. (See previous section on global warming and climate change.)

Indoor Air Pollution

A growing body of scientific evidence indicates that the air within homes and other buildings can be more seriously polluted than outdoor air in even the largest and most industrialized cities. Many indoor air pollutants and pollutant sources are thought to have an adverse effect on human health. These pollutants include asbestos; formaldehyde, which is associated with many consumer products, including certain wood products and aerosols; airborne pesticide residues; chloroform; perchloroethylene (associated particularly with dry cleaning); paradichlorbenzene (from mothballs and air fresheners); and many disease-causing microorganisms. (See figure 17.18.) Smoking is the most important air pollutant source in the United States in terms of human health. The Surgeon General estimates that 350,000 people in this country die each year from emphysema, heart attacks, strokes, lung cancer, or other diseases caused by tobacco smoking. Banning smoking probably would save more lives than would any other pollution-control measure.

A recent contributing factor to the concern about indoor air pollution is the weatherizing of buildings to reduce heat loss and save on fuel costs. In most older homes, there is a complete exchange of air every hour. This means that fresh air leaks in around doors and windows and through cracks and holes in the building. In a weatherized home, a complete air exchange may occur only once every five hours. Such a home is more energy efficient, but it also tends to trap air pollutants.

Even though we spend almost 90 percent of our time indoors, the movements to reduce indoor air pollution lag behind regulations governing outdoor air pollution. In the United States, the Environmental Protection Agency is conducting research to identify and rank the human health risks that result from exposure to individual indoor pollutants or mixtures of multiple indoor pollutants.

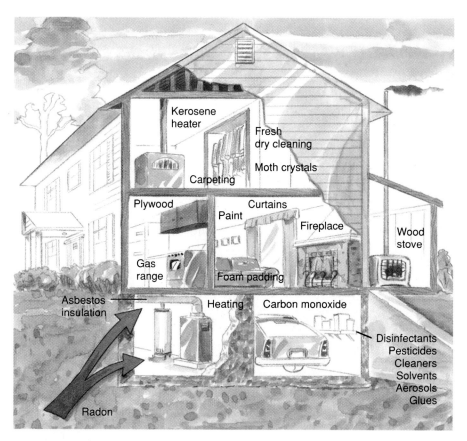

Figure 17.18

Sources of Indoor Air Pollution   In tightly sealed modern buildings, many small sources of air pollution can become a problem.

Source: Data from U.S. Environmental Protection Agency.

environmental
close-up

Radon

In 1985, the clothing worn by an engineer at the Limerick Nuclear Generating Station in Pottstown, Pennsylvania, registered a high radiation level. Initially, the generating station was believed to be the source of the radiation. However, subsequent studies indicated that radon 222 from the engineer's home was the source. Following this incident, there has been an increased interest in radon and its effects.

The source of radon is uranium 238, a naturally occurring element that makes up about three parts per million of the earth's crust. Uranium 238 goes through fourteen steps of decay before it becomes stable, nonradioactive lead 206. **Radon,** an inert radioactive gas having a half-life of 3.8 days, is one of the products formed during this process.

Since radon is an inert gas, it does not enter into any chemical reactions within the body, but it can be inhaled. Once in the lung it will undergo radioactive decay, producing other kinds of atoms called "daughters" of radon. These decay products (daughters) of radon—polonium 218, which has a three-minute half-life; lead 214, which has a twenty-seven minute half-life; bismuth 214, which has a twenty-minute half-life; and polonium 214, which has a millisecond half-life—are solid materials that remain in the lungs and are chemically active.

Increased incidence of lung cancer is the only known health effect associated with radon decay products. It is estimated that the decay products of radon are responsible for about 15,000 lung cancer deaths annually in the United States. This is about 10 percent of lung cancer deaths.

Some 4 million homes in the United States may have higher-than-acceptable levels of radon. Mining areas or areas with high amounts of igneous and metamorphic rocks are most likely to be affected. Uranium mining regions of Colorado and sections of Pennsylvania, New York, New Jersey, and Florida are the areas most affected in the United States.

As the radon gas is formed in the rocks, it usually diffuses up through the rocks and soil and escapes harmlessly into the atmosphere. It can also diffuse into groundwater. Radon usually enters a home through an open space in the foundation. A crack in the basement floor or the foundation, the gap around a water or sewer pipe, or a crawl space allows the radon to enter the home. It may also enter in the water supply from wells.

Radon risk evaluation chart

pCi/L	WL	Estimated lung-cancer deaths due to radon exposure (out of 1,000)	Comparable exposure levels	Comparable risk
200	1.0	440–770	One thousand times average outdoor level	More than sixty times nonsmoker risk
				Four-pack-a-day smoker
100	0.5	270–630	One hundred times average indoor level	
				Two thousand chest X rays per year
40	0.2	120–380		
				Two-pack-a-day smoker
20	0.1	60–120	One hundred times average outdoor level	
				One-pack-a-day smoker
10	0.05	30–120	Ten times average indoor level	
				Five times nonsmoker risk
4	0.02	13–50	Level at which EPA suggests remedial action	
				Two hundred chest X rays per year
2	0.01	7–30	Ten times average outdoor level	
				Nonsmoker risk of dying from lung cancer
1	0.005	3–13	Average indoor level	
				Twenty chest X rays per year
0.2	0.001	1–3	Average outdoor level	

Note: Measurement results are reported in one of two ways: (1) pCi/L (picocuries per liter) — Measurement of *radon gas,* or (2) WL (working levels) — Measurement of *radon decay products.*

Usually, the natural ventilation within a house allows air to escape through loose-fitting doors or windows or other openings. Thus, radon that enters a house can escape. Problems arise, however, as people make their homes more airtight in an attempt to save energy.

Only 10 percent of the homes in the United States have a potential radon problem. However, the increased publicity about radon has many people worried. In addition, in 1988, the Environmental Protection Agency and the U.S. Surgeon

Source: Data from Office of Air and Radiation Programs, U.S. Environmental Protection Agency.

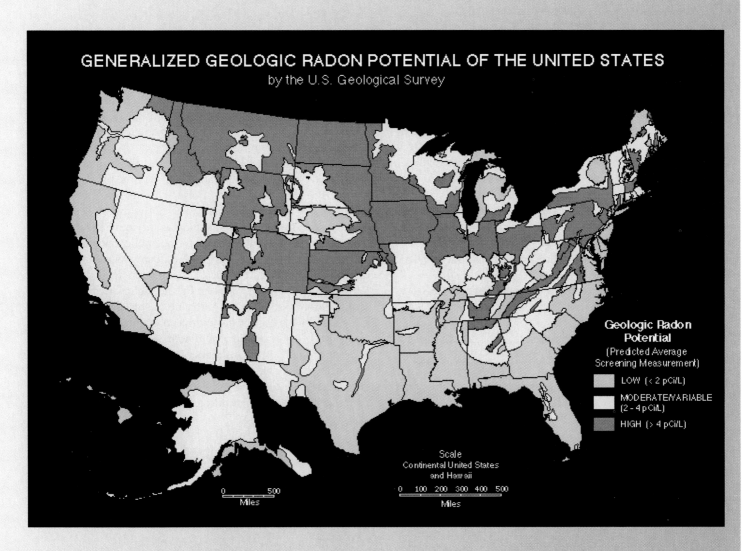

GENERALIZED GEOLOGIC RADON POTENTIAL OF THE UNITED STATES
by the U.S. Geological Survey

Geologic Radon Potential
(Predicted Average Screening Measurement)

LOW (< 2 pCi/L)

MODERATE/VARIABLE (2 – 4 pCi/L)

HIGH (> 4 pCi/L)

Scale
Continental United States and Hawaii

0   500
Miles

0   100   200   300   400   500
Miles

General recommended that all Americans (other than those living in apartment buildings above the second floor) test their home for radon. If the tests indicate that the level of radon is at or above 4 picocuries/liter, EPA recommends that the homeowner take action to lower the level. This is usually not expensive and consists of blocking the places where radon is entering or venting sources of radon to the outside. People who are concerned about radon should contact their state's public health department or environmental protection agency. These agencies can recommend self-testing kits to detect radon levels. The cost of these kits ranges from $15 to $35. Consumer frauds that prey upon people's fear of radon are increasing, so consumers should be wary. Two excellent books on radon—*Radon Reduction Methods, a Homeowner's Guide* and *Radon Reduction Methods for Detached Houses*—may be obtained free from the Environmental Protection Agency, 230 South Dearborn St., Chicago, Illinois 60604.

environmental
close-up

Noise Pollution

Noise is referred to as unwanted sound. However, noise can be more than just an unpleasant sensation. Research has shown that exposure to noise can cause physical, as well as mental, harm to the body. The loudness of the noise is measured by decibels (db). Decibel scales are logarithmic, rather than linear. Thus, the change from 40 db (a library) to 80 db (a dishwasher or garbage disposal) represents a ten-thousandfold increase in sound loudness.

The frequency or pitch of a sound is also a factor in determining its degree of harm. High-pitched sounds are the most annoying. The most common sound pressure scale for high-pitched sounds is the A scale, whose units are written "dbA." Hearing loss begins with prolonged exposure (eight hours or more) to 80 to 90 dbA levels of sound pressure. Sound pressure becomes painful at around 140 dbA and can kill at 180 dbA. (See the table.)

In addition to hearing loss, noise pollution is linked to a variety of other ailments, ranging from nervous tension

headaches to neuroses. Research has also shown that noise may cause blood vessels to constrict (which reduces the blood flow to key body parts), disturbs unborn children, and sometimes causes seizures in epileptics. The U.S. EPA has estimated that noise causes about 40 million U.S. citizens to suffer hearing damage or other mental or physical effects. Up to 64 million people are estimated to live in homes affected by aircraft, traffic, or construction noise.

The Noise Control Act of 1972 was the first major attempt made in the United States to protect the public health and welfare from detrimental noise. This act also attempted to coordinate federal research and activities in noise control, to set federal noise emission standards for commercial products, and to provide information to the public. Subsequent to the passage of the Noise Control Act, many local communities in the United States enacted their own noise ordinances. While such efforts are a step in the right direction, the United States is still controlling noise less than are many European countries. Several European countries have developed quiet construction equipment in conjunction with strongly enforced noise ordinances. The Germans and Swiss have established maximum day and night noise levels for certain areas. Regarding noise-pollution abatement, North America has much to learn from European countries.

Intensity of Noise

Source of Sound	Intensity in decibels
Jet aircraft at takeoff	145
Pain occurs	140
Hydraulic press	130
Jet airplane (160 meters overhead)	120
Unmuffled motorcycle	110
Subway train	100
Farm tractor	98
Gasoline lawn mower	96
Food blender	93
Heavy truck (15 meters away)	90
Heavy city traffic	90
Vacuum cleaner	85
Hearing loss after long exposure	85
Garbage disposal unit	80
Dishwasher	65
Window air conditioner	60
Normal Speech	60

International Air Pollution

Acid rain originates as a form of air pollution, but it may damage the environment in the form of water pollution. Also, air pollution that originates in one country may result in water pollution in another country. A particular nation may have stringent environmental controls within its own boundaries but have environmental problems because of the actions of a neighboring country.

Recently, the phenomenon of acid rain has underscored the need for international cooperation in dealing with various environmental concerns. For example, an estimated 56 percent of the acid rain falling in Sweden originates outside of that country. The main sources of the pollutants are Germany and the United Kingdom because the west coast of Sweden receives winds from these countries. When these industrialized countries release more sulfur dioxide into the atmosphere, the result is more acid rain in Sweden. Since the 1930s, lakes in western Sweden have become more acidic by a value of two pH units. Ten thousand lakes have a pH below 6.0, and 5,000 lakes have a pH below 5.0. A pH below 5.5 is too acidic for many species of fish.

In the 1930s, the United Kingdom initiated a program to reduce air pollution. This program has been successful in that the air in the United Kingdom has become cleaner. But one of the country's solutions was to build taller smokestacks. As a result, more of the pollutants from the United Kingdom are transported to Sweden.

- Should the United Kingdom be permitted to disperse air pollutants in a manner that damages the Swedish environment?
- Should there be a series of international agreements to control and regulate the movement of airborne pollutants across international boundaries?
- In addition to air pollutants, are there other forms of pollutants that can naturally be transported across international boundaries?

SUMMARY

The atmosphere has a tremendous ability to accept and disperse pollutants. Carbon monoxide, hydrocarbons, particulates, sulfur dioxide, and nitrogen compounds are the primary air pollutants. They can cause a variety of health problems. Lead and air toxics have also been identified as significant air pollutants.

Photochemical smog is a secondary pollutant formed when hydrocarbons and oxides of nitrogen are trapped by thermal inversions and react with each other in the presence of sunlight to form peroxyacetylnitrates and ozone. Elimination of photochemical smog requires changes in technology, such as more fuel efficient automobiles, special devices to prevent the loss of hydrocarbons, and catalytic converters to more completely burn hydrocarbons.

Acid rain is caused by emissions of sulfur dioxide and oxides of nitrogen in the upper atmosphere, which form acids that are washed from the air when it rains or snows. Direct effects of acid rain on terrestrial ecosystems are difficult to prove, but changes in many forested areas are suspected of being partly the result of additional stresses caused by acid rain. Recent evidence suggests that loss of calcium from the soil may be a major problem associated with acid rain. The effect of acid rain on aquatic ecosystems is easy to quantify. As waters become more acidic, the complexity of the ecosystem decreases, and many species fail to reproduce. The control of acid rain requires the use of scrubbers, precipitators, and filters or the removal of sulfur from fuels.

Currently, many are concerned about the damaging effects of greenhouse gases: carbon dioxide, methane, and chlorofluorocarbons. These gases are likely to be causing an increase in the average temperature of the earth and consequently are leading to major changes in the climate. Chlorofluorocarbons are also thought to lead to the destruction of ozone in the upper atmosphere, which results in increased amounts of ultraviolet light reaching the earth. Concern about the effects of chlorofluorocarbons has led to international efforts that have resulted in significant reductions in the amount of these substances reaching the atmosphere. Many commonly used materials release gases into closed spaces (indoor air pollution) where they cause health problems. The most important of these are associated with smoking.

REVIEW QUESTIONS

1. List the five primary air pollutants commonly released into the atmosphere and their sources.
2. Define *secondary air pollutant*, and give an example.
3. List three health effects of air pollution.
4. Why is air pollution such a large problem in urban areas?
5. What is photochemical smog? What causes it?
6. Describe three actions that can be taken to control air pollution.
7. What causes acid rain? List three possible detrimental consequences of acid rain.
8. Why is carbon dioxide (a nontoxic normal component of the atmosphere) called a "greenhouse gas"?
9. What would the consequences be if the ozone layer surrounding the earth was destroyed?
10. How does energy conservation influence air quality?

KEY TERMS

acid deposition   358
acid rain   358
carbon dioxide (CO_2)   362
carbon monoxide (CO)   351
carcinogenic   352

greenhouse effect   362
hydrocarbons (HC)   352
oxides of nitrogen (NO and NO_2)   353

ozone   364
particulates   352
photochemical smog   353
primary air pollutants   351

radon   366
secondary air pollutants   351
sulfur dioxide (SO_2)   352
thermal inversion   354

RELATED INTERNET SITES

U.S. EPA. Up-to-date news and information on air pollution and radiation.
   http://www.epa.gov/ARD-R5/
Ozone depletion facts.
   http://www.epa.gov/ARD-R5/cfc/cfc.htm
The health effects of environmental pollution. Great links. Indexed.
   http://charlotte.med.nyu.edu/HomePage.html
Award-winning site on environmental problems. Hundreds of links.
   http://charlotte.med.nyu.edu/outreach/index.html
International conference on global warming and climate change.
   http://www.islandnet.com/~skies/

The impact of ultraviolet radiation on the environmental and human health. Good links.
   http://www.islandnet.com/~see/uvb.htm
Pacific Northwest Pollution Prevention Resource Center.
   http://pprc.pnl.gov/pprc/
Award-winning site for pollution prevention, compliance, and policies. Great links.
   http://es.inel.gov/index.html
Pollution prevention technical library.
   http://clean.rti.org/larry/nav_in.html
Department of Energy site focusing on greenhouse gases and emissions.
   http://www.eia.doe.gov/oiaf/1605/frntend.html

chapter 18

Solid Waste Management and Disposal

Objectives

After reading this chapter, you should be able to:

- Explain why solid waste has become a problem throughout the world.
- Understand that the management of municipal solid waste is directly affected by economics, changes in technology, and citizen awareness and involvement.
- Recognize that the management of municipal solid waste in the future will require an integrated approach.
- Describe the various methods of waste disposal and the problems associated with each method.
- Understand the difficulties in developing new municipal landfills.
- Define the problems associated with incineration as a method of waste disposal.
- Describe some methods of source reduction.
- Describe composting and how it fits into solid waste disposal.
- List some benefits and drawbacks of recycling.

Chapter Outline

Introduction
The Disposable Decades
The Nature of the Problem
Methods of Waste Disposal
    Landfilling
    Incineration
    Source Reduction
Environmental Close-Up: *What You Can Do To Reduce Waste and Save Money*
    Recycling
Environmental Close-Up: *Recycling Is Big Business*
Environmental Close-Up: *Recyclables Market Basket*
Environmental Close-Up: *Resins Used in Consumer Packaging*
Environmental Close-Up: *Container Laws*
Issues and Analysis: *Corporate Response to Environmental Concerns*

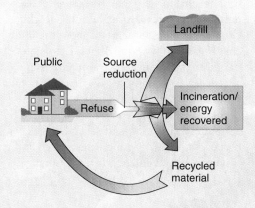

Introduction

In March 1987, a barge laden with 3,200 tons of garbage set out in search of a dump. The refuse had been turned away from a landfill in Islip, New York. The barge traveled 10,000 kilometers and stopped at several foreign ports, but found no one willing to accept its noxious load. The three-month odyssey took the barge to Mexico, Belize, and the Bahamas before it returned still fully loaded to New York. (See figure 18.1.) The futile voyage made headlines, giving many North Americans their first inkling of an impending crisis.

Lack of space for dumping solid waste has become a problem for many large metropolitan areas throughout the world. Communities are concerned about the increasing costs of waste disposal, possible hazards to groundwater, and maintaining air quality. In many areas, there is not suitable land available for new landfills. Problems with solid waste have increased dramatically over the past several decades because of population increases and an attitude that convenience is a very important part of the North American lifestyle.

The Disposable Decades

In 1955, *Life* magazine pictured a happy family in an article entitled "Throwaway Living." A disposable lifestyle was marketed as the wave of the future and as a way to cut down on household chores. "Use it once and throw it away" became a very popular advertising slogan in the 1950s.

The solid waste problems facing us today have their roots in the economic boom that followed World War II. Marketing experts set to work trying new tactics to get consumers to buy and toss or, as economists would say, to "stimulate consumption." In the mid-1950s, marketing consultant Victor Lebow wrote an emotional plea for "forced consumption" in the New York Journal of Retailing:

> Our enormously productive economy demands that we make consumption a way of life, that we convert the buying and use of goods into rituals, that we seek our spiritual satisfactions in consumption. . . . We need things consumed, burned up, worn out, replaced, and discarded at an ever-growing rate.

Consumers were quick to adapt to the lifestyle that Lebow envisioned. In fact, it was not long until a disposable, throwaway lifestyle was seen as a consumer's right. The idea was to sell "convenience" to the prosperous postwar consumers. What was initially a convenience was soon to become a "necessity"; at least, that was what the advertisements were telling people.

A good example of this way of thinking is the TV dinner, which was first marketed in 1953. The food-packaging industry was evolving into a major service, closely allied with marketing and advertising. From the first TV dinner, we progressed rapidly, so that by the late 1980s, microwave meals were available. One such meal consisted of 340 grams of edible material and six separate layers of packaging, five of them plastic.

Figure 18.1

Wandering Garbage In 1987, a barge filled with garbage similar to this barge traveled from New York to Mexico looking for a place to dispose of its cargo. This practice of shipping unwanted garbage to other countries continues today throughout the world.

Figure 18.2

Disposable Lifestyles The disposable lifestyle of the past four decades has created a shortage of space in landfills such as this. The development of new landfills is expensive, and in many areas there is not available land appropriate for development.

After nearly four decades of throw-away living, the disposable lifestyle may soon be changing. Cities and counties face a shortage of space in old landfills. Solid waste has become a major problem and, in a growing list of communities around the world, it has reached crisis proportions. (See figure 18.2.)

The Nature of the Problem

As we approach the 21st century, the world must take steps to solve the ever-increasing burden of garbage—or as the professionals say, **municipal solid waste.** We are all part of the problem, but we can also be part of the solution.

The United States produces about 160 million tons of municipal solid waste each year. This equates to about 1.5 kilograms of trash per person per day. The volume of municipal solid waste has increased by 80 percent since 1960 and is expected to grow by an additional 20 percent by 2000. At this rate, the U.S. annual garbage heap could reach close to 200 million tons by the end of the century. (See figure 18.3.)

Nations with high standards of living and productivity tend to have more municipal solid waste per person than less developed countries. (See figure 18.3c.) The United States and Canada, therefore, are world leaders in waste production. This high volume of solid waste and reliance on landfills for disposal have led to the problems we face today. For example, Toronto, Canada, is running out of places to put its municipal solid waste. With present methods, metropolitan Toronto's two landfill sites will be full by the beginning of the next century. The outlying districts are not willing to be the site of a new landfill for Toronto. Solutions to such a problem will not be easy; however, solutions must be found.

Until recently, the disposal of municipal solid waste did not attract much public attention. From prehistory through the present day, the favored means of disposal was simply to dump solid wastes outside of the city or village limits. Frequently, these dumps were in a wetlands adjacent to a river or lake. To minimize the volume of the waste, the dump was often burned. This method of waste disposal was state of the art until only recently. Unfortunately, this method is still being used in remote or sparsely populated areas in North America and the world today. (See figure 18.4.)

As better waste-disposal technologies were developed and as values changed,

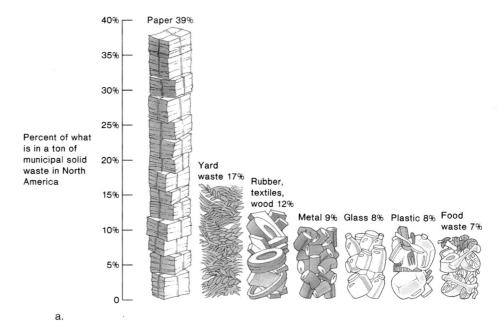

a.

U.S. per Capita Garbage Output, 1960, 1980, 1995, and 2000 (Projection), in Pounds Per Day

Note that 2.205 pounds equals 1 kilogram.				
Waste Material	1960	1980	1995	2000
Total nonfood and nonyard wastes:	1.65	2.57	3.18	3.38
Paper and paperboard	0.91	1.32	1.80	1.96
Glass	0.20	0.36	0.23	0.21
Metals	0.32	0.35	0.34	0.35
Plastics	0.01	0.19	0.40	0.43
Rubber and leather	0.06	0.10	0.10	0.11
Textiles	0.05	0.06	0.09	0.09
Wood	0.09	0.12	0.16	0.17
Other	0.01	0.07	0.06	0.06
Other wastes:				
Food wastes	0.37	0.32	0.28	0.27
Yard wastes	0.61	0.66	0.69	0.70
Miscellaneous organic wastes	0.03	0.06	0.06	0.06
Total waste	2.66	3.61	4.21	4.41

b.

Figure 18.3

The Changing Nature of Trash (*a* and *b*) While food waste was the primary waste product in the past, changes in lifestyle and packaging have led to a change in the nature of trash. Most of what is currently disposed of could be recycled.

Source: (b) Data from the Environmental Protection Agency.

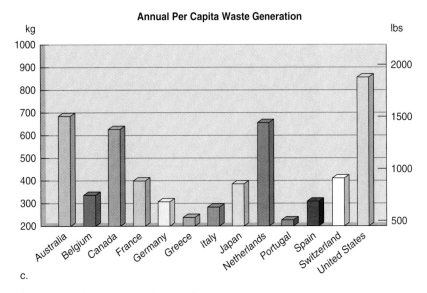

Annual Per Capita Waste Generation

Figure 18.3, *continued*

The Changing Nature of Trash (*c*) Solid waste generated per person in selected developed nations (note that volumes are for comparative purposes only; due to differences in methods of accounting, the annual waste generated per person in the United States does not precisely agree with the data given in 18.3 *a* and *b*.

Source: (c) Data compiled from World Resources Institute, World Resources 1992–93, 1992 Oxford University Press, New York.

Figure 18.4

Burning Landfills To minimize the volume of waste in landfills, the landfills were frequently burned. Waste is still being burned in sparsely populated areas in North America and other parts of the world today.

more emphasis was placed on the environment and quality of life. Simply dumping and burning our wastes is no longer an acceptable practice from an environmental or health perspective. While the technology of waste disposal has evolved during the past several decades, our options are still limited. Realistically, there are no ways of dealing with waste that have not been known for many thousands of years. Essentially, four techniques are used: (1) dumping it, (2) burning it, (3) converting it into something that can be used again, and (4) minimizing the volume produced in the first place.

Methods of Waste Disposal

Our present waste-management systems are failing, and changes are needed. It is increasingly important that each community adapt an integrated waste-management approach that combines the following four methods in a way best suited to local needs and capabilities: (1) landfills, (2) incineration, (3) source reduction, and (4) recycling.

Landfilling has historically been the primary method of waste disposal because it is the cheapest and most convenient, and because the threat of groundwater contamination was not initially recognized. (See figure 18.5.) However, the landfill of today is far different from a simple hole in the ground into which garbage is dumped.

Landfilling

A modern **municipal landfill** is typically a clay-lined depression in which each day's deposit of fresh garbage is covered with a layer of soil. Selection of modern landfill sites must be based on an understanding of groundwater geology, soil type, and sensitivity to local citizens' concerns. Once the site is selected, extensive construction activities are necessary to prepare it for use. New landfills have complex bottom layers to trap contaminant-laden water leaking through the buried trash. In addition, monitoring systems are necessary to detect methane gas production and groundwater contamination. In some cases, methane produced by rotting garbage is collected and used to generate electricity. The water that leaches through the site must be collected and treated. As a result, new landfills are becoming increasingly more complex and expensive. They currently cost up to $1 million per hectare to prepare. (See figure 18.6.)

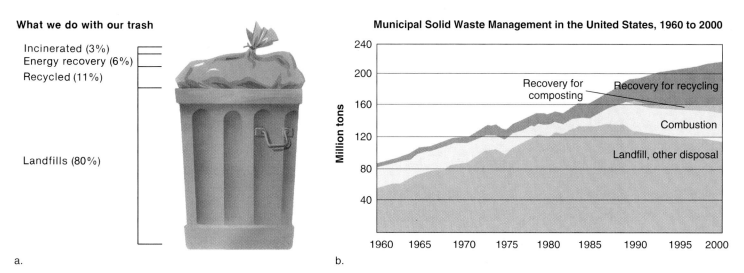

What we do with our trash

Incinerated (3%)
Energy recovery (6%)
Recycled (11%)

Landfills (80%)

a.

Municipal Solid Waste Management in the United States, 1960 to 2000

b.

Figure 18.5

Landfill (*a*) The landfill is still the primary method of waste disposal. Historically, landfills have been the cheapest means of disposal. This may not be the case in the future. As shown in (*b*), recycling has grown over the past decade while landfilling has declined somewhat.

Source: (a) Data from Franklin Associates, Ltd., INFORM Inc. (b) Data from EPA Journal.

How a modern landfill works

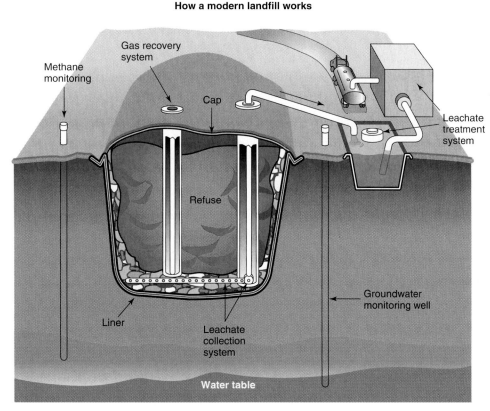

Figure 18.6

A Modern Well-Designed Landfill A modern sanitary landfill is far different from a simple hole in the ground filled with trash. A modern landfill is more of a "cell," which is sealed when filled. Methane gas and groundwater are continuously monitored.

Source: From National Solid Waste Management Association.

Today, almost 80 percent of municipal solid waste in North America goes into landfills, but this method is failing to handle the volume. For example, New York City's Fresh Kills Landfill on Staten Island is already standing 30 meters high and soon will be the sole recipient of the city's 27,000-ton daily trash load. Fresh Kills has been projected to reach a peak of 170 meters when it closes in a decade.

The problem New York City faces with the closing of its Fresh Kills Landfill is similar to what many communities will be facing in the near future. It is estimated that early in the next century there will be only 3,250 landfills in the United States compared to 18,500 landfills in 1979. (See figure 18.7.)

A prolonged public debate over how to replace lost landfill capacity is developing where population density is high and available land is scarce. Siting new landfills in locations like Toronto, New York, and Los Angeles is extremely difficult because of local opposition, which is commonly referred to as the NIMBY, or "not-in-my-backyard," syndrome. Resistance by the public comes from concern over groundwater contamination, odors, and truck traffic. Public officials look for alternatives to landfills to avoid problems with the public over landfill sitings. Although

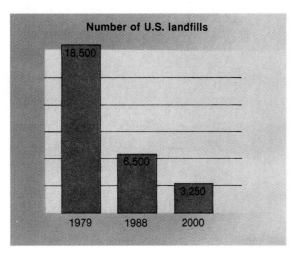

Figure 18.7

Reducing the Number of Landfills The number of landfills in the United States is declining because they are filling up or because their design and operation do not meet environmental standards.

Source: Data from U.S. Environmental Protection Agency, 1990.

siting a new landfill may be necessary, politicians are unwilling to take strong positions that might alienate their constituents. This is often called the NIMEY, or "not-in-my-election-year," syndrome.

Japan and many Western European countries have already moved away from landfilling as the primary method of waste disposal because of land scarcities and related environmental concerns. Sweden, Switzerland, and Japan dispose of 15 percent or less of their waste in landfills, compared to 80 percent in North America. Instead, recycling and incineration are the primary methods. In addition, the energy produced by incineration can be used for electric generation or heating.

Incineration

Incineration of refuse was quite common in North America and Western Europe prior to 1940. However, many incinerators were eliminated because of aesthetic concerns, such as foul odors, noxious gases, and gritty smoke, rather than for reasons of public health. Today, about 13 percent of the municipal solid waste in the United States is incinerated. Most incinerators are not used just to burn trash. The heat derived from the burning is converted into steam and electricity. There are approximately 115 operating waste-to-energy incinerators and about 60 more contracted or in the advanced-planning stage. (See figures 18.8 and 18.9.)

Most incineration facilities burn unprocessed municipal solid waste, which is not as efficient as some other technologies. About one-fourth of the incinerators use refuse-derived fuel—collected refuse that has been processed into pellets prior to combustion.

The newest means of incineration, a European concept, is called **mass burn.** In the mass-burn technique, municipal solid waste is fed into a furnace, where it falls onto moving grates and is burned at temperatures up to 1,300° C (2,400° F). The burning waste heats water, and the steam drives a turbine to generate electricity, which is solid to a utility.

Incinerators drastically reduce the amount of municipal solid waste—up to 90 percent by volume and 75 percent by weight. Primary risks of incineration, however, involve air-quality problems and the toxicity and disposal of the ash.

Though mass-burn technology works efficiently in Europe, the technology is not easily transferrable. North American municipal solid waste contains more plastic and toxic materials than European waste, thus creating air-pollution and ash-toxicity concerns. Modern incinerators have electrostatic precipitators, dual scrubbers, and fabric filters called baghouses; however, they still release small amounts of pollutants into the atmosphere, including certain metals, acid gases, and also classes of chemicals known as dioxins and furans, which have been implicated in birth defects and several kinds of cancer. The long-term risks from the emissions are still a subject of debate.

Ash from incineration is also a major obstacle to the construction of waste-to-energy facilities. Small concentrations of heavy metals are present both in the air emissions (fly ash) and residue (bottom ash) from these facilities. Because the ash contains lead, cadmium, mercury, and arsenic in varying concentrations from such items as batteries, lighting fixtures, and pigments, this ash may not need to be treated as hazardous waste. The toxic substances are more concentrated in the ash than in the original garbage and can seep into groundwater from poorly sealed landfills. Many cities have had difficulty disposing of incinerator ash, and there is still considerable debate about what is the best method of disposal.

The cost and siting of new incinerators are also major concerns facing many communities. Incinerator construction is often a municipality's single largest bond issue. Incinerator construction costs in North America in 1995 ranged from $35 million to $300 million, and the costs are not likely to decline.

Critics have argued that cities and towns have impeded waste reduction and recycling efforts by putting a priority on incinerators and committing resources to them. Proponents of incineration have been known to oppose source reduction. They argue that incinerators need large amounts of municipal solid waste to operate and that reducing the amount of waste generated makes incineration impractical. Many communities that have opposed incineration say that they support a vigorous waste-reduction and recycling effort.

Source Reduction

The most fundamental way to reduce waste is to prevent it from ever becoming waste in the first place. Examples of **source reduction** are using less material

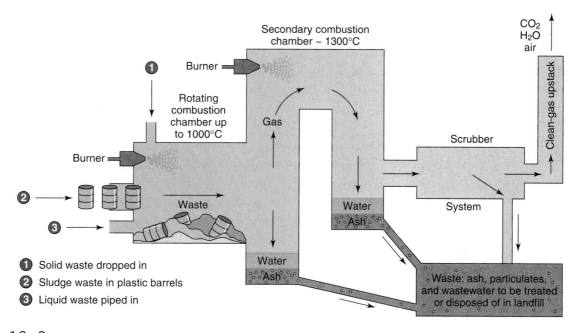

Secondary combustion chamber ~ 1300°C

Burner

Rotating combustion chamber up to 1000°C

Burner

Gas

Waste

Water Ash

Water Ash

CO₂
H₂O
air

Clean-gas upstack

Scrubber

System

Water Ash

Waste: ash, particulates, and wastewater to be treated or disposed of in landfill

❶ Solid waste dropped in
❷ Sludge waste in plastic barrels
❸ Liquid waste piped in

Figure 18.8

A high-temperature incineration system   The diagram shows how a high-temperature incinerator operates.

Figure 18.9

Incineration   Incineration of municipal solid waste reduces its weight and volume significantly. However, there are concerns about air-quality problems and the toxicity and disposal of the ash.

when making a product or converting from heavy packaging materials to lightweight ones. Some packagers have converted to lightweight aluminum and plastic by reducing the thickness of packages and thus the amount of packaging waste. Since the 2-liter soft-drink bottle was introduced in 1977, its weight has been reduced by more than 35 percent. Since 1965, aluminum cans have been reduced in weight by 35 percent. (See figure 18.10.)

Another way companies reduce waste is by making consumer products in concentrated form. These concentrated products can then be packaged in smaller containers. This approach, however, requires consumers to favor concentrates when making purchasing decisions.

Municipal composting is another source-reduction technique that could have substantial impact since 20 percent of material going into landfills in North America is yard waste. Building a compost pile is a popular practice with many home gardeners because it turns waste into a useful soil additive. (See figure 18.11.) Many communities require separate collection of such yard waste as grass clippings, leaves, and even discarded Christmas trees. Such waste is either composted or shredded into wood chips. The by-products can then be used for landscaping at city parks, schools, golf courses, and cemeteries. Preserving space in landfills by municipal composting is gaining in popularity because it is cost effective and substantially extends a landfill's useful life.

On an individual level, we can all attempt to reduce the amount of waste we

What You Can Do To Reduce Waste and Save Money

You can make a difference. While this phrase is sometimes overused, it does speak the truth when it comes to your ability to lessen the stream of solid waste being generated everyday. Here are a few ideas that are easy to follow, will save you money, and will help reduce waste.

- Buy things that last, keep them as long as possible, and have them repaired, if possible.
- Buy things that are reusable or recyclable, and be sure to reuse and recycle them.
- Buy beverages in refillable glass containers instead of cans or throwaway bottles.
- Use plastic or metal lunch boxes and metal or plastic garbage containers without throwaway plastic liners.
- Use rechargeable batteries.
- Skip the bag when you buy only a quart of milk, a loaf of bread, or anything you can carry with your hands.
- Use sponges and washable cloth napkins, dish towels, and handkerchiefs instead of paper ones.
- Don't use throwaway paper and plastic plates and cups, eating utensils, razors, pens, lighters, and other

disposable items when reusable or refillable versions are available.
- Buy recycled goods, especially those made by primary recycling, and then recycle them.
- Recycle all newspaper, glass, and aluminum, and any other items accepted for recycling in your community.
- Reduce the amount of junk mail you get. This can be accomplished by writing to Mail Preference Service, Direct Marketing Association, 11 West 42nd Street, P.O. Box 3681, New York, NY 10163-3861, or by calling (212) 768-7277. Ask that your name not be sold to large mailing-list companies. Of the junk mail you do receive, recycle as much of the paper as possible.
- Push for mandatory trash separation and recycling programs in your community and schools.
- Choose items that have the least packaging or, better yet, no packaging ("nude products").
- Compost your yard and food wastes, and pressure local officials to set up a community composting program.

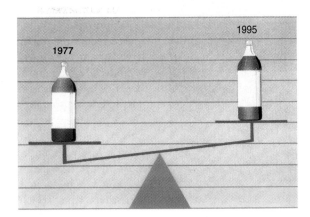

Figure 18.10

Source Reduction Plastic soft-drink bottles are now 35 percent lighter than in 1977. The weight of aluminum cans has also been reduced by 35 percent since 1965.

generate. Every small personal commitment from each of us could have the cumulative result of a significant reduction in municipal solid waste.

Recycling

About 10 percent of the waste generated in North America is handled through

recycling. The U.S. goal is to have 25 percent of municipal solid waste recycled by the year 2000. The goal for Ontario, Canada, is to reduce the amount of garbage going to the landfill sites by 50 percent by the year 2000. Recycling, along with source reduction, is a major part of the Ontario plan.

Recycling initiatives have grown rapidly in North America during the past several years. In the United States, mandatory recycling laws for all materials have passed in fifteen states. In 1990, 1,000 U.S. cities had curbside recycling programs. By 1996, the number had grown to over 7,000 cities. In Canada, Toronto and Mississauga, Ontario, have comprehensive recycling programs.

Even with the growth of recycling programs, however, North America recycles only a small percentage of the municipal solid waste generated. (See figure

Figure 18.11

Composting Household composting of yard waste is a popular practice with many home gardeners. Yard waste accounts for up to 20 percent of waste going into landfills. Composting can have a significant impact on the life of a landfill.

18.12.) In contrast, Japan recycles about 45 percent of its municipal solid waste, including half the paper, about 55 percent of glass bottles, and 66 percent of food and beverage containers. In the best-run Japanese programs, residents separate waste into several categories, which are collected on different days.

Benefits of Recycling

Some benefits of recycling are readily recognizable, such as conservation of resources and pollution reduction. These perceived benefits provide the primary motivation for participation in recycling programs. The following examples illustrate the resource-conservation and pollution-reduction benefits of recycling.

One Sunday edition of the New York Times consumes 62,000 trees. Currently, only about 20 percent of all paper used in North America is recycled.

The United States imports 91 percent of its aluminum and throws away over $400 million worth of aluminum annually.

There are no sources of tin within the United States; however, 2.7 kilograms of ultrapure tin can be reclaimed from each 900 kilograms of metal food cans.

Crushed glass (cullet) reduces the energy required to manufacture new glass by 50 percent. Cullet lowers the

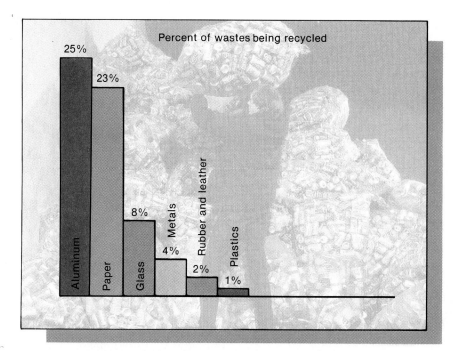

Figure 18.12

Recycling Even with the growth of recycling programs during the past several years, North Americans recycle only a small percentage of the municipal and solid waste generated.

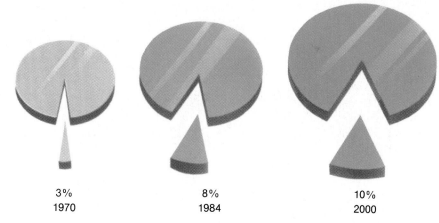

| 3% | 8% | 10% |
| 1970 | 1984 | 2000 |

Figure 18.13

Increasing Amounts of Plastics in Trash Plastics are a growing component of municipal solid waste in North America. Increased recycling of plastics could reverse this trend.

Source: Data from Franklin Associates, Ltd.

temperature requirements of the glass-making process, thus conserving energy and reducing air pollution.

Approximately 500 million barrels of oil were saved by recycling paper, metals, and rubber during 1990.

While recycling is a viable alternative to landfilling or the incineration of municipal solid waste, recycling does present several problems.

Recycling Concerns

Problems associated with recycling tend to be either technical or economic. Technical questions are of particular concern when recycling plastics. (See figure 18.13.)

environmental
close-up

Recycling Is Big Business

In 1996, Weyerhaeuser Company, one of the world's largest forest products producers, announced that it was going to build a new paper mill in Cedar Rapids, Iowa, an area of North America not known for its abundant forests. Even with a lack of local trees, the new mill will not be hurting for raw material to make paper and corrugated cardboard.

The mill will be supplied with old paper, including scrap boxes from K-Mart, which has signed an agreement with Weyerhaeuser to be a supplier. The fact that a supply of recycled paper can determine where new paper-making factories will be built indicates just how important the recycling business is becoming. It is projected that by the turn of the century over a third of the raw material that Weyerhaeuser uses for paper making will come from recycled materials. Weyerhaeuser is the fourth largest paper recycler and collector in the United States. It has 35 wastepaper processing plants in both Canada and the United States. According to a company spokesperson, the recycling business is becoming as significant as the millions of hectares of forests the company owns.

environmental
close-up

Recyclables Market Basket

Annually in North America, 12 percent of scrap used tires are recovered. Scrap tires are difficult to dispose of in landfills and waste incinerators. An estimated 2 billion to 4 billion are currently stockpiled. These stockpiles can provide convenient habitats for rodents, serve as breeding grounds for mosquitoes, and pose fire hazards. Of the scrap tires that are used, most are burned for energy. Scrap tires also are used for rubberized asphalt paving, molded rubber products, and athletic surfaces.

About 96 percent of automotive batteries are recovered each year in North America. Although these lead-acid batteries constitute a small portion of the waste stream, they contain metals that may be a concern when disposed of in landfills and incinerators. All three components of automotive batteries are recyclable: the lead, the acid, and the plastic casing.

Seventy percent of all used oil is recovered in North America. Only 10 percent of the amount generated by people who change their own motor oil is returned to collection programs. If disposed of improperly, such as being poured down sewage drains, used oil can contaminate soil, groundwater, and surface water. In some communities, used motor oil is collected at service stations, corporate or municipal collection sites, or at the curbside.

Pens and many other products can be made from recycled tires.

While the plastics used in packaging are recyclable, the technology to do so differs from plastic to plastic. There are many different types of plastic polymers. Since each type has its own chemical makeup, different plastics cannot be recycled together. In other words, a milk container is likely to be high-density polyethylene (HDPE), while an egg container is polystyrene (PS) and a soft-drink bottle is polyethylene terephthalate (PET).

Plastic recycling is still a relatively new field. Industry is researching new technologies that promise to increase the quality of plastics recycled and that will allow mixing of different plastics. Until such technology is developed, separation of different plastics before recycling will be necessary.

The economics of recycling are also a primary area of concern. The stepped-up commitment to recycling in many developed nations has produced a glut of certain materials on the market. Markets for collected materials fill up just like landfills. Unless the demand for recycled products keeps pace with the growing supply, recycling programs will face an uncertain future.

On the positive side, the markets for materials collected in recycling programs grew dramatically during the mid 1990s. By 1996 in North America, the price for collected aluminum was up nearly 100 percent; for corrugated containers and certain plastics it was up over 200 percent. The price for collected newspaper went from $10 a ton in 1990 to over $100 a ton by 1995. In 1995, the American Forest and Paper Association stated that it was planning to spend $10 billion by 2000 to retool and build new mills to produce recycled paper.

The long-term success of recycling programs is also tied to other economic incentives such as taxing issues and the development of and demand for products manufactured from recycled material. Government tax policy needs to be readjusted to encourage recycling efforts. Currently in the United States it is still cheaper to transport virgin material such as fresh cut pulp wood than to transport collected paper for recycling. Such taxing policy severely inhibits the cost-effectiveness of paper recycling. In addition, on an individual level, we can have an impact by purchasing products made from recycled materials. The demand for recycled products must grow if recycling is to succeed on a large scale. (See Environmental Close-Up: Recyclables Market Basket for photos of items made from recycled waste.)

environmental *close-up*

Resins Used in Consumer Packaging

Thermoplastics, which account for 87 percent of plastics sold, are the most recyclable form of plastics because they can be remelted and reprocessed, usually with only minor changes in their properties. Thermoplastic resins are commonly used in consumer packaging applications:

1. Polyethylene terephthalate (PET) is used extensively in rigid containers, particularly beverage bottles for carbonated beverages.
2. Polyethylene is the most widely used resin. High-density polyethylene (HDPE) is used for rigid containers, such as milk and water jugs, household-product containers, and motor oil bottles.
3. Polyvinyl chloride (PVC) is a tough plastic often used in construction and plumbing. It is also used in some food, shampoo, oil, and household-product containers.

4. Low-density polyethylene (LDPE) is often used in films and bags.
5. Polypropylene (PP) is used in a variety of areas, from snack-food packaging to battery cases to disposable diaper linings. It is frequently interchanged for polyethylene or polystyrene.
6. Polystyrene (PS) is best known as a foam in the form of cups, trays, and food containers. In its rigid form, it is used in cutlery.
7. Other. These usually contain layers of different kinds of resins and are most commonly used for squeezable bottles (for example, ketchup).

Currently, HDPE and PET are the two most commonly recycled resins. Efforts to recycle PS are also being explored, especially within the fast-food industry.

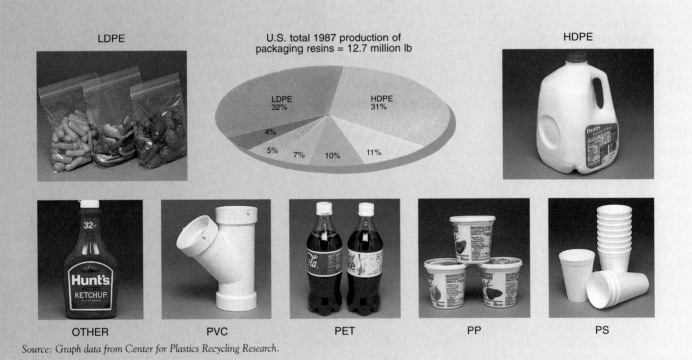

LDPE

U.S. total 1987 production of packaging resins = 12.7 million lb

HDPE

LDPE 32%
HDPE 31%
4%
5%
7%
10%
11%

OTHER   PVC   PET   PP   PS

Source: Graph data from Center for Plastics Recycling Research.

environmental close-up

Container Laws

In October 1972, Oregon became the first state to enact a "Bottle Bill." This statute required a deposit of two to five cents on all beverage containers that could be reused. It banned the sale of one-time-use beverage bottles and cans. The purpose of the legislation was to reduce the amount of litter. Beverage containers were estimated to make up about 62 percent of the state's litter. The bill succeeded in this respect: Within two years after it went into effect, beverage-container litter decreased by about 49 percent.

Those opposed to the bottle bill cited a loss of jobs that resulted from the enactment of the bill. In two small can-manufacturing plants, 142 persons did lose their jobs, but hundreds of new employees were needed to handle the returnable bottles. Other arguments against bottle bills focused on the "major" inconvenience to consumers who would need to return the bottles and cans to the store and on retailers' storage problems. Neither of these problems has proven to be serious. Nine other states—Vermont, Maine, Connecticut, New York, Iowa, Rhode Island, Michigan, Delaware, and California—have enacted legislation requiring deposits on returnable bottles, specifically beverage containers. Many states, such as New Jersey, Rhode Island, and California, are turning more toward mandating recycling laws. In 1991, Maine expanded its bottle law to all nondairy beverage containers, including all noncarbonated juice containers holding a gallon or less. Deposits are five cents, except on liquor and wine bottles, which carry a fifteen-cent deposit. Excluded are containers for milk and other dairy products, cough syrup, baby formula, soap, and vinegar.

At the present time, the amount of energy consumed in the United States is a major concern of many people. Can the United States afford to use the equivalent of 125,000 barrels of oil a day to allow for the convenience of throwaway containers? Many argue that a national bottle bill is long overdue. A national bottle bill would reduce litter, save energy and money, and create jobs. It would also help to conserve natural resources. A national container bill could save approximately 7 million metric tons of glass, 2 million metric tons of steel, and 0.5 million metric tons of aluminum each year.

In 1990, more than 50 legislative bills were introduced in the U.S. Congress addressing ways to reduce, revise, and recycle the nation's growing volume of solid waste. One of the bills would have required a minimum five-cent refundable deposit on most beverage containers nationwide. By the middle of the 1990s, the U.S. Congress had still failed to pass a national container law.

Corporate Response to Environmental Concerns

In 1990, McDonald's Corporation announced that it would switch from polystyrene to paper for packaging its food products. In making this announcement, McDonald's stated that it was responding to consumer pressure to become more environmentally conscientious. Is using paper to wrap fast food better for the environment than using polystyrene? The answer is not a simple yes or no.

Advocates for the switch say that polystyrene takes up space in landfills and does not decompose. They also argue that the burning of polystyrene foam in incinerators might release harmful air pollutants.

Opponents of the switch argue that polystyrene can be recycled into useful products, such as insulation board or playground equipment. They further argue that using paper means cutting forests and that, since the paper used to wrap the food is coated with wax, it cannot be recycled.

Many grocery chains now offer several choices of carryout bags. Some encourage reuse of bags by deducting a small amount from the customer's bill. Others provide a choice of paper or plastic.

- What are the pros and cons of paper and plastic?
- Which alternative to you prefer? Why?
- Is there an alternative to both?

Which is better?

Which do you use?

Is there an alternative?

SUMMARY

Beginning with the post-World War II era, increased consumption of consumer goods became a way of life. Products were designed to be used once and then thrown away. By the 1980s, a disposal lifestyle began to cause problems. There simply were no places to dispose of waste. Barges filled with municipal solid waste from the metropolitan areas along the eastern United States were traveling the world trying to dispose of their unwanted cargo.

Municipal solid waste is managed by landfilling, incineration, waste reduction, and recycling. Landfilling is the primary means of disposal; however, a contemporary landfill is significantly more complex and expensive than the simple holes in the

ground of the past. The availability of suitable landfill land is also a problem in large metropolitan areas.

About 13 percent of the municipal solid waste in the United States is incinerated. The number of incinerators in North America has declined over the past several decades. While incineration does reduce the volume of municipal solid waste, the problems of ash disposal and air quality continue to be major concerns.

The most fundamental way to reduce waste is to prevent it from ever becoming waste in the first place. Using less material in packaging, producing consumer products in concentrate form, and composting yard waste are all examples of source reduction. On an individual level, we can all attempt to reduce the amount of waste we generate.

About 10 percent of the waste generated in North America is handled through recycling. In contrast, Japan recycles about 45 percent of its municipal solid waste. Recycling initiatives have grown rapidly in North America during the past several years. As a result, the markets for some recycled materials have become glutted. Recycling of municipal solid waste will only be successful if markets exist for the recycled materials.

Another problem in recycling is the current inability to mix various plastics. The plastic industry is working on the development of a more universal plastic.

Future management of municipal solid waste will be an integrated approach involving landfilling, incineration, source reduction, and recycling. The degree to which any option will be used will depend on economics, changes in technology, and citizen awareness and involvement.

REVIEW QUESTIONS

1. How is lifestyle related to our growing municipal solid waste problem?
2. What four methods are incorporated under integrated waste management?
3. Describe some of the problems associated with modern landfills.
4. What are four concerns associated with incineration?
5. Describe examples of source reduction.
6. Describe the importance of recycling household solid wastes.
7. Name several strategies that would help to encourage the growth of recycling.

KEY TERMS

incineration   376
mass burn   376
municipal landfill   374
municipal solid waste   373
recycling   378
source reduction   376

RELATED INTERNET SITES

Solid waste management. Data sheets on topics from aluminum cans to cloth diapers.
    http://clean.rti.org/larry/sw.htm
Keep America beautiful site on solid waste management and litter prevention.
    http://www.kab.org/
up-to-date waste management databases in Canada.
    http://www.web.net/rco/3rsource.html
Solid waste management using environmentally sound waste management practices
    http://www.bluestem.org/
Repository for information on solid and hazardous waste.
    http://www.webvista.com/swix/

World Resource Foundation provides a worldwide database on sustainable waste management.
    http://www.wrfound.org.uk/
Papers on solid waste policy and often-neglected worldwide waste issues.
    http://www.yale.edu/pswp
Award-winning composting resource page.
    http://www.oldgrowth.org/compost/
Up-to-date information on recycling practices and legislation.
    http://www.cais.com/recycle/
Global Recycling Network. Great links and searchable.
    http://grn.com/grn/

chapter 19

Regulating Hazardous Materials

Objectives

After reading this chapter, you should be able to:
- Distinguish between hazardous substances and hazardous wastes.
- Distinguish between hazardous and toxic materials.
- Explain the complexity in regulating hazardous materials.
- Describe the four characteristics by which hazardous materials are identified.
- Describe the environmental problems of hazardous and toxic materials.
- Understand the difference between persistent and nonpersistent pollutants.
- Describe the health risks associated with hazardous wastes.
- Explain the problems associated with hazardous-waste dump sites and how such sites developed.
- Describe how hazardous wastes are managed, and list five technologies used in their disposal.
- Describe the importance of source reduction with regard to hazardous wastes.

Chapter Outline

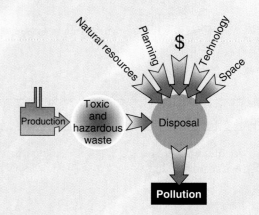

Hazardous and Toxic Materials in Our Environment

Our modern technological society makes use of a large number of substances that are hazardous and toxic. The benefits gained from using these materials must be weighed against the risks associated with their use.

- Pesticides thought to degrade in soils turn up in rural drinking-water wells.
- Underground plumes of toxic chemicals emanating from abandoned waste sites contaminate city water supplies.
- A gas leak at a chemical production plant in Bhopal, India, killed more than 2,000 people.
- Pesticides spilled into the Rhine River from a warehouse near Basel, Switzerland, destroyed a half million fish, disrupted water supplies, and caused considerable ecological damage.
- The collapse of an oil storage tank in Pennsylvania spilled over 2 million liters of oil into the Monongahela River and threatened the water supply of millions of residents.

Toxic and hazardous products and by-products are becoming a major issue of our time. At sites around the world, accidental or purposeful releases of hazardous and toxic chemicals are contaminating the land, air, and water. The potential health effects of these chemicals range from minor, short-term discomforts such as headaches and nausea to serious health problems such as cancers and birth defects (that may not manifest themselves for years) to major accidents that cause immediate injury or death. Today, names like Love Canal, New York, and Times Beach, Missouri, in the United States; Lekkerkerk in the Netherlands; Vac, Hungary; and Minamata Bay, Japan, are synonymous with the problems associated with the release of hazardous and toxic wastes into the environment.

Increasingly, governments and international agencies are attempting to con-trol the growing problem of hazardous substances in our environment. Controlling the release of these substances is difficult since there are so many places in their cycles of use at which they may be released. (See figure 19.1.)

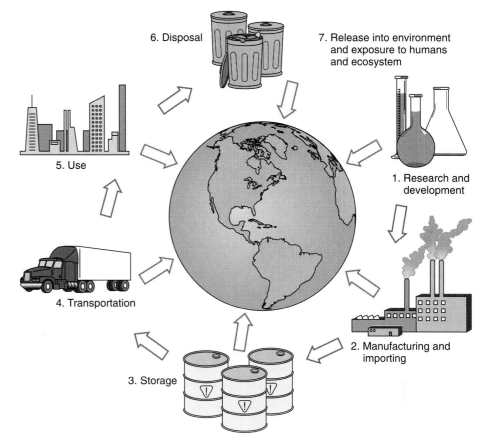

Figure 19.1

The Life Cycle of Toxic Substances Controlling the problems of hazardous substances is complicated because of the many steps involved in a substance's life cycle.

6. Disposal

7. Release into environment and exposure to humans and ecosystem

5. Use

1. Research and development

4. Transportation

2. Manufacturing and importing

3. Storage

Hazardous and Toxic Substances— Some Definitions

To begin, it is important to clarify various uses of the words *hazardous* and *toxic* as well as distinguish between things that are wastes and those that are not. **Hazardous substances** are those that can cause harm to humans or the environment. The U.S. Environmental Protection Agency (EPA) defines hazardous materials as having one or more of the following characteristics:

1. **Ignitability**—Describes materials that pose a fire hazard during routine management. Fires not only present immediate dangers of heat and smoke but also can spread harmful particles over wide areas. Common examples are gasoline, paint thinner, and alcohol.
2. **Corrosiveness**—Describes materials requiring special containers because of their ability to corrode standard materials, or requiring segregation from other materials because of their ability to dissolve toxic contaminants. Common examples are strong acids and bases.
3. **Reactivity** (or explosiveness)—Describes materials that, during routine management, tend to react spontaneously, to react vigorously with air or water, to be unstable to shock or heat, to generate toxic gases, or to explode. Common

examples are gunpowder, which will burn or explode; the metal sodium, which reacts violently with water; and nitroglycerine, which explodes under a variety of conditions.

4. **Toxicity**—Describes materials that, when improperly managed, may release toxicants (poisons) in sufficient quantities to pose a substantial hazard to human health or the environment. Almost everything that is hazardous is toxic in high enough quantities. For example, tiny amounts of carbon dioxide in the air are not toxic, but high levels are.

Some hazardous materials fall into several of these categories. Gasoline, for example, is ignitable, can explode, and is toxic. It is even corrosive to certain kinds of materials. Other hazardous materials meet only one of the criteria. Polychlorinated biphenyls (PCBs) are toxic but will not burn, explode, or corrode other materials. While the terms *toxic* and *hazardous* are often used interchangeably, there is a difference. **Toxic** commonly refers to a narrow group of substances that are poisonous and cause death or serious injury to humans and other organisms by interfering with normal body physiology. **Hazardous,** the broader term, refers to all dangerous materials, including toxic ones, that present an immediate or long-term human health risk or environmental risk.

Another important distinction that needs to be made is the difference between hazardous substances and hazardous wastes. Although the health and safety considerations regarding hazardous substances and hazardous wastes are similar, the legal and regulatory implications are quite different. Hazardous substances are materials that are used in business and industry for the production of goods and services. Typically, hazardous substances are consumed or modified in industrial processes. **Hazardous wastes** are by-products of industrial, business, or household activities for which there is no immediate use. These materials must be disposed of in an appropriate manner, and there are stringent regulations pertaining to their production, storage, and disposal.

Defining Hazardous Waste

The definition of hazardous waste varies from one country to another. One of the most widely used definitions, however, is contained in the U.S. **Resource Conservation and Recovery Act** of 1976 (RCRA). The RCRA considers wastes toxic and/or hazardous if they:

> cause or significantly contribute to an increase in mortality or an increase in serious irreversible, or incapacitating reversible, illness; or pose a substantial present or potential hazard to human health or the environment when improperly treated, stored, transported, disposed of, or otherwise managed.

This definition gives one an appreciation for the complexity of hazardous-waste regulation.

Working from the RCRA definition, the EPA compiled a list of hazardous wastes. Listing is the most common method for defining hazardous waste in European countries and in some state laws. The EPA has also proposed that a hazardous waste be identified by testing it to determine if it possesses any one of the four characteristics discussed earlier: ignitability, reactivity, corrosiveness, and toxicity. If it does, it is subject to regulation under RCRA.

Issues Involved in Setting Regulations

Whether a hazardous substance is a raw material, an ingredient in a product, or a waste, there are problems associated with determining regulations that pertain to it. In the industrialized countries of Europe and North America, chemical and petrochemical industries produce nearly 70 percent of all hazardous wastes; in developing countries, the figure is 50–66 percent. These industries produce many useful materials that are converted into the everyday products we use. Most toxic and hazardous wastes come from chemical and related industries that produce plastics,

soaps, synthetic rubber, fertilizers, medicines, paints, pesticides, herbicides, and cosmetics. (See figure 19.2.)

Identification of Hazardous and Toxic Materials

In attempting to regulate the use of toxic and hazardous substances and the generation of toxic and hazardous wastes, most countries simply draw up a list of specific substances for which there is sufficient scientific evidence linking them to adverse human health or environmental effects. However, since many potentially harmful chemical compounds have yet to be tested adequately, most lists include only the known offenders. Historically, we have often identified toxic materials only after their effects have shown up in humans or other animals. Asbestos was identified as a cause of lung cancer in humans who were exposed on the job, and DDT was identified as toxic to birds when robins began to die and eagles and other fish-eating carnivores failed to reproduce. Once these substances are identified as toxic, their use is regulated. Countries contemplating regulation of hazardous and toxic materials and wastes must consider not only how toxic each one is but also how flammable, corrosive, and explosive it is, and whether it will produce mutations or cause cancer.

Setting Exposure Limits

Even after a material is identified as hazardous or toxic, there are problems in determining appropriate exposure limits. Nearly all substances are toxic in sufficiently high doses. The question is, When does a chemical cross over from safe to toxic? There is no easy way to establish acceptable levels. For any new compounds that are to be brought on the market, extensive toxicology studies must be done to establish their ability to do harm. Usually these are tests on animals. (See Environmental Close-up: Exposure to Toxins.) Typically, the regulatory agency will determine the level of exposure at which none of the test animals is affected (**threshold level**) and then set the human exposure level lower

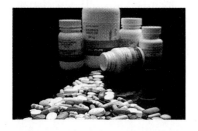

The products we use...	The potentially hazardous waste they generate...
Plastics	Organic chlorine compounds
Pesticides	Organic chlorine compounds, organic phosphate compounds
Medicines	Organic solvents and residues, heavy metals (mercury and zinc, for example)
Paints	Heavy metals, pigments, solvents, organic residues
Oil, gasoline, and other petroleum products	Oils, phenols, and other organic compounds, heavy metals, ammonia salts, acids, caustics
Metals	Heavy metals, fluorides, cyanides, acid and alkaline cleaners, solvents, pigments, abrasives, plating salts, oils, phenols
Leather	Heavy metals, organic solvents
Textiles	Heavy metals, dyes, organic chlorine compounds, solvents

Figure 19.2

Common Materials Can Produce Hazardous Wastes   Many commonly used materials can release toxic wastes if not properly disposed of.

to allow for a safety margin. This safety margin is important because it is known that threshold levels vary significantly among species, as well as among members of the same species. Even when concentrations are set, they may vary consider-

ably from country to country. For example, in the Netherlands, 50 milligrams of cyanide per kilogram of waste is considered hazardous; in neighboring Belgium, the toxicity standard is fixed at 250 milligrams per kilogram.

Acute and Chronic Toxicity

Regulatory agencies must look at both the effects of one massive dose of a substance (**acute toxicity**) and the effects of exposure to small doses over long periods (**chronic toxicity**). Acute toxicity is readily apparent because organisms respond to the toxin shortly after being exposed. Chronic toxicity is much more difficult to determine because the effects may not be seen for years. Furthermore, an acute exposure may make an organism ill but not kill it, while chronic exposure to a toxic material may cause death. A good example of this effect is alcohol toxicity. Consuming extremely high amounts of alcohol can result in death (acute toxicity and death). Consuming moderate amounts may result in illness (acute toxicity and full recovery). Consuming moderate amounts over a number of years may result in liver damage and death (chronic toxicity and death).

Another example of chronic toxicity involves lead. Lead has been used in paints, in gasoline, and in pottery glazes for many years, but researchers discovered that it has harmful effects. The chronic effects on the nervous system are most noticeable in children, particularly when children eat paint chips.

Synergism

Another problem in regulating hazardous materials is assessing the effects of mixtures of chemicals. Most toxicological studies focus on a single compound, even though industry workers may be exposed to a variety of chemicals, and in waste dumps, the compounds are usually found in mixtures. Although the materials may be relatively harmless as separate compounds, once mixed, they may become highly toxic and cause more serious problems than do individual pollutants. This is referred to as **synergism.** For example, all uranium miners are exposed to radioactive gases, but those who smoke tobacco and thus are exposed to the toxins in tobacco smoke have unusually high incidences of lung cancer. Apparently, the radioactive gases found in uranium mines interact synergistically with the carcinogens found in tobacco smoke.

Persistent and Nonpersistent Pollutants

The regulation of hazardous and toxic materials is also influenced by the degree of persistence of the pollutant. **Persistent pollutants** are those that remain in the environment for many years in an unchanged condition. Most of the persistent pollutants are human-made materials. An estimated 30,000 synthetic chemicals are used in the United States. They are mixed in an endless variety of combinations to produce all types of products used in every aspect of daily life. They are part of our food, transportation, clothing, building materials, home appliances, medicine, recreational equipment, and many other items. Our way of life is heavily dependent upon synthetic materials.

An example of a persistent pollutant is DDT. It was used as an effective pesticide worldwide and is still used in some countries because it is so inexpensive. However, once released into the environment, it accumulates in the food chain and causes death when its concentration is high enough. (See chapter 14 for a discussion of DDT as a pesticide.)

Another widely used group of synthetic compounds of environmental concern are polychlorinated biphenyls (PCBs). PCBs are highly stable compounds that resist changes from heat, acids, bases, and oxidation. These characteristics make PCBs desirable for industrial use but also make them persistent pollutants when released into the environment. At one time these materials were commonly used in transformers and electrical capacitors. Other uses included inks, plastics, tapes, paints, glues, waxes, and polishes. PCBs are harmful to fish and other aquatic forms of life because they interfere with reproduction. In humans, PCBs produce liver ailments and skin lesions. In high concentration, they can damage the nervous system, and they are suspected carcinogens. In 1970, PCB production was limited to those cases where satisfactory substitutes were not available.

In addition to synthetic compounds, our society uses heavy metals for many purposes. Mercury, beryllium, arsenic, lead, and cadmium are examples of heavy metals that are toxic. When released into the environment, they enter the food chain and become concentrated. In humans, these metals can produce kidney and liver disorders, weaken the bone structure, damage the central nervous system, cause blindness, and lead to death. Because these materials are persistent, they can accumulate in the environment even though only small amounts might be released each year. When industries use these materials in a concentrated form, it presents a hazard not found naturally.

A **nonpersistent pollutant** does not remain in the environment for very long. Most nonpersistent pollutants are biodegradable. Others decompose as a result of inorganic chemical reactions. Still others quickly disperse to concentrations that are too low to cause harm. A biodegradable material is chemically changed by living organisms and often serves as a source of food and energy for decomposer organisms, such as bacteria and fungi. Phenol and many other kinds of toxic organic materials can be destroyed by decomposer organisms.

Other toxic materials such as many of the insecticides are destroyed by sunlight or reaction with oxygen or water in the atmosphere. These include the "soft biocides." For example, organophosphates are a type of pesticide that usually decomposes within several weeks. As a result, organophosphates do not accumulate in food chains because they are pollutants for only a short period of time.

Other toxic and hazardous materials such as carbon monoxide, ammonia, or hydrocarbons can be dispersed harmlessly into the atmosphere (as long as their concentration is not too great) where they eventually react with oxygen.

Because persistent materials can continue to do harm for a long time (chronic toxicity), they are particularly important to regulate. Nonpersistent materials need to be kept below threshold levels to protect the public from acute toxicity. They are not likely to present a danger of chronic toxicity since they either disperse or decompose. Therefore, regulations must be aimed at eliminating acute toxicity.

Environmental Problems Caused by Hazardous Wastes

Hazardous wastes contaminate the environment in several ways. Many haz-

e n v i r o n m e n t a l
close-up

Exposure to Toxins

We are all exposed to materials that are potentially harmful. The question is, at what levels is such exposure harmful or toxic? One measure of toxicity is LD_{50}, the dosage of a substance that will kill (lethal dose) 50 percent of a test population. Toxicity is measured in units of poisonous substance per kilogram of body weight. For example, the deadly chemical that causes botulism, a form of food poisoning, has an LD_{50} in adult human males of 0.0014 milligrams per kilogram. This means that if each of 100 human adult males weighing 100 kilograms consumed a dose of only 0.14 milligrams—about the equivalent of a few grains of table salt—approximately 50 of them will die.

Lethal doses are not the only danger from toxic substances. During the past decade, concern has been growing over minimum harmful dosages, or threshold dosages, of poisons, as well as their sublethal effects.

The length of exposure further complicates the determination of toxicity values. Acute exposure refers to a single exposure lasting from a few seconds to a few days. Chronic exposure refers to continuous or repeated exposure for several days, months, or even years. Acute exposure usually is the result of a sudden accident, such as the tragedy at Bhopal,

India, mentioned at the beginning of the chapter. Acute exposures often make disaster headlines in the press, but chronic exposure to sublethal quantities of toxic materials presents a much greater hazard to public health. For example, millions of urban residents are continually exposed to low levels of a wide variety of pollutants. Many deaths attributed to heart failure or such diseases as emphysema may actually be brought on by a lifetime of exposure to sublethal amounts of pollutants in the air.

---

ardous materials are released directly to the environment. Many molecules that evaporate readily are vented directly to the atmosphere or escape from faulty piping and valves. These materials are often not even thought of as being hazardous waste. Once hazardous wastes are produced, they must be stored. Improper storage or even poor bookkeeping may inadvertently result in a release. Uncontrolled or improper incineration of hazardous wastes, whether on land or at sea, can contaminate the atmosphere and the surrounding environment. The discharge of hazardous substances into the sea or into lakes and rivers often kills fish and other aquatic life. Further, disposal on land in dumps that are later abandoned, or in improperly controlled landfills, can pollute both the soil and the groundwater as materials leach below the site.

Because most hazardous wastes are disposed of on or in land, the most serious environmental effect is contaminated groundwater. In the United States alone, an estimated 100,000 active industrial landfill sites may be possible sources of groundwater contamination, along with 200 special facilities for disposal of both liquid and solid hazardous wastes, and some 180,000 surface impoundments (ponds) for all types of waste. (See chapter 15.) Nearly 2 percent of North America's underground aquifers could be contaminated with such chemicals as chlorinated solvents, pesticides, trace metals, and PCBs. Once groundwater is polluted with hazardous wastes, the cost of reversing the damage is prohibitive. In fact, if an aquifer is contaminated with organic chemicals, restoring the water to its original state is seldom physically or economically feasible.

Health Risks Associated with Hazardous Wastes

Because most hazardous wastes are chemical wastes, controlling chemicals and their waste products is a major issue in most developed countries. Every year, roughly 1,000 new chemicals join the nearly 70,000 in daily use. Many of these hazardous chemicals are toxic, but they pose little threat to human health unless they are used or disposed of improperly. For example, many insecticides are extremely toxic to humans. However, if they are stored, used, and disposed of properly they do not constitute a human health hazard. Unfortunately, at the center of the hazardous-waste problem is the fact that the products and by-products of industry are often handled and disposed of improperly. Table 19.1 is a list of 20

environmental

close-up

Toxic Chemical Releases

In 1987, as the result of a new EPA requirement, industries in the United States had to report toxic chemicals released into the environment. Any industrial plant that released 23,000 kilograms or more of toxic pollutants was required to file a report. Industrial plants that released under 23,000 kilograms were not required to file; thus, the data are incomplete. In 1990, 20,612 reports were filed, covering 332 toxic chemicals.

About 2 billion kilograms of toxic chemicals were reported released into the environment by industry in 1988, compared with nearly 2.3 billion in 1987. Another 800 million kilograms—not counted in the overall figure—were sent

to municipal-waste treatment centers or private treatment and storage facilities.

According to the 1990 data, among the chemicals routinely emitted from industrial sources were 77 carcinogens. The most widely released cancer-causing chemical—52 million kilograms—was dichloromethane, a chemical often used as an industrial solvent and paint stripper. Industry released a total of 128 million kilograms of carcinogens, including such chemicals as arsenic, benzene, and vinyl chloride. Chemical and allied industries accounted for about 60 percent of the total releases.

Toxic Substances Commonly Released into the Environment

Name of Substance	Number of states that listed it in the top five
Methanol	37
Toluene	37
Ammonia	34
Xylene	22
Hydrochloric Acid	17

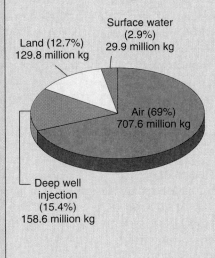

Toxic Releases to the Environment (U.S. 1994)
Environmental Compartment Affected

Land (12.7%) 129.8 million kg
Surface water (2.9%) 29.9 million kg
Air (69%) 707.6 million kg
Deep well injection (15.4%) 158.6 million kg

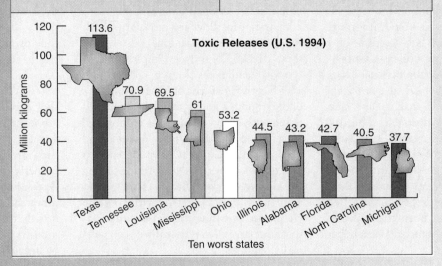

Toxic Releases (U.S. 1994)

Million kilograms

Ten worst states	
Texas	113.6
Tennessee	70.9
Louisiana	69.5
Mississippi	61
Ohio	53.2
Illinois	44.5
Alabama	43.2
Florida	42.7
North Carolina	40.5
Michigan	37.7

Lead and Mercury Poisoning

Lead and mercury are naturally present in the environment and probably have been a source of pollution for centuries. For example, the lead drinking and eating vessels used by the wealthy Romans may have caused many of their deaths. In 1865, when Alice in Wonderland was published, one of the characters was the Mad Hatter. At that period in history, mercury was widely used in the treatment of beaver skins for making hats. As a result of exposure to mercury, hat makers often suffered from a variety of mental problems; hence, the phrase "mad as a hatter."

In 1953, a number of physical and mental disorders in the Minamata Bay region of Japan were diagnosed as being caused by mercury: 52 people developed symptoms of mercury poisoning, 17 died, and 23 became permanently disabled. In 1970, an outbreak of mercury poisoning in North America was traced to mercury in the meat of swordfish and tuna. In both incidents, the toxic material was not metallic mercury but a mercurous compound, methylmercury. Metallic mercury is converted to methylmercury by

There are concerns about fish contamination and public health as a result of methylmercury.

bacteria in the water. Methylmercury enters the food chain and may become concentrated as the result of biological amplification. Sufficient amounts in humans can cause brain damage, kidney damage, or birth defects. Today, regulations reduce the release of mercury into the environment and set allowable levels in foods. The problem still persists, however, because it is impossible to eliminate the large amounts of mercury already present in the environment, and it is difficult to prevent the release of mercury in all cases. For example, burning coal releases 3,200 metric tons of mercury into the earth's atmosphere each year, and mercury is still "lost" when it is used for various industrial purposes.

Like mercury, lead is a heavy metal and has been a pollutant for centuries. Studies of the Greenland Ice Cap indicate a 1,500 percent increase in the lead content today as compared to 800 B.C. These studies reveal that the first large increase occurred during the Industrial Revolution and the second great increase occurred after the invention of the automobile. Oil companies added lead to gasoline to improve performance, and burning gasoline is a major source of lead pollution. There has been a reduction in airborne lead as a result of North America and Europe reducing lead content in gasoline. Another source of lead pollution is older paints. Prior to 1940, indoor and outdoor paints often contained lead.

There is still disagreement over what levels of lead and mercury can cause human health problems. Although ingested lead from paint can cause death or disability, such a strong correlation cannot be made for atmospheric lead.

top toxic materials as identified by the Agency for Toxic Substances and Disease Registry.

Establishing the medical consequences of exposure to toxic wastes is extremely complicated. The problem of linking a particular chemical or other hazardous waste to specific injuries or diseases is further compounded by the lack of toxicity data on most hazardous substances.

Although assessing environmental contamination from toxic wastes and determining health effects is extremely difficult, what little is known is cause for concern. Most older hazardous-waste dump sites, for example, contain danger-

ous and toxic chemicals along with heavy-metal residues and other hazardous substances.

Hazardous-Waste Dumps—A Legacy of Abuse

In the United States prior to the passage of the Resource Conservation and Recovery Act (RCRA) in 1976, hazardous waste was essentially unregulated. Similar conditions existed throughout most of the industrial nations of the world. The solution to hazardous-waste disposal was

simply to bury or dump the wastes without any concern for potential environmental or health risks. Such uncontrolled sites included open dumps, landfills, bulk storage containers, and surface impoundments. These sites were typically located convenient to the industry and were often in environmentally sensitive areas, such as floodplains or wetlands. Rain and melting snow soaked through the sites, carrying chemicals that contaminated underground waters. When these groundwaters reached streams and lakes, they were contaminated as well. When the sites became full or were abandoned, they were frequently left uncovered, thus increasing the likelihood of water pollution

Table 19.1

Top Twenty Hazardous Substances, 1995

Substance	Source	Toxic Effects
Lead	Lead-based paint Lead additives in gasoline	Neurological damage. Affects brain development in children. Large doses affect brain and kidneys in adults and children.
Arsenic	From elevated levels in soil or water	Multiple organ systems affected. Heart and blood vessel abnormalities, liver and kidney damage, impaired nervous system function.
Metallic Mercury	Air or water at contaminated sites	Permanent damage to brain, kidneys, developing fetus.
Vinyl Chloride	Plastics manufacturing Air or water at contaminated sites	Acute effects: dizziness, headache, unconsciousness, death. Chronic effects: liver, lung, and circulatory damage.
Benzene	Industrial exposure Glues, cleaning products, gasoline	Acute effects: drowsiness, headache, death at high levels. Chronic effects: damages blood forming tissues and immune system; also carcinogenic.
Polychlorinated biphenyls (PCBs)	Eating contaminated fish Industrial exposure	Probable carcinogens. Acne and skin lesions.
Cadmium	Released during combustion Living near a smelter or power plant Picked up in food	Probable carcinogen, kidney damage, lung damage, high blood pressure.
Benzo[a]pyrene	Product of combustion of gasoline or other fuels In smoke and soot	Probable carcinogen, possible birth defects.
Chloroform	Contaminated air and water Many kinds of industrial settings	Affects central nervous system, liver, and kidneys; probable carcinogen.
Benzo[b]fluoranthene	Product of combustion of gasoline and other fuels Inhaled in smoke	Probable carcinogen.
DDT	From food with low levels of contamination Still used as pesticide in parts of world	Probable carcinogen; possible long-term effect on liver; possible reproductive problems.
Aroclor 1260 (a mixture of PCBs)	From food and air	Probable carcinogens. Acne and skin lesions.
Trichloroethylene	Used as a degreaser, evaporates into air	Dizziness, numbness, unconsciousness, death.
Aroclor 1254 (a mixture of PCBs)	From food and air	Probable carcinogens. Acne and skin lesions.
Chromium (+6)	From food, water, and air Originates from combustion source	Ulcers of the skin, irritation of nose and gastrointestinal tract, also affects kidney and liver. Probable carcinogen.
Chlordane	Pesticide used to control termites Air in treated homes. Food and water	Gastrointestinal problems. Convulsions with high exposure. Headache, dizziness with low exposure.

Source: Data from Agency for Toxic Substances and Disease Registry.

Substance	Source	Toxic Effects
Dibenz[a,h]anthracene	Product of combustion of gasoline and other fuels Found in smoke and soot	Probable carcinogen.
Hexachlorbutadiene	Breathing in workplace Used in manufacture of rubber products	Kidney and liver damage.
DDD (a relative of DDT)	From food with low levels of contamination	Probable carcinogen; possible long-term effect on liver; possible reproductive problems.
Dieldrin	Previously used as pesticides Enters with food and drink	Acute effects: headache, dizziness, convulsions, death. Chronic effects: liver damage, immune system damage.

from leaching or flooding, and increasing the chances of people having direct contact with the wastes. At some sites, specifically the uncovered ones, the air was also contaminated as toxic vapors rose from evaporating liquid wastes or from uncontrolled chemical reactions. (See figure 19.3.) In North America alone, the number of abandoned or uncontrolled sites is over 25,000, and the list grows yearly. The costs involved in cleaning up the sites are high. Nearly all industrialized countries are faced with costly, even massive, cleanup bills.

As is to be expected from the amount of toxic wastes generated each year, the United States has the highest number of hazardous-waste dumps needing immediate attention. Europeans are also paying a heavy price for their negligence. Every country in Europe (except Sweden and Norway) is plagued by an abundance of toxic-waste sites—both old and new—needing urgent attention. Holland is a good example. Authorities estimate that up to 8 million metric tons of hazardous chemical wastes may be buried in Holland. Estimates for cleaning up those wastes run as high as $6 billion. In the republics of the former Soviet Union and Eastern Europe, many hazardous waste sites have been identified recently, and there is no money to pay for cleanup.

In the United States, the federal government has become the principal participant in the cleanup of hazardous-waste sites. The program that deals with the cleanup has popularly become known as **Superfund.** Superfund was established when Congress responded to public pressure to clean up hazardous-waste dumps and protect the public against the dangers of such wastes. The **Comprehensive Environmental Response, Compensation and Liability Act (CERCLA)** (Superfund) was enacted in 1980. CERCLA had several key objectives:

1. To develop a comprehensive program to set priorities for cleaning up the worst existing hazardous-waste sites.
2. To make responsible parties pay for those cleanups whenever possible.
3. To set up a $1.6 billion Hazardous Waste Trust Fund—popularly known as Superfund—to support the identification and cleanup of abandoned hazardous-waste sites.
4. To advance scientific and technological capabilities in all aspects of hazardous-waste management, treatment, and disposal.

A **National Priority List** of hazardous-waste dump sites requiring urgent attention was drawn up for Superfund action. The size of the list varies, however, depending on which governmental agency is making the list. The U.S. Office of Technology Assessment (OTA) estimates that 10,000 sites may eventually be placed on the National Priority List requiring Superfund cleanup. The OTA believes that cleaning up these hazardous dumps may take 50 years and cost up to $100 billion because of the complex mixture of contaminants in most sites. (See table 19.2.) The Government Accounting Office (GAO) believes that the National Priority List could reach more than 4,000 sites with cleanup costs of around $40 billion. The EPA has the shortest list of sites (2,500) that it says should be placed on the National Priority List at a cost of nearly $30 billion. Table 19.3 lists Superfund sites considered priority one.

By the mid-1990s, the Superfund program was still controversial. Millions of dollars have been spent by both the federal government and industry, but most of the money has involved litigation. The money has paid lawyers but has not paid for cleanup. One of the primary reasons for this is the way CERCLA was written. It provided that anyone who contributed to a specific hazardous-waste site could be required to pay for the cleanup of the entire site regardless of the degree to which it contributed to the

Figure 19.3

Toxic Chemical Storage   This is a site where toxic wastes were improperly stored. Local governments often assume the problems and costs of correcting bankrupt companies' errors.

problem. Since many industries that contributed to the problem had gone out of business or could not be identified, those who could be identified were asked to pay for the cleanup. Most businesses found it cost-effective to hire lawyers to fight their inclusion in a cleanup effort rather than to pay for the cleanup. Consequently, cleanup has been slow.

In 1997, the U.S. Congress will probably look at reauthorizing CERCLA. A key issue in this reauthorization will be an effort to change the law to allow for more realistic assignment of the costs of cleanup.

Managing Hazardous Wastes

In the past, the management of hazardous waste was always added on to the end of the industrial process. The effluents from pipes or smoke stacks were treated to reduce their toxicity or concentration. For 1993, the EPA reported that the United States produced about 235 million metric tons of hazardous waste. This was 42 million metric tons less than in 1991. In recent years it has become obvious that a better way to deal with the problem of hazardous waste is to

Table 19.2

Common Contaminants Found at Superfund Sites		
Chemical	Average occurrence (percent)	Average concentration (parts per billion)
Lead	51.4	309,000
Cadmium	44.7	2,185
Toluene	44.1	1,120,000
Mercury	29.6	1,379
Benzene	28.5	16,582
Trichloroethylene	27.9	103,000
Ethylbenzene	26.9	540,000
Benzo[a]anthracene	12.3	148,000
Bromodichloromethane	7.0	20
Polychlorinated biphenyls	3.9	128,000
Toxaphene	0.6	12,360

Source: Data from presentation by William Eckel and Donald Trees, Viar and Company, and Stanley Kovell, EPA, at the National Conference on Hazardous Wastes and Environmental Emergencies as appeared in Environmental Science & Technology, *20:28, 1986, American Chemical Society.*

Table 19.3

Proposed National List of Superfund Cleanup Sites in the United States (priority one)

State/City/County	Site name	State/City/County	Site name
Alabama		**New Jersey**	
Limestone and Morgan	Triana, Tennessee River	Bridgeport	Bridgeport Rent & Oil
Arkansas		Fairfield	Caldwell Trucking
Jacksonville	Vertac, Inc.	Freehold	Lone Pine Landfill
California		Gloucester Township	Gems Landfill
Glen Avon Heights	Stringfellow	Mantua	Helen Kramer Landfill
Delaware		Marlboro Township	Burnt Fly Bog
New Castle	Army Creek	Old Bridge Township	CPS/Madison Industries
New Castle County	Tybouts Corner	Pittman	Lipari Landfill
Florida		Pleasantville	Price Landfill
Jacksonville	Pickettville Road Landfill	**New York**	
Plant City	Schuylkill Metals	Oswego	Pollution Abatement Services
Indiana		Oyster Bay	Old Bethpage Landfill
Gary	Midco I	Wellsville	Sinclair Refinery
Iowa		**Ohio**	
Charles City	Labounty Site	Arcanum	Arcanum Iron & Metal
Kansas		**Oklahoma**	
Cherokee County	Tar Creek, Cherokee County	Ottawa County	Tar Creek
Maine		**Pennsylvania**	
Gray	McKin Company	Bruin Boro	Bruin Lagoon
Massachusetts		Grove City	Osborne
Acton	W.R. Grace	McAdoo	McAdoo
Ashland	Nyanza Chemical	**Rhode Island**	
East Woburn	Wells G&H	Coventry	Picillo Coventry
Holbrook	Baird & McGuire	**South Dakota**	
Woburn	Industri-Plex	Whitewood	Whitewood Creek
Michigan		**Texas**	
Swartz Creek	Berlin & Farro	Crosby	French, Ltd.
Utica	Liquid Disposal Inc.	Crosby	Sikes Disposal Pits
Minnesota		Houston	Crystal Chemical
Brainerd Baxter	Burlington Northern	La Marque	Motco
Fridley	FMC		
New Brighton/Arden	New Brighton		
St. Louis Park	Reilly Tar		
Montana			
Anaconda	Anaconda-Anaconda		
Silver Bow/Deer Lodge	Silver Bow Creek		
New Hampshire			
Epping	Kes-Epping		
Nashua	Sylvester, Nashua		
Somersworth	Somersworth Landfill		

Source: Data from U.S. Environmental Protection Agency, "Proposed National Priorities List: As Provided for in Section 105(8)(b) of CERCLA," December 20, 1982, Washington, DC.

not produce it in the first place. To this end, the EPA and regulatory agencies in other countries have emphasized pollution prevention and waste minimization. Strong regulatory control requires that industries report the hazardous wastes they produce and that the wastes be stored, transported, and disposed of properly.

The EPA now fosters a **pollution prevention hierarchy** that emphasizes reducing the amount of hazardous waste produced. This involves the following strategy:

First—reduce the amount of pollution at the source.

Second—recycle wastes wherever possible.

Third—treat wastes to reduce their hazard or volume.

Fourth—dispose of wastes on land or incinerate them as a last resort. (See figure 19.4.)

Pollution Prevention

Pollution prevention encourages changes in the operations of business and industry that prevent hazardous wastes from being produced in the first place. Many of these actions are simple to perform and cost little. Primary among them are activities that result in fewer accidental spills, leaks from pipes and valves, loss from broken containers, and similar mishaps. These reductions often can be achieved through better housekeeping and awareness training for employees at little cost. Many industries actually save money because they need to buy less raw material because less is being lost.

Waste Minimization

Waste minimization involves changes that industries could make in the way they manufacture products that would reduce the waste produced. For example, it may be possible to change a process so that a solvent that is a hazardous material is replaced with water, which is not a hazardous material. This is an example of source reduction: any change or strategy that reduces the amount of waste produced.

Another strategy is to use the waste produced in a process in another aspect of

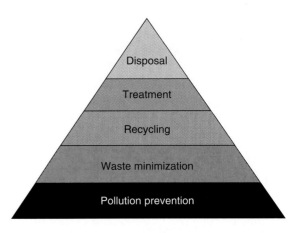

Figure 19.4

Pollution Prevention Hierarchy The simplest way to deal with hazardous wastes is to not produce them in the first place. The pollution prevention hierarchy stresses reductions in the amount of hazardous waste produced by employing several different strategies.

the process, thus reducing the amount of waste produced. For example, water used to clean equipment might be included as a part of the product rather than being discarded as a contaminated waste.

Another technique that can be used to reduce the amount of waste produced is to clean solvents used in processes. Using a still to purify solvents results in a lower total volume of hazardous waste being produced because the same solvent can be used over and over again.

The simple process of allowing water to evaporate from waste can reduce the total amount of waste produced. Obviously, the hazardous components of the waste are concentrated by this process.

Recycling of Waste

Often it is possible to use a waste for another purpose and thus eliminate it as a waste. Many kinds of solvents can be burned as a fuel in other kinds of operations. For example, waste oils can be used as fuels for power plants, and other kinds of solvents can be burned as fuel in cement kilns. Care needs to be taken that the contaminants in the oils or solvents are not released into the environment during the burning process, but the burning of these wastes destroys them and serves a useful purpose at the same time.

Similarly, many kinds of acids and bases are produced as a result of industrial

activity. Often these can be used by other industries that have a need for them.

Ash or other solid wastes can often be incorporated into concrete or other building materials and therefore do not require disposal. Thus the total amount of waste is reduced.

Treatment of Wastes

Wastes can often be treated in such a way that their amount is reduced or their hazardous nature is modified. Dangerous acids and bases can be reacted with one another to produce materials that are not hazardous.

Hazardous wastes that are biodegradable can be subjected to the actions of microorganisms that destroy the hazardous chemicals. Many kinds of organic molecules can be handled in this way.

Incineration (thermal treatment) can be used to treat a variety of kinds of wastes, although many people are suspicious of this technique because they feel that toxic materials may be escaping from the smokestacks. Wastes are heated in a flame-powered incinerator. Under controlled conditions, incineration can destroy 99.999 percent of organic wastes, and hazardous waste incinerators must burn 99.9999 percent of certain hazardous materials. Incineration accounts for the disposal of only about 2 percent of the hazardous wastes in North America. In Europe, the amount of hazardous wastes destroyed by incineration is higher, but still amounts to less than 50 percent. The relatively high costs of incineration (compared with landfills) and concerns for the safety of surrounding areas in case of accidents have kept incineration from becoming a major method of treatment or disposal.

Air stripping is sometimes used to remove volatile chemicals from water. Volatile chemicals, which have a tendency to vaporize easily, can be forced out of liquid when air passes through it. Steam stripping works on the same principle, except that it uses heated air to raise the temperature of the liquid and force out volatile chemicals that ordinary air would not. The volatile compounds can be captured and reused or disposed of.

Carbon absorption tanks contain specifically activated particles of carbon to treat hazardous chemicals in gaseous and liquid waste. The carbon chemically combines with the waste or catches hazardous particles just as a fine wire mesh catches grains of sand. Contaminated carbon must then be disposed of or cleaned and reused.

Precipitation involves adding special materials to a liquid waste. These bind to hazardous chemicals and cause them to precipitate out of the liquid and form large particles called floc. Floc that settles can be separated as sludge; floc that remains suspended can be filtered and the concentrated waste can be sent to a hazardous waste landfill.

Land Disposal

When all other options have been exhausted, any remaining hazardous wastes are typically disposed of on land. (See table 19.4.) For over 80 percent of their hazardous wastes, North America, Europe, and Japan still rely principally on six methods of disposal:

1. Deep-well injection into porous geological formations or salt caverns.
2. Discharge of treated and untreated liquids into municipal sewers, rivers, and streams.
3. Placement of liquid wastes or sludges in surface pits, ponds, or lagoons.

Table 19.4

Hazardous-Waste Management Methods, North America

Management method	Share of total waste managed (percent)
Land disposal[1]	67
Discharge to sewers, rivers, streams	22
Distillation for recovery of solvents	4
Burning in industrial boilers	4
Chemical treatment by oxidation	1
Land treatment of biodegradable waste	1
Incineration	1
Recovery of metals through ion exchange	less than 1
Total	**100**

**Hazardous Waste
Treatment Methods U.S. 1993
(Million tons of waste)**

Recovery 8.096
Incineration 2
Land disposal 2.435
Used as fuel 1.7
Deep well injection 24
Wastewater treatment 166

Source: Data from U.S. Environmental Protection Agency

[1]Includes injection wells (25 percent of total), surface impoundments (19 percent), hazardous-waste landfills (13 percent), and sanitary landfills (10 percent).

Source: Data from U.S. Congressional Budget Office, Hazardous Waste Management, Recent Changes and Policy Alternatives, U.S. Government Printing Office, 1990, Washington, D.C.

4. Storage of solid wastes in specially lined dumps covered by soil.
5. Storage of liquid and solid wastes in underground caverns and abandoned salt mines.
6. Sending wastes to sanitary landfills not designated for toxic or hazardous wastes.

There are techniques that reduce the chance that hazardous materials will escape from these locations and become a problem for the public. Immobilizing a waste puts it into a solid form that is easier to handle and less likely to enter the surrounding environment. Waste immobilization is useful for dealing with wastes, such as certain metals, that cannot be destroyed. Once a waste has been immobilized, it can be disposed of in a hazardous-waste landfill. Two popular methods of immobilizing waste are fixation and solidification. Engineers and scientists mix materials such as fly ash or cement with hazardous wastes. This either "fixes" hazardous particles, in the sense of immobilizing them or making them chemically inert, or "solidifies" them into a solid mass. Solidified waste is sometimes made into solid blocks that can be stored more easily than can a liquid.

Love Canal

Love Canal, near Niagara Falls, New York, was constructed as a waterway in the nineteenth century. It was subsequently abandoned and remained unused for many years. In the 1930s, it became an industrial dump. The Hooker Chemical Company purchased the area in 1947 and used it as a burial site for 20,000 metric tons of chemicals.

Hooker then sold the property to the local government for one dollar. A housing development and an elementary school were constructed on the site. Soon after the houses were constructed, people began to complain about chemicals seeping into their basements. In 1978, 80 different chemicals were found in this seepage. Approximately one dozen probable carcinogens were identified among these chemicals.

As a result of these findings, $27 million in government funds were appropriated in 1978 to purchase homes and permanently relocate 237 families. The funds also provided for the construction of a series of ditches to contain the chemicals and a clay cap to prevent the fumes from entering the atmosphere. But the problems continued.

The remaining 710 families in the Love Canal area were not satisfied with the government's approach to the problem. They cited the fact that women in the area had a 50 percent higher rate of miscarriages. Of 17 reported pregnancies in the area during 1979, two children were born normal, nine had defects, two were stillborn, and four ended in miscarriage. In addition to the abnormalities in birth, there are other biological problems in the Love Canal area.

Neurologists determined that the speed of the nerve impulses in 37 residents who were examined were slower than normal. They stated that chemical exposure could have caused this damage. In 1980, the EPA released the findings of a study that found that 11 out of 36 residents tested in the Love Canal area had broken chromosomes, which are linked to cancer and birth defects. As a result, the federal government released $5 million to temporarily relocate Love Canal residents to motels or other quarters.

By 1990, a $150 million cleanup effort had sealed off the leaky dump, demolished 238 homes nearest the chemical graveyard, and scoured toxins from neighborhood storm sewers and streams (the two major sources of danger to area homes).

In early 1991, some families began to move back into the area. The Love Canal Area Revitalization Agency is planning to sell 236 homes in Love Canal (renamed Black Creek Village) by 1996. One of the incentives is that the houses can be purchased at a low cost. The families are confident that the houses are safe, but many environmental groups oppose them.

- Who is responsible for providing treatment for the physical and mental problems experienced by the residents in this community?

- There is a question no one ever seems to ask. Why were permits ever awarded to construct a thousand-unit housing development on top of a site known to contain 20,000 metric tons of toxic wastes?

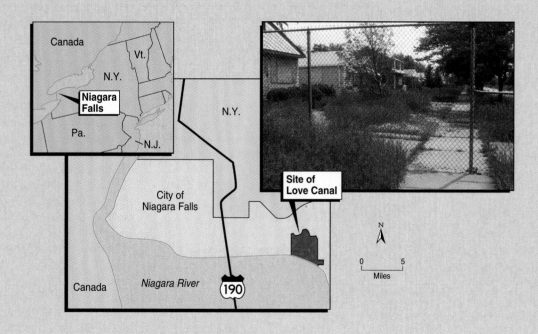

environmental
close-up

A Contaminated Community

"We don't know of any other community in the nation that has ever been contaminated like this one." This statement was made by a spokesperson from the U.S. Centers for Disease Control in Atlanta, Georgia, about the 600 residents of Triana, Alabama. Triana is located 25 kilometers from Huntsville.

From 1947 to 1970, the Olin Chemical Corporation manufactured DDT on a site leased from the nearby Redstone Arsenal. During the time of production, about 4,000 metric tons of DDT residue accumulated in the soil of an adjacent marsh. Today, the DDT from this soil is filtering into Indian Creek, which flows through Triana. The fish in this stream contain 40 times the federal standard of allowable DDT. Because the people from Triana depend upon catfish from the stream as a source of food, they have accumulated large amounts of DDT in their bodies.

The average person in the United States contains 8 to 10 ppm of DDT in his or her body fat. People from Triana contain up to 60 times that amount. The older residents have higher levels of DDT than the younger members of the community, perhaps because the older people have bioaccumulated the DDT for a longer time.

A large portion of the community's income was earned from the sale of catfish. Because these fish are not suitable for sale, this income has been lost. The wildlife from nearby Wheeler National Wildlife Refuge also suffer from the effects of DDT. Mallard ducks have been found to contain 480 ppm DDT. A wintering population of thousands of double-crested cormorants once used the refuge, but they are now absent. A heron rookery that contained three hundred nests has disappeared from the area. These and other environmental changes will occur in Triana and the surrounding area as long as the DDT-contaminated soil remains in the marsh.

As yet, the only move toward a solution has been an allotment of $1.5 million by the federal government for a study to determine what to do about the problem. However, many residents feel that, regardless of the solution, the town will never be the same because of the problems caused by the DDT.

Hazardous-Waste Management Choices

Today, the two most common methods for disposing of hazardous wastes are land disposal and incineration. The choice between these two methods involves both economic decisions and acceptance by the public. In North America there is abundant land available for land disposal, making it the most economical and most widely used method. In Europe and Japan where land is in short supply and is expensive, incineration is more economical and is a major method for dealing with hazardous waste. Because of concerns about the emissions from incinerators, significant amounts of hazardous waste are incinerated at sea on specially designed ships. (See table 19.5)

Laws and regulations dealing with hazardous-waste disposal are driving industrial behavior toward pollution prevention and waste minimization. In addition, because the costs of safe disposal are mounting, waste-handling firms—both private and public—are looking for better and cheaper ways to treat and dispose of hazardous wastes. Strong, enforceable laws have eliminated the economic incentives to pollute. Unfortunately, as strong laws have been enacted, some small companies that were unable or, more likely, unwilling to properly dispose of their wastes have turned to illegal nighttime dumping.

The environmental costs of not managing hazardous wastes, as witnessed in virtually every industrialized country, are astronomical. And because major generators of hazardous wastes remain liable for past mistakes, economic and regulatory incentives for complying with hazardous-waste regulations should continue to encourage responsible management.

Table 19.5

Average Amounts of Hazardous Wastes Incinerated in Selected Countries

Country	Amount incinerated (metric tons)	
	On land	At sea
Denmark	circa 32,000	0
Belgium	0	10,000
France	400,000	10,000
Germany	875,000	60,000
Netherlands	66,000	20,000
Switzerland	120,000	5,000
United Kingdom	80,000	3,500
United States	2,700,000*	0
Norway	0	8,000

*This large figure is only 1 percent of the total amount of hazardous wastes generated in the United States. In contrast, Germany incinerates 15 percent of its total amount of hazardous wastes. The U.S. Congress, Office of Technology Assessment, estimates that 10–20 percent of all U.S. hazardous wastes could, in theory, be incinerated.

Source: Data adapted from U.S. Congress, Office of Technology Assessment (OTA), Ocean Incineration: Its Role in Managing Hazardous Waste (OTA), August 1986, pp.198–201, Washington, D.C.

SUMMARY

Public awareness of the problems of hazardous substances and hazardous wastes is relatively recent. The industrialized countries of Europe and North America began major regulation of hazardous materials only during the past 20 years, and most developing countries exercise little or no control over such substances. As a result, many countries are living with serious problems from prior uncontrolled dumping practices, while current systems for management of hazardous and toxic waste remain incomplete and incapable of even identifying all hazardous waste.

A number of fundamental problems are involved in hazardous-waste management. First, there is no agreement as to what constitutes a hazardous waste. Moreover, little is known about the amounts of hazardous wastes generated throughout the world. The issue is further complicated by our limited understanding of the health effects of most hazardous wastes and the fact that large numbers of potentially hazardous chemicals are being developed faster than their health risks can be evaluated.

Hazardous-waste management must move beyond burying and burning. Industries need to be encouraged to generate less hazardous waste in their manufacturing processes. Although toxic wastes cannot be entirely eliminated, technologies are available for minimizing, recycling, and treating wastes. It is possible to enjoy the benefits of modern technology while avoiding the consequences of a poisoned environment. The final outcome rests with governmental and agency policymakers, as well as with an educated public.

REVIEW QUESTIONS

1. Explain the problems associated with hazardous-waste dump sites and how such sites developed.
2. Distinguish between acute and chronic toxicity.
3. Give two reasons why regulating hazardous wastes is difficult.
4. In what ways do hazardous wastes contaminate the environment?
5. Describe how hazardous wastes contaminate groundwater.
6. Why is there often a problem in linking a particular chemical or hazardous waste to a particular human health problem?
7. Describe what is meant by the U.S. National Priority List.
8. Describe five technologies for managing hazardous wastes.
9. What is meant by pollution prevention and waste minimization?
10. Describe the pollution prevention hierarchy.
11. What are RCRA and CERCLA? Why is each important for managing hazardous wastes?

KEY TERMS

acute toxicity   389
chronic toxicity   390
Comprehensive Environmental
    Response, Compensation
    and Liability Act
    (CERCLA)   395
corrosiveness   387

hazardous   388
hazardous substances   387
hazardous wastes   388
ignitability   387
incineration   398
LD_{50} 391
National Priority List   395

nonpersistent pollutant   390
pollution prevention
    hierarchy   398
persistent pollutant   390
reactivity   387
Resource Conservation and
    Recovery Act   388

Superfund   395
synergism   390
threshold levels   388
toxic   388
toxicity   388
waste minimization   398

RELATED INTERNET SITES

Hazardous Materials magazine, comparing over 800 products and services.
    http://www.hazmatmag.com/
Communities against toxics, a database of concerned groups and incidents of toxic pollution.
    http://www.gn.apc.org/cats/

An on-line newsletter covering issues on nuclear waste, land use, and military toxics.
    http://www.downwinders.org

chapter 20

Environmental Policy and Decision Making

Objectives

After reading this chapter, you should be able to:

- Explain how the executive, judicial, and legislative branches of the U.S. government interact in forming policy.
- Describe the forces that led to changes in environmental policy in the United States during the past three decades.
- Understand the history of the major U.S. environmental legislation.
- Understand what is meant by "Green" politics.
- Describe the reasons why environmentalism is a growing factor in international relations.
- Understand the factors that could result in "ecoconflicts."
- Understand why it is not possible to separate politics and the environment.
- Explain how citizen pressure can influence governmental environmental policies.

Chapter Outline

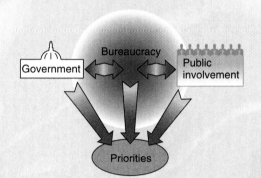

Politics in the United States

Government affects our lives in many ways. Food, air, and water quality are regulated by government. Individually, each of us supports the government by paying a variety of taxes. Even though not all governmental policies have been perceived as wise, many of them have helped to solve complex environmental problems. We need a knowledge of the political process and institutional structure of society before we can effectively influence public policies.

The government of the United States is structured into three separate branches: the legislative, judicial, and executive. Each of these branches has specific functions, and each is a check on the other two to ensure that the assigned responsibilities are fulfilled for the smooth operation of the whole system.

The **legislative branch** of the U.S. government is the Congress, composed of the House of Representatives and the Senate. A primary function of Congress is to develop and approve policy. **Policy** is a planned course of action on a question or topic. Congress generally supports policy by passing bills or acts that establish a government agency or by instructing an existing agency to take on new tasks or programs. (See figure 20.1.)

Another function of the legislative branch is to oversee the various agencies of the executive branch. Congress looks at how well the laws are being carried out and how effectively the agencies are spending their appropriations. An important input associated with the development, funding, and oversight of agencies is that concerned individuals and groups exert pressure to develop policies and to fund programs that are favorable to them. These pressure groups are called lobbies, and they sometimes have tremendous influence on members of Congress. They can be a source of shortsighted legislative action, especially during an election year. (See figure 20.2.)

The **judicial branch** of the U.S. government is a complex and layered series of courts, ranging from local traffic courts to

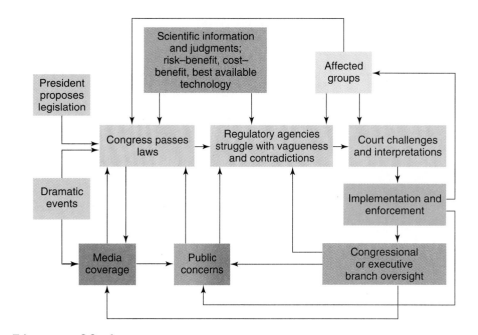

Figure 20.1

U.S. Environmental Policy The diagram shows the primary forces involved in making environmental policy at the federal level in the United States.

the Supreme Court. The courts interpret the several different categories of laws, such as constitutional law, statutory law, and administrative law, and specific courts deal with specific areas of interpretation. Generally, statutory and administrative law are the areas involving the environment.

Federal agencies, such as the Environmental Protection Agency, have developed their own methods of establishing regulations that are as binding as laws passed by Congress. These regulations have the force of law because of the statutes establishing the agency, and the enforcement of their regulations falls under the category of statutory and administrative law.

The powers of the courts are generally less than the powers of the chief executive, the president. The powers of the **executive branch** come from the Constitution, but many additional powers have been assumed over the years by executives with particular personalities and priorities. Presidential powers are really permissions given to the office. When a president wants to gain support for programs or legislation, personality can be either an

important asset or a serious liability. The powers of the chief executive when dealing with environmental issues include support for the development and funding of agencies and the budgeting of monies for existing environment-related agencies. The major environmental role of the executive branch is to lead members of Congress and the people they represent toward respect for and appreciation of quality in their surroundings. (See figure 20.3.)

Enactment of policies into law may be only the beginning. Relatively few statutes are "self-executing" in the sense that they can be put into effect immediately by existing agencies. Many require funding, if only in the form of new personnel to perform the investigatory services or the enforcement called for in the statute. In such cases, the appropriation process following submission of the budget serves as the second consideration of the issues addressed by the statute. Thus, the policy must be considered in light of its monetary ramifications. Once money has been both authorized and appropriated, the agency or program can spend the money. The agency is now a part of the executive branch and can implement policy.

The legislative process

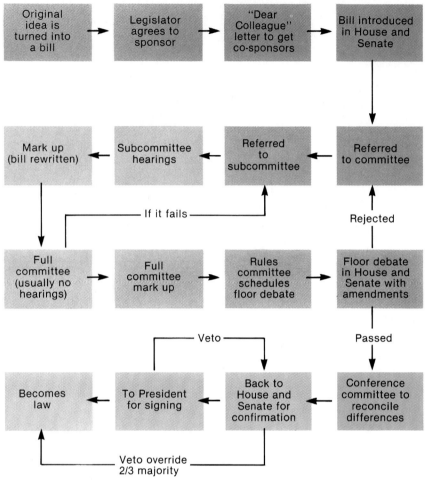

Figure 20.2

Passage of a Law   This figure illustrates the path of a bill from organization to becoming a law. As we can see, the process is not a quick one.

The Development of Environmental Policy in the United States

U.S. President Teddy Roosevelt declared 80 years ago that nothing short of defending your country in wartime "compares in leaving the land even a better land for our descendants than it is for us." Environmental issues that Roosevelt strongly believed in, however, did not become major political issues until the early 1970s.

April 22, 1970, was the first Earth Day, and in the opinion of many, it was the event that gave birth to the modern environmental movement. In 1970, as a result

of mounting public concern over environmental deterioration—cities clouded by smog, rivers on fire, waterways choked by raw sewage—many nations, including the United States, began to address the most obvious, most acute environmental problems. (See table 20.1)

Earth Day was the largest organized demonstration in U.S. history. Thousands of schools, colleges, and univerities took part, and millions of ordinary citizens demonstrated their desire to work toward environmental goals. Congress adjourned for the day. New York City's Fifth Avenue was closed, and hundreds of ecology fairs were held.

Public opinion polls indicate that a permanent change in national priorities

followed Earth Day 1970. When polled in May 1971, 25 percent of the U.S. public declared protecting the environment to be an important goal—a 2,500 percent increase over 1969.

During the early 1970s, the U.S. Congress tackled many environmental problems. Important pieces of legislation were passed that called for setting air-quality standards, cleaning up rivers, and lakes, protecting coastal areas, regulating pesticides, protecting endangered species, and safeguarding drinking water. Many of the identified environmental problems were so immediate, so obvious, that it was relatively easy to see what had to be done and to summon the political will to do it.

Just as it was beginning to gain momentum, however, the environmental movement began to decline. When the energy crisis threatened to stall the North American economy in the early 1970s, environmental concerns quickly faded. By 1974, U.S. President Gerald Ford had proposed an acceleration of his administration's leasing program for offshore gas and oil drilling. A turnaround in environmental policy was even more pronounced during the 1980s. Former Vice President Walter Mondale was fond of noting that President Reagan "would rather take a polluter to lunch than to court."

During the mid-1980s, the Reagan administration embarked on a policy course that radically altered the environmental agenda in the United States. The guiding principles, set out by the President's Council on Environmental Quality, were: (1) regulatory reform including the use of cost-benefit analysis to help determine the value of environmental regulations and programs; (2) reliance, as much as possible, on the free market to allocate resources; and (3) decentralization—shifting responsibility for environmental protection to state and local governments whenever feasible. These goals reversed much of the environmental policy developed in the early 1970s.

Despite the wavering of political will toward environmental concerns in the period from 1970 to 1990, there were some very tangible accomplishments. Among the most visible and quantifiable is the expansion of protected areas. Since 1970, federal parklands in the United States—

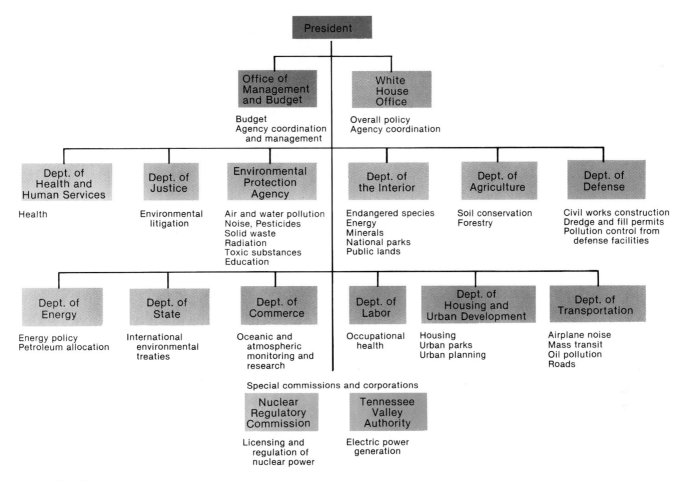

Figure 20.3

Major Agencies of the Executive Branch   Major agencies of the executive branch are shown with their environmental responsibility.

excluding Alaska—increased 800,000 hectares, to 10.5 million. In Alaska, 18.3 million additional hectares have been protected, bringing the state's total to 21.8 million hectares. Also, the extent of the waterways included in the National Wild and Scenic Rivers System increased by more than 12 times, to about 15,000 kilometers.

By the late 1980s, however, a new environmental awareness and concern began to surface as a major political issue. Once again, the public reacted by organizing and putting pressure on the political system, and as in 1970, the politicians began to respond. For the first time in the history of the United States, the environment became a key issue in a presidential campaign. In 1988, the environmental records of the two major candidates were hotly debated. Environmentalism was evolving as a major public issue. By the 1992 elec-

tion, the environment was firmly established as a major campaign issue.

April 22, 1990, was the twentieth anniversary of Earth Day and, in the opinion of many, the beginning of a new decade of environmental concern in the United States. In many respects, the environmental movement of the 1970s and 1980s has come of age in the 1990s. Putting the bands of politics and science, or emotionalism and logic, together in a new environmental movement represents a significant integration of human thinking.

It has been said that politics have always forged science. Prioritization of issues and political will determine where money will be spent. By the early 1990s, it appeared that the United States political will to address environmental concerns was on the rise. A 1990 Gallup Poll indicated that 79 percent of people ages 30 to 49 called themselves environmen-

talists. Two out of three Americans said that pollution was a "very serious threat," according to the poll. A National Wildlife Federation study in 1990 documented a heightened public concern for the environment. In that study, 84 percent of the population called the threat of freshwater pollution "very serious," up from 48 percent in 1970; 73 percent called air pollution a "very serious" threat, up from 46 percent in 1970.

A 1992 survey revealed that most Americans would prefer a federal government more actively involved in environmental protection. Ninety-two percent desired stronger active participation by leaders in government and business in environmental concerns. Forty-one percent believed the general public should bear the primary responsibility for a clean environment, followed by industry at 34 percent and government at 22 percent.

Table 20.1

Major U.S. Environmental and Resource Conservation Legislation

Wildlife conservation

Anadromous Fish Conservation Act of 1965

Fur Seal Act of 1966

National Wildlife Refuge System Act of 1966, 1976, 1978

Species Conservation Act of 1966, 1969

Marine Mammal Protection Act of 1972

Marine Protection, Research, and Sanctuaries Act of 1972

Endangered Species Act of 1973, 1982, 1985, 1988

Fishery Conservation and Management Act of 1976, 1978, 1982

Whale Conservation and Protection Study Act of 1976

Fish and Wildlife Improvement Act of 1978

Fish and Wildlife Conservation Act of 1980 (Nongame Act)

Land use and conservation

Taylor Grazing Act of 1934

Wilderness Act of 1964

Multiple Use Sustained Yield Act of 1968

Wild and Scenic Rivers Act of 1968

National Trails System Act of 1968

National Coastal Zone Management Act of 1972, 1980

Forest Reserves Management Act of 1974, 1976

Forest and Rangeland Renewable Resources Act of 1974, 1978

Federal Land Policy and Management Act of 1976

National Forest Management Act of 1976

Soil and Water Conservation Act of 1977

Surface Mining Control and Reclamation Act of 1977

Antarctic Conservation Act of 1978

Endangered American Wilderness Act of 1978

Alaskan National Interests Lands Conservation Act of 1980

Coastal Barrier Resources Act of 1982

Food Security Act of 1985

Coastal Development Act of 1990

General

National Environmental Policy Act of 1969 (NEPA)

International Environmental Protection Act of 1983

Environmental Education Act of 1990

Energy

National Energy Act of 1978, 1980

Water quality

Water Quality Act of 1965

Water Resources Planning Act of 1965

Federal Water Pollution Control Acts of 1965, 1972

Ocean Dumping Act of 1972

Safe Drinking Water Act of 1974, 1984

Clean Water Act of 1977, 1987

Great Lakes Critical Programs Act of 1990

Oil Spill Prevention and Liability Act of 1990

Air quality

Clean Air Act of 1963, 1965, 1970, 1977, 1990

Noise control

Noise Control Act of 1965

Quiet Communities Act of 1978

Resources and solid-waste management

Solid Waste Disposal Act of 1965

Resources Recovery Act of 1970

Resource Conservation and Recovery Act of 1976

Waste Reduction Act of 1990

Toxic substances

Toxic Substances Control Act of 1976

Resource Conservation and Recovery Act of 1976

Comprehensive Environmental Response, Compensation, and Liability (Superfund) Act of 1980, 1986, 1990

Nuclear Waste Policy Act of 1982

Pesticides

Federal Insecticide, Fungicide, and Rodenticide Control Act of 1972, 1988

Only 3 percent believe all three bear equal responsibility. The rising figures in the polls were being matched by rising membership in nongovernmental environmental organizations.

By 1994, however, it was becoming evident that the dramatic increase in membership in environmental organizations was beginning to level off and even decline in some instances. Public interest in the environment, while still very high, began to decrease somewhat by the middle 1990s. A USA Today poll conducted in late 1994 found that the 10 largest environmental groups in the United States collectively lost 6.5 percent of their membership between 1990 and 1994. (See table 20.2.)

A major part of the problem with the larger environmental groups may be that they have actually grown into larger bureaucracies and in the process lost the trust of many grassroots environmentally concerned individuals. For example, the National Wildlife Federation has an annual budget of around $100 million. This includes profits on merchandise of nearly $26 million and nearly $12 million spent on magazines. By any definition, the National Wildlife Federation is a large bureaucracy.

Table 20.2

Major Environmental Advocacy Groups

Organization	Membership
National Wildlife Federation	1,720,000
Greenpeace Inc.	1,600,000
World Wildlife Fund	1,180,000
Nature Conservancy	719,400
National Audubon Society	542,000
Sierra Club	535,000
Ducks Unlimited	530,000
National Parks and Conservation Association	349,000
Wilderness Society	300,000
Environmental Defense Fund	250,000
Natural Resources Defense Council	95,000
Izaak Walton League	50,000
Environmental Policy Institute/Friends of the Earth	42,000
Environmental Action	20,000
Total	7,932,400

Source: Data from Capital Research Center, Washington, D.C.

While many of the larger, established environmental organizations are declining somewhat, many newer, smaller, local, and grassroots organizations are being created or are expanding, often in response to environmental threats in their own communities. Estimates are that some 6,500 local environmental organizations are active in the United States. On college campuses, interest in the environment is growing. Environmental studies programs are expanding at both the undergraduate and graduate level. Students are not just concerned with local recycling programs or nuclear power plants but are focusing on the broader issues to be faced in the next century. The primary issue is the achievement of long-term sustainability and the fundamental changes in society that this will entail.

By 1995, an anti-environmental backlash was beginning throughout the United States. Many people who felt that their interests were threatened by proposals for restricted resource use, wilderness preservation, or pollution control were organizing to resist such change. This anti-environmental movement rallies supporters under the banner of "wise use" or "multiple use" of public lands. Ranchers, loggers, miners, industralists, and land developers who believe their access to resources is threatened or who face expensive changes in the way they do business form the core of the group. Many people join simply because they believe that the social changes are occurring too rapidly or going too far. Some individuals, for instance, question the morality of liberal environmentalists, equating biocentric ecology with heathenism and social change with communism.

The largest numbers of adherents to the wise-use movement are generally found in the Western states, where public land ownership is high, rugged individualism and independence are valued, and the federal government is regarded with some suspicion and in some cases hostility. Their central policy statement is contained in *The Wise Use Agenda*, written by conservative activists Ron Arnold and Alan Gottlieb. It lists 25 goals, including opening all public lands, including wilderness areas and national parks, to mineral, energy, and timber production; eliminating protection of "nonadaptive species" such as spotted owls and California condors; and allowing motorized recreation anywhere on all public lands.

Wise-use groups are often generously funded by oil, mineral, and timber corporations. Workers join in because they believe that their jobs are jeopardized by environmental restrictions. A few violent types threaten to form vigilante posses, turning the tactics of intimidation and "direct action" used by radical environmentalists back on their adversaries. So far, this movement is a fringe group without much national power. Those concerned with the environment should not underestimate the anger, alienation, and resistance that underlie this reactionary force.

A 1996 Roper poll, however, revealed that U.S. citizens do not believe that economic achievements must come at the expense of environmental or social needs. The poll found that 66 percent of respondents say that economic development, environmental protection, and health and happiness of people can go hand in hand. The poll revealed that safety from crimes was the most important characteristic for a community (71 percent) followed closely by healthy air and water (68 percent).

The International Environmental Movement

In the United States and in other industrialized countries, the nongovernmental organizations (NGOs) that work on behalf of the environment range from small to large, as we have noted. In the developing world, most of these groups are community-based, relatively small, and their chances of failure far outweigh their chances of success. For many groups in developing countries, survival depends on the courage and persistence of a few individuals. In some developing countries, members of such organizations must contend with the active hostility of the government; in a few cases, members may be risking their lives.

Nevertheless, nongovernmental organizations are growing, both in numbers and in influence, especially in the developing countries of the Southern Hemisphere. The reasons for this growth are

Farm Policy Legislation in the United States

In 1985, the U.S. Congress faced the question of whether farm programs should be consistent with the nation's environmental goals. The outcome was a major piece of legislation: the 1985 Food Security Act, better known as the "Farm Bill." The Farm Bill addressed many environmental issues, such as groundwater and surface-water quality, wetlands protection, soil erosion, and pesticide residues.

Before 1985, U.S. federal farm programs too often encouraged negative environmental behavior—the draining of wetlands and the farming of erosion-prone soil. By passing two key conservative programs—the Conservation Compliance Program and the Conservation Reserve Program—in 1985, Congress made major strides toward reversing this trend. Unfortunately, each of these programs had loopholes, and the programs were not always enforced.

In 1990, a new Farm Bill was passed by Congress. The following environmental issues were addressed by the 1990 Farm Bill:

1. The amount of money in the budget for agricultural price supports was lowered by reducing the amount of land eligible for income support payments by 15 percent. Farmers now have 15 percent of their agricultural land exempted from farm support payments. This encourages farmers to take out of production marginal land that is most subject to degradation by being farmed.

2. An Office of Environmental Quality was established within the U.S. Department of Agriculture.

3. The bill encourages farmers to set aside high-quality wetlands by initiating a 30-year Wetland Reserve program that pays farmers for not using wetlands for agricultural purposes.

4. The Conservation Reserve Program, which pays farmers to take highly erodible land out of production, now includes windbreaks, shelterbelts, and marginal pasturelands. This encourages farmers to preserve current shelterbelts and windbreaks, rather than to remove them to allow for more farmland. Some farmers may be encouraged to plant new windbreaks.

5. The Conservation Compliance Program, which requires farmers to use approved farming techniques on erodible land, has been made stronger by increasing the number of benefits a farmer would lose if U.S. Department of Agriculture recommendations are not followed.

The 1990 Farm Bill is the platform that is going to govern U.S. agricultural policy into the next century. It appears that environmental concerns will be a growing component of agricultural practices.

---

complex. Local groups often form in response to specific needs such as the need to improve water supplies. National groups may form to fulfill a specific need such as environmental protection, often with the support of groups in the Northern Hemisphere. In some cases, ineffective or nonexistent government programs create a vacuum that private organizations step into; in other cases, resource-poor governments use private organizations as a means to extend their authority.

As the links among environment, development, and population become clearer, organizations devoted to a single issue see the advantages of working with other groups; development NGOs, for example, may join with NGOs advancing entrepreneurship or environmental protec-

tion or human rights. This trend is fueled by the fast-growing number of national and international NGO networks and coalitions, and in turn is helped by rapid advances in communications technology.

Is Environmental Quality a Right?

When Thomas Jefferson, John Hancock, Ben Franklin, and 52 other delegates wrote the U.S. Constitution in 1787, they never considered addressing the right to a clean and healthy environment. They did not include an environmental guarantee because they never envisioned that the land and water would ever be so

extensively utilized; nor could they foresee the environmentally destructive by-products of modern industrialized society.

There were eight attempts between 1968 and 1972 to amend the U.S. Constitution to include environmental-quality rights. Although none of these proposals survived congressional committees, their introduction indicated the popularity of such an idea.

A total of 21 states already have adopted language in their constitutions that guarantees certain environmental rights. In addition, the constitutions of eight countries—including China, Canada, and Switzerland—also address environmental issues.

In 1989, the National Wildlife Federation passed a resolution calling for the

U.S. Constitution to be amended to include an Environmental Bill of Rights. According to the Federation, the principles that an Environmental Quality Amendment should embody are as follows:

> The people have a right to clean air, pure water, productive soils, and to the conservation of the natural, scenic, historic, recreational, esthetic, and economic values of the environment. America's natural resources are the common property of all people, including generations yet to come. As trustee of these resources, the United States government shall conserve and maintain them for the benefit of all people.

In spite of support for an Environmental Quality Amendment, considerable opposition exists. Other nations also experienced opposition, and yet the environmental quality conditions were incorporated into similar constitutions. Perhaps the U.S. Constitution will also incorporate environmental rights in the future.

Environmental Policy and Regulation

Environmental laws are not a recent phenomenon. As early as 1306, London adopted an ordinance limiting the burning of coal because of the degradation of local air quality. Such laws become more common as industrialization created many sources of air and water pollution.

To date, much environmental law has reflected the perception that environmental problems are localized in time, space, and media (i.e., air, water, soil). For example, many hazardous-waste sites in the United States have been "cleaned" by simply shipping the contaminated dirt someplace else, which not only does not solve the problem, but creates the danger of incidents during removal and transportation. Environmental regulation has focused on specific phenomena and adopted the so-called "command-and-control" approach, in which restrictive and highly specific legislation and regulation are implemented by centralized authorities and used to achieve narrowly defined ends.

Such regulations generally have very rigid standards, often mandate the use of specific emission-control technologies, and generally define compliance in terms of "end-of-pipe" requirements. Examples in the United States include the Clean Water Act (which applied only to surface waters), the Clean Air Act (urban air quality), and the Comprehensive Environmental Response, Compensation, and Liability Act (Superfund), which applied to specific landfill sites.

If properly implemented, command-and-control methods can be effective in addressing specific environmental problems. For example, rivers such as the Potomac and Hudson in the United States are much cleaner as a result of the Clean Water Act. Moreover, where applied against particular substances, such as the ban on tetraethyl lead in gasoline in the United States, the command-and-control approach has clearly worked well. Such regulation, however, is characterized by a number of problems. These include (1) a growth of mandatory requirements, (2) a relative unconcern with economic efficiency, (3) a focus on the manufacturing stage of industrial activity rather than the life cycle of materials or products, (4) a tendency toward adversarial relations between the regulated community and regulators (which varies significantly by country) and, (5) because specific technologies are prescribed, a strong bias against technological innovation. It has also proven very difficult to modify such regulations to reflect advances in scientific understanding.

Learning from the Past

For the past quarter century, the basic pattern of environmental protection in economically developed nations has been to react to specific crises. Institutions have been established, laws passed, and regulations written in response to problems that already were posing substantial ecological and public health risks and costs, or that already were causing deep-seated public concern.

The United States is no exception. The U.S. Environmental Protection Agency has focused its attention almost exclusively on present and past problems. The political will to establish the agency grew out of a series of highly publicized, serious environmental problems, like the fire on the Cuyahoga River in Ohio, smog in Los Angeles, and the near extinction of the bald eagle. During the 1970s and 1980s, Congress enacted a series of laws intended to solve these problems, and the EPA, which was created in 1970, was given the responsibility for enforcing most environmental laws.

Despite success in correcting a number of existing environmental problems, there has been a continuing pattern of not responding to environmental problems until they pose immediate and unambiguous risks. Such policies, however, will not adequately protect the environment in the future. People are recognizing that the agencies and organizations whose activities affect the environment must begin to anticipate future environmental problems, and then take steps to avoid them. One of the most important lessons learned during the past quarter century of environmental history is that the failure to think about the future environmental consequences of prospective social, economic, and technological changes may impose substantial and avoidable economic and environmental costs on future generations.

Thinking about the Future

Thinking about the future is more important today than ever before, because the accelerating rate of change is shrinking the distance between the present and the future. Technological capabilities that seemed beyond the horizon just a few years ago are now outdated. Scientific developments and the flow of information are accelerating. For example, who would have envisioned cellular phones, voice-activated computers, or a widely used Internet only 10 years ago? Similarly, the environmental effects of changes in global economic activity are being felt

Earth Day 1970–1995

On April 22, 1970, citizens of all ages celebrated Earth Day by planting trees, cleaning up vacant lots, starting gardens, and learning about the delicacy of the natural world.

The spontaneity of Earth Day 1970 and the grassroots-powered energy it developed were felt in the political arena for many years. Plans for an American SST (a controversial supersonic transport plane) were scrapped, Congress enacted the Clean Air Act, new in-plant pollution laws were passed, and the U.S. government banned the military's controverisal use of toxic defoliants in Indochina. Earth Day 1970 marked the beginning of political activism for many individuals and awakened public sentiment regarding the health of our environment.

Many saw Earth Day 1970 as a chance to reflect on the environment of the 21st century. While many of the issues are the same today as they were 20 years ago, scientific knowledge and public awareness have grown. We are also beginning to recognize that environmental problems in one country cross national borders, affecting the planet on which all of us must live. While Earth Day 1970 was primarily a North American event, Earth Day 1995 was international in scope.

Public concern about the environment took many forms on Earth Day, both in 1970 and in 1990.

more rapidly by both nations and individuals. Examples include the collapse of the cod fishing industry in the maritimes of Canada and the northeast United States, the severe air-pollution problems in Mexico, and the shortage of safe drinking water supplies in Russia.

Initiating thought and analysis well in advance of anticipated change can shorten the time needed to respond to such change or allow us to avoid the problem entirely. Because some damage is irreversible, response time may be critical.

Thinking about the future also is valuable because the cost of avoiding a problem is often far less than the cost of solving it later. The U.S. experience with hazardous-waste disposal provides a compelling example. Some private companies and federal facilities undoubtedly saved money in the short term by disposing of hazardous wastes inadequately, but those savings were dwarfed by the cost of cleaning up hazardous-waste sites years later. In that case, foresight could have saved private industry, insurance companies, and the federal government (i.e., taxpayers)

The organizers of Earth Day 1995 were working toward specific accomplishments, including:

A worldwide ban on emissions of chlorofluorocarbons, which destroy the ozone layer and contribute to global warming, to be fully implemented within five years

Slowing the rate of global warming through dramatic, sustained reductions in carbon dioxide emissions, including higher standards for automobile fuel efficiency and the rapid adoption of a transportation system not powered by fossil fuels

Preservation of old-growth forests, in both temperate and tropical areas

A ban on packaging that is neither recyclable nor biodegradable, and the implementation of strong, effective recycling programs in every community

A swift transition to renewable energy resources

Dramatic increases in residential and industrial energy efficiency

A comprehensive hazardous-waste minimization program, emphasizing source reduction

Heightened protection for endangered species and habitats

New protections for marine resources, including marine mammals and fisheries

A new sense of responsibility for the protection of the planet by individuals, communities, and nations

The adoption by all countries of strategies to stabilize their populations within sustainable limits using environmentally sustainable agricultural and industrial processes

World population in billions in 1970: 3.7. 1995: 5.8.

Number of whales killed worldwide in 1970: 42,480. In 1994: 741.

U.S. population in millions in 1970: 205. 1995: 261.

North American population of breeding mallard ducks in 1970: 10,379,000. Estimate in 1994: 7,000,000.

Number of people in millions that visited a national park in 1970: 172. In 1993: 273.

Number of national wildlife refuges in the United States in 1970: 331. In 1994: 500.

Total U.S. energy consumption (including wood) in quadrillion BTUs in 1970: 66.4. In 1993: 83.8.

Amount of glass recycled in 1970: 1 percent. In 1993: 25 percent.

Cost of solar electricity per watt in 1970: $22/watt. In 1994: $5/watt.

Looking Back: Earth Day at 25

Amount of aluminum cans recycled in 1970: 3 percent. In 1993: 62 percent.

Millions of tons of sulfur dioxides emitted into America's air in 1970: 31.3. In 1992: 22.7.

Amount of waste generated per person, per day in the United States in 1970: 3.3 pounds. In 1993: 4.4 pounds.

Thousands of tons of lead polluting America's air in 1970: 219. In 1992: 5.2.

billions of dollars, while reducing exposure to pollutants and public anxieties in the affected communities.

Thinking about the future has another value, one that goes beyond the immediate costs and benefits of environmental protection. Environmental foresight can preserve the environment for future generations. When one generation's behavior necessitates environmental remediation in the future, an environmental debt is bequeathed to future generations just as surely as unbalanced government budgets bequeath a burden of financial debt. By anticipating environmental problems, and by taking steps now to prevent them, the present generation can minimize the environmental and financial debts that its children will incur.

Today, we face new classes of environmental problems that are more diffuse than those of the past and thus demand different approaches. Since the first Earth Day in 1970, the vast majority of the most significant "point" sources of air and water pollution—large industrial facilities and municipal sewage systems—that once spewed untreated wastes into the air, rivers, and lakes have been controlled. The most important remaining sources of pollution are diffuse and widespread: sediment, pesticides, and fertilizers that run off farmland; oil and toxic heavy metals that wash off city streets and highways; and air pollutants from automobiles, outdoor grills, and woodstoves. Pollution from these sources cannot be controlled with sewage treatment plants or the same regulatory techniques used to check emissions from large industries. To make matters more complicated, we now have environmental problems on a global scale—biodiversity loss, ozone depletion, and climate change. These problems will require cooperative international responses. We are also recognizing that controlling pollutants alone, no matter how successful, will not achieve an environmentally sustainable economy, since many global concerns are related to the size of the human population and the unequal distribution of resources.

Defining the Future

We are progressing from an environmental paradigm based on cleanup and control to one including assessment, anticipation, and avoidance. Expenditures to develop technologies that prevent environmental harm are beginning to pay off. Agricultural practices are becoming less wasteful and more sustainable, manufacturing processes are becoming more efficient in the use of resources, and consumer products are being designed with the environment in mind. The infrastructures that supply energy, transportation services, and water supplies are becoming more resource efficient and environmentally benign. Remediation efforts are cleaning up a large portion of existing hazardous-waste sites. Our ability to respond to emerging problems is being aided by more advanced monitoring systems and data analysis tools that continually assess the state of the local, regional, and global environment. Finally, we are developing effective ways of restoring or re-creating severely damaged

ecosystems to preserve the long-term health and productivity of our natural resource base.

This trend will continue if we are willing to develop strong environmental policies, develop new strategies, and closely coordinate the actions of the public and private sectors toward a set of shared, long-term goals. Such a strategy should include three aims. The first is to articulate a vision of the future and the role environmental technology will play in shaping that future. This vision must be built on an understanding of the strengths and weaknesses of past policies and actions. The second is to define the roles of the many individuals and organizations who are needed to implement the goals. The third is to chart a course by offering suggestions for strategic goals for all the partners in this endeavor.

In the long run, environmental quality is not determined solely by the actions of government, regulated industries, or nongovernment organizations. It is largely a function of the decisions and behavior of individuals, families, businesses, and communities everywhere. Consequently, the extent of environmental awareness and the strength of environmental institutions will be two critical factors driving changes in environmental quality in the future.

A concerned, educated public, acting through responsive local, national, and international institutions, will serve as effective agents for avoiding future environmental problems, no matter what they are. Environmental institutions, strengthened by informed public support, will play a critical role in devising and implementing effective national and international responses to emerging issues.

The Greening of Geopolitics

The dramatic increase in concern about environmental issues led pollster Louis Harris to predict that, within the decade of the 1990s, national leaders would be chosen and elected with a pro-environmental stance as a primary identification factor. Harris further stated that the "issue

of the environment has become an explosive and decisive cutting edge in mainstream politics." While the environment was not one of the top four issues in the 1996 presidential election, the issue was still carefully watched by pollsters.

Environmental or "green" politics has emerged from minority status and become a political movement in many nations. Such events as the massive destruction that resulted from a chemical spill on the Rhine River and the nuclear disaster at Chernobyl have served to bolster the emergence of green politics. Even in the former Soviet Union and the rest of the newly enfranchised Eastern Europe, the public has demanded more environmental protection, and leaders are beginning to respond.

In October 1988, 300,000 Lithuanians signed a petition against a nuclear power plant. In Poland, where 80 percent of the Vistula River is too polluted for even industrial use, concern about the environment helped spark the 1989 Polish revolution. Norway has volunteered to contribute 0.1 percent of its gross national product, about $65 million, to an international climate fund if other countries are willing to join. At the economic summit in Paris in 1989, the leaders of the seven largest industrial democracies devoted a third of their final statement to an appeal for "decisive action" to "understand and protect the earth's ecological balance." A month prior to the summit, the Organization for Economic Cooperation and Development, representing the major economic powers, had called upon "all relevant national, regional, and international organizations" of its 24 member states to take a "vigilant, serious, and realistic" look at "balancing long-term environmental costs and benefits against near-term economic growth."

Concern about the environment is not limited to developed nations. A 1989 treaty signed in Switzerland limits what poorer nations call toxic terrorism: use of their lands by richer countries as dumping grounds for industrial waste. In 1990, more than one hundred nonaligned nations called for a "productive dialogue with the developed world" on "protection of the environment." Shortly after this

The Greens

In all Western countries, the growing environmental movement has been an important political development over the past 20 years. Numerous environmental organizations have been created. Indeed, not just Western nations but virtually all countries now have environmental organizations that are helping to shape the perspective of governments on environmental issues.

As environmental organizations have developed, they have focused the question whether to work with existing political parties or to create their own. While organizations in the United States generally work through existing structures, in West Germany, a new political party—the Greens—is closely identified with the environmental agenda. As distinguished from environmental organizations, the need for a "Green" party is not evident in every country. Nevertheless, the German Greens has been modeled in many countries, so that it is possible to speak of a "Green phenomenon."

In the West German elections in April 1983, the Green Party gained international attention by breaking through the 5 percent vote barrier of the nation's proportional representation system and placing 27 people in the national parliament. Since then, Greens in seven other Western European countries have entered parliament with between 5 and 8 percent of the vote, and Greens have a voice in the all-European parliament that meets in Strasbourg.

In the early summer of 1989, Greens in Great Britain astounded the experts, and themselves, by getting 15 percent of the vote, a figure that unfortunately for them does not translate into parliamentary representation, given the simple majority, winner-take-all system in that country. There are organized Green political tendencies in Poland, Hungary, and the former Soviet Union. Japan has a small but very vigorous Green Party, as do Australia, Brazil, and Canada.

While Greens exist in the United States and Canada, there is little sign that they will be able to elect representatives in significant numbers. They are ultimately victims of the political system within which they work. Also, the need for a specifically Green Party is arguably less in North America than elsewhere because of a strong tradition of freedom of association and the corresponding characteristic for forming political interest and pressure groups. Thus, environmental issues are pushed by a long list of local pressure groups and by strong national environmental lobbies. If the Greens continue to grow in Western Europe, there may come a time when they will also become a significant political force in North America.

statement, Japan pledged $2.25 billion to tackle pollution in the Third World. As was covered in chapter 2, the Earth Summit in Rio brought together nearly 180 governments to address world environmental concerns.

Environmentalism is also a growing factor in international relations. Many world leaders see the concern for environment, health, and natural resources as entering the policy mainstream. A new sense of urgency and common cause about the environment is leading to cooperation in some areas. Ecological degradation in any nation is now understood almost inevitably to impinge on the quality of life in others. Drought in Africa and deforestation in Haiti have resulted in large numbers of refugees, whose migrations generate tensions both within and between nations. From the Nile to the Rio Grande, conflicts flare over water rights. The growing megacities of the Third World are areas of potential civil unrest. Sheer numbers of people overwhelm social services and natural resources. The government of the Maldives has pleaded with the industrialized nations to reduce their production of greenhouse gases, fearing that the polar ice caps may melt and inundate the island nation.

Economic progress in the Third World also could bring the possibility for environmental peril and international tension. China, which accounts for 21 percent of the world's population, has the world's third largest recoverable coal reserves. If China's current "modernization" campaign succeeds, the boom will be fueled by that coal, to the detriment of the planet as a whole. Some experts estimate that the developing world, which today produces one-fourth of all greenhouse-gas emissions, could be responsible for nearly two-thirds by the middle of the next century. Third World nations have repeatedly indicated that they are not prepared to slow down their own already weak economic growth to help compensate for decades of environmental problems caused largely by the industralized world.

Some developing countries may resist environmental action because they see a chance to improve their bargaining leverage with foreign aid donors and international bankers. Where before the poor nations never had a strategic advantage, they now may have an ecological edge. Ecologically, there could be more parity than there ever was economically or militarily.

In 1991, the U.S. ambassador to the United Nations stated that, just as the cold war between the East and West seems to be winding down, "ecoconflicts" between the industrialized North and the developing South may pose a comparable challenge to world peace. National security may no longer be about fighting forces and weaponry alone. It relates increasingly to watersheds, croplands, forests, climate, and other factors rarely considered by military experts and political leaders, but that, when taken together, deserve to be viewed as equally crucial to a nation's security as are military factors.

The increased attention to the environment as a foreign policy and national security issue is only the beginning of what will be necessary to avert problems in the future. The most formidable obstacle may be the entrenched economic and political interests of the world's most advanced nations. If the United States, for example, asks others not to cut their forests, then it will have to be more judicious about cutting its own. If North Americans wish to stem the supply of hardwood from a fragile jungle or furs from endangered species,

then they will have to stem demand for fancy furniture and fur coats. If they wish to preserve wilderness from the intrusions of the oil industry, then they will have to find alternative sources of energy and use all fuels more efficiently.

What may be needed is self-discipline on the part of the world's haves and increased assistance to the have-nots. In the world today, a billion people live in a degree of poverty that forces them to deplete the environment without regard to its future. Their governments often are too crippled by international debt to afford the short-term costs of environmental safeguards.

William Ruckelshaus, the former administrator of the U.S. Environmental Protection Agency, believes a historical watershed may be at hand. If the industrialized and developing countries did everything they should, he says, the resulting change would represent a"a modification of society comparable in scale to the agricultural revolution of the Neolithic age and to the Industrial Revolution of the past two centuries."

Just as environmentalism began as a local movement, it needs to continue to grow at the local level, even as it gathers force in the parliaments and legislative bodies of the world. Individuals and nations alike should learn to think on a far broader scale about themselves, their needs, and their interests before a global catastrophe forces them to do so.

International Environmental Policy

If there must be a war, let it be against environment contamination, nuclear contamination, chemical

contamination, against the bankruptcy of soil and water systems; against the driving of people away from the lands as environmental refugees. If there must be war, let it be against those who assault people and other forms of life by profiteering at the expense of nature's capacity to support life. If there must be war, let the weapons be your healing hands, the hands of the world's [youth] in defence of the environment.

Mustafa Tolba
Secretary General
United Nations
Environmental Program

Examples of international cooperation on issues other than war are unfortunately somewhat limited. Institutional coordination and political resolve, however, are necessary to preserve and protect the global environment. A major step in that direction was the 1972 United Nations Conference in Stockholm, Sweden. This was the first international conference specifically dealing with environmental concerns. Out of that conference was born the U.N. Environmental Program, a separate department of the United Nations that deals with environmental issues.

In 1982, the Third United Nations Conference on the Law of the Sea produced a comprehensive convention that addressed many of the issues concerning jurisdiction over ocean waters and use of ocean resources. This treaty is viewed by many as a model for international environmental protection. Positive results are already evident from application of the agreements on pollution control, marine mammal protection, navigation safety, and other aspects of the marine environ-

Table 20.3

Factors Affecting International Environmental Laws

1. **Severity of the problem.** Once a problem is widely acknowledged as critical, it is easier to make progress.
2. **Science.** Good data are needed on the extent of the problem and possible solutions.
3. **Geography.** The extent to which the problem is transboundary and where its effects are felt.
4. **Law.** Whether countries have laws protecting the environment and whether foreigners have access to court and administrative proceedings to enforce those laws.
5. **Domestic interests and pressures.** Who favors and who opposes action on the issue in each country.
6. **Formal institutions and policies.** Whether there is a mechanism in place for cooperative action among the interested countries.
7. **History.** Whether there is a tradition of cooperation or conflict among the countries.
8. **Relative economic strength, military power, and population.** Large disparities in size or strength may hinder agreement unless the stronger country depends on the weaker for the resource.
9. **Outside influences.** Third parties can influence negotiations positively or negatively.
10. **Timing.** It may be easier to reach agreement before various interests are entrenched or at times when other changes in societies or economies are occurring.

Page 182 from World Resources 1987 *by* World Resources Institute *and* International Institute for Environment and Development. *Copyright © 1987 by the International Institute for Environment and Development and the World Resources Institute. Reprinted by permission of BasicBooks, a division of HarperCollins Publishers, Inc.*

ment. The issue of deep ocean mining, primarily of manganese nodules, has to date kept many industrial nations, such as Germany and the United States, from ratifying the treaty.

There have been several other successful international conventions and treaties. The Antarctic Treaty of 1961 reserves the Antarctic continent for peaceful scientific research and bans all military activities in the region. The 1979 Convention on Long-Range Transboundary Air Pollution was the first multilateral agreement on air pollution and the first environmental accord involving all the nations of Eastern and Western Europe and North America. The 1989 Accord on Chlorofluorocarbon Emissions addresses the threat to the stratospheric ozone shield.

There is no international legislature with authority to pass laws; nor are there international agencies with power to regulate resources on a global scale. An international court at the Hague in the Netherlands has no power to enforce its decisions. Nations can simply ignore the court if they wish. However, a growing network of multilateral environmental organizations have a greater sense of their roles and a greater incentive to work together. These include not only the United Nations Environment Program but also the Environment Committee of the Organization for Economic Cooperation and Development and the Senior Advisors on Environmental Problems of the Economic Commission for Europe. Such institutions perform unique functions that cannot be carried out by governments acting alone or bilaterally. Table 20.3 lists factors that affect international environmental laws.

Is the goal of an environmentally healthy world realistic? There is growing optimism that the community of nations is slowly maturing with regard to our common environment. We have all suffered a loss of innocence about "earth management." Laissez-faire may be good economics, but it can be a prescription for disaster in ecology.

This environmental "coming of age" is reflected in the broadening of intellectual perspective. Governments used to be preoccupied with domestic environmental affairs. Now, they are beginning to broaden their scope to confront problems that cross international borders, such as transboundary air and water pollution, and threats of a planetary nature, such as stratospheric ozone depletion and climatic warming. It is becoming increasingly evident that only decisive mutual action can secure the kind of world we seek.

Eco-Terrorism

The effects of war on the natural environment are frequently catastrophic. By its nature, war is destructive, and the environment is an innocent victim. The 1990–91 war in the Persian Gulf, however, witnessed a new role for the environment in war: its degradation was used as a weapon.

The phrase "eco-terrorism" was applied to Iraqi leader Saddam Hussein's use of oil as a weapon in the war. Iraq dumped an estimated 1.1 billion liters of crude oil into the Persian Gulf from Kuwait's Sea Island terminal before the United States made bombing raids on the pumps feeding the facility. The oil spill that resulted from the dumping was the world's largest—over 30 times the size of the 1989 Exxon Valdez disaster in Alaska.

Saddam Hussein may have engineered the spill to block allied plans for an amphibious invasion of Kuwait, but he was also probably trying to shut down seaside desalination plants that provide much of the freshwater for Saudi Arabia's Eastern Province. Whatever the military or political effect, the environmental effect was much greater.

The Persian Gulf is a sensitive ecosystem. The water is very salty, and temperatures vary widely from one season to another. As a result, indigenous animals and plants are finely attuned to specialized conditions. The gulf is also rather isolated, with only one narrow outlet—the Strait of Hormuz, just 55 kilometers across. The Gulf takes up to five years to flush out.

The Persian Gulf waters, shores, and islands are dotted with coral reefs, mangrove swamps, and beds of sea grass, alive with many unique species of birds, fish, and marine mammals. Many of the region's species, such as the bottlenose dolphin, dugong, green turtle, and caspian tern were already classified as threatened. This complex ecosystem, already pushed to the limits of survival by years of pollution, was again being threatened by a deliberate human act.

Another act of eco-terrorism that took place during the Iraq-Kuwait war was the intentional burning of hundreds of oil wells in Kuwait. The air pollution resulting from the burning oil covered a large area. It took several months to extinguish the fires following the war. Using environmental damage as a weapon of war is a frightening concept. The environmental consequences of such acts of eco-terrorism will not be fully understood for years. Though the war in the Persian Gulf lasted less than one year, the environmental effects on some species could be irreversible.

A Guide to International Organizations with Environmental Influence

Global Environment Facility—Established in 1990 through cooperation among the World Bank, UNEP, and UNDP, the GEF is an experiment in providing low- or no-interest loans for programs in four areas: protection of the ozone layer, reduction of greenhouse gases, protection of international water resources, and protection of biodiversity.

International Development Association—An affiliate of the World Bank, IDA is a lending agency intended to finance development projects in the poorer member countries for the same general purpose as the World Bank.

International Monetary Fund—The IMF is a specialized agency of the United Nations that aims to promote international monetary cooperation and stabilization of currencies, to facilitate the expansion and balanced growth of world trade, and to help member countries meet temporary difficulties in foreign payments.

Organization for Economic Cooperation and Development—OECD promotes economic and social welfare in member countries and harmonious development of the world economy. Members of OECD are the industrialized countries of North America and Western Europe, and Australia, Japan, and New Zealand.

United Nations Development Program—UNDP is the United Nations' central agency for funding economic and social development projects around the world. It is intended to help developing countries increase the wealth-producing capabilities of their natural and human resources.

United Nations Environment Program—UNEP is the U.N. agency intended to cover the major environmental issues facing both the developed and the developing areas of the world. UNEP also is responsible for promoting environmental law and education and training for the management of the environment.

World Bank—Officially named the International Bank for Reconstruction and Development, the World Bank is the leading organization in the field of multilateral financing of investment and technical assistance. Beginning operations in 1946, this specialized agency of the United Nations originally was concerned with reconstruction of Europe after World War II and now provides assistance to developing nations and the underdeveloped areas of the Western world.

Sources: International Organizations: A Dictionary and Directory *by Giuseppe Schiavone (St. James Press, 1983)*; World Directory of Environmental Organizations *by Thaddeus Tryzna and Robert Childes (California Institute of Public Affairs, 1992).*

Environmental Degradation in Eastern Europe

The environmental legacy of four decades of Communist rule is one of stark environmental degradation. While the causes of such degradation are many, four appear to be primary: (1) burning of high-sulfur brown coal, (2) uncontrolled auto tail-pipe emissions, (3) little or no control equipment for industrial smokestacks, and (4) intensive use of pesticides in a monoculture-based agriculture.

In many of the countries, environmental regulations are thought to be adequate but poorly enforced. Environmental institutions, official and nonofficial, exist. Environmental knowledge is believed considerable, but more theoretical than practical. Many environmental activists view their efforts as part and parcel of greater political and economic reform. For over a decade, West Germany paid East Germany to take its toxic wastes and other garbage. The new Germany will bear the responsibility of cleaning them up.

Cleaning up the problems in Eastern Europe will not be easy. Some suggest that the United States demand all loans to the area from the World Bank and other such institutions be conditioned upon tight environmental standards and that private loans from banks uphold basic environmental standards. As Eastern Europe leaders find that foreign investment is scarce and prospects for growth are dimming, however, they may be tempted to accept dirty technologies from the West: plants and production lines that fail Western environmental standards but are a bit above those now in the East.

OVERVIEW OF ENVIRONMENTAL PROBLEMS IN FOUR COUNTRIES

Former Czechoslovakia

Major air pollutant is sulfur dioxide from uncontrolled industries and power plants. Sixty percent of energy comes from burning brown coal. About 70 percent of the rivers are polluted from mining wastes and agricultural runoff, and 40 percent of all sewage is untreated. Seventy-five percent of toxic waste is dangerously stored. Poor agricultural practices and heavy metals from open-cast mining have contaminated soil.

Former East Germany

Produces most sulfur dioxide in Eastern Europe. Uncontrolled burning of brown coal provides 70 percent of energy. Sixty-six percent of rivers are polluted, and sewage is poorly treated. Toxic waste, stored for the West, is poorly controlled. Agricultural land is reduced as a result of heavy fertilizer use and monoculture agriculture.

Hungary

One-third of the population lives in areas where air pollution exceeds international standards. Primary pollutant is sulfur dioxide. Burning of brown coal produces 20 percent of energy; oil, gas, and power from one nuclear power plant provide another 60 percent of energy needs. Hundreds of villages and towns have water that is unfit for drinking. Adequate sewage treatment exists for less than half of the population. Toxic waste, transported from Western Europe, is stored inadequately. Soil erosion is a problem in some areas.

Poland

Twenty-seven areas have been declared ecological danger zones; five regions have been declared disaster areas. Sulfur dioxide is the major air pollutant, especially in Upper Silesia and Krakow. The burning of brown coal supplies almost all energy needs. Nearly all rivers are polluted; water from one-third is unfit for industrial use. Seventy percent of sewage is untreated. Of the food produced in the Krakow area, 60 percent may be unfit for human consumption because of heavy metal contamination.

Sources: U.S. Environmental Protection Agency and World Bank.

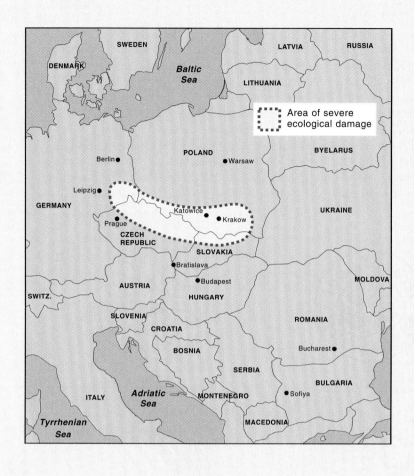

Eco-Labels

Green consumerism is the concept of rational consumption of our scarce resources for the benefit of the environment and future generations. The old saying "the world is enough for everyone's need but not for everyone's greed" calls for a change in our behavior and lifestyle in favour of a sustainable future. Eco-labels have been introduced in a number of countries to help consumers choose products with a proven environmental edge, determined by the product's choice of raw materials, production process, product life cycle, and associated disposal problems. Eco-labels provide evidence that products have met the safety, quality, and environmental protection requirements of the authority that issues the label.

The first eco-label was introduced in Germany in 1978. This distinctive Blue Angel logo was followed by other environmental certification, including the White Swan of Northern European countries, Environmental Choice of Canada, Green Label of Singapore, and Environmental Label of China. By 1995, over a dozen countries had adopted this system of providing consumers with information enabling them to become responsible green consumers. The eco-label of the European Union is significant because it is the world's first regional scheme to apply the same minimum standards across national markets.

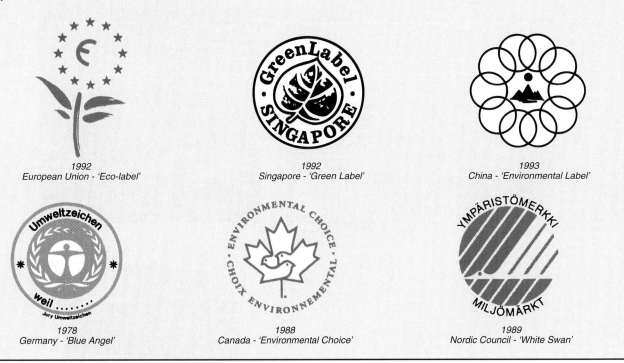

1992
European Union - 'Eco-label'

1992
Singapore - 'Green Label'

1993
China - 'Environmental Label'

1978
Germany - 'Blue Angel'

1988
Canada - 'Environmental Choice'

1989
Nordic Council - 'White Swan'

Source: From ECCO, Bulletin of the Environmental Campaign Committee, Issue 49, October 1995.

Keeping Tabs on the Global Environment

The United Nations Environment Program (UNEP) describes its mandate as "bridging the gap between awareness and action." Since its founding in 1972, UNEP has worked to coordinate the environmental activities of other UN organizations and to develop new relationships among scientists, decision-makers, engineers and financiers, industrialists and environmental activitists.

Some of UNEP's most substantial achievements include:

GEMS

An acronym for the Global Environmental Monitoring Service, GEMS is a system for gathering and distributing data on the Earth's vital signs. Data are collected via land-based stations and remote-sensing satellites.

The system involves the collaboration of hundreds of national and international organizations spread out over 142 nations. The network monitors changes in atmospheric composition and the climate system, freshwater and coastal pollution, air pollution, food contamination, deforestation, ozone-layer depletion, the build-up of greenhouse gases, acid rain, the extent of global ice cover, and numerous issues related to biological diversity.

INFOTERRA

INFOTERRA provides access to the vast pool of environmental data collected by governments, industry, and researchers. Some 139 countries, representing 99 percent of the world's population, contribute to the system.

REGIONAL SEAS PROGRAM

Getting Arabs and Israelis to sit around a negotiating table—no matter what the subject—must be considered a major accomplishment. Through the Regional Seas Program, just such an event was a milestone in bringing about the Barcelona Convention that has helped stem water quality decline in the Mediterranean Sea. The Mediterranean program is one of five regional-cooperation groups orchestrated through this UNEP program. The latest is a plan being set up now in the northwest Pacific between both North and South Korea along with several other neighboring nations.

Coastal zone management and water quality analysis are elements of the Regional Seas Program that are gaining in importance as coastal population currently lives near seacoasts.

INDUSTRY AND ENVIRONMENT OFFICE (IEO)

Opened in 1975, IEO was one of the first international organizations to break the traditional adversarial mold between industry and environmental activists. The IEO's primary message to business is that pollution prevention pays.

Long a promoter of clean technology, its latest program has set up working groups of experts to gather information on clean technology relevant to specific industries. A computer-based information-exchange system called the International Cleaner Production Information Clearinghouse is now operational.

SUMMARY

Politics and the environment cannot be separated. In the United States, the government is structured into three separate branches, each of which impacts environmental policy.

April 22, 1970, was the first Earth Day, and in the opinion of many, it was the event that initiated the modern environmental movement. By 1990, it appeared that the United States political will to address environmental concerns was increasing. Earth Day II on April 22, 1990, was seen as the beginning of a new era of environmental concern.

The late 1980s and early 1990s witnessed a new international concern about the environment, both in the developed and developing nations of the world. Environmentalism is also seen as a growing factor in international relations. This concern is leading to international cooperation where only tension had existed before.

While there exists no world political body that can enforce international environmental protection, the list of multilateral environmental organizations is growing.

It remains too early to tell what the ultimate outcome will be, but progress is being made in protecting our common resources for future generations. Several international conventions and treaties have been successful. In the final analysis, however, each of us has to adjust our lifestyle to clean up our own small part of the world.

REVIEW QUESTIONS

1. What are the major responsibilities of each of the three branches of the U.S. government?
2. In which branch of the government would a group find it most effective to apply pressure to develop programs and policies favorable to that group?
3. How has public opinion in the United States changed concerning the protection of the environment in the past 10 years?
4. Why is environmentalism a growing factor in international relations?
5. Give some examples of some international environmental conventions and treaties.

KEY TERMS

executive branch   405
judicial branch   405
legislative branch   405
policy   405

RELATED INTERNET SITES

Research on environmental policy and sustainability for all states.
   http://www.brocku.ca/epi/
Corporate Conservation Council, working toward sustainable economic development and "greening" business.
   http://www.nwf.org/ccc/
Public finance policies based on the democratic right of all people to land and resources.
   http://www.envirolink.org/orgs/earthrights/
U.S. environmental laws and policies.
   http://www.webcom.com/~staber/welcome.html
Award-winning site on policies, frameworks, and decision-making necessary for sustainable development.
   http://iisd1.iisd.ca/

Multimedia resource for environment and development policy makers.
   http://www.mbnet.mb.ca/linkages/
Green Plans as the path to a sustainable environment and economy.
   http://www.rri.org/
Sustainable living on a personal level.
http://condor.stcloud.msus.edu/~dmichael/eco/
Permaculture and related solutions to sustainable living.
   http://www.uea.ac.uk/~e415/home.html

APPENDIX ONE

Is an Environmental Career in your Future?

As interest in the environment grows, so does the need for qualified individuals to pursue the many career opportunities available in the environmental field. A person can work for a nonprofit environmental organization, a business, a school, a park, or the government. Environmental issues are involved in most fields.

In order to prepare yourself for an environmental career, you first need to ask yourself some questions: 1. What do I enjoy doing and why? Do I like writing, computer programming, or science? 2. What area of the environment interests me and why? For example, am I interested in recycling or am I curious about certain endangered animals? 3. What problem in the environment would I like to see solved? Do I like local or international issues?

In addition to asking yourself questions, you can decide if you are interested in an environmental career by talking to people who already work in the field and have a job you think is interesting. These people can answer questions and tell you how they prepared to enter the environmental arena.

You can prepare yourself either by combining a liberal arts background with some course work dealing with environmental science or policy or by pursuing a scientific approach to your education by taking more technical courses. The most important thing is to pursue something that you enjoy and that interests you. Most academic degrees can be developed into an environmental career. For example, a person with a communication degree can work in an environmental organization's public relations office; a person with a math degree can collect environmental statistics and conduct research; a person with an English degree can write environmental regulations; a business degree can be utilized in corporate environmental management.

One of the best ways to see if this field is for you is to get involved now! You can read about environmental issues, start a recycling program in your community, or get a summer job in an environmental field. Volunteering or having an internship with an environmental agency or business is a valuable learning experience! It allows you to get practical work experience while you investigate the field.

The environmental field is growing rapidly both domestically and internationally. With a sincere interest in the environment, a college degree, and hands-on experience, you can be on your way to a rewarding career where you can make a difference in the world we all share.

The following list of resource books, journal articles, and contacts provide information about environmental careers. Contact your school or local library for assistance in locating these items:

- *Careers for Nature Lovers and Other Outdoor Types* by Louise Miller. Lincolnwood, Ill.: VGM Career Horizons, 1992.
- *Careers in Conservation and Environmental Protection: Fish and Wildlife Management* compiled by Institute for Research. Chicago: Institute for Research, 1991.
- *Careers in Hazardous Waste Management: A Job Hunter's Guide to the Hazardous Waste Management Field* by Jacalyn Spiszman.

Sacramento, Calif.: Environment Employment Clearinghouse, 1989.

- *Careers in Natural Resources and Environmental Law* by Percy R. Luney. Chicago, Ill.: American Bar Association, 1987.
- *Complete Guide to Environmental Careers* by The CEIP Fund. Washington, D.C.: Island Press, 1993.
- *Environmental Careers Guide: Job Opportunities with the Earth in Mind* by Nicholas Basta. New York: Wiley, 1991.
- "Environmental Careers: A Garbage Primer for Ecoeds" by Amy Martin. In *Garbage: The Practical Journal for the Environment,* Vol. 4, No. 1, January/February 1992, pp. 24–31.
- *Environmental Jobs for Scientists and Engineers* by Nicholas Basta. New York: Wiley, 1992.
- *Fixing the Environment: Guide to Science/Engineering Careers in Environmental Conservation* by Nicholas Basta. New York: Wiley, 1992.
- *The Job Seekers Guide: Opportunities in Natural Resource Management for the Developing World* by Timothy M. Resch and Mary G. Porter. Washington, D.C.: U.S. Department of Agriculture, Forest Service and Office of International Cooperation and Development, Forestry Support Program, 1990.
- *Massachusetts Environmental Industry/Education Resource Directory: A Guide to Schools, Careers, and Environmental Companies* written and compiled by Fenna Hanes. Boston, Mass.: Bays State Skills Corporation, 1992.
- *Occupational Outlook Handbook 1996–1997* by Neal H. Rosenthal and Ronald E. Kutscher. Washington, D.C.: U.S. Department of Labor, Bureau of Labor Statistics, 1996.
- *Opportunities in Environmental Careers* by Odom Fanning. Lincolnwood, Ill.: VGM Careers Horizons, 1991.

APPENDIX TWO

Environmental Work, Study, and Travel Opportunities

Access
50 Beacon Street, Boston, MA 02108
(617) 720-JOBS

Gathers information on public interest and nonprofit internships and career opportunities.

American Friends Service Committee (AFSC)
1501 Cherry Street,
Philadelphia, PA 19102
(215) 421-7000

AFSC is a Quaker organization that provides material assistance to grassroots groups in Asia, Africa, Latin America, the Middle East, and poor communities in the United States. AFSC also coordinates cross-cultural education programs, foreign policy programs, and human rights campaigns.

American Geological Institute
4220 King Street,
Alexandria, VA 22302
(703) 379-2480

Publishes *Directory of Geoscience Departments and Careers in Geology*.

Amigos de las Americas
5618 Star Lane, Houston, TX 77057
(800) 231-7796 or (713) 782-5290

Volunteers lead public health projects in Latin America and the Caribbean. Each volunteer receives training and raises the funds needed for his or her project during the school year and then works with a partner in a village for six to eight weeks during the summer. Projects include teaching dental hygiene, distributing eyeglasses, helping to build latrines, and providing immunization against disease.

Canadian Organization for Development through Education
321 Chapel Street
Ottawa, Ontario K1N 7Z2
Canada
(613) 232-3569

Supports sister school projects between Canada and Uganda, Kenya, the Caribbean and the Philippines.

Center for Global Education
c/o Augsburg College
731 21st Avenue South
Minneapolis, MN 55454
(612) 330-1159

Coordinates travel seminars designed to introduce participants to the realities of life in Mexico, Central America, the Caribbean, and the Philippines. Costs range between $900 and $2,000. Some scholarships are available.

Community Jobs
50 Beacon Street,
Boston, MA 02108
(617) 720-5627

Provides listings of more than 400 jobs each month; college edition each year lists summer employment opportunities for students. Published by ACCESS: Networking in the Public Interest.

Conservation Directory
1400 16th Street NW,
Washington, D C 20036
1 (800) 432-6564 or(202) 797-6800

Lists organizations, agencies, and personnel engaged in conservation work and natural resources use and management at state, national, and international levels. Also lists college and universities in the United States and Canada that have conservation studies programs. Published yearly by the National Wildlife Federation.

Cool It!
National Wildlife Federation
1400 16th Street NW
Washington, D C 20036
(202) 797-5435

Regional Cool It! coordinators work with students over the phone and conduct on-site workshops to help establish models of environmentally sound practices. Other services include a newsletter, job bank, and speakers' bureau.

Earth Day Resources
116 New Montgomery Street,
Suite 530
San Francisco, CA 94105
(800) 727-8619 or (415) 495-5987

Maintains a network of grassroots organizations committed to environmental solutions at the local level. Offers materials such as lesson plans, fact sheets, and guidance for recycling and rideshare programs.

Earthwatch
P.O. Box 403
Watertown, MA 02272
(617) 926-8200

Earthwatch links volunteers with scientists and scholars on academic research expeditions in the United States and abroad. You might help to study rock art in Italy or help to make a documentary movie on Brazilian festivals. Costs range between $600 and $2,500.

Eco-Net
18 DeBoom Street, Dept. GM
San Francisco, CA 94107
(415) 442-0220

An online computer network of more than 100 electronic bulletin boards that deal with environmental issues and job opportunities.

The Environmental Careers Organization, Inc.
286 Congress Street, Dept. GM
Boston, MA 02210
(617) 426-4375

Places college students and recent graduates in short-term paid internships with environmental groups and publishes a guide to environmental careers.

Environmental Careers World
22 Research Drive,
Hampton, VA 23666
(804) 865-0605

Lists over 500 jobs each month. Offers career advice, career news, and interviews with employers, and a career networking calendar. Published twice a month by the Environmental Career Center.

The Environmental Consortium for Minority Outreach
1001 Connecticut Avenue NW,
Washington, D C 20036
(202) 331-8387

Recruits minorities who are interested in working for environmental organizations.

Environmental Job Opportunities
550 N. Park Street,
15 Science Hall,
Madison, WI 53706
(608) 263-1815

Published every five weeks by the Institute for Environmental Studies at the University of Wisconsin–Madison.

Environmental Opportunities
P.O. Box 788,
Walpole, NH 03608
(603) 756-4553

Lists permanent, seasonal, and internship opportunities with nonprofit organizations and governmental agencies. Published monthly.

Environmental Project on Central America (EPOCA)
Earth Island Institute
300 Broadway, Suite 28
San Francisco, CA 94133-3312
(415) 788-3666

Sponsors reforestation brigades in Nicaragua. Brigadistas plant tens of thousands of trees to provide wind blocks and to prevent erosion and flooding on Nicaragua's farmland.

Environmental Studies and Resource Management
730 Polk Street,
San Francisco, CA 94109

An annotated guide to universities and government training programs in the United States. Published by the Sierra Club.

Foundation for International Training
200-1262 Don Mills Road, Don Mills, Ontario M3B 2W7, Canada
(416) 449-8838

The Foundation sponsors projects in community development, industry administration, and management in Third World countries. Participants should have some overseas experience as well as experience teaching a technical skill. Assignments are for six to eight weeks. All expenses are paid and a small honorarium is offered.

Gemquest
Global Exchange Motivators, Inc.
Montgomery County Intermediate
Unit Building
Montgomery Avenue and
Paper Mill Road
Erdenheim, PA 19118
(215) 233-9558

Organizes study and travel programs that promote cultural understanding through intensive contact between travelers and their hosts. Includes homestays and the possibility for university study abroad.

Global Action Plan
84 Ferry Hill Road
Woodstock, NY 12498
(914) 679-4830; fax (914) 679-4834

Provides workbook and support for the formation of small groups, or "Eco Teams," that meet to learn about simple and effective ways to reduce their impact on the environment. Each group quantifies their results, such as water saved each month, and adds the data to the results of other Eco Teams around the world.

Global Exchange
2940 16th Street, Room 307
San Francisco, CA 94103
(415) 255-7296

Creates tours designed to help people become more involved in Third World development efforts. Tours go to Central and South America, the Caribbean, and southern Africa.

Green Corps
1724 Gilpin Street
Denver, CO 80218
(303) 355-1881

An intensive training program in grassroots organizing, fund-raising, media, and campaign skills, followed by a year using your skills in a real life situation.

Habitat for Humanity
419 West Church Street
Americus, GA 31709
(912) 924-6935

Volunteers donate labor, money, and materials in this nonprofit Christian housing ministry to build and renovate homes in the United States and abroad.

Highlander Research and Education Center
Route 3, Box 370
New Market, TN 37820
(615) 933-3443

The Highlander Center organizes exchanges between community activists in Appalachia and the Third World, focusing on education, labor rights, health care, land use, toxics, and economic development.

International Bicycle Fund (IBF)
4247 135th Place SE
Bellevue, WA 98006-1319

Sponsors bicycle tours to Africa to encourage people-to-people contact and to learn about cultures, histories, and economies of the peoples visited. Costs are about $1,000 plus airfare.

International Development Exchange (IDEX)
777 Valencia Street
San Francisco, CA 94110
(415) 621-1494

IDEX brings together school and church groups in the United States with grassroots development projects in the Third World.

Job Opportunities in the Environment 1994
P.O. Box 2123,
Princeton, NJ 08543
(800) 338-3282

Provides information on over 1,100 of the fastest-growing and most successful employers in the environmental industry. Published by Peterson's Guide, Inc.

MADRE
121 West 27th Street, Room 301
New York, NY 10001
(212) 627-0444

Madre works to build friendships between women in Central America, the Caribbean, and the United States.

Mennonite Central Committee
21 South 12th Street
Akron, PA 17501
(717) 859-1151

Volunteers work on health, education, social services, and community development projects in over 50 countries.

National Association of Environmental Professionals
P.O. Box 15210,
Alexandria, VA 22309
(703) 660-2384

Holds an annual conference and publishes a newsletter and journal. Has local chapters and student chapters.

Opportunities
130 Azalea Drive,
Roswell, GA 30075
(404) 594-9367

Lists teacher/naturalist, camp counselor, and internship positions. Published six times annually by the Natural Science for Youth Foundation.

Overseas Development Network (ODN)
P.O. Box 1430
Cambridge, MA 02238
(617) 868-3002

A network of college activists, ODN organizes educational and fund-raising events dealing with Third World development issues.

Partners for Global Justice
4920 Piney Branch Road NW
Washington, DC 20011
(202) 723-8273

Partners for Global Justice empowers U.S. citizens to influence public policy effectively. Volunteers spend one year in a Third World country and one year in the United States working with cooperatives and community organizations.

Pax World Services
1111 16th Street NW, Suite 120
Washington, D C 20036
(202) 293-7290

A nonprofit organization that initiates and supports projects that encourage international understanding, reconciliation, and sustainable development. Sponsors hand-on development trips that work on community projects and educational, fact-finding trips.

Plowshares Institute
P.O. Box 243
Simsbury, CT 06070
(203) 651-9675, (203) 658-6645

Travel seminars to Africa, Asia, and Australia. Participants are asked to do advanced reading and preparation and to share experiences with others after the trip.

Public Interest Research Groups
The Fund for Public Interest Research
29 Temple Place
Boston, MA 02111
(617) 292-4800

The PIRGs organize student groups, campaign on a variety of environmental and social justice issues, and provide numerous canvassing and fund-raising jobs to recent college graduates.

SERVAS International
11 John Street, Suite 706
New York, NY 10038
(212) 267-0252

SERVAS is an international network of people interested in peace issues. SERVAS does not organize or lead trips. Instead, the organization compiles national directories of people in over 70 countries who are interested in meeting travelers for a short homestay, or just for coffee.

Student Conservation Association
1800 North Kent Street, Suite 1260
Arlington, VA 22209
(703) 524-2441

Volunteers work with stewardship and conservation programs on public land and in national parks. For example, you might help with a botany research project in Yellowstone National Park.

Student Environmental Action Coalition (SEAC)
P.O. Box 1168
Chapel Hill, NC 27514
(919) 967-4600

SEAC publishes a monthly newsletter for its members and sponsors over 70 gatherings across the country for students interested in environmental justice. Through the Campus Ecology Project, students can assess environmental quality and create strategies for change on their own campuses.

Youth Ambassadors of America
P.O. Box 5273
Bellingham, WA 98227
Exchange programs, conferences, and summits for young people.

Volunteers for Peace International Workcamps
Tiffany Road
Belmont, VT 05730
(802) 259-2759

Opportunities for young people of all nationalities to work together on two- and three-week hands-on projects in over 800 camps in 36 countries.

Witness for Peace
P.O. Box 567
Durham, NC 27702-0567
(919) 688-5049

A grassroots, faith-based organization committed to changing U.S. policy toward Central America through nonviolent action. WFP sponsors both long-term and short-term work delegations.

World Environmental Directory
P.O. Box 1068,
Silver Spring, MD 20910

Includes over 40,000 company, organization, agency, institution, and personal listings worldwide. Published by Business Publishers, Inc.

APPENDIX THREE

Environmental Organizations

This is a sampling of nongovernmental national and international environmental organizations. For the more complete, annually updated Conservation Directory, send $25 and $5.25 S&H to the National Wildlife Federation, 1400 Sixteenth Street NW, Washington, D C 20036. The directory includes many professional associations and state or local organizations not listed here. Your local library may have a copy. Most of the organizations publish a magazine or newsletter.

- Acid Rain Foundation, 1410 Varsity Drive, Raleigh, NC 27606. A nonprofit, nonpartisan organization that educates the public on air-quality issues, including acid rain, air pollutants, and global climate changes.
- African Wildlife Foundation, 1717 Massachusetts Avenue NW, Washington, D C 20036 (202-265-8393). Finances and operates wildlife conservation projects in Africa.
- American Farmland Trust, 1920 N Street NW, Suite 400, Washington, D C 20036 (202-659-5170). Works for preservation of family farms and for soil conservation.
- American Museum of Natural History, Central Park West at 79th Street, New York, NY 10024 (212-769-5100). Conducts natural history research and publishes educational material.
- American Rivers, Inc., 801 Pennsylvania Avenue SE, Suite 400, Washington, D C 20003 (202-547-6900). Works for preservation of American rivers.

- Appalachian Mountain Club, 5 Joy Street, Boston, MA 02108 (617-523-0636). Sponsors trail maintenance, outdoor education, and recreational hikes and climbs. Operates a mountain hut system on the Appalachian trail.
- Bat Conservation International, P.O. Box 162603, Austin, TX 78716 (512-327-9721). Works to preserve bats and to change people's perceptions of bats.
- CARE, 660 First Avenue, New York, NY 10016 (212-686-3110). Offers technical and material assistance to developing countries focusing on health, nutrition, agriculture, environment, small business support, and emergency aid.
- Center for Environmental Education, 1725 DeSales Street NW, Suite 500, Washington, D C 20036 (202-429-5609). A nongovernmental, nonprofit organization that sponsors programs for protection of endangered species.
- Center for Science in the Public Interest, 1875 Connecticut Avenue NW, Suite 300, Washington, D C 20009 (202-332-9110). National consumer advocacy organization that focuses on health and nutrition.
- Cenozoic Society, P.O. Box 455, Richmond, VT 05477 (802-434-4077). Founded in 1991 as a splinter group from Earth First!, the Cenozoic Society publishes *Wild Earth*, a journal of conservation biology and wildlands activism.

- Citizen's Clearinghouse for Hazardous Wastes, P.O. Box 6806, Falls Church, VA 22040 (703-237-2249). Collects information on hazardous-waste effects, disposal, and cleanup.
- Clean Water Fund, 1320 18th Street NW, Washington, D C 20036 (202-457-0336). Works for clean water and for protection of natural resources. Conducts voter education and public awareness projects through 14 state groups.
- Conservation International, 1015 18th Street NW, Suite 1000, Washington, D C 20036 (202-429-5660). Buys land or trades foreign debt for land set aside for nature preserves in developing countries.
- Cousteau Society, Inc., 870 Greenbriar Cir., Suite 402, Chesapeake, VA 23320 (804-523-9335). Produces television films, lectures, books, and research on ocean quality and other resource issues.
- Cultural Survival, 53-A Church Street, Cambridge, MA 02138 (617-495-2562). Helps small, vulnerable societies survive the encroachments of governments and industrialized society.
- Defenders of Wildlife, 1101 14th St., NW, Suite 1400, Washington, D C 20005 (202-682-9400). Seeks to protect and restore native species, habitats, and ecosystems.
- Ducks Unlimited, Inc., 1 Waterfowl Way, Memphis, TN 38120 (901-758-3825). Perpetuates waterfowl by purchasing and protecting wetland habitat.
- Earth Island Institute, 300 Broadway, Suite 28, San Francisco, CA 94133 (415-788-3666). A clearinghouse for international information on environmental and resource issues. Founded by David Brower to bring together sources of conservation action and news.
- Earthwatch, P.O. Box 403N Mt. Auburn St., Watertown, MA 02272 (800-776-0188). Sponsors scientific field research worldwide. Recruits volunteers for a wide variety of research expeditions.

- Environmental Careers Organization, 286 Congress St., 3rd Floor, Boston, MA 02210 (617-426-4375). Promotes environmental careers through consulting and publications.
- Environmental Defense Fund, Inc., 257 Park Avenue South, New York, NY 10010 (212-505-2100). Protects environmental quality and public health through litigation and administrative appeals.
- Environmental Law Institute, 1616 P Street NW, Suite 200, Washington, D C 20036 (202-328-5150). Sponsors research and education on environmental law and policy.
- Food and Agriculture Organization of the United Nations, Via delle Terme di Caracalla, Rome 00100 Italy (Tele: 57971). A special agency of the United Nations established to improve food production, nutrition, and the health of all people.
- Friends of the Earth, 1025 Vermont Ave, NW, Suite 300, Washington, D C 20005 (202-783-7400). Affiliated with groups in 32 countries around the world. Works to form public opinion and to influence government policies to protect nature.
- Fund for Animals, Inc., 200 West 57th Street, New York, NY 10019 (212-246-2096). Advocacy group for humane treatment of all animals.
- Greater Yellowstone Coalition, P.O. Box 1874, 13 S. Willson, Bozeman, MT 59715 (406-586-1593). A small but influential organization dedicated to the preservation and protection of the Greater Yellowstone ecosystem.
- Green Party USA/The Greens, P.O. Box 30208, Kansas City, MO 64112 (816-931-9366). Headquarters of the green political movement in the U.S.
- Greenpeace USA, 1436 U Street NW, Washington, D C 20009 (202-462-1177). Worldwide organization that works to halt nuclear weapons testing, to protect marine animals,

and to stop pollution and environmental degradation.
- Humane Society of the United States, 2100 L Street NW, Washington, D C 20037 (202-452-1100). Dedicated to the protection of both domestic and wild animals.
- Institute for Food and Development Policy, 398 60th St., Oakland, CA, 94618 (570-654-4400). An international research and education center working for social justice and environmental protection.
- Institute for Social Ecology, P.O. Box 89, Plainsfield, VT 85667. Offers courses for credit through Goddard College and workshops on social ecology, ecofeminism, urban design, and ecological planning.
- International Alliance for Sustainable Agriculture, 1701 University Avenue SE, Minneapolis, MN 55414 (612-331-1099). An alliance of organic farmers, researchers, consumers, and international organizations dedicated to sustainable agriculture.
- International Society for Ecological Economics, P.O. Box 1589, Solomons, MD 20688 (410-326-0794). Publishes the journal *Ecological Economics*.
- International Union for Conservation of Nature and Natural Resources (IUCN), US, 1400 16th St., NW, Washington, D C 20036 (202-797-5454). Promotes scientifically based action for the conservation of wild plants, animals, and resources.
- International Wolf Center, 1396 Hwy 169, Ely, MN 55731 (218-365-4695). Offers classes and tours and maintains displays and a captive wolf pack for conservation education and research.
- Izaak Walton League of America, Inc., Conservation Center, 707 Conservation Lane, Gaithersburg, MD 20878. Educates the public on land, water, air, wildlife, and other conservation issues.

- Land Institute, 2440 E. Water Well Road, Salina, KS 67401 (913-823-5376). Carries out research on perennial species, prairie polycultures, and sustainable agriculture. Has training programs and conferences.
- League of Conservation Voters, 1707 L Street NW, Suite 750, Washington, D C 20036 (202-785-8683). A nonpartisan national political campaign committee that strives to elect environmentally responsible public officials. Publishes an annual evaluation of voting records of Congress.
- National Audubon Society, 700 Broadway, New York, NY 10003 (212-979-3000). One of the oldest and largest conservation organizations, Audubon has many educational and recreational programs as well as an active lobbying and litigation staff.
- National Parks and Conservation Association, 1776 Massachusetts Ave. NW, Suite 200, Washington, D C 20036 (202-223-6722). A private nonprofit organization dedicated to preservation, promotion, and improvement of our national parks.
- National Wildlife Federation, 1400 Sixteenth Street NW, Washington, D C 20036 (202-797-6800). Specializes in wildlife conservation but recognizes the importance of habitat and other resources to all living things. More than 5 million members.
- Natural Resources Defense Council, Inc., 40 W. 20th Street, New York, NY 10011 (212-727-2700). An environmental organization that monitors government agencies and brings legal action to protect the environment.
- (The) Nature Conservancy, 1815 North Lynn Street, Arlington, VA 22209 (703-841-5300). Works with state and federal agencies to identify ecologically significant natural areas. Manages a system of over 1,000 nature sanctuaries nationwide.
- Planet Drum Foundation, P.O. Box 31251, San Francisco, CA 94131 (415-285-6556). Promotes bioregionalism, wise use of resources, and new attitudes toward nature.
- Population Reference Bureau, 1875 Connecticut Ave NW, Suite 540, Washington, D C 20009 (202-483-1100). Gathers, interprets, and publishes information on social, economic, and environmental implications of world population dynamics. Excellent data source.
- Rainforest Action Network, 450 Sansome St., Suite 700, San Francisco, CA 94111 (415-398-4404). Focuses on actions designed to save rainforest and to defend the rights of indigenous people around the world.
- Resources for the Future, 1616 P Street NW, Washington, D C 20036 (202-328-5000). Conducts research and provides education about natural resource conservation issues.
- Rodale Institute, 222 Main Street, Emmaus, PA 18098 (610-967-8509). A leading research institute for organic farming and alternative crops. Publishes magazines, books, and reports on regenerative farming.
- Save-the-Redwoods League, 114 Sansome Street, Room 605, San Francisco, CA 94104 (415-362-2352). Buys land, plants trees, and works with state and federal agencies to save redwood trees.
- Scientists' Institute for Public Information, 355 Lexington Avenue, New York, NY 10017 (212-661-9110). Enlists scientists and other experts in public information programs and public policy forums on a variety of environmental issues.
- Sea Shepherd Conservation Society, P.O. Box 628, Venice, CA 90294 (310-301-7325). An international marine conservation action program. Carries out field campaigns to call attention to and stop wildlife destruction and resource misuse.
- Sierra Club, 730 Polk Street, San Francisco, CA 94109 (415-776-2211). Founded in 1892 by John Muir and others to explore, enjoy, and protect the wild places of the earth. Conducts outings, educational programs, volunteer work projects, litigation, political action, and administrative appeals. Has one of the most comprehensive programs of any conservation organization.
- Society for Conservation Biology, Stanford University, Department of Biological Sciences, Stanford, CA 94305 (415-725-1852). A professional society for research and information. Publishes Conservation Biology.
- Soil and Water Conservation Society, 7515 NE Ankeny Rd., Ankeny, IA 50021 (515-289-2331). Promotes soil and water conservation.
- Student Conservation Association, Inc., Route 12A, River Road, Charlestown, NH 03603 (603-543-1700). Coordinates environmental internships and volunteer jobs with state and federal agencies and private organizations for students and adults.
- United Nations Environment Programme, P.O. Box 30552, Nairobi, Kenya; and United Nations, Rm, DC2-0803, New York, NY 10017 (212-963-8138). Coordinates global environmental efforts with United Nations agencies, national governments, and nongovernmental organizations.
- Waldebridge Ecological Centre, Worthyvale Manor Farm, Camelford, Cornwall P132 9TT England (Tele: 0840 212711). Conducts research and education on a variety of environmental issues. Publishes The Ecologist, an excellent global environmental journal.
- Wilderness Society, 900 17th Street NW, Washington, D C 20006 (202-833-2300). Dedicated to preserving wilderness and wildlife in America.

- World Resources Institute, 1709 New York Avenue NW, Washington, D C 20006 (202-638-6300). A policy research center that publishes excellent annual reports on world resources.
- Worldwatch Institute, 1776 Massachusetts Avenue NW, Washington, D C 20036 (202-452-1999). A nonprofit research organization concerned with global trends and problems. Publishes excellent periodic reports and annual summaries.
- World Wildlife Fund, 1250 24th Street NW, Washington, D C 20037 (202-293-4800). Seeks to protect the world's endangered wildlife and the habitat they need to survive.
- Zero Population Growth, Inc., 1400 16th Street NW, Suite 320, Washington, D C 20036 (202-332-2200). Advocates of worldwide population stabilization.

United States Government Agencies

- Bureau of Land Management, 1620 L St. NW, Washington, D C 20240 (202-205-3801). Administers about half of all public lands, mainly in the western United States. Follows policy of multiple use for maximum public benefit.
- Bureau of Reclamation, Room 7654, Department of the Interior, Washington, D C 20240. Builds and operates federal water projects, mostly in the western states with the Army Corps of Engineers.
- Bureau of Sports Fisheries and Wildlife (formerly U.S. Fish and Wildlife Service), Washington, D C 20240 (202-208-4717). Carries out wildlife research and management. Enforces game, fish, and endangered species laws. Administers the national wildlife refuges.
- Council on Environmental Quality, 722 Jackson Place NW, Washington, D C 20503 (202-395-5750). Advises the president on environmental matters.
- Department of Agriculture, 14th Street and Independence Avenue SW, Washington, D C 20250 (202-720-8732). Manages national forests and grasslands and oversees farm prices, farm policies, and soil conservation. (http://www.usda.gov)
- Department of Health and Human Services: Food and Drug Administration, 5600 Fishers Lane, Rockville, MD 20857. Enforces laws requiring that foods and drugs be pure, safe, and wholesome.
- Department of the Interior, C Street between 18th and 19th NW, Washington, D C 20240 (202-208-3100). Administers national parks, monuments, wildlife refuges, and public lands. (http://www.usgs.gov.doi)
- Environmental Protection Agency, 401 M Street SW, Washington, D C 20460 (202-260-2090). Enforces clean air and clean water laws. Identifies, regulates, and purifies toxic and hazardous materials. (http://www.epa.gov)
- Forest Service, P.O. Box 96090, Washington, D C 20013 (202-250-0951). Administers national forests and grasslands.
- National Oceanic and Atmospheric Administration, Herbert C. Hoover Bldg., Rm 5128, 14th and Constitution Ave. NW, Washington, D C 20230 (202-482-3384). Part of the Department of Commerce. Promotes global environmental stewardship and conducts atmospheric and oceanic research. Good source of information on atmosphere and climate. (http://www.gov.doc)
- National Park Service, Interior Bldg, P.O. Box 37127, Washington, D C 20013 (202-208-4717). Administers the national park system, monuments, and wild and scenic rivers. Possible source of seasonal jobs.
- National Resource Conservation Service (formerly Soil Conservation Service), 14th and Independence Ave. SW, P.O. Box 2890, Washington, D C 20013 (202-447-4543). Provides technical and educational assistance for soil conservation and watershed protection.

APPENDIX FIVE

Metric Unit Conversion Tables

The Metric System

Standard Metric Units		Abbreviations
Standard unit of mass	Gram	g
Standard unit of length	Meter	m
Standard unit of volume	Liter	l

Common Prefixes		Examples
Kilo	1,000	A kilogram is 1,000 grams.
Centi	0.01	A centimeter is 0.01 meter.
Milli	0.001	A milliliter is 0.001 liter.
Micro (μ)	One-millionth	A micrometer is 0.000001 (one-millionth) of a meter.
Nano (n)	One-billionth	A nanogram is 10^{-9} (one-billionth) of a gram.
Pico (p)	One-trillionth	A picogram is 10^{-12} (one-trillionth) of a gram.

Units of Length

Unit	Abbreviation	Equivalent
Meter	m	Approximately 39 in
Centimeter	cm	10^{-2} m
Millimeter	mm	10^{-3} m
Micrometer	μm	10^{-6} m
Nanometer	nm	10^{-9} m
Angstrom	Å	10^{-10} m

Length Conversions

1 in = 2.5 cm	1 mm = 0.039 in
1 ft = 30 cm	1 cm = 0.39 in
1 yd = 0.9 m	1 m = 39 in
1 mi = 1.6 km	1 m = 1.094 yd
	1 km = 0.6 mi

To Convert	Multiply By	To Obtain
Inches	2.54	Centimeters
Feet	30	Centimeters
Centimeters	0.39	Inches
Millimeters	0.039	Inches

Units of Volume

Unit	Abbreviation	Equivalent
Liter	l	Approximately 1.06 qt
Milliliter	ml	10^{-3} (1 ml = 1 cm^3 = 1 cc)
Microliter	μl	10^{-6} l

Volume Conversions

1 tsp = 5 ml	1 ml = 0.03 fl oz
1 tbsp = 15 ml	1 l = 2.1 pt
1 fl oz = 30 ml	1 l = 1.06 qt
1 cup = 0.24 l	1 l = 0.26 gal
1 pt = 0.47 l	
1 qt = 0.95 l	
1 gal = 3.8 l	

To Convert	Multiply By	To Obtain
Fluid ounces	30	Milliliters
Quarts	0.95	Liters
Milliliters	0.03	Fluid ounces
Liters	1.06	Quarts

Units of Weight

Unit	Abbreviation	Equivalent
Kilogram	kg	10^3 g (approximately 2.2 lb)
Gram	g	Approximately 0.035 oz
Milligram	mg	10^{-3} g
Microgram	μg	10^{-6} g
Nanogram	ng	10^{-9} g
Picogram	pg	10^{-12} g

Weight Conversions

1 oz = 28.3 g	1 g = 0.035 oz
1 lb = 453.6 g	1 kg = 2.2 lb
1 lb = 0.45 kg	

To Convert	Multiply By	To Obtain
Ounces	28.3	Grams
Pounds	453.6	Grams
Pounds	0.45	Kilograms
Grams	0.035	Ounces
Kilograms	2.2	Pounds

Temperature Conversions

$$°C = \frac{(°F - 32) \times 5}{9}$$

$$°F = \frac{°C \times 9}{5} + 32$$

Some Equivalents

0°C = 32°F

37°C = 98.6°F

100°C = 212°F

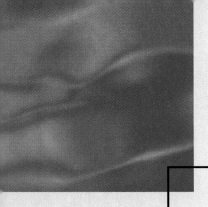

APPENDIX SIX

What *You* Can Do to Make the World A Better Place in which to Live

It is easy to complain; in fact, we all tend to do so almost on a daily basis. Complaining, however, never corrects a problem or helps to resolve a dispute. There are actions you can take that *will* help improve the environment. The following list offers a few suggestions, and others will probably come to mind once you have thought about it. Try to expand the list with others in your class, and give your suggestions to your instructor. Have your instructor send your suggestions to the authors at Delta College, University Center, Michigan 48710 or Huxley College of Environmental Studies, Western Washington University, Bellingham, Washington 98226. We will expand the list in the next edition of the textbook. Remember, you do not have to start big to help!

1. **Continue your education.** Learning should not stop when you leave school. Become informed about issues; then you can begin to bring about change.

2. **Do not feel responsible for every problem in the world.** You cannot do everything. Concentrate on issues that you feel strongly about and that you can do something about. Focus your energy.

3. **Think about the consequences of your profession and your lifestyle.** If they are damaging to other people or to the environment, adjust your behavior accordingly. Try to persuade friends, family, and coworkers to do the same. Make environmental awareness a family affair.

4. **Work with others.** Attend meetings of your local government and ask officials about their plans to prevent pollution. Often, officials are very responsive to visits of this kind. Being part of a group of people with similar interests gives you support and increases your effectiveness. If you cannot find an appropriate group, start one of your own.

5. **Become active in your community.** Organize a community conference to discuss positive approaches to pollution prevention. Invite public officials, industry and labor representatives, other interested groups, and individual citizens. Get all the facts, and then try to get the appropriate action programs initiated.

6. **Learn about the ecology of your bioregion.** Develop a sense of place that puts you in contact with your local physical environment. Learn about the unique environmental features of your area. What are the most urgent environmental problems?

7. **Vote.** You cannot improve your world by not voting. If you do not like the choices available, work to get individuals on the ballot who represent your interests.

8. **Think globally and act locally.** You need to be aware of global conditions, but you should also work to improve your own particular place.

9. **Do not be discouraged.** It is important to face facts honestly and to be realistic about the state of the world, but it does not help to wallow in despair. Do not dwell on negatives. Do what you can to improve the world, and take pleasure and pride in the small victories and elements of success.

10. **Try to leave things better than you found them.** Pick up a piece of litter on a beach or on your street, plant a tree, recycle your papers . . . the list goes on. Don't wait for the next person to begin—you can make a difference!

GLOSSARY

A

abiotic factors Nonliving factors that influence the life and activities of an organism.

abyssal ecosystem The collection of organisms and the conditions that exist in the deep portions of the ocean.

acid Any substance that, when dissolved in water, releases hydrogen ions.

acid deposition The accumulation of potential acid-forming particles on a surface.

acid mine drainage A kind of pollution, associated with coal mines, in which bacteria convert the sulfur in coal into compounds that form sulfuric acid.

acid rain (acid precipitation) The deposition of wet acidic solutions or dry acidic particles from air.

activated sludge sewage treatment Method of treating sewage in which some of the sludge is returned to aeration tanks, where it is mixed with incoming wastewater to encourage degradation of the wastes in the sewage.

activation energy The initial energy input required to start a reaction.

active solar system A system that traps sunlight energy as heat energy and uses mechanical means to move it to another location.

acute toxicity A serious effect, such as a burn, illness, or death, that occurs shortly after exposure to a hazardous substance.

age distribution The comparative percentages of different age groups within a population.

agricultural products Any output from farming; milk, grain, meat, etc.

agricultural runoff Surface water that carries nutrients, such as phosphate and nitrates, as it runs off agricultural land to lakes and streams.

air stripping The process of pumping air through water to remove volatile materials dissolved in the water.

alpha radiation A type of radiation consisting of a particle with two neutrons and two protons.

alpine tundra The biome that exists above the tree line in mountainous regions.

aquiclude An impermeable layer in an artesian aquifer.

aquifer A layer of earth material that can transmit water sufficient for water supply purposes.

aquitard A permeable layer in an artesian aquifer.

artesian aquifer The result of a pressurized aquifer intersecting the surface or being penetrated by a pipe or conduit, from which water gushes without being pumped.

atom The basic subunit of elements, composed of protons, neutrons, and electrons.

auxin A plant hormone that stimulates growth.

B

base Any substance that, when dissolved in water, removes hydrogen ions from solution; forms a salt when combined with an acid.

benthic Describes organisms that live on the bottom of marine and freshwater ecosystems.

benthic ecosystems A type of marine or freshwater ecosystem consisting of organisms that live on the bottom.

beta radiation A type of radiation consisting of electrons released from the nuclei of many fissionable atoms.

bioaccumulation The buildup of a material in the body of an organism.

biochemical oxygen demand (BOD) The amount of oxygen required to destroy organic molecules in aquatic ecosystems.

biocide A kind of chemical that kills many different types of living things.

biodegradable Able to be broken down by natural biological processes.

biogenetic law The generally accepted concept that living things can only be produced by other living things. Life cannot be created from non-life.

biological amplification The increases in the amount of a material at successively higher trophic levels.

biomass Any accumulation of organic material produced by living things.

biome A kind of plant and animal community that covers major geographic areas. Climate is a major determiner of the biome found in a particular area.

biotic factors Living portions of the environment.

biotic potential The inherent reproductive capacity.

birthrate The number of individuals born per thousand individuals in the population per year.

black lung disease A respiratory condition resulting from the accumulation of large amounts of fine coal dust particles in miners' lungs.

boiling-water reactor (BWR) A type of light-water reactor in which steam is formed directly in the reactor, which is used to generate electricity.

boreal forest A broad band of mixed coniferous and deciduous trees that stretches across northern North America (and also Europe and Asia); its northernmost edge is integrated with the arctic tundra.

C

carbamate A class of soft pesticides that work by interfering with normal nerve impulses.

carbon absorption The use of carbon particles to treat chemicals by having the chemicals attach to the carbon particles.

carbon cycle The cyclic flow of carbon from the atmosphere to living organisms and back to the atmospheric reservoir.

carbon dioxide (CO_2) A normal component of the earth's atmosphere that in elevated concentrations may interfere with the earth's heat budget.

carbon monoxide (CO) A primary air pollutant produced when organic materials, such as gasoline, coal, wood, and trash, are incompletely burned.

carcinogen A substance that causes cancer.

carcinogenic The ability of a substance to cause cancer.

carnivores Animals that eat other animals.

carrying capacity The optimum number of individuals of a species that can be supported in an area over an extended period of time.

catalyst A substance that alters the rate of a reaction but is not itself changed.

chemical bond The physical attraction between atoms that results from the interaction of their electrons.

chemical weathering The process of breaking up large particles by chemical reactions that convert insoluble materials to soluble materials and then removing the particles.

chlorinated hydrocarbon A class of pesticide consisting of carbon, hydrogen, and chlorine, which are very stable.

chronic toxicity A serious effect, such as an illness or death, that occurs after prolonged exposure to small doses of a toxic substance.

clear-cutting A forest harvesting method in which all the trees in a large area are cut and removed.

climax community Last stage of succession; a relatively stable, long-lasting, complex, and interrelated community of plants, animals, fungi, and bacteria.

combustion The process of releasing chemical bond energy from fuel.

commensalism The relationship between organisms in which one organism benefits while the other is not affected.

community Interacting groups of different species.

competition An interaction between two organisms in which both require the same limited resource, which results in harm to both.

composting A waste disposal system whereby organic matter is allowed to decay to a usable product.

compound A kind of matter composed of two or more different kinds of atoms.

Comprehensive Environmental Response, Compensation, and Liability Act (CERCLA) The 1980 U.S. law that addressed the issue of cleanup of hazardous-waste sites.

confined aquifer An aquifer that is bounded on the top and bottom by confining layers.

conservation To use in the best possible way so that the greatest long-term benefit is realized by society.

conservation ethic An environmental ethic that stresses a balance between total development and absolute preservation.

consumers Organisms that rely on other organisms for food.

contour farming A method of tilling and planting at right angles to the slope, which reduces soil erosion by runoff.

controlled experiment An experiment in which two groups are compared. One, the control, is used as a basis of comparison and the other, the experimental, has one factor different from the control.

coral reef ecosystem A tropical, shallow-water, marine ecosystem dominated by coral organisms that produce external skeletons.

corporation A business structure that has a particular legal status.

corrosiveness Ability to degrade standard materials.

cost-benefit analysis A method used to determine the feasibility of pursuing a particular project by balancing estimated costs against expected benefits.

cover A term used to refer to any set of physical features that conceals or protects animals from the elements or their enemies.

D

death phase The portion of the population growth curve that shows the population declining.

death rate The number of deaths per thousand individuals in the population per year.

decommissioning Decontaminating and disassembling a nuclear power plant and safely disposing of the radioactive materials.

decomposers Small organisms, like bacteria and fungi, that cause the decay of dead organic matter and recycle nutrients.

demand Amount of a product that consumers are willing and able to buy at various prices.

demographic transition The hypothesis that economies proceed through a series of stages, resulting in stable populations and high economic development.

demography The study of human populations, their characteristics, and changes.

denitrifying bacteria Bacteria that convert nitrogen compounds in the soil into nitrogen gas.

density-dependent limiting factors Those limiting factors that become more severe as the size of the population increases.

density-independent limiting factors Those limiting factors that are not affected by population size.

desert A biome that receives less than 25 centimeters of precipitation per year.

desertification The conversion of arid and semiarid lands into deserts by inappropriate farming practices or overgrazing.

detritus Organic material that results from fecal waste material or the decomposition of plants and animals.

development ethic Philosophy that states that the human race should be the master of nature and that the earth and its resources exist for human benefit and pleasure.

dispersal Migration of organisms from a concentrated population into areas with lower population densities.

domestic water Water used for domestic activities, such as drinking, air conditioning, bathing, washing clothes, washing dishes, flushing toilets, and watering lawns and gardens.

E

ecology A branch of science that deals with the interrelationship between organisms and their environment.

economic costs Those monetary costs that are necessary to exploit a natural resource.

economic growth The perceived increase in monetary growth within a society.

ecosystem A group of interacting species combined with the physical environment.

ectoparasite A parasite that is adapted to live on the outside of its host.

electron The lightweight, negatively charged particle that moves around the nucleus of an atom.

element A form of matter consisting of a specific kind of atom.

emergent plants Aquatic vegetation that is rooted on the bottom but has leaves that float on the surface or protrude above the water.

emigration Movement out of an area that was once one's place of residence.

endangered species Those species that are present in such small numbers that they are in immediate jeopardy of becoming extinct.

endoparasite A parasite that is adapted to live within a host.

energy The ability to do work.

energy cost The amount of energy required to exploit a resource.

environment Everything that affects an organism during its lifetime.

environmental costs Damage done to the environment as a resource is exploited.

environmental justice Fair application of laws designed to protect the health of human beings and ecosystems; that no groups suffer unequal environmental harm.

Environmental Protection Agency (EPA) U.S. government organization responsible for the establishment and enforcement of regulations concerning the environment.

environmental resistance The combination of all environmental influences that tend to keep populations stable.

environmental science An interdisciplinary area of study that includes both applied and theoretical aspects of human impact on the world.

enzyme Protein molecules that speed up the rate of specific chemical reactions.

erosion The loss of soil because it is carried away by running water or wind.

estuaries Marine ecosystems that consist of shallow, partially enclosed areas where freshwater enters the ocean.

ethics A discipline that seeks to define what is fundamentally right and wrong.

euphotic zone The upper layer in the ocean where the sun's rays penetrate.

eutrophication The enrichment of water (either natural or cultural) with nutrients.

eutrophic lake A usually shallow, warm-water lake that is nutrient rich.

evolution A change in the structure, behavior, or physiology of a population of organisms as a result of some organisms with favorable characteristics having greater reproductive success than those organisms with less favorable characteristics.

executive branch The office of the President of the United States.

experiment An artificial situation designed to test the validity of a hypothesis.

exponential growth phase The period during population growth when the population increases at an ever-increasing rate.

external costs Expenses, monetary or otherwise, borne by someone other than the individuals or groups who use a resource. .

extinction The death of a species; the elimination of all the individuals of a particular kind.

F

first law of thermodynamics A statement about energy that says that under normal physical conditions, energy is neither created nor destroyed.

fixation A form of waste immobilization in which materials, such as fly ash or cement, are mixed with hazardous waste.

floodplain Lowland area on either side of a river.

floodplain zoning ordinances Designations that restrict future building in floodplains.

food chain The series of organisms involved in the passage of energy from one trophic level to the next.

food web Intersecting and overlapping food chains.

fossil fuels The remains of plants, animals, and microorganisms that lived millions of years ago.

free-living, nitrogen-fixing bacteria Bacteria that live in the soil and can convert nitrogen gas (N_2) in the atmosphere into forms that plants can use.

freshwater ecosystem Aquatic ecosystems that have low salt content.

friability The ability of a soil to crumble.

fungicide A pesticide designed to kill or control fungi.

G

gamma radiation A type of electromagnetic radiation that comes from disintegrating atomic nuclei.

gas-cooled reactor (GCR) A type of reactor that uses graphite as a moderator and carbon dioxide or helium as a coolant.

geothermal energy The heat energy from the earth's molten core.

grasslands Areas receiving between 25 and 75 centimeters of precipitation per year. Grasses are the dominant vegetation, and trees are rare.

greenhouse effect The property of carbon dioxide (CO_2) that allows light energy to pass through the atmosphere but prevents heat from leaving; similar to the action of glass in a greenhouse.

gross national product (GNP) An index that measures the total goods and services generated annually within a country.

groundwater Water that infiltrates the soil and is stored in the spaces between particles in the earth.

groundwater mining Removal of water from an aquifer faster than it is replaced.

H

habitat An identifiable region in which a particular kind of organism lives.

habitat management The process of changing the natural community to encourage the increase in populations of certain desirable species.

hard pesticide A pesticide that persists for long periods of time; a persistent pesticide.

hazardous-waste dump A disposal site for hazardous waste in a dump, landfill, or surface impoundment without any concern for potential environmental or health risks.

hazardous wastes Substances that could endanger life if released into the environment.

heavy-water reactor (HWR) A type of reactor that uses the hydrogen isotope deuterium in the molecular structure of the coolant water.

herbicide A pesticide designed to kill or control plants.

herbivores Primary consumers; animals that eat plants.

high-temperature, gas-cooled reactor (HTGCR) A type of reactor that uses graphite as a moderator and helium as a coolant.

horizon A horizontal layer in the soil. The top layer (A horizon) has organic matter. The lower layer (B horizon) receives nutrients by leaching. The C horizon is partially weathered parent material.

host The organism a parasite uses for its source of food.

humus Soil organic matter.

hydrocarbons (HC) Group of organic compounds consisting of carbon and hydrogen atoms that are evaporated from fuel supplies or are remnants of the fuel that did not burn completely, and that act as a primary air pollutant.

hydrologic cycle Constant movement of water from surface water to air and back to surface water.

hydroxyl ion A negatively charged particle consisting of a hydrogen and an oxygen atom, commonly released from materials that are bases.

hypothesis A logical guess that explains an event or answers a question.

I

ignitability Characteristic of materials that results in their ability to combust.

immigration Movement into an area in which one has not previously resided.

incineration Method of disposing of solid waste by burning.

industrial ecology A concept that stresses cycling resources rather than extracting and eventually discarding them.

Industrial Revolution A period of history during which machinery replaced human labor.

industrial uses Uses of water for cooling and for dissipating and transporting waste materials.

insecticide A pesticide designed to kill or control insects.

in-stream uses Use of a stream's water flow for such purposes as hydroelectric power, recreation, and navigation.

integrated pest management A method of pest management in which many aspects of the pest's biology are exploited to control its numbers.

interspecific competition Competition between numbers of different species for a limited resource.

intraspecific competition Competition among members of the same species for a limited resource.

ion An atom or group of atoms that has an electric charge because it has either gained or lost electrons.

irrigation Adding water to an agricultural field to allow certain crops to grow where the lack of water would normally prevent their cultivation.

isotope Atoms of the same element that have different numbers of neutrons.

J

judicial branch That portion of the U.S. government that includes the court system.

K

kinetic energy Energy of moving objects.

kinetic molecular theory The widely accepted theory that all matter is made of small particles that are in constant movement.

K-strategists Large organisms that have relatively long lives, produce few offspring, provide care for their offspring, and typically have populations that stabilize at the carrying capacity.

L

lag phase The initial stage of population growth during which growth occurs very slowly.

land The surface of the earth not covered by water.

landfill A method of disposing of solid wastes that involves burying the wastes in specially constructed sites.

land-use planning The construction of an orderly list of priorities for the use of available land.

law A hypothesis that has survived repeated examination by many investigators.

LD$_{50}$ A measure of toxicity; the dosage of a substance that will kill (lethal dose) 50 percent of a test population.

leaching The movement of minerals from the A horizon to the B horizon by the downward movement of soil water.

legislative branch That portion of the U.S. government that is responsible for the development of laws.

light-water reactor A reactor that uses ordinary water as a coolant.

limiting factor The one primary condition of the environment that determines the success of an organism.

limnetic zone Region that does not have rooted vegetation in a freshwater ecosystem.

liquefied natural gas Natural gas that has been converted to a liquid by cooling to $-162°C$.

liquid metal fast-breeder reactor (LMFBR) Nuclear fission reactor using liquid sodium as the moderator and heat transfer medium; produces radioactive plutonium 235, which can be used as a nuclear fuel.

littoral zone Region with rooted vegetation in a freshwater ecosystem.

loam A soil type with good drainage and good texture that is ideal for growing crops.

M

macronutrient A nutrient, such as nitrogen, phosphorus, and potassium, that is required by plants in relatively large amounts.

mangrove swamp ecosystems Marine shoreline ecosystems dominated by trees that can tolerate high salt concentrations.

marine ecosystems Aquatic ecosystems that have high salt content.

marsh Area of grasses and reeds that is either permanently flooded or flooded for a major part of the year.

mass burn A method of incineration of solid waste.

matter Substance with measurable mass and volume.

mechanical weathering The process of breaking up large particles by physical forces that cause the particles to collide or rub against one another. Wind, moving water, and glaciers are common causes of mechanical weathering.

megalopolis A large, regional urban center.

micronutrient A nutrient needed in extremely small amounts for proper plant growth; examples are boron, zinc, and magnesium.

migratory birds Birds that fly considerable distances between their summer breeding areas and their wintering areas.

mixture A kind of matter consisting of two or more kinds of matter intermingled with no specific ratio of the kinds of matter.

moderator Material that absorbs the energy from neutrons released by fission.

molecule Two or more atoms combined to form a stable unit.

monoculture A system of agriculture to which large tracts of land are planted in the same crop.

morals Predominant feeling of a culture about ethical issues.

mortality The number of deaths per year.

multiple land use Land uses that do not have to be exclusionary, so that two or more uses of land may occur at the same time.

municipal landfill An area used for the containment of solid wastes.

municipal solid waste All the waste produced by the residents of a community.

mutualism The association between organisms in which both benefit.

N

natality The number of individuals added to the population through reproduction.

National Priority List A listing of hazardous-waste dump sites requiring urgent attention as identified by the Superfund.

natural resources Those structures and processes that can be used by humans for their own purposes, but cannot be created by them.

natural selection A process that determines which individuals within a species will reproduce more effectively.

nature centers Teaching institutions that provide a variety of methods for people to learn about and appreciate the natural world.

negligible risk A point at which there is no significant health or environmental risk.

neutralization Reacting acids with bases to produce relatively safe end products.

neutron Neutrally charged particle located in the nucleus of an atom.

niche The total role an organism plays in a habitat.

nitrogen cycle The series of stages in the flow of nitrogen in ecosystems.

nitrogen-fixing bacteria Bacteria that are able to convert the nitrogen gas (N_2) in the atmosphere into forms that plants can use.

nonpersistent pollutants Those pollutants that do not remain in the environment for long periods and are biodegradable.

nonpoint source Diffuse pollutants, such as agricultural runoff, road salt, and acid rain, that are not from a single, confined source.

nonrenewable energy sources Those energy sources that are not replaced by natural processes within a reasonable length of time.

nonrenewable resources Those resources that are not replaced by natural processes, or those whose rate of replacement is so slow as to be noneffective.

nontarget organism An organism whose elimination is not the purpose of pesticide application.

northern coniferous forest See boreal forest.

nuclear breeder reactor Nuclear fission reactor designed to produce radioactive fuel from nonradioactive uranium and at the same time release energy to use in the generation of electricity.

nuclear chain reaction A continuous process in which a splitting nucleus releases neutrons that strike and split the nuclei of other atoms, releasing nuclear energy.

nuclear fission The decomposition of an atom's nucleus with the release of particles and energy.

nuclear fusion The union of smaller nuclei to form a heavier nucleus accompanied with the release of energy.

nuclear reactor A device that permits a controlled nuclear fission chain reaction.

nucleus The central region of an atom that contains protons and neutrons.

O

observation Ability to detect events by the senses or machines that extend the senses.

oligotrophic lakes Deep, cold, nutrient-poor lakes that are low in productivity.

omnivores Organisms that eat both plants and animals.

organophosphate A class of soft pesticides that work by interfering with normal nerve impulses.

outdoor recreation Recreation that uses the natural out-of-doors for leisure-time activities.

overburden The layer of soil and rock that covers deposits of desirable minerals.

oxides of nitrogen (NO and NO$_2$) Primary air pollutants consisting of a variety of different compounds containing nitrogen and oxygen.

ozone (O$_3$) A molecule consisting of three atoms of oxygen, which absorb much of the sun's ultraviolet energy before it reaches the earth's surface.

P

parasite An organism adapted to survival by using another living organism (host) for nourishment.

parasitism A relationship between organisms in which one, known as the parasite, lives in or on the host and derives benefit from the relationship while the host is harmed.

parent material Material that is weathered to become the mineral part of the soil.

particulates Small pieces of solid materials, such as smoke particles from fires, bits of asbestos from brake linings and insulation, dust particles, or ash from industrial plants, that are dispersed into the atmosphere.

passive solar system A design that allows for the entrapment and transfer of heat from the sun to a building without the use of moving parts or machinery.

patchwork clear-cutting A forest harvest method in which patches of trees are clear-cut among patches of timber that are left untouched.

peat The first stage in the conversion of organic material into coal.

pelagic Those organisms that swim in open water.

pelagic ecosystem A portion of a marine or freshwater ecosystem that occurs in open water away from the shore.

periphyton Attached organisms in freshwater streams and rivers, including algae, animals, and fungi.

permafrost Permanently frozen ground.

persistent pesticide A pesticide that remains unchanged for a long period of time; a hard pesticide.

persistent pollutant A pollutant that remains in the environment for many years in an unchanged condition.

pest An unwanted plant or animal that interferes with human activity.

pesticide A chemical used to eliminate pests; a general term used to describe a variety of different kinds of pest killers, such as insecticides, fungicides, rodenticides, and herbicides.

pH The negative logarithm of the hydrogen ion concentration; a measure of the number of hydrogen ions present.

pheromone A chemical produced by one animal that changes the behavior of another.

photochemical smog A yellowish-brown haze that is the result of the interaction of hydrocarbons, oxides of nitrogen, and sunlight.

photosynthesis The process by which plants manufacture food. Light energy is used to convert carbon dioxide and water to sugar and oxygen.

photovoltaic cell A means of directly converting light energy into electricity.

phytoplankton Free-floating, microscopic, chlorophyll-containing organisms.

pioneer community The early stages of succession that begin the soil-building process.

plutonium-239 (Pu-239) A radioactive isotope produced in a breeder reactor and used as a nuclear fuel.

point source Pollution from a single pipe or series of effluent pipes.

policy Planned course of action on a question or a topic.

pollution Waste material that people produce in such large quantities that it interferes with their health or well-being.

pollution costs The private or public expenditures undertaken to avoid pollution damage once pollution has occurred and the increased health costs and loss of the use of public resources because of pollution.

pollution prevention To prevent either entirely or partially the pollution that would otherwise result from some production or consumption activity.

population density A measure of how close organisms are to one another, generally expressed as the number of organisms per unit area.

postwar baby boom A large increase in the birthrate immediately following World War II.

potable waters Unpolluted freshwater supplies.

potential energy The energy of position.

prairies Grasslands.

precipitation Removal of materials by mixing with chemicals that cause the materials to settle out of the mixture.

predator An organism that kills and eats another organism.

preservation To keep from harm or damage; to maintain in its original condition.

preservation ethic Philosophy that considers nature to be so special that it should remain intact.

pressurized-water reactor (PWR) A type of light-water reactor in which the water in the reactor is kept at high pressure and steam is formed in a secondary loop.

prey An organism that is killed and eaten by a predator.

primary consumer An organism that eats plants (producers) directly.

primary air pollutants Types of unmodified materials that, when released into the environment in sufficient quantities, are considered hazardous.

primary sewage treatment Process that removes larger particles by filtering raw sewage through large screens into ponds or lagoons.

primary succession Succession that begins with bare mineral surfaces or water.

probability A mathematical statement about how likely it is that something will happen.

producer An organism that can manufacture food from inorganic compounds and light energy.

profitability The extent to which economic benefits exceed the economic costs of doing business.

proton The positively charged particle located in the nucleus of an atom.

public resources Those parts of the environment that are owned by everyone.

R

radiation Energy that travels through space in the form of waves or particles.

radioactive Describes unstable nuclei that release particles and energy as they disintegrate.

radioactive half-life The time it takes for half of the radioactive material to spontaneously decompose.

radon Radioactive gas emitted from certain kinds of rock; can accumulate in very tightly sealed buildings.

range of tolerance The ability organisms have to succeed under a variety of environmental conditions. The breadth of this tolerance is an important ecological characteristic of a population.

reactivity The property of minerals that indicates the degree to which a material is likely to react vigorously to water or air, or to become unstable or explode.

recycling The process of reclaiming a resource and reusing it for another or the same structure or purpose.

reforestation The process of replanting areas after the original trees are removed.

rem A measure of the biological damage to tissue caused by certain amounts of radiation.

renewable energy sources Those energy sources that can be regenerated by natural processes.

renewable resources Those resources that can be formed or regenerated by natural processes.

repeatability An important criterion in the scientific method, it requires that independent investigators be able to repeat an experiment and get the same results.

replacement fertility The number of children per woman needed to replace just the parents.

reserves The known deposits from which materials can be extracted profitably with existing technology under present economic conditions.

Resource Conservation and Recovery Act (RCRA) The 1976 U.S. law that specifically addressed the issue of hazardous waste.

resource exploitation The use of natural resources by society.

resources Naturally occurring substances that are potentially feasible to extract under prevailing conditions.

respiration The process that organisms use to release chemical bond energy from food.

ribbon sprawl Development along transportation routes that usually consists of commercial and industrial building.

risk assessment The use of facts and assumptions to estimate the probability of harm to human health or the environment that may result from exposures to specific pollutants, toxic agents, or management decisions.

risk management Decision-making process that uses input such as risk assessment, technological feasibility, economic impacts, public concerns, and legal requirements.

rodenticide A pesticide designed to kill rodents.

r-strategist Typically, a small organism that has a short life span, produces a large number of offspring, and does not reach a carrying capacity.

runoff The surface water that enters a river system.

S

savanna Tropical biome having seasonal rainfall of 50 to 150 centimeters per year. The dominant plants are grasses, with some scattered fire- and drought-resistant trees.

science A method for gathering and organizing information that involves observation, hypothesis formation, and experimentation.

scientific method A way of gathering and evaluating information. It involves observation, hypothesis formation, hypothesis testing, critical evaluation of results, and the publishing of findings.

secondary consumers Organisms that eat animals that have eaten plants.

secondary air pollutants Pollutants produced by the interaction of primary air pollutants in the presence of an appropriate energy source.

secondary recovery Techniques used to obtain the maximum amount of oil or natural gas from a well.

secondary sewage treatment Process that involves holding the wastewater until the organic material has been degraded by bacteria and other microorganisms.

secondary succession Succession that begins with the destruction or disturbance of an existing ecosystem.

second law of thermodynamics A statement about energy conversion that says that, whenever energy is converted from one form to another, some of the useful energy is lost.

selective harvesting A forest harvesting method in which individual high-value trees are removed from the forest, leaving the majority of the forest undisturbed.

septic tank Underground holding tank into which sewage is pumped and where biological degradation of organic material takes place; used in places where sewers are not available.

seral stage A stage in the successional process.

sere A stage in succession.

sex ratio Comparison between the number of males and females in a population.

soft pesticide A nonpersistent pesticide that breaks down into harmless products in a few hours or days.

soil A mixture of mineral material, organic matter, air, water, and living organisms; capable of supporting plant growth.

soil profile The series of layers (horizons) seen as one digs down into the soil.

soil structure Refers to the way that soil particles clump together. Sand has little structure because the particles do not stick to one another.

soil texture Refers to the size of the particles that make up the soil. Sandy soil has large particles, and clay soil has small particles.

solidification The conversion of liquid wastes to a solid form to allow for more safe storage or transport.

solid waste Unusable or unwanted solid products that result from human activity.

source reduction Reducing the amount of solid waste generated by using less, or converting from heavy packaging materials to lightweight ones.

speciation The process of developing a new species.

species A group of organisms that can interbreed and produce offspring capable of reproduction.

stable equilibrium phase The phase in a population growth pattern in which the deathrate and birthrate become equal.

standard of living The necessities and luxuries essential to a level of existence that is customary within a society.

steam stripping The use of heated air to drive volatile compounds from liquids.

steppe A grassland.

storm-water runoff Stormwater that runs off of streets and buildings and is often added directly to the sewer system and sent to the municipal waste-water treatment facility.

strip farming The planting of crops in strips that alternate with other crops. The primary purpose is to reduce erosion.

submerged plants Aquatic vegetation that is rooted on the bottom and has leaves that stay submerged below the surface of the water.

subsidy A gift given to private enterprise by government when the enterprise is in temporary economic difficulty but is viewed as being important to the public.

succession Regular and predictable changes in the structure of a community, ultimately leading to a climax community.

successional stage A stage in succession.

sulfur dioxide (SO₂) A compound containing sulfur and oxygen produced when sulfur-containing fossil fuels are burned. When released into the atmosphere, it acts as a primary pollutant.

Superfund The common name given to the U.S. 1980 Comprehensive Environmental Response, Compensation, and Liability Act, which was designed to address hazardous waste sites.

supply Amount of a good or service available to be purchased.

supply/demand curve The relationship between available supply of a commodity or service and its price.

surface impoundment Pond created to hold liquid materials. Some may hold only water, while others may be used to contain polluted water or liquid contaminants.

surface mining (strip mining) A type of mining in which the overburden is removed to procure the underlying deposit.

sustainable development Using renewable resources in harmony with ecological systems to produce a rise in real income per person and an improved standard of living for everyone.

swamp Area of trees that is either permanently flooded or flooded for a major part of the year.

symbiosis A close, long-lasting physical relationship between members of two different species.

symbiotic nitrogen-fixing bacteria Bacteria that grow within a plant's root system and that can convert nitrogen gas (N₂) from the atmosphere to nitrogen compounds that the plant can use.

synergism The interaction of materials or energy that increases the potential for harm.

T

taiga Biome having short, cool summers and long winters with abundant snowfall. The trees are adapted to winter conditions.

target organism The organism a pesticide is designed to eliminate.

technological advances Increasing use of machines to replace human labor.

temperate deciduous forest Biome that has a winter–summer change of seasons and that receives 100 centimeters or more of relatively evenly distributed precipitation throughout the year.

terrace A level area constructed on steep slopes to allow agriculture without extensive erosion.

tertiary sewage treatment Process that involves a variety of different techniques designed to remove dissolved pollutants left after primary and secondary treatments.

theory A unifying principle that binds together large areas of scientific knowledge.

theory of evolution The widely accepted concept that populations of living things can change genetically over time and that this change can lead to a population that is very well adapted to its environment.

thermal inversion The condition in which warm air in a valley is sandwiched between two layers of cold air and acts like a lid on the valley.

thermal pollution Waste heat that industries release into the environment.

thermal treatment A form of hazardous-waste destruction involving heating waste. Such incineration can destroy 99 percent of the organic wastes.

threatened species Those species that could become extinct if a critical factor in their environment were changed.

threshold levels The minimum amount of something required to cause measurable effects.

total fertility rate The number of children born per woman per lifetime.

toxicity A property of materials which, in sufficient quantity, can cause danger to human health or the environment.

toxic waste Substances that are poisonous and cause death or serious injury to humans and animals when released into the environment.

tract development The construction of similar residential units over large areas.

transpiration Transportation of water to leaves and its evaporation from the surfaces of plants.

trophic level A stage in the energy flow through ecosystems.

tropical rainforest A biome with warm, relatively constant temperatures where there is no frost. These areas receive more than 200 centimeters of rain per year in rains that fall nearly every day.

tundra A biome that lacks trees and has permanently frozen soil.

U

unconfined aquifer An aquifer that usually occurs near the land's surface and may be called a water table aquifer.

underground mining A type of mining in which the deposited material is removed without disturbing the overburden.

underground storage tank Tank located below ground level for the storage of materials, such as oil, gasoline, or other chemicals.

uranium-235 (U-235) A naturally occurring radioactive isotope of uranium used as fuel in nuclear reactors.

urban sprawl Unplanned suburban growth.

W

waste destruction Destruction of a portion of hazardous waste with harmful residues still left behind.

waste immobilization Putting hazardous wastes into a solid form that is easier to handle and less likely to enter the surrounding environment.

waste separation Either separating one hazardous waste from another, or separating hazardous waste from nonhazardous material that it has contaminated.

water diversion The physical process of transferring water from one area to another.

water table The top of the layer of water in an aquifer.

waterways Low areas that water normally flows through.

weathering The physical and chemical breakdown of materials; involved in the breakdown of parent material in soil formation.

weed An unwanted plant.

wetlands Areas that include swamps, tidal marshes, coastal wetlands, and estuaries.

wilderness Designation of land use for the exclusive protection of the area's natural wildlife; thus, no human development is allowed.

windbreak The planting of trees or strips of grasses at right angles to the prevailing wind to reduce erosion of soil by wind.

Z

zero population growth The stabilized growth stage of human population during which births equal deaths and equilibrium is reached.

zoning Type of land-use regulation in which land is designated for specific potential uses, such as agricultural, commercial, residential, recreational, and industrial.

zooplankton Weakly swimming microscopic animals.

CREDITS

Photographs

Part Openers

1: © Carl Rogers Photography
2: © McGraw-Hill Higher Education Group Inc.,/ Carlyn Iverson, photographer
3: © Blanche Haning
4: © Blanche Haning
5: Kennan Ward © 1997

Chapter 1

1.2: AP/Wide World Photos; p.8 left and right: © Leonard Lee Rue, III/Animals Animals/Earth Scenes; 1.4a: © Steve McCutcheon/Visuals Unlimited; 1.4b: © Frank Hanna/Visuals Unlimited; 1.4c: © John Shaw/Tom Stack and Associates; 1.4d: © Wm. W. Knaver II/Visuals Unlimited; 1.5a-b: © Bob Coyle; 1.5c: © Larry Miller/Photo Researchers, Inc.; 1.5d: © Link/ Visuals Unlimited; 1.6a: © Sylvan Wittwer/Visuals Unlimited; 1.6b: © John Cunningham/Visuals Unlimited; 1.6c: © David Halpern/Photo Researchers, Inc.; 1.6d: © Gary Lado/Photo Researchers, Inc.; 1.6e: © Science Vu/Visuals Unlimited; 1.7a: © Milton Rand/Tom Stack and Associates; 1.7b: © Bob Pool/Tom Stack and Associates; 1.7c: © William J. Weber/Visuals Unlimited; 1.7d: © Joe McDonald/Tom Stack and Associates; 1.8a: © Richard Choy/Peter Arnold, Inc.; 1.8b: © Louis Goldman/Photo Researchers, Inc.; 1.8c: © Matt Bradley/Tom Stack and Associates; 1.8d, 1.9a: © Bob Coyle; 1.9b: © Malcolm S. Kirk/Peter Arnold, Inc.; 1.9c: © Science VU/Visuals Unlimited; 1.9d: © Brian Parker/Tom Stack and Associates; 1.9e: © Martin G. Miller/Visuals Unlimited; 1.10b: © Eastcott/Momatiuck/The Image Works

Chapter 2

2.1: NASA; p.20 left: © The Granger Collection; p.20 middle and right: © The Bettman Archive; p.21 left and right: © AP/Wide World Photos; 2.1 2.2a: © Terry Donnelly/Tom Stack and Associates; 2.2b: © J. Lotter/Tom Stack and Associates; 2.2c: © Bob Pool/Tom Stack and Associates; 2.2d: © Chase Swift/ Tom Stack and Associates; p.23 left: © Bob Daemmrich/The Image Works; p.23 right: © Wendy Shattil/Tom Stack and Associates; p.24 & 25: © Courtesy General Motors; 2.3a: © Toni Michaels; 2.3b: © Stephen Frisch/Stock Boston; 2.3c: © McGraw-Hill Higher Ed., Inc.;

2.4a: © Greg Vaughn/Tom Stack and Associates; 2.4b: AP Wide World Photos; 2.5a: © E.R. Denninger/Animals Animals; 2.5b: © G.I. Bernard/Animals Animals; 2.5c: © L.L.T. Rhodes/Animals Animals; 2.5d: © Bruce Davidson/Animals Animals; 2.5e: © E.R. Degginger/Animals Animals; 2.5f: © Dr. Nigel Smith/Animals Animals; p.31: © AP/Wide World Photos; p.32 left: © Anna Zuckerman/Tom Stack and Associates; p.32 right: © Johnny Johnson/ Animals Animals/Earth Scenes;

Chapter 3

p.40 top and bottom: © Times Mirror Higher Education Group, Inc./Bob Coyle, Photographer; 3.8: © Carl Purcell/Photo Researchers, Inc.

Chapter 4

4.4: © L. West/ Photo Researchers, Inc.; 4.6: © Doug Sokell/Visuals Unlimited; 4.7: © M. W. Tweedie/Photo Researchers, Inc.; 4.8: © Stephen Krasemann/Photo Researchers, Inc.; 4.9: © Fritz Polking GDT/Peter Arnold, Inc.; 4.10a: © Dr. Tom McHugh/SPL/Photo Researchers, Inc.; 4.10b: © Manfred Kage/Peter Arnold, Inc.; 4.11: © Tom McHugh/Photo Researchers, Inc.; 4.12: © Dr. J. Burgess/Photo Researchers, Inc.; p.60 bottom left: © Sinclair Stammers/SPL/Photo Researchers, Inc.; p.60 top: © K. Maslowski/Visuals Unlimited; p.60 bottom right: © Charles Sykes/Visuals Unlimited; 4.17b: Beans: © Alexander Lowry/Photo Researchers, Inc.; 4.17c: Fawn: © Maresa Pryor/Animals Animals/Earth Scenes; 4.17d: © Paul Souders/Tony Stone Images; 4.17e: Dead Deer: © Knolan Benfield/Visuals Unlimited; 4.17f: Bacteria Right: © Kessel - G. Shih/Visuals Unlimited; 4.17g: Bacteria Left: © David M. Phillips/Visuals Unlimited; 4.17h: Bacteria Top: Fred Hossler/Visuals Unlimited; p.67: © Courtesy Marc Blouin, National Biological Survey, U.S. Dept. of The Interior; p.68: © Richard P. Smith/Tom Stack and Associates

Chapter 5

5.1: © William E. Ferguson; 5.4: © Larry Mellichamp/Visuals Unlimited; p.78 left and middle: © Harold Hunger-ford/University of S. Illinois at Carbondale; p.78 right: © Carl Bollwinkel/University of Northern Iowa; 5.9b: © Len Rue, Jr./Photo Researchers, Inc.; 5.10b: © William E. Ferguson. Photo by Stephanie Ferguson.; p.81: © Michael J. Balick/Peter Arnold,

Chapter 16

16.5b: © William E. Ferguson; **16.5c:** © Mark Antman/The Image Works; **16.5d:** © Thomas Kitchin/Tom Stack and Associates; **16.5e:** © McGraw-Hill Higher Education/Bob Coyle, Photographer; **16.5f:** © Toni Michaels; **16.5g:** © Jon Feengersh/Tom Stack & Associates; **16.5h:** © Joe Sohm/The Image Works; **16.5I:** © Scott Blackman/Tom Stack and Associates; **16.7a:** © C. Max Dunham/Photo Researchers, Inc.; **16.7b:** © Didier Givois/Photo Researchers, Inc.; **16.7c:** © Chris Caswell/Photo Researchers, Inc.; **16.7d:** © Clyde H. Smith/Peter Arnold, Inc.; **16.7e:** © Jan Halaska/The Image Works; **16.7f:** © George E. Jones, III/Photo Researchers, Inc.; **16.8b:** AP/Wide World Photos; **16.9b:** © Runk/Schoenberger/Grant Heilman Photography; **16.9c:** © Bob Daemmrich/The Image Works; p.346 left: © C. Allan Morgan/Peter Arnold, Inc.; p.346 right: © John R. MacGregor/Peter Arnold, Inc.

Chapter 17

17.4: © John D. Cunningham/Visuals Unlimited; **17.5:** © Mark Antman/The Image Works; **17.6a-b:** © Peggy Yoram Kahana/Peter Arnold, Inc.; p.357: © James Shaffer; **17.9a-b:** © Toni Michaels; **17.11:** © Don and Pat Valenti/Tom Stack and Associates; **17.12:** © John Shaw/Tom Stack and Associates; **17.13:** Canapress Photo Service, Inc.; p.364: © Robert W. Ginn/Unicorn Stock Photos; p.368: © Bruce M. Wellman/Stock Boston

Chapter 18

18.1: © R. Maiman/Sygma; **18.2:** © Ray Pfortner/Peter Arnold, Inc.; **18.12b:** © Steve Elmore/Tom Stack and Associates; **18.4:** © Rapho Division/Photo Researchers, Inc.; **18.9:** © David M. Dennis/Tom Stack and Associates; **18.11:** © Marie Mills/Unicorn Stock Photos; p.380: © David Putnam, Courtesy of Weyerhaeuser; p.381 © Pat Toth-Smith; p.382 all, 384 all: © McGraw-Hill Higher Education/Bob Coyle, Photographer;

Chapter 19

19.2b: © McGraw-Hill Higher Education/Bob Coyle; **19.2c:** © John Colwell/Grant Heilman Photography; **19.2d:** © Larry Kolvoord/The Image Works; **19.2e-f:** © Toni Michaels; **19.2g:** © Benelux/Photo Researchers, Inc.; **19.2h:** © Bob Daemmrich/The Image Works; **19.2I:** © Joseph Nettis/Photo Researchers, Inc.; p.391: © Kerry T. Givens/Tom Stack and Associates; p.393: © Lowell Georgia/Photo Researchers, Inc.; **19.3:** © Gary Milburn/Tom Stack and Associates; p.397: © C. William Campbell/Peter Arnold, Inc.; p.400: © The Bettmann Archives

Chapter 20

p.412 top: © Paul Fusco/Magnum Photos; p.412 bottom: © Robert Winslow/Tom Stack and Associates; p.415: © Craig Newbauer/Peter Arnold, Inc.; p.418 left: © Peter Turnley/Black Star; p.418 right: © P. Durand/Sygma; p.421 top: © H. R. Bramaz/Peter Arnold, Inc.; p.421 top middle: © T. Okapia/Photo Researchers, Inc.; p.421 bottom middle: © Craig Newbauer/Peter Arnold, Inc.; p.421 bottom: © Horst Schafer/Peter Arnold, Inc.

INDEX

hazardous waste disposal
environmental problems caused by
toxic waste, 390–91, 392
health risks associated with, 391, 393,
394–95, 400
historical abuses, 393, 395–96
incineration, 398–99, 401–2
land disposal, 399–402
managing, 396, 398–402
national priority list (U.S.), 395, 397
necessity of proper, 389
pollution prevention, 398
recycling, 398
Superfund, 395–96
toxic chemical releases, 392
treatment of wastes, 398–99
waste minimization, 398
hazardous wastes
defined, 388
environmental effects, 390–91, 392
health risks associated with, 391, 393,
394–95, 400
regulation, 388–90, 391
heavy-water reactors (HWR), 172
Helsinki Declaration (1989), 363
herbicides, 265, 268, 270, 271
herbivores, 59
herring gulls, 281
high-density polyethylene (HDPE),
381, 382
high-level radioactive waste disposal,
183, 184
high-temperature, gas-cooled reactors
(HTGCR), 172
Hiroshima (Japan), atomic bomb
dropped on, 169
Holy Ganges River (India) cleanup, 320
Homer, 265
horizon, 245, 246
host, 57
household chemicals, 40
human-accelerated extinction,
211–12, 213
human impact on ecosystems
animals that benefit from human
activity, 219
aquatic ecosystems, 201–5
areas with minimal impact,
193–94, 195
changing role of human impact, 189
ecosystems modified by human use,
194–208
human-managed ecosystems replacing
natural ecosystems, 208, 210
rangelands, 199–201
wildlife, modifying ecosystems to
manage, 205–8
human population
age distribution, 96, 97
carrying capacity, 103–4
current trends, 110
fluctuations, 102
social aspects of, 104
and standard of living, 110–11, 112, 121
ultimate size limitations, 104–5
urbanization of, 119
U.S., 118–20
world, 110
human population growth
in Canada, 123
causes, 112–17
changes with continued growth, 124
demographic transition, 117–18
governmental effects on, 116–17
growth curve, 102–3
limiting factors, 103–4
in Mexico, 123
political factors affecting, 116–17
and poverty, 121
social reasons for, 113–16

and standard of living, 110–11, 112, 121
ultimate size limitations, 104–5
in U.S., 118–20
world problems, as contributing factor
to, 111–12, 120–24
humus, 242, 264
Hungary, environmental degradation in,
420–21
hunting, 205–6
Hussein, Saddam (President of Iraq), 418
Hyde Park (London), 233
hydrocarbons
as air pollutant, 352
efforts to reduce, 355, 356
hydroelectric power, 153–55
in-stream use of water for, 292–93, 294
James Bay power project (Canada),
164–65
hydrogen
in hydrocarbons, 352
vehicles, 137
hydrologic cycle, 285–88
hydroxyl ion, 41
hypotheses, 37, 38

I

ignitability, 387
imidacloprid, 267
immigration, 96
policy in U.S., 119–20
pressure on developing countries,
116–17
incineration
of hazardous wastes, 398–99,
401–2
solid waste, 374, 376, 377
indoor air pollution, 365–68
industrial
energy use, 135
water pollution, 299–300
water use, 293–94
industrial ecology, 27
Industrial Revolution, 130–31
Industry and Environment Office
(IEO), 423
information programs, 334
INFOTERRA, 423
inorganic matter, 41
insecticides
beneficial insects killed by, 276
chlorinated hydrocarbons, 265, 267
defined, 265
historical overview of use, 265
kinds, 269
new generation of, 267
organophosphates and carbamates,
265, 267–68
resistance to, 275–76
insects, 244–45
control of, 265
killing of beneficial, by insecticides, 276
use of beneficial, 278–80
institutional commitment, 342
in-stream use of water, 292–93, 294
integrated pest management, 278–80, 282
interaction of organisms
carrying capacity and, 99, 103–4
in communities, 59
competition, 56–57
predation, 56
symbiotic relationships, 57–58
interdependence, 342
Intergovernmental Panel on Climate
Change (IPCC), 360–61
international air pollution, 369
International Boundary and Water
Commission, 295
International Cleaner Production
Information Clearinghouse, 423

International Development Association
(IDA), 419
International Environmental Education
Programme (IEEP), 7
international environmental policy,
409–10, 416–23
International Fund for Agricultural
Development (IFAD), 248
International Joint Commission, 7, 285
International Monetary Fund (IMF), 419
International Union for the
Conservation of Nature and
Natural Resources (IUCN), 214–15
International Whaling Commission
(IWC), 217
interspecific competition, 57
intraspecific competition, 56, 57
iodine, 277
ions, 41
irrigation, 289, 292, 293
California Water Plan and, 318
Isle Royal (Lake Superior), 106
isotopes, 41, 169
ivory, ban on sale of, 101

J

James Bay power project (Canada),
164–65
Japan
atomic bombs dropped on, 169
recycling in, 379
Japan Economic Research Center, 337
Journals, 20
judicial branch of U.S. government, 405

K

Kemp's ridley sea turtles, 346
Kettlewell, H.B.D., 55
killer fog, 353
kinetic energy, 43, 45
Kirtland's warblers, 206–7
Krasnoyarsk (Russia), 181
K-strategists, 100, 102
kudzu vine, 104–5
Kuwait, eco-terrorism and, 418

L

labeling, eco-labels, 422
labor-intensive agriculture, 261, 262
ladybugs, 279
lag phase, 97
Lake Victoria (Uganda, Kenya, Tanzania)
environmental damage and
conservation efforts, 306–7
lakes
acid rain and, 359, 360, 361
characteristics, 90, 91
heated water discharged into, 301
pollution of, 294, 298
land
defined, 242
as nonrenewable resource, 225
soil and, 242
land capability, 250–51
classification, 254–55
land disposal of hazardous waste, 399–402
landfills. *See also* solid waste disposal
burning, 373, 374
groundwater pollution from, 305
in municipal waste disposal programs,
374–76
lands, state of world's, 227
land use
forests, state of world's, 228
historical overview, 225–28
lands, state of world's, 227
multiple land use, 232

nonfarm land, uses of, 257–58
recreational land use (*see* recreational
land use)
suburbia, rise of, 228–30
water and, 225, 226
land-use planning
defined, 236
malls, planning decisions surrounding
development of, 239
principles, 236–39
regional planning, 236–37
urban transportation planning, 237–38
laterite soil, 81
latitude, and vegetation distribution,
85, 86
laws, passage of, 405, 406
laws, scientific, 38
LD_{50}, 391
leaching, 245, 247
lead
as air pollutant, 354–55
in drinking water, 290–91
poisoning, 393
toxicity, 390
Lebow, Victor, 372
legislative branch of U.S. government,
405
legumes, 65, 81
Leopold, Aldo, 18–19, 20, 21
lichens, 71, 242
Life magazine, 372
lifestyles, environmental impact of, 29, 30
lighting, improvements, 135
light-water reactors (LWR), 171
lignite, 146, 148
Limerick Nuclear Generating Station
(Pennsylvania), 366
limestone, acid rain and, 358
limiting factors, 50, 53
phosphates as, 299
on population growth, 99, 103–4
limnetic zone, 90
liquefied natural gas, 153
liquid metal fast-breeder reactors
(LMFBR), 173
litter, 190
littoral zone, 90
loam, 243
loggerhead sea turtles, 346
London killer fog (England, 1952), 353
Los Alamos National Laboratory (New
Mexico), mining heat, 156–57
Love Canal (New York), 400
low-density polyethylene (LDPE), 382
low-level radioactive waste disposal, 183

M

macronutrients, 264
malls, planning decisions surrounding
development of, 239
manatees, 219
manganese, 264
mangrove swamp ecosystems, 89
marine ecosystems
benthic, 86–89
defined, 86
pelagic, 86, 87
marine oil pollution, 301–2
market-based instruments (MBIs),
334–35
marshes, 91
mass burn, 376
mass transit, 237–38
matter, 38–42, 43, 44
acids, bases and pH, 41
atomic structure, 38–39, 41
chemical reactions, 41–42
inorganic and organic, 41
living things, chemical reactions in, 42

DATE DUE

WITHDRAWN

MR 12 '99			
JUN 2 3 2000			
JUL 2 0 2000			
MAY 0 1 2001			
AUG 2 9 2002			
1/16/04			
GAYLORD			PRINTED IN U.S.A